Introduction to

COMPLEXITY AND COMPLEX SYSTEMS

Robert B. Northrop

CRC Press
Taylor & Francis Group
Boca Raton   London   New York

CRC Press is an imprint of the
Taylor & Francis Group, an **informa** business

CRC Press
Taylor & Francis Group
6000 Broken Sound Parkway NW, Suite 300
Boca Raton, FL 33487-2742

First issued in paperback 2017

© 2011 by Taylor and Francis Group, LLC
CRC Press is an imprint of Taylor & Francis Group, an Informa business

No claim to original U.S. Government works

ISBN 13: 978-1-138-11333-6 (pbk)
ISBN 13: 978-1-4398-3901-0 (hbk)

Library of Congress Cataloging-in-Publication Data

Northrop, Robert B.
  Introduction to complexity and complex systems / Robert B. Northrop.
     p. ; cm.
  Includes bibliographical references and index.
  ISBN 978-1-4398-3901-0 (Hardcover : alk. paper)
    1. Complexity (Philosophy) 2. Biocomplexity. 3. Chaotic behavior in systems. I. Title.
    [DNLM: 1. Biophysical Phenomena. 2. Biomedical Technology. 3. Models, Theoretical. QT 34]

  Q175.32.C65N67 2010
  570.1'1--dc22   2010026038

Visit the Taylor & Francis Web site at
http://www.taylorandfrancis.com

and the CRC Press Web site at
http://www.crcpress.com

I dedicate this text to my complex family.

Contents

Preface

This introductory textbook is intended for use in a one-semester course to acquaint biomedical engineers, biophysicists, systems physiologists, ecologists, biologists, and other scientists, in general, with *complexity* and *complex systems*. I have focused on biochemical, genomic, and physiological complex systems, and I have also introduced the reader to the inherent complexity in economic systems.

Reader background: Readers should have had college courses in algebra, calculus, ordinary differential equations, and linear algebra, and, hopefully, engineering systems analysis. They should also have had basic college courses in chemistry, biochemistry, cell biology, and ideally even in human physiology and anatomy. This is the broad background that is required in the interdisciplinary fields of biomedical engineering, biophysics, systems physiology, and economics.

Overview: In this book, I address ways in which scientists can describe the characteristics of certain complex systems, model them, and use these models to try and predict unanticipated behavior. All scientists need to be familiar with complex systems, thereby minimizing *unintended consequences*.

One often hears the words *complex (adj. and n.)*, *complexity (n.)*, and *complicated (adj.)*, particularly in the media, to describe systems and situations. Complex, complexity, and complicated all come from the Latin word, *complexus*, meaning twisted together, or entwined. Complexity is difficult to define; it can be defined in several ways, depending on the type of system being studied, how it is being studied, and who is studying it. Broadly stated, we consider that *complexity* is a subjective measure of the difficulty in *describing* and *modeling* a system (thing or process), and thus being able to predict its behavior. We may also view the complexity of a system or dynamic process as an increasing function of the degree to which its components engage in structured, organized interactions. Complexity has also been seen as a global characteristic of a system that represents the gap between component and parameter knowledge and knowledge of overall behavior.

Complexity lies in the eyes of the beholder: what is complex to one observer may not be complex to another, based on the observer's knowledge and skills. It is safe to say, in general, that complexity is relative, graded, and system- and observer-dependent. Unfortunately, there is no single, simple definition of complexity.

Biomedical engineering is an interdisciplinary science that is characterized by noisy, nonstationary signals acquired during the course of diagnosis, therapy, and research, and it deals with the interlinked, nonlinear, time-variable, complex systems from which these signals arise. Conventional engineering systems analysis, as practiced by electrical engineers, has generally focused on linear, time-invariant (LTI) systems that are easily described by linear ordinary differential equations (ODEs) and linear algebraic techniques (state variables), and it considers signals that are stationary and ergodic, which can be analyzed using Fourier analysis. They can be characterized by descriptors using time averages such as auto- and cross-correlograms, auto- and cross-power spectrograms, coherence, etc.

Early bioengineers, biophysicists, and systems physiologists tried to characterize certain physiological regulators as linear and stationary. Initially, linear systems analysis was inappropriately applied to certain complex, physiological regulators and control systems (e.g., pupil regulation and eye movement control), which resulted in "black-box" closed-loop models in which linear transfer function modules were connected to a nonlinear module in a single feedback loop. These were phenomenological I/O models that gave little insight into the physiology and complexity of the systems.

Living organisms, by their very nature, emit signals that are noisy and nonstationary; that is, the parameters of nonlinear systems giving rise to the signals change with time. As several physiological systems are implemented in parallel at the cellular level for robustness, one is likely to pick up the "cross-talk" from other channels as noise while recording from one "channel" of one system (e.g., in EMG and EEG recordings). There are many reasons for nonstationarity: circadian rhythms affect gene expression, hormone levels, certain physiological processes, and behavior; and exogenous drugs alter physiological processes in a concentration-dependent manner. Another cause of nonstationarity involves endogenous, periodic rhythms such as those associated with breathing, or the heart beating, and still other nonstationarities can be associated with natural processes such as the digestion of food or motor activity.

Thus, we see that all physiological systems are complex, nonstationary, and nonlinear *regardless of scale*—that is, they are complex, nonlinear systems (CNLSs). Nonstationarity can generally be ignored if the changes are slow compared to the time over which data are acquired. Nonlinearity can arise from the concatenated chemical reactions underlying physiological system functions. (There are no negative concentrations or gene expression rates.) The coupled ODEs of mass-action kinetics are generally nonlinear, which makes biochemical system characterization a challenge. Other nonlinearities may arise from the signal processing properties of the nervous system and the all-or-no regulation of gene expression.

RATIONALE

Purpose of the book: This book is intended for use in a classroom course or extended seminar designed to introduce advanced students (juniors, seniors, and graduates) majoring in biomedical engineering, biophysics, or physiology to complexity and complex systems. It could also serve as a reference book for cell biologists, ecologists, medical students, and economists interested in the topics.

The interdisciplinary fields of biomedical engineering, biophysics, and systems physiology are demanding in that they require a thorough knowledge of not only certain mathematical (e.g., calculus, linear algebra, and differential equations) and engineering (electronic, photonic, mechanical, materials) skills, but also a diversity of material in the biological sciences (e.g., molecular biology, biochemistry, physiology, and anatomy).

In summary, this book was written to aid biomedical scientists understand how to deal with biological complexity and complex systems. Analysis and modeling of complex systems presents a particular challenge if one is to be able to predict system behavior and avoid the effects of the *law of unintended consequences.*

DESCRIPTION OF THE CHAPTERS

This book is organized into 12 chapters, as summarized in the following:

Chapter 1, "Introduction to Complexity and Complex Systems," provides an overview of the properties of complex systems. "Complexity" is defined as being relative, graded, and system- and observer-dependent. Examples of simple, complicated, and complex systems are given, and the law of unintended consequences is explained. The chapter also describes human shortcomings in dealing with complexity and complex systems, and the efforts made by academia to implement courses on complex systems analysis, modeling, and engineering.

Chapter 2, "Introduction to Large Linear Systems," reviews the classical approaches to linear, dynamic systems analysis (superposition, convolution, transforms, linear algebra, state variables, etc.). Signal flow graphs are introduced as useful tools for the analysis of LTI systems as well as the topological description of CNLSs.

Chapter 3, "Introduction to Biochemical Oscillators and Complex Nonlinear Biochemical Systems," introduces some basic properties of nonlinear systems and stresses the fact that all living systems are nonlinear. The phase plane is introduced as a means of describing the behavior of CNLSs. The destabilizing effect of transport lags in closed-loop system stability is described. The chapter also provides a comprehensive description of oscillating solutions of nonlinear ODEs, and analyzes some well-known chemical and biochemical oscillators. It describes models for the synchronization of multiple, intracellular, biochemical oscillators and introduces autoinducer (AI) molecules.

Chapter 4, "Modularity, Redundancy, Degeneracy, Pleiotropy, and Robustness in Complex Biological Systems," describes how these properties of complex living systems help them survive at the intracellular, biochemical, and organ-system levels. It introduces the Red Queen, and describes operons, homeobox genes, and gene regulation and introduces system sensitivities as a means of finding branch gains in signal flow graph (SFG) models of CNLSs.

Chapter 5, "Evolution of Biological Complexity: Invertebrate Immune Systems," discusses the complex issues inherent in natural selection and Darwin's theory of evolution, the origin of species, including speciation by horizontal gene transfer and transposons. As a first detailed example of physiological complexity, we give a comparative description of invertebrate immune systems, including invertebrate and bacterial antimicrobial proteins (AMPs), and their potential applications in biomedicine.

Chapter 6, "Complex Adaptive and Innate Human Immune Systems," deals with the two major components of the complex human immune system (hIS): the human adaptive IS (hAIS) and the innate IS, including its cellular and complementary branches. It describes the roles of the various cellular components of the hAIS, including antigen presentation, and provides detailed information on the many immunocytokines of the hAIS. It also discusses the function of the human complement system (hCS) and its membrane attack complex (MAC) and introduces catalytic antibodies.

Chapter 7, "Complexity in Quasispecies and MicroRNAs," examines the complexity issues associated with viral and bacterial evolution in terms of viral quasispecies (VQS) and bacterial quasispecies (BQS). It describes the role of miRNAs in the complex regulation of gene expression and in the etiology of diseases.

Chapter 8, "Introduction to Physiological Complexity: Examples of Models of Some Complex Physiological Systems," introduces the important concept of homeostasis, and describes three important homeostatic, physiological regulatory systems: glucoregulation, the hIS vs. HIV, and the hIS vs. cancer. It covers the normal human glucoregulatory system and its major hormones and develops a parsimonious, dynamic, nonlinear model using mass-action kinetics; simulation results are also shown. Of considerably greater complexity, models for the hIS vs. the HIV virus are derived and evaluated. Two models for the hIS vs. cancer cells are explained and tested. Finally, some common conditions caused when complex physiological regulators fail are elaborated upon, including hIS dysfunctions and shock.

Chapter 9, "Quest for Quantitative Measures of Complexity," describes various measures by which one can quantify and compare system complexity. These include intuitive measures, sensitivities, structural (topological) measures, and measures based on information theory and statistics.

In Chapter 10, " 'Irreducible' and 'Specified Complexity' in Living Systems," we consider the faith-based, pseudoscientific approaches supporting irreducible complexity and specified complexity. We provide examples from evolution showing that irreducible complexity is in fact not irreducible, and specified complexity is a product of unsubstantiated, ad hoc mathematical formulas.

Chapter 11, "Introduction to Complexity in Economic Systems," introduces the reader to classical, steady-state supply-and-demand microeconomics, as well as many economic "schools of thought." We stress on Forrester's call for the use of dynamic models of economic systems in 1956. One reason for the complexity of economic systems is shown to be human behavior in response to information. Agent-based modeling (ABM) is also introduced in this chapter.

In Chapter 12, "Dealing with Complexity," we examine the works of two, little-known (in the United States) German complexity scientists: Dietrich Dörner and Fredrick Vester. General guidelines on characterizing and regulating complex systems are presented by Dörner, and Vester describes an heuristic "impact matrix" in which a medium-sized CNLS can be examined for the relative impact of one node in its SFG to another node, a first step toward determining edge gains or sensitivities.

For MATLAB® and Simulink® product information, please contact:

The MathWorks, Inc.
3 Apple Hill Drive
Natick, MA, 01760-2098 USA
Tel: 508-647-7000
Fax: 508-647-7001
E-mail: info@mathworks.com
Web: www.mathworks.com

Author

Robert B. Northrop majored in electrical engineering at MIT, Cambridge, Massachusetts, graduating with a bachelor's degree in 1956. He received his master's degree in systems engineering from the University of Connecticut (UConn) in 1958. As a result of a long-standing interest in physiology, he entered a PhD program in physiology at UConn, conducting research on the neuromuscular physiology of molluscan catch muscles. He received his PhD in 1964.

In 1963, he rejoined the UConn Electrical Engineering Department as a lecturer, and was hired as an assistant professor in 1964. In collaboration with his PhD advisor, Dr. Edward G. Boettiger, he secured a 5-year training grant in 1965 from the National Institute of General Medical Sciences (NIGMS) (NIH), and started one of the first, interdisciplinary biomedical engineering graduate training programs in New England. UConn currently awards MS and PhD degrees in this field of study, as well as BS degrees in engineering under the BME area of concentration.

Throughout his career, Dr. Northrop's research interests have been broad and interdisciplinary and have been centered on biomedical engineering and physiology. He has conducted research (sponsored by the U.S. Air Force Office of Scientific Research [AFOSR]) on the neurophysiology of insect and frog vision, and devised theoretical models for visual neural signal processing. He has also conducted sponsored research on electrofishing and developed, in collaboration with Northeast Utilities, effective working systems for fish guidance and control in hydroelectric plant waterways on the Connecticut River at Holyoke, Massachusetts, using underwater electric fields.

Still another area of his research (sponsored by NIH) has been in the design and simulation of nonlinear, adaptive, digital controllers to regulate *in vivo* drug concentrations or physiological parameters, such as pain, blood pressure, or blood glucose in diabetics. The development of mathematical models for the dynamics of the human immune system was an outgrowth of this research, and they were used to investigate theoretical therapies for autoimmune diseases, cancer, and HIV infection.

Dr. Northrop and his graduate students have also conducted research on biomedical instrumentation: An NIH grant supported studies on the use of the ocular pulse to detect obstructions in the carotid arteries. Minute pulsations of the cornea from arterial circulation in the eyeball were sensed using a no-touch, phase-locked, ultrasound technique. Ocular pulse waveforms were shown to be related to cerebral blood flow in rabbits and humans.

More recently, Dr. Northrop addressed the problem of noninvasive blood glucose measurement for diabetics. Starting with a Phase I Small Business Initiative Research (SBIR) grant, he developed a means of estimating blood glucose by reflecting a beam of polarized light off the front surface of the lens of the eye and measuring the very small optical rotation resulting from glucose in the aqueous humor, which in turn is proportional to the concentration blood glucose. As an offshoot of techniques developed in micropolarimetry, he developed a magnetic sample chamber for glucose measurement in biotechnology applications. The water solvent was used as the Faraday optical medium.

Dr. Northrop has written 10 textbooks and is currently working on two more: *Analog Electronic Circuits: Analysis and Applications*; *Introduction to Instrumentation and Measurements*; *Introduction to Instrumentation and Measurements* (2nd edn.); *Endogenous and Exogenous Regulation and Control of Physiological Systems*; *Dynamic Modeling of Neuro-Sensory Systems*; *Signals and Systems Analysis in Biomedical Engineering*; *Signals and Systems Analysis in Biomedical Engineering* (2nd edn.); *Noninvasive Instrumentation and Measurements in Medical Diagnosis*; *Analysis and Application of Analog Electronic Circuits to Biomedical Instrumentation*; *Introduction to Molecular Biology, Genomics and Proteomics for Biomedical Engineers* (with Anne N. Connor). He is currently working on *Analysis and Application of Analog Electronic Circuits in Biomedical Engineering* (2nd edn.) (MS in preparation, November 2010).

Dr. Northrop was on the electrical and computer engineering faculty at UConn until his retirement in June 1997. Throughout this time, he was director of the Biomedical Engineering Graduate Program. As emeritus professor, he teaches graduate courses in biomedical engineering, writes texts, sails, and travels. His current research interest lies in complex systems.

1 Introduction to Complexity and Complex Systems

The significant problems we face cannot be solved at the same level of thinking we were at when we created them. A quotation attributed to Albert Einstein.

Albert Einstein Quotes
S.F. Heart 2008

1.1 INTRODUCTION TO COMPLEXITY

This book is targeted toward readers with educational backgrounds in biomedical engineering, physiology, biophysics, molecular biology, and other life sciences who wish to learn how to better understand, describe, model, and manipulate complex systems. In this book I have stressed the complexity associated with biochemical, genomic, physiological, medical, ecological, and other living systems, and I also have introduced the reader to the complexity inherent in economic systems. This focus reflects the education, experience, interests, and bias of the author. Also, we should all recognize the complexity inherent in energy, weather, political, and sociological systems, to mention a few of many other complex systems we deal with on a near-daily basis. Hopefully, the material presented in this book will introduce those with interests in these systems to the tools and mind-set they need to effectively tackle complex systems.

To extract maximum information from this book, the reader should have had college courses in algebra, calculus, cell biology, and introductory differential equations, as well as basic courses in physics and chemistry. A background course in engineering systems analysis is also important.

In the fast-paced, twenty-first century world we live in, we are often challenged with having to project how our actions (or inactions) will affect events (e.g., the outputs) in a complex system. In order not to be blindsided by unexpected results, we need a systematic, comprehensive way of analyzing, modeling, and simulating complex systems in order to predict nonanticipated outcomes. Sadly, comprehensive models of complex systems are generally unattainable.

This chapter addresses how we can describe the important characteristics of certain complex systems, model them, and use the models to try to predict unanticipated behavior. We want to be able to work competently with complex systems in order to create desired results while simultaneously avoiding the pitfalls of the *Law of Unintended Consequences* (LUC). Our mathematical models of complex systems, in general, will be imperfect. Certainly, one of the major attributes of a complex system is the difficulty we have in modeling it. Even so, our imperfect models may still be able to capture the essence of complex nonlinear systems' (CNLSs') behavior.

We use the noun, *system*, a lot in this chapter. We define a system as a group of interacting, interrelated, or interdependent *elements* (also agents, entities, parts, states) forming or regarded as forming a collective entity. There are many definitions of systems that are generally context-dependent. (The preceding definition is very broad and generally acceptable.) There are many kinds of systems, some are simple, others complicated, and many are complex. Systems can also be classified as being linear or nonlinear. These are described below.

1.1.1 When Is a System Complex?

We often hear the words *complex* (*adj.* and *n.*), *complexity* (*n.*), and *complicated* (*adj.*) used, particularly in the media, to describe systems and situations. (Just count the number of times you hear or see the word, *complex*, during a day on television news or in the printed media.) Complex, complexity, and complicated all come from the Latin word, *complexus*, meaning twisted together, or entwined. Complexity is difficult to define; hence you will find many definitions given for it, depending on who is studying it, the type of system being studied, and how it is being studied. Broadly stated, we consider that *complexity* is a subjective measure of the difficulty in describing and modeling a system (thing or process), and thus being able to predict its behavior. Or we might view the complexity of a system or dynamic process as some increasing function of the degree to which its components engage in structured, organized interactions. Complexity also has been seen as a global characteristic of a system that represents the gap between component and parameter knowledge and knowledge of overall behavior.

Corning (1998) suggested that complexity generally has three attributes: (1) A complex system has many parts (or items, units, or individuals). (2) There are many relationships/interactions between the parts. (3) The parts produce combined effects (synergies) that are not easily foreseen and may often be novel or surprising (chaotic).

Lloyd (2001) compiled 45 different definitions of complexity [!], underscoring the plasticity of the definition problem. Quantitative measures of complexity necessarily must rely on the structure of mathematical models of complex systems, most of which are nonlinear and time-variable (nonstationary).

One condition that has been proposed for a system to be called complex is that its overall behavior is destroyed or markedly altered by trying to simplify it by removing subsystems. However, a simple system can also be easily altered by removing components or subsystems, rendering this criterion invalid. No one will dispute the complexity of the human brain. Yet it often appears to function with minimal disturbance following moderate injury, surgery, stroke, etc. This is because the central nervous system (CNS) is a complex adaptive system (CAS), capable of limited reorganization of its local structures and their functions. It possesses the property of robustness.

Edmonds (1999a,b) argued "…that complexity is not a property *usefully* attributed to natural systems but only to our models of such systems." Edmonds proposed the definition: "Complexity is that property of models which make it difficult to formulate its overall behaviour in a given language of representation, even when given almost complete information about its components and their inter-relations." The use of our models of natural complex systems to measure complexity has the advantage of working with quantifiable formalisms such as graphs and graph theory, sets of ordinary differential equations (ODEs), linear algebra, transfer functions, algebraic nonlinearities, etc. For model validation, we can derive and measure sensitivities from models and compare them with those measured on the natural system. Note that when we consider a complex natural system (such as the human glucoregulatory system) and a mathematical model formulated from it, we might be tempted to view the model as complicated by itself. This does not mean that we should infer that the natural system can therefore be demoted to complicated status, it only means that the model is parsimonious; it sacrificed much detail while preserving the essence of one or more aspects of complex natural behavior.

Rosser (2008) gave a definition of *dynamic complexity* based on that of Day (1994): "…systems are dynamically complex if they [their phase-plane trajectories] fail to converge to either a point, a limit cycle, or an exponential expansion or contraction due to endogenous causes. The system generates irregular dynamic patterns of some sort, either sudden discontinuities, aperiodic chaotic dynamics subject to sensitive dependence on initial conditions, multi-stability of basins of attraction, or other such irregular patterns."

A measure of organismic complexity was cited by Hintze and Adami (2008): "The complexity of an organism can be estimated by the amount of information its genome encodes about the environment within it thrives." Presumably, "about the environment" excludes proteins having housekeeping functions and includes proteins having I/O functions, such as transmembrane proteins. For example, the *information content* I of a molecular sequence s of length L encoding the bases 0, 1, 2, 3 can be estimated by $I = L - H(s)$, where the *entropy* of the sequence $H(s)$ is approximated by the sum of the per-site entropies $H(s) \approx \sum_{x=1}^{L} H(x)$, having a per-site entropy, $H(x) = -\sum_{i=0}^{3} p_i \log_4(p_i)$ (Hintze and Adami 2008). p_i is the probability of the ith base at a given site in s.

Suppose a natural CNLS is analyzed exhaustively by Team B of systems analysts. They assign all variables, including hidden ones, to nodes in a graph model, and *all* of the directed branches between the nodes are described, and a very large, complete digraph of the CNLS is made. Thus the CNLS is described mathematically by a very large set of equations so that Team B's knowledge of the model of the CNLS's structure and behavior is exhaustively complete. Is this well-described/explained CNLS model still complex? Now Team A views the same natural CNLS. It appears complex to them. So is complexity an inherent property of a CNLS, or does it require our ignorance and confusion about the system to exist?

In summary, complexity must lie in the eyes of the beholder; what is complex to one observer may be complicated to another, based on observer knowledge and skills. It is safe for us to say that, in general, complexity is relative, graded, and system- and observer-dependent. You will know it when you see it.

1.1.2 EXAMPLES

There are many examples of complex systems—all are characterized by having many parameters or states that are functionally interconnected, generally leading to nonintuitive system behavior. Complex systems have been subdivided into those with *fixed* (time-invariant or static) *structures, dynamic complexity, evolving complexity,* and *self-organizing, complex adaptive systems* (Lucas 2006). All of the complex systems we will consider in this book are dynamic, and thus rely on the use of differential or difference equations for their model descriptions. Complex systems are generally multiple-input, multiple-output (MIMO) systems. However, others are multiple-input, single-output (MISO), and single-input, multiple-output (SIMO); a few even have a single-input, single-output (SISO) architecture.

Some examples of simple systems include, but are not limited to: A flashlight, a mechanical wristwatch, a gyroscope, a power lawn mower, an electronic oscillator, an orbiting satellite, a pendulum, the van der Pol equation, etc. A simple system can be nonlinear and/or time variable, but its behavior is always predictable and easily modeled mathematically.

Some examples of complicated systems include, but are not limited to: A laptop computer, a modern automobile, an automatic dishwasher, a microwave oven, an iPod, an income tax form, etc. Complicated systems can generally be reduced to component subsystems and successfully analyzed and modeled. They generally do not exhibit unexpected, chaotic behavior.

Some examples of complex systems include, but are not limited to: Cell biology systems including intracellular biochemical (metabolic) pathways and genomic regulatory pathways (with emphasis on embryonic development). Physiological systems (especially those based on various multifunction organs—e.g., the CNS, the kidneys, the liver, the immune system, etc.). Also on the list are: Self-organizing and adaptive biological and man-made neural networks, ecosystems, economic systems, epidemics (HIV, H1N1, etc.), the stock market, energy market systems, meteorological systems, oceanographic systems, transportation systems, the World Wide Web, social systems (e.g., family groups, sports teams, legislative bodies); political parties, governments, management of multinational corporations, wars against insurgencies, health-care legislation, etc. A chess game played between two humans (as opposed to between a human and a computer) can be complex because of human behavior. Often we see ecological systems grouped with economic systems as examples of CNLSs. A major challenge in understanding the behavior of a complex system is how to formulate a mathematical model of that system that will allow us to predict the system's behavior for novel inputs. What are the system's relevant parameters (states), how are they functionally interconnected, what are the system's inputs and outputs, and what other systems do they interface with? Putting such information together into an effective, validated, and verified model can be daunting.

The boundaries between simple and complicated, and complicated and complex system designations are *fuzzy* (Zadeh et al. 1996) and debatable, even using quantitative measures of complexity. This is because humans draw the boundaries or set the criteria. Try ranking the complexity of human physiological systems. A candidate for the most complex is the human (h)CNS; the human immune system (hIS) might be second. A larger challenge is to name the least complex human physiological system.

Generally, a well-described complex system can be modeled by a graph composed of *nodes* (*vertices*) and *branches* (*edges*). The nodes represent the complex system's states or variables; the edges represent some sort of causal connections between the nodes. Nonlinearity and time variability are generally present in the relationships (branches, pathways, edges, or arcs) interconnecting the parameters or states (at vertices or nodes) of complex systems' models. Once modeled by a graph, a complex system's structure can be described by a variety of objective mathematical functionals used in graph theory, and its dynamic behavior can be studied by computer simulation.

Another, more complicated means of modeling a CNLS is by *agent-based simulation* (ABS). An agent inputs information and signals, and outputs behavior, signals, and information. They are autonomous decision-making units with diverse characteristics. A network of information flow can exist between neighboring agents. Some agents also can have "memories," and are adaptive and can learn. Computer models of "human" agents figure in *agent-based models* (ABMs) of political, economic, contagion, terrorism, and social CNLSs. In economic ABMs, human agent models can be programmed to produce, consume, buy, sell, bid, hoard, etc. They can be designed to be influenced by the behavior of other adjacent agents. ABMs can exhibit emergent, collective behavior. Dynamic ABMs have been used to model bacterial chemotaxis (Emonet et al. 2005) and have application in hIS modeling (immune cell trafficking). (See Section 11.2.8 for more on ABM.)

1.1.3 Properties of Complex Systems: Chaos and Tipping Points

As we have noted, complex systems, by definition, are hard to describe and their input/output relationships are difficult to understand. A large, dynamic, nonlinear, complex system is always a challenge to model. Generally, many, many coupled, nonlinear differential equations are required. Numerical rate constants in the equations are often poorly known or unknown and must be estimated. Often the rate constants vary parametrically; that is, their values are functions of certain variables or states in the complex system, and these parametric relationships must also be estimated.

A complex system may exhibit abrupt switches in its overall behavior (e.g., switch from a stable, I/O behavior to a limited oscillatory one, or from one oscillatory mode to another, or to unbounded (saturated) output(s)). The study of such labile, unstable, periodic, and aperiodic behaviors in nature and in dynamic models of CNLSs is part of *chaos theory*. These abrupt switches in complex system behavior are called *tipping points*; that is, a small change in one or more inputs, parameters, or initial conditions will trigger a major, I/O behavioral mode change. Such tipping point behavior has often been called the "butterfly effect," where ideally, a butterfly flapping its wings in Brazil many miles away from a weather system supposedly can trigger an abrupt change in the system, such as the formation of a tornado in Texas. With due respect to Edward Lorenz, the butterfly effect is actually a poor metaphor. Although all kinds of real-world complex systems, including social and economic systems (Gladwell 2002) can exhibit tipping points, a certain minimum increment in the level of a signal, energy, momentum, or information is required to trigger a rapid, global change in their behavior. In the natural case of weather systems, air is a viscous, lossy medium. The gentle air currents caused by a butterfly's wings are quickly attenuated with increasing distance from the insect. From the same dissipative property, sound dies out with distance, and it is impossible to blow out a candle from across a room. The so-called butterfly effect is more aptly applied to the behavior of our mathematical models than real-world CNLSs. In fact, it was in weather system simulations that meteorologist Lorenz (1963) in 1961 found that a change in the fourth significant figure of an initial condition value could trigger "tipping" and radical changes in the outputs of his model. In fact, the change was ca. $0.025\% = (100 \times 0.000127)/0.506127$. (See the description of the "Lorenz equations" in Section 3.4.3.)

Mathematical models of noise-free, deterministic, complex, nonlinear systems can exhibit chaotic behavior, in which output variables in the system oscillate or jump from level to level in what appears to be a random manner. However, on a microscale, this chaotic behavior is in fact, deterministic; it involves thresholds or tipping points. Its occurrence may at first appear random because it appears as the result of one or more minute changes in the CNLS's inputs, initial conditions, and/or parameters. Under certain conditions, a chaotic CNLS may exhibit bounded, steady-state periodic *limit cycle oscillations* of its states in the *phase plane*, or unbounded (saturating) responses. In certain chaotic CNLSs, a combination of initial conditions and/or inputs can trigger an abrupt transition from one nonoscillatory, steady-state, behavioral mode to oscillatory limit cycle behavior, then a further perturbation of an input or parameter can cause the system to go into a new limit cycle, or a new global behavioral mode.

One example of a natural, physiological, chaotic system is *atrial fibrillation* in the human heart. Some subtle factor(s) cause the normal, organized electrical conduction between atrial heart muscle cells to fail, and the atrial muscles contract weakly and asynchronously in local groups in response to the S-A pacemaker signal. Atrial fibrillation (AFib) results in poor filling of the ventricles, hence reduced cardiac output and a propensity to form emboli. In idiopathic AFib, the exact cause or trigger of the condition is not known, it appears suddenly.

Untreated, AFib can lead to embolic strokes from blood clots that form in the atria. AFib can be treated by surgery, and a variety of drugs.

Ecologists modeling the variation of a species' population as it interacts with its environment (niche) have found that chaotic behavior can be observed under certain circumstances that mimic data from field studies. In some cases the models do not show the "chaos" observed in field studies, underscoring that the art and science of mathematically modeling complex ecosystems is challenging (Zimmer 1999).

Much of the research on chaotic behavior has used mathematical models consisting of systems of coupled, nonlinear ODEs. Some of the examples of these complex mathematical systems entering chaotic behavior have been traced to round-off errors in numerical simulation of the ODEs. That is, the tipping point for the system was so sensitive that it responded to equivalent digital (quantization, round-off) noise in the computations.

The author has found from personal experience that the choice of integration routine is very important when simulating a CNLS described by a set of nonlinear, *stiff* ODEs. A poor choice of integration routine was seen to lead to chaos-like noise in plots of system output variables (Northrop 2000, 2001). The noise disappeared when simple rectangular integration with a very small Δt was used.

If one plots the derivative of the kth output vs. the kth output of a self-oscillating CNLS in the steady state (i.e., $\dot{x}_k$ vs. x_k), the resulting 2-D phase plane plot will contain a closed path because x_k is periodic. This steady-state, closed, phase-plane trajectory is called a *simple attractor*. CNLSs may also exhibit strange attractors in which $x_k(t)$ is still periodic, but its SS phase trajectory contains multiple, reentrant loops, rather than a simple closed path. The Lorenz weather system is a good example of a system with a strange attractor (see Section 3.4.3).

In examining the apparently chaotic behavior of real-world CNLSs, chaos theorists have the problem of separating behavior that is intrinsic (caused by noise-free system dynamics), and behavior that is the result of random noise entering the system from within or without. Such studies rely heavily on statistical analysis of output time series, and also make use of the *Lyapunov exponent* to analyze the fine structure of trajectories near their attractors in the phase plane. The Lyapunov exponent is a logarithmic measure of whether trajectories are approaching or diverging from an attractor (see the Glossary).

1.1.4 LAW OF UNINTENDED CONSEQUENCES

As noted above, the nonintuitive, unexpected, chaotic behavior of imperfectly described, complex nonlinear systems has led to the consensual establishment of the LUC. An unintended consequence is where a new action (input) to a CNLS (or an incremental change in an input) results in a behavior that is not intended, or not expected. The unintended outcome can be adverse, fortuitous, or have zero impact.

There are several circumstances that can lead to unanticipated, unforeseen, or unintended consequences: (1) Actions taken (or not taken) when there is *ignorance* of a CNLS's complete structure (part of the definition of a CNLS). (2) Actions taken (or not taken) when there is a *lack of information about* the system's *sensitivities.* (3) Actions taken (or not taken) by humans based on greed, fear, anger, faith, and other cognitive biases, rather than facts. (4) Actions taken (or not taken) in systems with time lags. For example, a corrective input produces no immediate effects, so the operator interprets the initial lack of response as a need for a stronger action, resulting in a delayed, excessive output, which in turn promotes another stronger, corrective input action, etc., leading to system–operator instability, or a very long settling time. "Actions" refer to such things as the excessive administration of drugs (such as antibiotics to factory-farmed poultry and animals), chronic application of pesticides, weed killers, and fertilizers to the environment, the passage of new laws regulating human behavior, application of new economic regulations and taxes, and the introduction of new technologies (e.g., the World Wide Web, cameras, and global positioning systems [GPS] in cell phones, text messaging, and fish-locating sonar).

A sociological corollary to the LUC is called *Campbell's law* (CL) (Campbell 1976, Gehrman 2007). Campbell stated: "The more any quantitative social indicator is used for social decision-making, the more subject it will be to corruption pressures and the more apt it will be to distort and corrupt the social processes it is intended to monitor." Yes, regulations applied to social and economic systems can have some unintended consequences; however, an absence of regulation can often lead to many more unintended consequences. To regulate or not to regulate, that is the question (Tabarrok 2008), especially in economic systems; also, how to regulate optimally? (The control of CNLSs was initially considered by Ashby 1958.)

The LUC and CL are not strict physical or chemical laws that have mathematical descriptions like the "law of gravity," "Newton's laws," or the "laws of thermodynamics." The LUC and CL are more in the realm of Murphy's law. The LUC is often, but not always, manifested as part of the behavior of CNLSs. Examples of the expression of the LUC are demonstrated by certain recent ecological, political, social, and economic events, for example, the increase in food prices caused by the diversion of a large fraction of the U.S. corn crop into the production of ethanol for a gasoline additive, and soy beans to manufacture biodiesel fuel. (Because of system complexity, food price increases are not solely due to crop diversion.)

Introduction of xenospecies into ecosystems can have adverse, unexpected effects. For example, rabbits were introduced into Australia for hunting sport and rapidly became a crop pest, competing with sheep and cattle for forage. Kudzu vine was introduced into southeastern United States for erosion control and rapidly spread, choking out native species of trees and bushes. The Eastern Snakehead Fish was introduced into several U.S. southeastern lakes and waterways in the early twenty-first century. This invasive species grows rapidly; it is a hardy, voracious, apex predator that feeds on all native fish species (catfish, bass, trout, etc.) and destroys these native gamefish. The Snakehead is an East Asian delicacy, and was imported illegally to the United States as a specialty food. Unfortunately, some were released into local ponds and streams in Maryland and Virginia.

Ecosystems are complex; species' populations are held in balance by predation, weather, available food, pathogens, etc. There were few natural predators to eat the Australian rabbits, and they had plenty of food and a rapid reproductive rate. Eventually, to control their population, a lethal virus specific for rabbits (*Myxoma*) was introduced that caused myxomatosis. Myxomatosis symptoms range from skin tumors (in cottontail rabbits), to puffiness around the head and genitals, acute conjunctivitis leading to blindness; there is also loss of appetite and fever, as well as secondary infections (in European rabbits).

In ecosystems, another way we have seen the LUC manifested is by the creation of insecticide-resistant crop pests as the result of growing genetically modified (GM) crops with built-in genes to make *Bacillus thuringiensis* (BT), a natural insecticidal protein designed to protect that particular crop. Another ecosystem example of the LUC also involves GM crops: The genes for a crop plant were modified to make the plants resistant to a specific herbicide (e.g., Roundup™). It was subsequently found that the herbicide resistance was transferred via the crop plants' pollen to weeds related to the crop, complicating the chemical weeding process of the crop (Northrop and Connor 2008). The LUC also occurs in complex physiological systems. For example, one side effect of taking the drug Viagra™ in some individuals is impaired vision. The metabolism of the neurotransmitter, nitric oxide (NO), may be altered in the retina, as well as in the penis.

Still another example of the untoward effects of the LUC was seen in the late 1930s when the "dust bowl" destroyed Midwestern U.S. agriculture and drove over a million people out of the states where it occurred (Texas, Oklahoma, New Mexico, Colorado, Kansas, and Nebraska). Grassland prairie had been a stable ecosystem for millennia, resistant to floods and droughts. It was a major energy source for grazing prairie animals including bison and antelope. In the 1930s, land speculators and the United States Department of Agriculture (USDA) encouraged the sale of this cheap grassland for farming money-making grain crops. The sod was "busted" by the newly available, powerful diesel tractor gang plows, and grain crops planted. They did well at first because of above-average rainfall. No one bothered to ask what would happen if drought occurred. When several years of severe drought did occur in the mid-1930s, crops failed catastrophically, and the soil, unprotected by sod, suffered incredible wind-caused erosion. An area of the size of Pennsylvania was turned to desert; the topsoil blew away (Condon 2008). Clearly agriculture and weather systems are closely linked. It has taken years of hard work to ecologically reclaim most of this wasted land.

The Prohibition Amendment to the U.S. Constitution in the 1920s was intended to curb alcohol abuse and public intoxication. It also energized an entire industry of moonshining, smuggling, and speakeasies. Addiction proved to be a more powerful motivation to produce and consume alcohol than the governmental proscription of alcohol sales and consumption, hence the rise of organized crime in the alcohol business. The unintended consequence of the United States "War on Drugs" is the spawning of violent criminal organizations devoted to growing, smuggling into, distribution, and sale of drugs in the United States. Aircraft, and now boats and "submarines," are used for the importation of drugs by sea, and clever schemes have been used to bring them in by land routes (tunnels on the

U.S./Mexico border). Again, addiction, a very powerful motivator, was not used in the decision-making model.

Note that the LUC can demonstrate fortuitous results, as well as adverse ones. In 1928, bacteriologist Alexander Fleming "accidentally" discovered the antibiotic penicillin by noting the bactericidal effect of unwanted *Penicillium* sp. mold growing in old cultures of *Staphylococcus* sp. in his lab. Fleming went on to isolate the active agent from the mold and named it penicillin. He found it was bactericidal for a number of Gram-positive bacterial pathogens, for example, those causing scarlet fever, pneumonia, gonorrhea, meningitis, and diphtheria, but not typhoid or paratyphoid. His discovery led to the eventual development of many important antibiotics, some natural and others synthetic. Fleming shared a Nobel prize for his discovery in 1945.

Ironically, over-medication with penicillin and related antibiotics has also unleashed the dark side of the LUC. That is, through the evolution of antibiotic-resistant bacterial strains, particularly in hospitals, leading to life-threatening nosocomial infections including methicillin-resistant *Staphococcus aureus* (MRSA). The chronic use of penicillin and tetracycline in cattle feed in factory farms raising hogs, cattle, and chickens has also led to the emergence of MRSA in the farms, and in about 40% of hog farm workers (Wallinga and Mellon 2008).

Another example of the good side of the LUC involves common aspirin, long used for mitigation of headaches and joint pain. It was soon discovered that aspirin has the beneficial effect of inhibiting blood clotting, an unintended consequence leading to the use of low-dose aspirin as prophylactic against coronary heart attacks and embolic strokes. An overdose of aspirin, however, can lead to hemorrhage and hemorrhagic stroke (again, the dark side).

Often two or more complex systems can interact in an undesired manner. A recent example involves the planetary weather system and the world energy resource system (Searchinger et al. 2008). Because the global demand for petroleum is exceeding the supply, prices of crude oil steadily increased in 2008, topping over 147 US$/bbl. on July 11, 2008. Following this inflationary increase, the light crude price steadily decreased to below $50/bbl. at the end of November 2008. It has risen again slowly and irregularly as demand trumps supply and the recession. Sure enough, light crude was over 82 US$/bbl on March 30, 2010.

To satisfy the insatiable world demand for gasoline, ethanol distilled from fermented sugarcane and corn is used as a fuel additive in concentrations ranging from 10% to 85% in order to stretch the petroleum-based gasoline supply. This ethanol is called a *carbon-neutral* (green) biofuel because it comes from yearly plant crops that use CO_2 to grow.

In the United States, the demand for ethanol has resulted in farmers planting more corn (instead of say, soybeans, or vegetables). Sadly, increased carbon-neutral ethanol production from corn has resulted in reduced production of corn for food use, the reduction of arable land available to grow other food crops as well as an increase in the price of corn/bu. An obvious unintended consequence of this shift in corn usage has been increased food prices, especially beef, pork, and chicken, traditionally fed on corn. (The residue of the corn fermentation process is fed to animals, but most of the carbohydrate energy in the residue has been lost to the fermentation process while making ethanol.) Still another unintended consequence of biofuel production is the consumption of significant amounts of fresh water for irrigation, depleting public water supplies in areas already having low rainfall.

Another fuel substitute that is carbon-neutral is biodiesel made from natural plant oils. New, arable land is being reclaimed from forests in order to grow soybeans for both food and biodiesel in Indonesia, Brazil, China, and India. Biodiesel is also derived from oil seed crops such as canola, corn, cottonseed, crambe, flaxseed, mustard seed, palm oil, peanuts, rapeseed, and safflower, to name a few sources of plant oils (Faupel and Kurki 2002, Ryan 2004).

Where the LUC comes into play in the production of "green" alternate fuels, it has to do with CO_2 emissions that affect the planetary weather system, in particular, global warming. In their recent paper, Searchinger et al. (2008) pointed out that the extensive, new, arable land reclamation required to grow carbon-neutral fuel crops is not without cost. The heavy machinery used for land clearing, tillage, planting, and harvesting all burn fossil fuels and emit CO_2 over and above the CO_2 that will be formed when the ethanol and biodiesel are burned. The industrial production of fertilizers and herbicides for these new crops also adds to the CO_2 burden. Searchinger et al. have calculated the time required for a new, carbon-neutral fuel crop to achieve a net reduction in CO_2 emissions, considering the carbon cost of crop production. Net reduction times were found to range from 167 to 423 years, depending on the fuel crop and its location!

One way to solve the unintended consequence of this excess CO_2 emission is to derive ethanol from crop waste, compost, garbage, and switch grass crops that require minimum energy for tillage and processing. GM bacteria and yeasts will be required to improve fermentation efficiency. Another way to reduce CO_2 emissions is very obvious; burn less fuel by mandating increased fuel efficiency of vehicles, use more public mass transportation, also improve insulation efficiency of buildings and rely more on noncarbon energy sources (solar, wind, hydro, tidal, nuclear). The United States is sadly lagging in these efforts.

Genetic engineering may eventually save the day in the direct production of petroleum substitutes. Researchers at Yale headed by Prof. Scott Strobel have recently found a species of fungus that produces compounds found in diesel oil (Labossiere 2009). They are attempting to sequence the fungus' genome and locate the genes that regulate the biosynthesis of these compounds. Rather than trying to extract the compounds directly from the slowly growing fungus as an industrial process, we expect genetic engineers to transfer the fungal genes for these compounds into GM bacteria such as *Escherichia coli* that can be grown in huge batches under industrial conditions. The diesel components will then be extracted from the GM bacteria. But what will be used as an energy source for the GM *E. coli*, and what will it cost?

Like every scientific discipline, the study of complex systems has evolved its own jargon, acronyms, and abbreviations; many of these will be found in the

Glossary of this book, along with selected entries for cellular and molecular biology, physiology, and economic systems.

1.1.5 COMPLEX ADAPTIVE SYSTEMS

CAS are dynamic CNLSs that possess an internal adaptive capability that endows them with robustness in the face of internal and external (environmental) disturbances (Schuster 2005). More specifically, their adaptive capability comes from their architecture; they monitor certain of their internal states as well as their outputs, and use this information to adjust ("tune") internal branch gains in order to maintain normal functions. This parametric regulation tends to be highly dispersed and decentralized. CAS's adaptive behavior is the result of many decisions made continuously by many individual *agents*. Holland (1992, 1995) stated that a CAS behaves according to three main principles: (1) Their order is emergent as opposed to predetermined; that is, it evolves. (2) The CAS's history is irreversible. (3) The system's future states are often unpredictable (chaos). *Agents* scan their environment and develop schema representing interpretive and action rules.

Adaptive artificial neural networks used in pattern recognition are one manifestation of CASs. Another class of CAS is *social systems*, that is, the system agents or entities are humans, ants, termites, bees, etc. In their text on social CASs, Miller and Page (2007) treat such topics as complexity in social worlds, emergence, agent-based models, banks, social cellular automata, segregation, city formation, etc. In an interesting review paper, Lansing (2003) considered CASs in the framework of evolutionary biology, automata, ecology, economics, and anthropology. He gave examples of adaptive landscapes, including NK (automata) systems. However, in this chapter, we are primarily interested in biological CASs.

Probably the best-known biological CAS is the human central nervous system (brain) (hCNS). The hCNS is very large (ca. 10^{11} neurons [ca. 10^4 different types], 10^{12} glial cells, and 10^{13-14} dendrites and synapses), has modular anatomy (pons, cerebellum, medulla, corpus callosum, various cortical regions, limbic system and its components, thalamus, hypothalamus, hippocampus, amygdala, various nuclei, etc.), and has redundant elements capable of plastic behavior that allows new neural circuits to be made that can replace damaged elements and modules. While new neurons generally cannot arise, new connections (synapses, dendrites) can grow in the rewiring process. Synapses can transmit neural information chemically (by neurotransmitter release from dendrite terminals [boutons]), or electrotonically through intimate conductive coupling of neuron–neuron membranes (Kandel et al. 1991). The number and strength of synaptic connections can be modulated by neurotransmitters and hormones.

A second example of a biological CAS is the adaptive arm of the hIS. You will see that through the processes of phagocytosis, antigen presentation, and clonal selection, populations of specific antibodies (Abs) can be produced, and the information for that Ab clone stored in the genome of special "memory cells."

1.2   WHY STUDY COMPLEX SYSTEMS?

If you are a biomedical engineer, a biologist (of any subdiscipline), a physiologist, a physician, an agronomist, ecologist, pharmacologist, economist, politician, stock market speculator, politician, etc., you have already encountered complex systems, and have had to make decisions concerning how to interact with it. We have already commented on the propensity of complex systems to respond in unexpected or unanticipated ways to our inputs and disturbances, and we have cited the LUC (the generation of both untoward and beneficial, unanticipated "side-effects").

With complex systems, some things happen way beyond our collective control (e.g., tides, climate, rainfall, monotonically increasing world population growth), other things we have marginal control over (e.g., total CO_2 emissions, petroleum consumption), and other parameters we can manipulate as inputs to try to manipulate output variables (e.g., the federal funds prime interest rate that banks charge each other set by the Federal Reserve; the acreage of corn planted for ethanol production; subsidies paid to farmers to grow corn for ethanol; subsidies paid by governments to support food agriculture; Chinese government subsidies supporting Chinese gasoline production that keeps pump prices artificially low and encourages consumption and the consequent air pollution.)

As we have remarked, complex systems are called that because they are difficult to describe, analyze, and model. They cannot be understood by inspection or intuition. Thus if we are to deal with them intelligently, we need a set of "tools" to describe, analyze, and model them in order to be able to manipulate certain of their output parameters with minimum unintended consequences. For example, hIS is an adaptive, complex, nonlinear, molecular biological/physiological system. We can approximately model certain aspects of its behavior using the ODEs for chemical mass-action and diffusion (see Sections 8.4.3 and 8.4.4). All of the immune system's relevant cells, cytokines, autacoids, and proteins must figure in the model, as well as other neural, hormonal, and exogenous drug inputs. Such equations are gross approximations of what really goes on at the molecular level, yet they give us some quantitative as well as qualitative insight into what will happen, for example, if a patient were to be treated with new drugs, or certain (exogenous) immune system cytokines.

We have learned to manipulate the living hIS in various ways: We can suppress certain components of it to control tissue transplant rejection and autoimmune disease symptoms. More specifically, we can immunize (cause the hIS to create specific antibodies) against specific pathogens, and desensitize it to specific exogenous allergens. In other words, medical science has learned to manipulate some features of the immune system with a certain degree of confidence. Still on our immune system to-do list, however, is causing it to attack and kill specific cancer cells, and totally destroy parasites, chronic infectious pathogens (e.g., syphilis, malaria, and Lyme disease), and persistent viruses such as HIV.

1.3   HUMAN RESPONSES TO COMPLEXITY

In general, human thinking has not evolved to deal effectively with complexity and complex systems. Modern humans have descended from short-lived, hunter-gatherers whose greatest threats were sudden physical attacks by apex predators, other humans, and diseases. These threats were generally simple in their origins. Today, the threats to the sustainability of the human race include a spectrum of relatively slowly happening adverse conditions involving the complex systems that now surround us, including (1) a slow, steady increase in the world's population; (2) slow climate change due to rapidly rising greenhouse gas concentrations (causing global warming, storms, floods, droughts, and food shortages); (3) a slow decline in the availability of agricultural and potable water due to climate change and pollution from agriculture and industry; (4) slow depletion of petroleum and natural gas availability while demand for energy is rapidly growing; (5) the relatively rapid evolution of antibiotic-resistant bacteria, pesticide-resistant insect pests, and herbicide-resistant weeds; (6) the slow decay of metropolitan infrastructures, including highway bridges, metropolitan underground water distribution systems and sewer systems; and (7) the overloaded capacity of electrical power grids and the rising need for more electrical power. (Electric power is replacing fossil fuels in transportation systems and industries, because it is perceived as more "green.")

In general, if something bad happens slowly enough, it fails to trigger a meaningful, *organized* human response to combat it until martyrs are made. It appears that the human brain has evolved the ability to remember and predict the timing and location of dangers to individuals before they occur. But we are handicapped by the high threshold in our ability to detect the rate of change of the cues that trigger meaningful, *mutual, cooperative, corrective responses*. For example, it appears that we need to have interstate highway bridge collapses and lives lost before our attention is drawn to the need for costly prophylactic bridge repairs and scheduled maintenance. (See the article on the slowly degrading infrastructure in the United States by Petroski [2009].)

Our innate ability to deal with complex systems, and our predictive and correlative abilities as a species are limited, especially in detecting and correcting slow disasters. We need specific training in the area of dealing with complex systems.

We humans generally tend to oversimplify our analyses of complex systems and their models. We tend to use intuition instead of scientific method and verifiable data, and we apply *reductionism*, often assigning *single causes (or cures)* for any unwanted outputs or conditions (e.g., for the steadily increasing price of gasoline). I call this the *single-cause mentality*. Such single-cause thinking is generally based on hunches and poor extrapolations rather than on observed facts and detailed, quantitative models. It can lead to fixing the blame on a single individual (a scapegoat), such as a political leader, when something goes wrong in a complex system (i.e., unemployment), or when supposed corrective actions produce unanticipated delayed results.

We often rely on the manipulation of a single input to try to attain the desired effect in a complex system (e.g., adjusting the prime rate to stimulate the national economy). Complexity science has indicated that it is often more effective to try multiple inputs and select those that prove more effective in obtaining the desired result. In other words, we need to find the CNLS's *gain sensitivities* for the output states under attention. Gain sensitivities can be estimated by simulation, and of course, measured on the actual CNLS whenever possible. However, such measurements are only valid on a CNLS around a stable, steady-state *operating point*. That is, the system is not oscillating or responding to a time-varying input.

Another shortcoming of humans in interacting with coupled, complex systems is the "not in my box" *mentality* (Smith 2004). This is a myopic approach to manipulating a complex system in which the manipulator has been restricted to dealing with a circumscribed part of a CNLS, (e.g., a module) neglecting the effects of his or her manipulations on the balance of the system. ("It's not my responsibility.") The module boundary may have been set arbitrarily by the manipulator, or by poor planning and external rules. The "not in my box" mentality provides a shortcut to the results of the LUC.

A recent illustration of "not in my box" behavior was seen in the recent attempted terrorist downing of a Detroit-bound airliner with an explosive (PETN) on December 25, 2009 by a Nigerian student Umar Farouk Abdulmutallab. Abdulmutallab's father, concerned by his son's association with religious radicals in Yemen, visited the U.S. Embassy in Abuja, Nigeria, on November 19, 2009 and reported his son's behavior. The embassy staff, following protocol under a security program called *Visa Viper,* promptly sent information on to the National Counterterrorism Center in Washington for entry into their database. According to Johnson (2009), "…neither the State Department nor the NCTC ever checked to see if Abdulmutallab had ever entered the United States or had a valid entry visa; information readily available in separate consular and immigration databases. 'It's not for us to review that,' a State Department Official said" (my underline). Such checking was evidently "outside their box." According to the U.S. Homeland Security Department Abdulmutallab had twice obtained U.S. visas and had actually visited the United States twice before (in 2005 and 2008), and had once been denied a U.K. visa. Obviously, the complex communications network involved in U.S. counterterrorism efforts still need "tuning."

In Section 1.5, we make the case for a formal educational background (i.e., courses) on how to analyze, model, and manipulate complex systems, and avoid the pitfalls of reductionist thinking, the single cause, and "my box" mentalities.

We see many examples every day of reductionism, the single cause, and "my box" mentalities in the way people deal with complexity. For example, consider the U.S. economy, including energy costs, recession, and inflation. We were told that manipulation of the *federal funds lending rate* by the Federal Reserve System would dampen inflation and the symptoms of recession. The theory had merit, but certain effects of this lending rate reduction are only felt after a destabilizing *time delay* of ca. 12 months. The recession of 2008–2009 occurred anyway.

The U.S. government also tried to stimulate the economy by modest tax refund payments in May 2008. There was uncertainty whether this cash injection would be effective in reducing the symptoms of recession. Economists debated whether this cash would be used to pay for local goods and services including gasoline, spent to pay off credit card and mortgage debts, or saved. Clearly, little research on human economic behavior had been done on how the refunds would be used before the decision to provide them was made. In retrospect, the United States officially entered a recession (as of November 2008), in spite of the refunds. Were they too little, too late, or the wrong input, or all of the above?

Consider the complex issues in the corn-for-ethanol debate. In an Op-Ed article, "Food-To-Fuel Failure" by Lester Brown and Jonathan Lewis that appeared in the *Hartford Courant* on April 24, 2008, some of the factors affecting this energy/economic system were discussed. This article rightly lamented the failure of the congressional "food-to-fuel" mandate. The authors pointed out that although ca. one quarter of the U.S. corn crop is now used for ethanol production to add to automotive gasoline, there has been only a 1% reduction in the country's oil consumption. (Most gasoline sold in the United States is now 10% ethanol, in Brazil, 85% ethanol is common.) One of the goals of adding ethanol to gasoline was to increase the available fuel supply, so only a 1% reduction in oil consumption is not surprising; in fact, this small reduction may have been due to the high and monotonically rising cost of gasoline through October 2008.

Also, according to Brown and Lewis, the food-to-fuel mandates have caused "in large part" the current escalation in food prices. However, the riots that have taken place in Haiti and Egypt are because of food shortages and high prices, and are not directly linked to U.S. ethanol production. Food is a finite resource, and global food shortages are caused by a number of factors, including but not limited to bad weather (drought in Australia wheat farms in 2007), insect pests, as well as excessive demand linked to steady world population growth, and standard of living increases in China and India. Yes, there is a close causality between higher U.S. food prices and corn crop diversion from food use. However, the high prices of diesel fuel and gasoline must also be factored into the prices of foods that are transported to U.S. markets, some from Mexico, Central, and South American countries in the winter. Farmers must pay more to run their tractors for plowing and harvesting, adding to food costs. Americans expect many of their foods to be wrapped in plastic on foam trays, and take their food purchases home in plastic bags, all made from expensive petroleum derivatives. Also, expensive fossil fuel energy is required to produce the fertilizers, herbicides, and insecticides used in modern agriculture. Diversion of arable land to grow more corn for ethanol has caused farmers to plant less wheat. Projected shortages of wheat and corn have led to a spate of speculative buying of these grain futures, further driving up their cost, and consequently the cost of foods made from, or fed from, corn and wheat. These foods not only include those for human consumption, but also those consumed as cattle and poultry foods. In a kind of chain reaction, as shortages appear in corn and wheat, prices of alternate grains (including rice, oats, barley, etc.) are also driven up by international investors, speculators in grain futures, and market demand. We expect that higher-yield, GM grain crops will be planted and consumed, regardless of the ecological and health consequences. (In modeling the CNLS of the world food supply, how do you use agents to model greed and paranoia-including grain hoarding?)

To further illustrate the single-cause mentality, the inability of otherwise thoughtful persons to handle complex system thinking, consider the matter of Lyme disease and how it is spread. The Lyme disease pathogen, a prokaryote spirochete not unlike that which causes syphilis, is spread by the deer tick, *Ixodes scapularis*. Infected *Ixodes* ticks were initially found in southcentral Connecticut and are now found in rural areas throughout Connecticut. Lyme disease now ranges from southern Maine, southern Massachusetts, Cape Cod and the islands, Long Island, Connecticut, S.E. New York, New Jersey, Maryland, S.E. Pennsylvania and strangely, western Wisconsin. Reservoirs for the *Borrelia burgdorferi* pathogen include small ground-dwelling mammals including voles, mice, shrews, moles, rabbits, and chipmunks and perhaps squirrels, as well as whitetail deer. Ground-feeding migratory birds are also infected *Ixodes* hosts, and provide rapid, long-range transit for the ticks to spread the disease (Gylfe et al. 2000). When infected *Ixodes* adults (and their larvae and nymphs) feed on human blood, they infect their hosts with Lyme disease that is now a serious epidemic in rural southern New England, New Jersey, Maryland, and Long Island. The treatment of chronic Lyme disease is now controversial; one line of treatment that uses very long-term administration of antibiotics is apparently not based on objective clinical or laboratory evidence of infection as a diagnostic criterion (Federer et al. 2007).

In March 23, 2008, an editorial appeared in the *Hartford Courant* (newspaper) entitled, "Taking Aim at Lyme Disease." The editorial writer advocated severely reducing whitetail deer populations in the tick-infested regions as a means of Lyme disease epidemic remediation. Hunting by man presumably would be the method employed. This editorial is an excellent example of the single-cause mentality applied to a complex ecosystem. N. Zyla responded on March 25, 2008 to this editorial with a letter in the *Courant* entitled: "Are Deer the Biggest Culprit?" Zyla cited an article in the January 22, 2008 *New England Journal of Medicine* (*NEJM*) in which the point was made that Lyme disease… "infects more than a dozen vertebrate species, any one of which could transmit the pathogen [*Borrelia*] to feeding ticks and increase the density of infected ticks and Lyme disease risk." The research found that deer, in particular, "are poor reservoirs" for *Borrelia*. The *NEJM* article correctly commented: "Because several important host species [mice, chipmunks and shrews] influence Lyme disease risk, interventions directed at multiple host species will be required to control this epidemic." Yes, deer are mobile hosts in terms of spreading the ticks, but eliminating them will not eliminate the reservoirs of *Borrelia*-infected small mammals and birds, or the ticks. Zyla correctly urged looking "at the bigger picture."

Another letter to the editors in the March 25, 2008 *Hartford Courant* from S. Jakuba, observed that the ticks not on animal hosts have predators in the form of ground-nesting birds (e.g., pheasants, partridges, wild turkeys, chickens, etc.) (but to what extent?). He also observed that coyotes prey on these birds (and their eggs). (So do foxes, weasels, and hawks.) Jakuba advocated getting rid of the coyotes, as well as reducing the deer population. We note that coyotes also feed on mice, voles, moles, shrews, rabbits, and chipmunks; so eliminating coyotes in the Lyme zone could activate the LUC by contributing to a population increase of these *Borrelia* reservoir animals.

We note in closing this section that mature *Ixodes* ticks climb up onto the leaves of bushes and tall grasses where they are relatively safe from predation by ground birds, and there they can easily be brushed onto passing humans, dogs, and deer. Ecosystems are complex; there are no simple fixes for the spread of Lyme disease.

1.4   COMPLEX SYSTEMS ENGINEERING

Our confrontation as a species with complexity and complex systems has recently started an accelerating trend in various college and university curricula of adding courses on complex systems analysis and engineering. For an overview and a call to study complex systems science, see the web essay edited by Bourgine and Johnson (2006). The online papers by Sheard and Mostashari (2008) and Sheard (2007) entitled that by *Principles of Complex Systems for Systems Engineering* also describe complex systems and CASs, and treat three examples of complex systems engineering: (1) INCOSE: (Guide to the systems engineering body of knowledge), accessed on December 3, 2009 at: http://www.incose.org/practice/guidetose-bodyofknow.aspx, 2006. (2) The systems engineering process within a company. (3) The National Air Traffic Control System. Note that Sheard and Sheard and Mostashari consider CAS examples that have human agents. They also consider *Systems-of-Systems* (SoS), of great interest to physiologists and cell biologists.

In 2001, Ferreira called for the establishment of university courses in *Complex Systems Engineering*. Indeed, at present (2009), one can find online descriptions of many academic courses dealing with complexity. Several U.S. universities now have created complex systems study programs; for example, Arizona State University has an interdisciplinary study program on Complexity that lists many courses. See: www.asu.edu/clas/csdc/courses.html. The Binghamton University Bioengineering program offers three complexity courses: BE 410 *Complexity in Biological Systems*, BE 450 *Complex Systems and Evolutionary Design*, and BE 451 *Complex Systems Engineering*. (See http://buweb.binghamtom.edu/bulletin/courses.asp? program_id=1016/ accessed November 06, 2009.) Brandeis University has the Benjamin and Mae Volen National Center for Complex Systems that focuses on cognitive neuroscience. See www.bio.brandeis.edu/volen/. Brown University's computer science department has the Von Neuman Research Program in Biological Complex Systems: See www.cs.brown.edu/~sorin/lab/pages/proj_vnrpbcs.html. Duke University has a Center for Nonlinear and Complex Systems (CNCS 2008) that encompasses complexity-related topics in biology, ECE, math, physics, philosophy, ME, computer science, etc. The computer science department at Iowa State University has a Complex Adaptive Systems Group; See www.cs.iastate.edu/~honavar/alife.isu.html. Johns Hopkins University offers courses: 605.716—*Modeling and Simulation of Complex Systems*—See http://ep.jhu.edu/course-homepages/viewpage.php?homepage_id=2651. Also at JHU: 645.742 *Management of Complex Systems*—see http://ep.jhu.edu/courses/645/742. Northwestern University has the Northwestern Institute on Complex Systems (NICO); see www.northwestern.edu/nico/. The University of Alaska offers an interdisciplinary program on complex systems; courses are specified. See www.uaa.alaska.edu/complexsystems/. UCLA offers an interdisciplinary minor degree in Human Complex Systems. See http://hcs.ucla.edu. The University of California at Davis listed potential CSE courses on February 26, 2005: *Advanced Complex Systems I & II*, a 1 year sequence in *Engineering Complex Systems*, and a 1 year sequence in *Complex Earth Systems*. (See http://cse.ucdavis.edu/public/news/cse-teaching-plans-potential-cse-courses/ Accessed November 6, 2009.) The University of New Hampshire also has a Complex Systems Research Center (UNHCSRC 2005) focusing on the study of Earth, oceans, and space. The University of Michigan Rackham School of Graduate Studies listed nine graduate courses in complex systems. See http://www.umich.edu/~bhlumrec/acad_unit/rackham/degree_req/www.rackham.umich.edu/... Accessed November 6, 2009. The University of Michigan's Center for the Study of Complex Systems can be reached at: www.cscc.umich.edu/education/grad/CSCS-courses/cscs-f09.html/. The University of Vermont has established a Complex Systems Center in their College of Engineering and Mathematical Sciences. The UVM program lists courses from biology, computer science, and mathematics. See, for example, the UVM course: 300—*Principles of Complex Systems*, at URL: www.uvm.edu/~pdodds/teaching/courses/2009-08UVM-300/index.html.

Many overseas universities have also heeded the call to implement complex systems study programs: A consortium of Australian universities in 2004 formed the *ARC Centre for Complex Systems*. (These include University of Queensland, Griffith University, Monash University, and The University of New South Wales.) See http://en.wikipedia.org/wiki/ARC_Centre_for_Complex_Systems. The University of Gothenburg, Sweden, has a Master's program in *Complex Adaptive Systems*. The Technical University of Denmark has a Biophysics and Complex Systems Group; see www.dtu.dk/centre/BioCom.aspx. In 2003, the University of Oxford (U.K.) established a CABDyN Complexity Centre (CABDyN is the acronym for Complex Agent-Based Dynamic Networks). This is a broad, interdisciplinary group with interests in proteomics, ecology, social systems, political science, physics, communications, networks, etc. See http://sbs-net.sbs.ox.ac.uk/complexity/complexity-about.asp. The University of Western Sydney offers an online Masters Program on *Complexity, Chaos & Creativity*; see http://clad-home-page.fcpages.com/UWS/MAchaos.html. There are a number universities in the United Kingdom that offer PhDs in complexity science, and which have MA programs; see a number of hot links to these programs at: www.complexity.ecs.soton.ac.ul/courses.php/ (accessed November 6, 2009). Another page of hot links to complexity courses in the United Kingdom can be seen at: http://css.csregistry.org/tiki-index.php?page=EPSRC+Courses+Complex+Sys tems&bl=y (accessed November 6, 2009). Chalmers University of Technology (Sweden) has a doctoral program in *Complex Systems*. See the description at: www.chalmers.se/en/selections/education/doctoral_programmes/graduate-schools/ (accessed November 6, 2009). The Medical University of Vienna has a *Complex Systems Research Group*; see www.complex-systems.meduniwien. ac.at/about/contact.php.

In addition to academic courses focusing on complexity in various areas, several complexity "think tanks" and universities now offer short courses and seminars on complex systems (Complexity 2008, CSS 2008). Some examples: In 2003, the University of Tennessee at Knoxville offered a short course in the area of biological complexity: *Optimal Control Theory in Application to Biology*. In Fall, 2007, Florida State University offered a seminar in complexity theory: *GEO-5934 Seminar in Complexity Theory*. The Ecole Normale Supérieure de Lyon offered a short course: *Models for Complex Systems in Human and Social Sciences* (S2004). (See for example Complexity Links at: www.psych.lse.ac.uk/complexity/links.html.) The University of Wisconsin at Madison has offered a seminar series in *Chaos and Complex Systems* since 1994. (See http://sprott. physics.wisc.edu/Chaos-Complexity [accessed November 6, 2009].)

The New England Complex Systems Institute (NECSI), 238 Main St., Cambridge, MA 02142 (617-547-4100, http://necsi.org) was founded in 1996 by faculty from New England-area academic institutions (e.g., MIT, Harvard, Brandeis) to further international research and understanding of complex systems. Some of the complex areas considered by NECSI faculty are: networks, evolution and ecology, systems biology, engineering, health care, ethnic violence, management, education, military conflict, negotiation, and development. NECSI offers short courses on complex systems.

The Santa Fe Institute (SFI) is a private, not-for-profit, independent research center focusing on complex systems, formed in 1984 (http://www.santafe.edu). The scope of their research is broad, dealing with the emergence, organization and dynamics of living systems, including origin, synthesis and form of life, the emerging ecology of living systems, the emergence of social ecosystems, and HIV propagation and treatment. Also, SFI deals with emergence and robustness in evolutionary systems, including innovation in biological systems, innovation in technological systems, innovation in markets, and robustness in biological and social systems. In addition to the topics above, the SFI also considers the physics of complex systems, information processing and computation in complex systems, and dynamics and quantitative studies in human behavior. The SFI has a resident faculty and many visiting faculty and scholars.

We would like to see courses dealing with complexity appear regularly in the specialized graduate curricula of such disciplines as automata theory, biomedical engineering, ecology, economics, energy systems, evolutionary biology, genetics, genomics, meteorology, microbiology, molecular and cellular biology, physiology (all disciplines), systems engineering, to name some important areas.

1.5 SUMMARY

In this chapter, we have introduced the concepts of systems, and simple, complicated, and complex systems. Also introduced are some general properties of CNLSs: Limit cycle oscillations, chaos, tipping points, the LUC, and Campbell's law. We also examined and gave examples for human responses to complexity, the dangers of reductionism, the *single cause mentality*, and the *not in my box* mind-set. We have seen that it is dangerous to apply *Occam's razor* to understanding the input/output relationships of CNLSs. The "shaving" process may, in fact, ignore those internal dynamic pathways that fully characterize the behavior of the CNLS.

We also have listed many academic programs and university courses dealing with general and specific kinds of CNLSs, and directed readers to visit the Web sites of the Santa Fe Institute and the New England Complex Systems Institute.

PROBLEMS

1.1 Give three examples each of *simple systems, complicated systems*, and *complex systems*. Do not use any systems cited in this chapter. Consider only nonliving examples of complex systems.

1.2 Give a real-world example of *Campbell's law.*

1.3 It is reasonable to identify 12 major physiological systems in humans: (1) cardiovascular (circulatory); (2) CNS/PNS; (3) endocrine; (4) gastrointestinal (digestive); (5) glucoregulatory; (6) immune; (7) integumentary; (8) muscular (striated, smooth, cardiac); (9) renal/ urinary; (10) reproductive; (11) respiratory (including O_2 and CO_2 transport in blood); and (12) skeletal (including bone marrow and its functions). Distinctions between these systems can be blurry; they all interact to some degree (some more than others). Which do you consider the *least complex* human physiological system? Give reasons for your choice.

1.4 Give three examples of CNLSs with *tipping points* (aka *regime shift, bifurcation point*, Norberg and Cumming 2008) (excluding cardiac fibrillation). Describe what happens in each case. (Consider *mechanical, fluid dynamic, economic, social/behavioral, political, climate, ecological, physiological, etc.* systems.)

1.5 What do you think the *unintended consequences* will be as the result of the recent 5:4 U.S. Supreme Court decision (in January, 2010) allowing corporations and labor unions to pay for political advertisements at any time? (The rationale for this decision was that it permits free speech as defined by the First Amendment to the U.S. Constitution for corporate entities, as well as individuals.) In particular, how will the decision affect the functioning of legislative branch of the U.S. government?

1.6 In a letter to the editor of *The Hartford Courant* of December 1, 2008, M.J. Paul advocated converting all school buses from gasoline or diesel engines to purely electric motors. He said: "There would be savings on fuel and engine and transmission parts. Buses would be quieter and lighter, and would produce fewer fumes."

(A) Comment on the economics of this plan; both the consequences and possible unintended consequences. Consider the efficiency of electricity production and transmission, including transformers and power lines, as well as the efficiency of charging batteries and running electric motors, also the cost in labor and materials in producing appropriate storage batteries, electric motors, and control units. Compare this with the efficiency and cost of a diesel engine and the costs of producing oil, refining fuel, and its distribution. What about the cost of converting existing buses from fossil fuel engines to electric motors and their transmissions? What kind of batteries would you use? Comment on the carbon "footprint" of an electric school bus vs. that of a similar, gasoline-powered vehicle.

(B) Would converting the buses to biodiesel be a cheaper and greener solution than all-electric buses?

1.7 Animal behavior can be considered to be a CAS with which a species maintains fitness. Give an example of complex, adaptive animal behavior.

1.8 Recent media attention to a student fatality in a school bus crash has led to the call for mandatory seat belts in all school busses in Connecticut. Discuss any unintended consequences of such an action, if made law. What technical features of school bus seat belts would you recommend that would make them more effective and minimize the LUC? [*Hint*: How would you insure that each student is buckled in, and does not unbuckle to "fool around"?]

1.9 The U.S. government may place a value-added tax (VAT) on all manufactured goods. What will the unintended consequences of this tax?

1.10 Proteomics, the study of proteins, is a complex science. Misfolded prion proteins are the cause of mad cow disease, scrapie in sheep, chronic wasting disease in cervids, and Creutzfeld-Jakob disease (CJD) in humans. Are there any other mammalian diseases associated with misfolded proteins?

1.11 Describe the *intended* and *unintended consequences* of the use of the pesticide DDT in the 1940–1960s.

1.12 What might be the adverse unintended consequences of relying heavily on wind turbines and solar cell arrays to generate electric power?

2 Introduction to Large Linear Systems

2.1 INTRODUCTION

In order to better understand the complexity of large, *nonlinear* systems, we will first describe the important attributes of large, dynamic, *linear, time-invariant* (LTI) systems. Large LTI systems are generally *complicated*, not necessarily complex. They generally do not exhibit tipping points or emergent behavior. They are modeled by sets of linear ODEs, difference equations, or linear algebraic equations. They can generally be analyzed by using reductionist methods; by describing their subsystems or modules, you can understand the system as a whole. This is because the important property of *superposition* applies to LTI systems. Below, we examine large linear systems and review the more important mathematical formalisms used to analyze them.

Fortunately, there are many mathematical tools available for analyzing large LTI systems that are not applicable to CNLSs. Or conversely, the many mathematical tools of linear system analysis unfortunately cannot be applied to CNLSs. In this chapter, we review the art of solving the ODEs that describe the input/output behavior of continuous, LTI systems. Section 2.3 treats the mathematical tools used to characterize the LTI systems.

Reviewed below are the concepts of *system impulse response, real convolution, transient response,* and *steady-state sinusoidal frequency response,* including *Bode* and *Nyquist plots.* Matrix algebra and matrix operations are covered in Section 2.4, and the state-variable (SV) formalism for describing the behavior of high-order, continuous dynamic systems is introduced in Section 2.4.2.3. In Section 2.5, we describe the factors affecting the stability of LTI systems.

2.2 LINEARITY, CAUSALITY, AND STATIONARITY

Even though the real world is fraught with thousands of examples of nonlinear systems, engineers and applied mathematicians have almost exclusively devoted their attention to developing mathematical tools to describe and analyze *linear systems.* One reason for this specialization appears to be that certain mechanical systems and many electrical/electronic circuits can be treated as linear systems. Also, many nonlinear electronic circuits can be easily linearized around a steady-state *operating point,* and thus the powerful mathematical tools of linear system analysis are easy to apply in circuit and mechanical system analysis.

A system is said to be linear if it obeys all of the following properties (Northrop 2000): By logical exclusion, *nonlinear systems do not obey one or more of the following properties*: (Here $\mathbf{x}(t)$ is the input, $\mathbf{y}(t)$ is the output, and $\mathbf{h}(t)$ is the *impulse response* or *weighting function* of the linear, single-input/single-output [SISO] system.)

$$\xrightarrow{a_1\mathbf{x}_1+a_2\mathbf{x}_2}\{LS\}\xrightarrow{\mathbf{y}=a_1\mathbf{y}_1+a_2\mathbf{y}_2} \quad superposition \qquad (2.1A)$$

$$\mathbf{x}_1 \rightarrow \{LS\} \rightarrow \mathbf{y}_1 = \mathbf{x}_1 \otimes \mathbf{h} \quad real\ convolution\ (a\ result\ of\ superposition) \quad (2.1B)$$

if

$$\xrightarrow{\dfrac{\mathbf{x}_2}{a_2\mathbf{x}_2}}\{LS\}\xrightarrow{\dfrac{\mathbf{y}_2=\mathbf{x}_2\otimes\mathbf{h}}{\mathbf{y}_2'=a_2(\mathbf{x}_2\otimes\mathbf{h})}} \quad scaling \qquad (2.1C)$$

then

$$\xrightarrow{\dfrac{\mathbf{x}_1(t-t_1)}{+\mathbf{x}_2(t-t_2)}}\{LS\}\xrightarrow{\dfrac{\mathbf{y}=\mathbf{y}_1(t-t_1)}{+\mathbf{y}_2(t-t_2)}} \quad shift\ invariance \qquad (2.1D)$$

Note that a dynamic system can still be linear and have time-variable (T-V) coefficients in its describing ODEs. This nonstationarity adds to the system's complexity. That is,

$$\dot{\mathbf{x}} = \mathbf{A}(t)\mathbf{x} + \mathbf{B}(t)\mathbf{u} \qquad (2.2)$$

where

$\dot{\mathbf{x}}$ is an **n**-element column matrix of the time derivates of the system's states
$\mathbf{x}$ is a column matrix of the system's **n** dependent variables
$\mathbf{u}$ is a column vector of **p** independent inputs
$\mathbf{A}(t)$ is a square $(\mathbf{n} \times \mathbf{n})$ matrix of the system's time-variable coefficients
$\mathbf{B}(t)$ is an $\mathbf{n} \times \mathbf{p}$ matrix of the time-variable coefficients describing how the **p** inputs, **u**, enter the system

Solution of such T-V linear systems is tedious and is best done by simulation (Northrop 2000).

Causality is a property of reality. A noncausal system can respond to an input before it occurs in time! Probably the only time you may encounter noncausal systems is in the derivation and implementation of *matched filters* and *Wiener filters* for time signals. In these cases, they must be approximated by causal filters (Schwartz 1959). A linear *spatial filter* (used to process 2-D optical data) can have a weighting function $\mathbf{h}(x, y)$ that does respond for x and/or $y < 0$, however.

Stationarity is a term generally used to describe random processes and noise, but it can be extended to the description of linear systems as well. In describing noise, we say a noisy voltage (or parameter) is *nonstationary* if the underlying physical process generating it is changing in time. This change causes the noise statistics to change in time (e.g., *probability density function, auto-power spectrum*, etc.). A system is said to be nonstationary if it is time variable and can be characterized by matrices, $\mathbf{A}$(t) and $\mathbf{B}$(t), as in Equation 2.2. A signal is generally nonstationary if it arises from a nonstationary system.

Interestingly, biochemical and physiological systems are all nonlinear and nonstationary. Life is characterized by changes. Biochemical and physiological systems are *causal*, however. We still can use the linear system theory to analyze and describe certain of them under appropriate limiting conditions, using linearization strategies, and by assuming short-term stationarity.

2.3 LTI SYSTEM TOOLS: IMPULSE RESPONSE, FREQUENCY RESPONSE, AND REAL CONVOLUTION

2.3.1 TRANSIENT RESPONSE OF SYSTEMS

There are three major classifications of *inputs to systems* which are used to characterize their input/output characteristics including sensitivities. These are: (1) *transient inputs* (including impulses, steps, ramps, pulses, and wavelets); (2) *steady-state sinusoidal inputs*; (3) *stationary random inputs* (*noise*) (viewed in the steady state).

Often we are interested in the *settling time* of a system when it is suddenly disturbed by an input. Such *transient inputs* include the *impulse* (*delta*) *function* (practically, a narrow pulse of known area whose duration is much less than the shortest time constant in the system's weighting function). The LTI system's output in response to a *unit impulse* input is, by definition, the system's *weighting function*, $\mathbf{h}$(t), which tells us all about the system's dynamic behavior. *Step functions* (U(t)) are also used to characterize the dynamic behavior of LTI systems. By definition, the step response of an LTI system is the integral of the system's weighting function. A *ramp* input (x(t) = at), starting at t = 0, can also be used experimentally to characterize the error properties of closed-loop control systems. The system's ramp response is the time integral of its step response. The *inverse Laplace transform* is widely used to calculate linear system outputs, given the system's *transfer function* and the Laplace transform of the system's input. This process is covered in detail in Northrop (2003), Lathi (1974), and Kuo (1967).

If a high-order system is described at the level of its component ODEs or state equations, it is more expedient to simulate the system's transient response with MATLAB® and Simulink® or Simnon™ than to derive its transfer function and then use Laplace, Fourier, or $\mathbf{z}$-transforms.

2.3.2 REAL CONVOLUTION

Real convolution is a direct result of the property of superposition. Real convolution is an integral (summing) operation that allows us to calculate the output of an LTI system at some time $t_1 > 0$, given the system's continuous input $\mathbf{x}$(t) for $t \geq 0$ and the system's impulse response, $\mathbf{h}$(t).

To derive the *real convolution integral*, let the system's input be a continuous function, $\mathbf{x}$(t), defined for $t \geq 0$. This input signal is approximated by a continuous

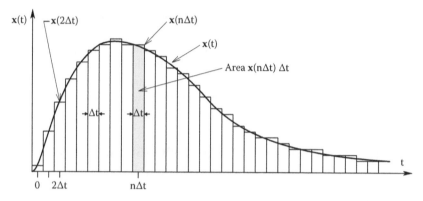

FIGURE 2.1   The continuous function, $\mathbf{x}(t)$, is approximated by a continuous train of rectangular pulses of width Δt and height $\mathbf{x}(n\Delta t)$. This figure is important in the derivation of the real convolution integral. (From Northrop, R.B., *Signals and Systems in Biomedical Engineering*, 2nd edn., CRC Press, Boca Raton, FL, 2010.)

train of rectangular pulses, as shown in Figure 2.1. That is, each pulse has a height $\mathbf{x}(n\Delta t)$ and occurs at time $t = n\Delta t$, $n = 0$, 1, 2, ..., ∞. Thus, each rectangular pulse has an area of $[\mathbf{x}(n\Delta t)\Delta t]$. If we let $\Delta t \to 0$ in the limit, the nth pulse produces a *delayed impulse response* given by

$$\mathbf{y}(n\Delta t) = \lim_{\Delta t \to 0} [\mathbf{x}(n\Delta t)\Delta t]\mathbf{h}(t - n\Delta t), \quad t \geq n\Delta t \tag{2.3}$$

That is, $\mathbf{y}(n\Delta t)$ is the impulse response to an (approximate) impulse of area $[\mathbf{x}(n\Delta t)\Delta t]$ occurring at time $t = n\Delta t$.

At a particular time, $t_1 > n\Delta t$, this response to a single input impulse is

$$\mathbf{y}_1(n\Delta t) = \lim_{\Delta t \to 0} [\mathbf{x}(n\Delta t)\Delta t]\mathbf{h}(t_1 - n\Delta t), \quad t_1 \geq n\Delta t \tag{2.4}$$

Now by superposition, the *total response* at time t_1 is given by the sum of the responses to each component (equivalent) impulse making up the input approximation, $\mathbf{x}(n\Delta t)$:

$$\mathbf{y}(t) = \lim_{\Delta t \to 0} \sum_{n=0}^{n\Delta t = t_1} [\mathbf{x}(n\Delta t)\Delta t]\mathbf{h}(t_1 - n\Delta t) \tag{2.5}$$

Because the system is causal, input pulses occurring beyond $t = t_1$ have no effect on the response at t_1. In the limit as $\Delta t \to 0$, the discrete summation approaches the real convolution integral:

$$\mathbf{y}(t_1) = \int_0^{t_1} \mathbf{x}(t)\mathbf{h}(t_1 - t)\, dt \tag{2.6}$$

It is less confusing to replace the time variable of integration in Equation 2.6 with τ and to replace t_1 with a general time, $t \geq 0$, at which we view the output. Thus, we have the general form for the *real convolution* process:

$$y(t) = \int_0^t x(\tau)h(t - \tau)\, d\tau \tag{2.7}$$

Some properties of real convolution are as follows:

1. *Real convolution is commutative.* That is,

$$f_1(t) \otimes f_2(t) = f_2(t) \otimes f_1(t) \tag{2.8A}$$

This may be proved by writing

$$f_1(t) \otimes f_2(t) = \int_{-\infty}^{\infty} f_1(\tau)f_2(t - \tau)\, d\tau \tag{2.8B}$$

Now let us substitute $(t - \mu)$ for τ. We can now write

$$f_1(t) \otimes f_2(t) = \int_{-\infty}^{\infty} f_2(\mu)f_1(t - \mu)\, d\mu = f_2(t) \otimes f_1(t) \tag{2.8C}$$

For a linear system with input x and output y, this means that

$$y(t) = x(t) \otimes h(t) = h(t) \otimes x(t) \tag{2.8D}$$

2. Convolution is also *distributive.* That is,

$$f_1(t) \otimes [f_2(t) + f_3(t)] = f_1(t) \otimes f_2(t) + f_1(t) \otimes f_3(t) \tag{2.8E}$$

3. It is also *associative*, sic

$$f_1(t) \otimes [f_2(t) \otimes f_3(t)] = [f_1(t) \otimes f_2(t)] \otimes f_3(t) \tag{2.8F}$$

A first example of a graphical interpretation of the convolution process helpful in understanding the abstract notation given by Equation 2.7 is shown in Figure 2.2. In this example, the system's impulse response $h(t) = B$ for $0 \leq t \leq 2T$, and 0 for $t < 0$ and $t > 2T$. The input, $x(t)$, is a triangular pulse of height A and duration T. Analytically, $x(t)$ can be expressed as

$$x(t) = A - \left(\frac{A}{T}\right)t = A\left(1 - \frac{t}{T}\right) \tag{2.9}$$

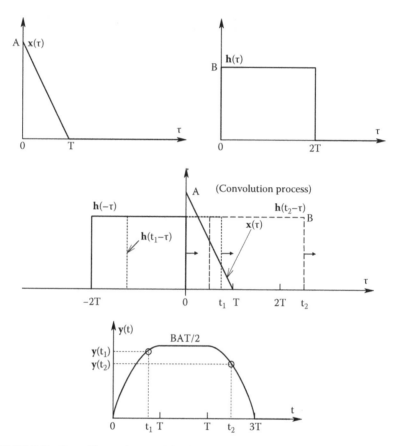

FIGURE 2.2 Steps illustrating graphically the process of continuous, real convolution between an input $x(t)$ and an LTI system with a rectangular impulse response, $h(t)$. In this example, $h(\tau)$ is reversed in time and displaced as $h(t - \tau)$ and slid past $x(\tau)$. Where $h(t - \tau)$ overlaps $x(\tau)$, their product is integrated to form the system output, $y(t)$. (From Northrop, R.B., *Signals and Systems in Biomedical Engineering*, 2nd edn., CRC Press, Boca Raton, FL, 2010.)

for $0 \le t \le T$, and 0 elsewhere. These waveforms are shown as functions of the age variable, τ. In the middle figure, $h(t - \tau)$ is plotted as $h(\tau)$ reversed in time τ and shifted to the right by $t > 0$ so that $h(t - \tau)$ overlaps the input pulse $x(\tau)$. The value of the convolution integral for a particular shift, t_1, is the integral of the product $x(\tau)h(t_1 - \tau)$, *where the functions overlap*. In this example, the integral can be divided into three regions: $0 \le t < T$, $T \le t < 2T$, and $2T \le t \le 3T$. In the first region,

$$y_1(t) = \int_0^t BA\left(1 - \frac{\tau}{T}\right)d\tau = BA\left(t - \frac{t^2}{2T}\right), \quad 0 \le t < T \qquad (2.10)$$

In the second region, $\mathbf{h}(t-\tau)$ overlaps the triangle, $\mathbf{x}(\tau)$ and the integral is simply the triangle's area times B, a constant. By inspection,

$$\mathbf{y}_2(t) = \frac{BAT}{2}, \quad T \le t < 2T \tag{2.11}$$

In the third region, the trailing edge of $\mathbf{h}(t-\tau)$ sets the lower limit of the integral to $t-2T$, and the upper limit is set by $\mathbf{x}(\tau)$ to T, sic:

$$\mathbf{y}_3(t) = \int_{t-2T}^{T} BA\left(1 - \frac{\tau}{T}\right) d\tau = BA\left[\tau - \frac{\tau^2}{2T}\right]\Bigg|_{t-2T}^{T}$$

$$= BA\left[T - \frac{T}{2}\right] - BA\left[(t-2T) - \frac{(t-2T)^2}{2T}\right]$$

$$= \frac{BAT}{2} - BA\,t + BA2T + BA\frac{t^2}{2T} - BA\frac{4tT}{2T} + BA\frac{4T^2}{2T}$$

$$= \left(\frac{9}{2}\right)BAT - 3BA\,t + BA\left(\frac{t^2}{2T}\right) \tag{2.12}$$

When $t = 2T$, $\mathbf{y}_3(t) = BAT/2$, and when $t \ge 3T$, $\mathbf{y}_3(t) = 0$, which agrees with the limits seen in the figure.

For a second graphical example of real convolution, consider the signals shown in Figure 2.3. The input is a rectangular pulse of duration T and amplitude A. The system weighting function is $\mathbf{h}(\tau) = Be^{-b\tau}$. In this case, the convolution integral can be broken up into two regions over which $\mathbf{x}(\tau)$ and $\mathbf{h}(t-\tau)$ overlap to form a nonzero product.

$$\mathbf{y}_1(t) = \int_0^t AB\exp[-b(t-\tau)]d\tau = AB\exp(-bt)\int_0^t \exp(b\tau)d\tau$$

$$= AB\exp(-bt)\exp\frac{(bt)}{b} - AB\exp\frac{(-bt)}{b} \tag{2.13}$$

$\downarrow$

$$\mathbf{y}_1(t) = \left(\frac{AB}{b}\right)[1 - \exp(-bt)], \quad 0 \le t \le T$$

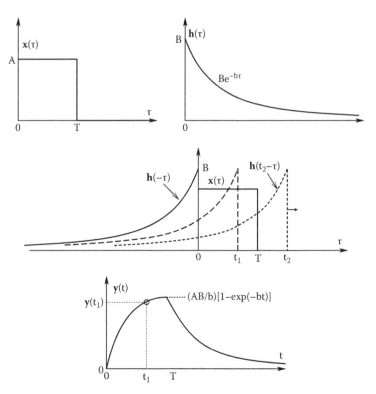

FIGURE 2.3 A second illustration of real convolution. The system's input, $x(\tau)$, is a rectangular pulse; the system's weighting function is a simple exponential decay, characteristic of a first-order, low-pass filter. The output, $y(t)$, rises to a peak, then decays. (From Northrop, R.B., *Signals and Systems in Biomedical Engineering*, 2nd edn., CRC Press, Boca Raton, FL, 2010.)

The *second region* exists for $T < t \leq \infty$. The value of the convolution integral in this region is

$$y_2(t) = \int_0^T AB\exp[-b(t-\tau)]\,d\tau = AB\exp(-bt)\int_0^T \exp(b\tau)\,d\tau$$

$$= AB\exp(-bt)\exp\frac{(bT)}{b} - AB\exp(-bt)\left(\frac{1}{b}\right)$$

(2.14)

$\downarrow$

$$y_2(t) = \left(\frac{AB}{b}\right)\exp(-bt)[\exp(bT)-1], \quad T < t \leq \infty$$

Thus, the peak $y(t)$ occurs at $t = T$ and is: $y(T) = (AB/b)[1 - \exp(-bT)]$.

Real convolution can also be performed on *discrete (sampled) signals*. In this case, the input can be written as $\mathbf{x}(nT)$, $\mathbf{x}(n)$, or $\mathbf{x_n}$. T is the sampling period and n is the sample number. Similarly, the system's impulse response is also expressed in numerical (sampled) form, $\mathbf{h}(nT)$ or $\mathbf{h}(n)$. Following the development above, we can write the discrete system output $\mathbf{y}(nT)$ as

$$y(n) = \sum_{k=0}^{\infty} x(k)\,h(n-k) = \sum_{k=0}^{\infty} h(k)x(n-k) \qquad (2.15)$$

All of the properties of convolution with continuous variables apply to discrete convolution.

An important use of convolution is *deconvolution*, or signal recovery. As an example, an acoustical signal, $\mathbf{x}(t)$ (such as heart valve sounds), propagates through the lungs and chest walls to the skin surface. Acoustic power at certain frequencies is attenuated and delayed in the internal tissues. The sound, $\mathbf{y}(t)$, picked up at the chest surface is convolved (*deconvolved*) with an *inverse filter* which essentially compensates for the effects of the propagation of $\mathbf{x}(t)$ through the tissues, $\mathbf{h}(t)$, then an estimate of the original sound, $\mathbf{x}(t)$, can be recovered (Widrow and Stearns 1985).

2.3.3 STEADY-STATE SINUSOIDAL FREQUENCY RESPONSE OF LTI SYSTEMS

The sinusoidal steady-state (SSS) *frequency response* is often used as a figure of merit for sound processing (audio) equipment (amplifiers, microphones, loudspeakers, CD players, etc.). It is also used to characterize the performance of other electronic equipment (RF amplifiers, op amps, fiber-optic cables, cell phones, etc.). Because the SSS frequency response can be related to LTI system transient response and stability, it finds wide application, even for nonlinear systems.

If a sinusoidal input, $\mathbf{x}(t) = A\sin(\omega t)$, $t \geq 0$, has been applied to a linear system for a time that is very long compared to the longest time constant in the system's transient response, the system is said to be in the SSS. In the steady state, the exponential transient terms of the homogeneous solutions to the system's ODEs have died out to zero. In SSS excitation, the linear system's output $\mathbf{y}(t)$ will be of the same frequency as the input but generally will have a different phase and amplitude; thus for a system input, $\mathbf{x}(t) = A\sin(2\pi ft)$, the output is $\mathbf{y}(t) = B\sin(2\pi ft + \phi)$. In phasor (vector) notation, $\mathbf{y}(t)$ is summarized by the 2-D (polar) vector, $\mathbf{Y} = B\angle\phi$.

By definition, the LTI system's *steady-state frequency response function*, $\mathbf{H}(f)$, can be expressed as a 2-D (polar) vector with magnitude (B/A) and phase angle, ϕ. That is, $\mathbf{H}(f) = (\mathbf{Y}/\mathbf{X}) = (B/A)\angle\phi$. Note that the frequency response function can be written as a function of Hz frequency, f, or radian frequency, $\omega = 2\pi f$. Both (B/A) and ϕ are functions of frequency. For reasons that will be made clear in the next section, engineers and scientists generally plot the decibel (dB) frequency response, $20\log_{10}|\mathbf{H}(f)|$, on a linear vertical scale vs. f on a logarithmic frequency scale. $\phi(f)$ is plotted on a linear vs. log frequency scale, as well. Such plots are called *Bode plots* (Northrop 2003). *Polar plots* of the frequency response can also be made where $20 \log_{10} |\mathbf{H}(f)|$ is plotted vs. $\phi(f)$ on polar coordinates for a set of f values (see Appendix D).

To find a linear system's frequency response from its defining ODE, we first write the nth-order system's ODE in general form:

$$\dot{y}^n + a_1\dot{y}^{(n-1)} + a_2\dot{y}^{(n-2)} + \cdots + a_{n-1}\dot{y} + a_n y = b_0\dot{u}^m + b_1\dot{u}^{(m-1)} + \cdots + b_{m-1}\dot{u} + b_m u$$

(2.16)

Next, we write the ODE using *operator notation*, where $\mathbf{p}y = (dy/dt)$ and $\mathbf{p}^n y = (d^n y/dt^n)$, etc. Thus, we have

$$\mathbf{p}^n y + \mathbf{p}^{n-1}ya_1 + \mathbf{p}^{n-2}ya_2 + \cdots + \mathbf{p}ya_{n-1} + ya_n = \mathbf{p}^m ub_0 + \mathbf{p}^{m-1}ub_1 + \cdots + \mathbf{p}ub_{m-1} + ub_m$$

(2.17)

Equation 2.17 can also be written in a factored form

$$y[\mathbf{p}^n + a_1\mathbf{p}^{n-1} + a_2\mathbf{p}^{n-2} + \cdots + a_{n-1}\mathbf{p} + a_n] = u[b_0\mathbf{p}^m + b_1\mathbf{p}^{m-1} + \cdots + b_{m-1}\mathbf{p} + b_m]$$

(2.18)

It can be shown that if we replace $\mathbf{p}$ by the imaginary number, $j\omega$, where ω is the radian frequency of the sinusoidal input, $\mathbf{u}(t) = A \sin(\omega t)$, then $y(t)$ and $u(t)$ can be treated as 2-D vectors (phasors), $\mathbf{Y}$ and $\mathbf{U}$, respectively. The angle of the input, $\mathbf{U}$, is by definition, 0, and its magnitude is A. $\mathbf{Y}$ has magnitude B and angle ϕ. Thus the linear, nth-order SISO system's frequency response is given in general by the ratio of two vectors:

$$B$$

$$\frac{\mathbf{Y}}{\mathbf{U}}(j\omega) = \frac{b_0(j\omega)^n + b_1(j\omega)^{n-1} + \cdots + b_{n-2}(j\omega)^2 + b_{n-1}(j\omega) + b_n}{(j\omega)^m + a_1(j\omega)^{m-1} + \cdots + a_{m-2}(j\omega)^2 + a_{m-1}(j\omega) + a_m} = \mathbf{H}(j\omega) \quad (2.19)$$

$$A$$

Note that even powers of $(j\omega)$ in the numerator and denominator are $\pm$ real numbers and odd powers of $(j\omega)$ are imaginary with either $\pm j$ factors. Recall also that the ratio of two vectors, $\mathbf{Q}$ and $\mathbf{R}$, can be written in general using polar notation as

$$\mathbf{P}\frac{\mathbf{Q}}{\mathbf{R}} = \frac{|\mathbf{Q}|\angle\theta_q}{|\mathbf{R}|\angle\theta_r} = \left(\frac{Q}{R}\right)\angle(\theta_q - \theta_r) \quad (2.20)$$

In this case, $\phi(\omega) = (\theta_q - \theta_r)$ and $(Q/R) = (A/B)$. Since terms in the numerator and denominator are either real or imaginary, we can group them accordingly:

$$H(j\omega) = \frac{\alpha(\omega) + j\beta(\omega)}{\gamma(\omega) + j\delta(\omega)} \tag{2.21}$$

The denominator can be made real by multiplying the denominator and numerator by the *complex conjugate* of the denominator, which is $R^* = [\gamma(\omega) - j\delta(\omega)]$. Thus,

$$H(j\omega) = \frac{[\alpha(\omega)\gamma(\omega) + \beta(\omega)\delta(\omega)] + j[\beta(\omega)\gamma(\omega) - \alpha(\omega)\delta(\omega)]}{\gamma^2(\omega) + \delta^2(\omega)} \tag{2.22}$$

Note that the angle of **H** is given by

$$\phi(\omega) = \angle H(j\omega) = \tan^{-1}\left\{\frac{[\beta(\omega)\gamma(\omega) - \alpha(\omega)\delta(\omega)]}{[\alpha(\omega)\gamma(\omega) + \beta(\omega)\delta(\omega)]}\right\} \tag{2.23}$$

As a first example of calculating the frequency response of a simple system, let us find the SSS frequency response of the pharmacokinetic (PK) system described below. This system describes the rate that a drug is removed from the *blood volume compartment.* [D] is the drug concentration in the blood in µg/L, K_L is the loss rate constant (units: L/min), V_c is the total blood volume in liters, and R_D is the rate of drug injection into the blood in µg/min. The system is modeled by the ODE:

$$[\dot{D}] = -K_L[D] + \left(\frac{R_D}{V_c}\right)U(t) \quad (\mu g/L)/\min \tag{2.24}$$

Rewriting the ODE in *operator notation,*

$$[D](\mathbf{p} + K_L) = \frac{R_D(t)}{V_c} \tag{2.25}$$

Normally, for an intravenous (IV) drip, R_D is a constant (a step). Because we wish to examine the sinusoidal response of this system, it is tempting to let $R_D = R_{D_o}\sin(\omega t)$. However, the sine wave is negative over half of each cycle, which is physically impossible in a PK system (there are no negative drug infusions, injections, or concentrations). Thus, we must add a constant level to $R_D(t)$ *so it does not go negative.* Thus, the input to the PK system will be

$$R_D(t) = R_{D_o}[1 + \sin(\omega t)]U(t) \geq 0 \tag{2.26}$$

where $U(t)$ is the unit step function: $U(t)=0$, $t<0$; $U(t)=1$, $t\geq 0$. ($U(t)$ is used as a mathematical switch here.) Thus the input is actually the sum of a step and a sinusoid for $t\geq 0$. From superposition, the steady-state, constant step response is

$$[D]_{ss} = \frac{R_{Do}}{K_L V_c} \mu g/L \qquad (2.27)$$

To find the SSS frequency response, as we have shown above, let $\mathbf{p} \rightarrow j\omega$ in Equation 2.25. Thus, the PK system's frequency response is

$$\frac{[\mathbf{D}]}{\mathbf{R_D}}(j\omega) = \frac{1/V_c}{j\omega + K_L} = \frac{1/V_c}{\sqrt{\omega^2 + K_L^2}} \angle \phi \qquad (2.28A)$$

$$\phi = -\tan^{-1}\left(\frac{\omega}{K_L}\right) \qquad (2.28B)$$

Thus, the sinusoidal component of drug concentration in the blood *lags* the input injection sinusoid by ϕ degrees, and ϕ falls off with frequency to reach $-90°$ as $\omega \rightarrow \infty$ (the system behaves like a first-order low-pass filter). The output sinusoidal response sits on the dc level, $R_{Do}/(K_L V_c)$, required to make $u(t)$ non-negative.

In a second example, we will illustrate finding the SSS frequency response of a second-order, time-domain, SV system given by

$$\dot{x} = \mathbf{Ax} + \mathbf{Bu} \qquad (2.29A)$$

$$y(t) = x_1(t) \qquad (2.29B)$$

The **A** and **B** matrices are

$$\mathbf{A} = \begin{bmatrix} 0 & 1 \\ -2 & -3 \end{bmatrix}, \quad \mathbf{B} = \begin{bmatrix} 0 \\ 1 \end{bmatrix} \qquad (2.30)$$

and $\mathbf{u}(t) = A\sin(\omega t)$. Using **p** as the differential operator, the SSS frequency response of this simple, linear SISO system is given by

$$\mathbf{px} = \mathbf{Ax} + \mathbf{Bu}$$

$$\downarrow$$

$$\mathbf{px} - \mathbf{Ax} = \mathbf{Bu} \qquad (2.31)$$

$$\downarrow$$

$$(\mathbf{pI} - \mathbf{A})\mathbf{x} = \mathbf{Bu}$$

Assume $\mathbf{u} = u(t) = A\sin(\omega t)$. Replace the operator $\mathbf{p}$ with the imaginary vector, $j\omega$, and solve for the state phasors:

$$\mathbf{X} = (j\omega\mathbf{I} - \mathbf{A})^{-1}\mathbf{B}\mathbf{U} \tag{2.32}$$

The inverse matrix $\mathbf{\Phi}(j\omega) \equiv (j\omega\mathbf{I} - \mathbf{A})^{-1}$ has complex elements. It is found by first calculating

$$(j\omega\mathbf{I} - \mathbf{A}) = \left(j\omega \begin{bmatrix} 1 & 0 \\ 0 & 1 \end{bmatrix} - \begin{bmatrix} 0 & 1 \\ -2 & -3 \end{bmatrix} \right) = \begin{bmatrix} j\omega & -1 \\ 2 & j\omega+3 \end{bmatrix} \tag{2.33}$$

The inverse of $(j\omega\mathbf{I} - \mathbf{A}) \equiv \mathrm{adj}(j\omega\mathbf{I} - \mathbf{A})/\det(j\omega\mathbf{I} - \mathbf{A}) = \mathbf{\Phi}(j\omega)$. $\det(j\omega\mathbf{I} - \mathbf{A}) = (j\omega)^2 + 3j\omega + 2$. The adjoint of $(j\omega\mathbf{I} - \mathbf{A})$ is found to be

$$\mathrm{adj}(j\omega\mathbf{I} - \mathbf{A}) = \begin{bmatrix} j\omega+3 & 1 \\ -3 & j\omega \end{bmatrix} \tag{2.34}$$

Thus the $\mathbf{\Phi}(j\omega)$ is

$$\mathbf{\Phi}(j\omega) = \frac{\begin{bmatrix} j\omega+3 & 1 \\ -3 & j\omega \end{bmatrix}}{(j\omega)^2 + 3j\omega + 2} \tag{2.35}$$

The system output is $y(t) = x_1(t)$, or in phasor notation, $\mathbf{Y} = \mathbf{X}_1(j\omega)$. Thus the system's frequency response function is

$$\mathbf{X}(j\omega) = \frac{\begin{bmatrix} j\omega+3 & 1 \\ -3 & j\omega \end{bmatrix}\begin{bmatrix} 0 \\ 1 \end{bmatrix}\mathbf{U}}{(j\omega)^2 + 3j\omega + 2} \tag{2.36}$$

And after some algebra, we have

$$\frac{\mathbf{Y}}{\mathbf{U}}(j\omega) = \frac{1}{(j\omega)^2 + 3j\omega + 2} = \frac{1}{(2 - \omega^2) + 3j\omega} = \frac{1}{\sqrt{(2 - \omega^2)^2 + 9\omega^2}} \angle\phi \tag{2.37A}$$

where

$$\phi = -\tan^{-1}\left(\frac{3\omega}{2 - \omega^2} \right) \tag{2.37B}$$

Note that there are easier pencil-and-paper means of finding a linear system's frequency response or transfer function, given its state equations. Finding $\Phi(j\omega)$ for a quadratic (2×2) SV system is relatively easy, but for $n \geq 3$, it becomes tedious and subject to algebraic errors. A simpler, pencil-and-paper approach is to construct a *signal flow graph* (SFG) from the state equations and reduce it using *Mason's rule* (see Section 2.6), which is far more direct algebraically than crunching $\Phi(j\omega)$ for $n \geq 3$. (A review of Bode and Nyquist frequency response plotting is found in Appendices C and D.)

2.4 SYSTEMS DESCRIBED BY LARGE SETS OF LINEAR ODEs

2.4.1 INTRODUCTION

Multicompartmental, dynamical, PK systems (Godfrey 1983, Northrop 2000) are excellent examples of a class of biomedical systems that are generally describable by sets of coupled, first-order, linear ODEs. If N *compartments* are interconnected, there will be in general N linear ODEs, and N roots in the overall system's characteristic equation. The roots will generally be real and have negative real parts for stable systems. Below we introduce some of the techniques used in manipulating large sets of linear ODEs.

As a first example, consider a forced, linear second-order system modeled by the second-order ODE:

$$\ddot{x} + \dot{x}b + xc = u \qquad (2.38)$$

In order to express this second-order system as two interconnected first-order systems, we now define *two state variables:* $x_1 \equiv x$ and $x_2 \equiv \dot{x}_1 = \dot{x}$. Now the ODE of Equation 2.43 can be rewritten as

$$\dot{x}_1 = x_2 \qquad (2.39A)$$

$$\dot{x}_2 = -x_1 c - x_2 b + u \qquad (2.39B)$$

where x_1 and x_2 are the state variables of the second-order system. As you will see in the next section, these two first-order ODEs can be put in SV form using vector matrix notation.

As a second example, consider a general nth-order system with one input:

$$\frac{dx^n}{dt^n} + a_{n-1}\frac{dx^{n-1}}{dt^{n-1}} + \cdots + a_1\frac{dx}{dt} + a_0 x = b_0 u(t) \qquad (2.40)$$

As above, we define $x_1 \equiv x$, $x_2 \equiv \dot{x} = \dot{x}_1$, $x_3 \equiv \dot{x}_2, \ldots, x_n \equiv \dot{x}_{n-1}$, from which we can find the set of n first-order ODEs:

$$\dot{x}_1 = x_2$$

$$\dot{x}_2 = x_3$$

$$\dot{x}_3 = x_4$$

$$\vdots$$

$$\dot{x}_{n-1} = x_n$$

$$\dot{x}_n = -a_0 x_1 - a_1 x_2 - a_2 x_3 \cdots - a_{n-2} x_{n-1} - a_{n-1} x_n + b_n u(t)$$

(2.41)

PK systems with $N \geq 3$ can give complicated sets of equations (Godfrey 1983). Solution of these ODEs is made easier by the *SV formalism* and, in some cases, by using linear SFGs to describe the system. When solving a system with pencil and paper, *Mason's rule* can be easily used with SFGs to find a high-order PK system's transfer function without the need for *matrix inversion*. Once the transfer function is known, the outputs can be readily found, given the inputs.

2.4.2 INTRODUCTION TO MATRIX ALGEBRA AND STATE VARIABLES

2.4.2.1 Matrix Algebra

Before reviewing *state variables*, we will examine some basic properties of *matrices*. A matrix is simply a collection of numbers or algebraic *elements* arranged in a rectangular or square array of rows and columns. The elements can be the coefficients of a set of simultaneous, linear algebraic equations. A matrix can have one row, one column, or n rows and m columns, or be *square* with n rows and n columns. A matrix is said to be of *order* (n, m) if it has n rows of elements and m columns of elements. A matrix is not a *determinant*, although a determinant can be calculated from the elements of a square (n × n) matrix. A square 3 × 3 matrix is shown below in Equation 2.42.

$$\mathbf{A} = \begin{bmatrix} a_{11} & a_{12} & a_{13} \\ a_{21} & a_{22} & a_{23} \\ a_{31} & a_{32} & a_{33} \end{bmatrix}$$

(2.42)

The determinant of $\mathbf{A}$ is a number or algebraic quantity:

$$\det \mathbf{A} = a_{11}(a_{22}a_{33} - a_{32}a_{23}) - a_{21}(a_{13}a_{33} - a_{32}a_{13}) + a_{31}(a_{12}a_{23} - a_{22}a_{13})$$ (2.43A)

or

$$\det \mathbf{A} = a_{11}(a_{22}a_{33} - a_{32}a_{23}) - a_{12}(a_{21}a_{33} - a_{31}a_{23}) + a_{13}(a_{21}a_{32} - a_{31}a_{22})$$ (2.43B)

and so forth.

Some terms: A *diagonal matrix* is a square matrix in which the diagonal elements, a_{kk}, are nonzero, and all the other a_{jk} elements $(j \neq k)$ are zero.

A *unit or identity matrix*, **I**, is defined as a diagonal matrix in which all $a_{kk} = 1$.

A *singular matrix* is a square matrix whose determinant $= 0$. Singularity is the result of nonindependence of the n simultaneous equations from which the matrix elements were derived. *For example*, consider the three, simple, linear algebraic equations below:

$$x_1 + 2x_2 + 3x_3 = 0 \tag{2.44A}$$

$$8x_1 - x_2 - x_3 = 0 \tag{2.44B}$$

$$7x_1 - 3x_2 - 4x_3 = 0 \tag{2.44C}$$

In matrix form, these equations can be written as

$$\mathbf{Ax} = \begin{bmatrix} 1 & 2 & 3 \\ 8 & -1 & -1 \\ 7 & -3 & -4 \end{bmatrix} \mathbf{x} = 0 \tag{2.45}$$

where **x** is the column matrix $\begin{bmatrix} x_1 \\ x_2 \\ x_3 \end{bmatrix}$.

The determinant of **A** is det $\mathbf{A} = 1(4-3) - 8(-8+9) + 7(-2+3) = 0$. In this example, singularity was the result of the third equation being derived by subtracting the top equation from the middle equation, causing it not to be independent.

2.4.2.2   Review of Some Matrix Operations

The transpose of a matrix is found by exchanging its columns for rows. For example, $\mathbf{B}^T$ is the transpose of **B**. Let

$$\mathbf{B} = \begin{bmatrix} 1 & 4 \\ 2 & 5 \\ 3 & 6 \end{bmatrix}, \quad \text{then } \mathbf{B}^T = \begin{bmatrix} 1 & 2 & 3 \\ 4 & 5 & 6 \end{bmatrix} \tag{2.46}$$

The *adjoint matrix* of the square matrix **A** is adj $\mathbf{A} = (\text{cof } \mathbf{A})^T$. That is, it is the *transpose* of the *cofactor matrix* of **A**. These operations are best shown by example. Given a square **A**,

$$\mathbf{A} = \begin{bmatrix} a_{11} & a_{12} & a_{13} \\ a_{21} & a_{22} & a_{23} \\ a_{31} & a_{32} & a_{33} \end{bmatrix} \tag{2.47}$$

$$\text{cof } \mathbf{A} = \begin{bmatrix} (a_{22}a_{33} - a_{32}a_{23}) & -(a_{21}a_{33} - a_{31}a_{23}) & (a_{21}a_{32} - a_{31}a_{22}) \\ -(a_{12}a_{33} - a_{32}a_{13}) & (a_{11}a_{33} - a_{31}a_{13}) & -(a_{11}a_{32} - a_{31}a_{12}) \\ (a_{12}a_{23} - a_{22}a_{13}) & -(a_{11}a_{23} - a_{21}a_{13}) & (a_{11}a_{22} - a_{21}a_{12}) \end{bmatrix} \quad (2.48)$$

Now the *transpose* of cof **A** gives the desired adj **A**:

$$\text{adj } \mathbf{A} = \begin{bmatrix} (a_{22}a_{33} - a_{32}a_{23}) & -(a_{12}a_{33} - a_{32}a_{13}) & (a_{12}a_{23} - a_{22}a_{13}) \\ -(a_{21}a_{33} - a_{31}a_{23}) & (a_{11}a_{33} - a_{31}a_{13}) & -(a_{11}a_{23} - a_{21}a_{13}) \\ (a_{21}a_{32} - a_{31}a_{22}) & -(a_{11}a_{32} - a_{31}a_{12}) & (a_{11}a_{22} - a_{21}a_{12}) \end{bmatrix} \quad (2.49)$$

Matrix addition: Two matrices can be added algebraically if they are of the same $(n \times m)$ order. Sic:

$$\mathbf{A} + \mathbf{B} = \mathbf{D} \quad (2.50)$$

The associative and commutative laws of addition and subtraction of real numbers holds for matrices of equal order. That is,

$$(\mathbf{A} + \mathbf{B}) + \mathbf{C} = \mathbf{A} + (\mathbf{B} + \mathbf{C}) \quad \text{(Associative law)} \quad (2.51)$$

$$\mathbf{A} + \mathbf{B} + \mathbf{C} = \mathbf{B} + \mathbf{C} + \mathbf{A} = \mathbf{C} + \mathbf{A} + \mathbf{B} \quad \text{(Distributive law)} \quad (2.52)$$

Matrix multiplication: A necessary condition to multiply two matrices together is that they be *conformable*. That is, the number of columns of **A** must equal the number of rows of **B**. Equation 2.58 below illustrates how matrix multiplication is done for a 3×2 times a 2×3 matrix:

$$\mathbf{AB} = \begin{bmatrix} a_{11} & a_{12} & a_{13} \\ a_{21} & a_{22} & a_{23} \end{bmatrix} \begin{bmatrix} b_{11} & b_{12} \\ b_{21} & b_{22} \\ b_{31} & b_{32} \end{bmatrix}$$

$$= \mathbf{C} = \begin{bmatrix} c_{11} & c_{12} \\ (a_{11}b_{11} + a_{12}b_{21} + a_{13}b_{31}) & (a_{11}b_{12} + a_{12}b_{22} + a_{13}b_{32}) \\ (a_{21}b_{11} + a_{22}b_{21} + a_{23}b_{31}) & (a_{21}b_{12} + a_{22}b_{22} + a_{23}b_{32}) \\ c_{21} & c_{22} \end{bmatrix} \quad (2.53)$$

Note that each element in the left-hand matrix's top row is multiplied by the corresponding element in the right-hand matrix's first (left-hand) column and the products of elements are added together to form the c_{11} element of the product matrix, **C**. Next, each element of the top row of **A** is multiplied by the corresponding element of the right-hand column of **B**, and the element products are added together to form c_{12} of **C**.

Now observe what happens when the order of multiplication is reversed:

$$\mathbf{BA} = \begin{bmatrix} b_{11} & b_{12} \\ b_{21} & b_{22} \\ b_{31} & b_{32} \end{bmatrix} \begin{bmatrix} a_{11} & a_{12} & a_{13} \\ a_{21} & a_{22} & a_{23} \end{bmatrix}$$

$$= \mathbf{D} = \begin{bmatrix} (b_{11}a_{11} + b_{12}a_{21}) & (b_{11}a_{12} + b_{12}a_{22}) & (b_{11}a_{13} + b_{12}a_{22}) \\ (b_{21}a_{11} + b_{22}a_{21}) & (b_{21}a_{12} + b_{22}a_{22}) & (b_{21}a_{13} + b_{22}a_{23}) \\ (b_{31}a_{11} + b_{32}a_{21}) & (b_{31}a_{12} + b_{32}a_{22}) & (b_{31}a_{13} + b_{32}a_{23}) \end{bmatrix} \qquad (2.54)$$

It is clear that $\mathbf{AB} \neq \mathbf{BA}$, and except for some special cases, *matrix multiplication is not commutative*. That is, matrix multiplication is order dependent. For example:

$$(\mathbf{A} + \mathbf{B})\mathbf{C} = \mathbf{AC} + \mathbf{BC} \qquad (2.55A)$$

and

$$\mathbf{C}(\mathbf{A} + \mathbf{B}) = \mathbf{CA} + \mathbf{CB} \qquad (2.55B)$$

As a *final example* of matrix multiplication, consider the product

$$\begin{bmatrix} 2 & 1 & 3 \end{bmatrix} \begin{bmatrix} 2 \\ 1 \\ 5 \end{bmatrix} = [20] \qquad (2.56)$$

An *identity matrix*, $\mathbf{I}$, is a (square) diagonal matrix in which all the a_{kk} elements are 1s, and all the other (a_{jk}, $j \neq k$) elements are zeros. Note that an exception to the $\mathbf{AB} \neq \mathbf{BA}$ rule is

$$\mathbf{AI} = \mathbf{IA} = \mathbf{A} \qquad (2.57)$$

Note that the order of $\mathbf{A}$ must agree with that of $\mathbf{I}$ for this to happen. That is, the number of elements in the rows of the left-hand matrix must equal the number of elements in the columns of the right-hand matrix. As an example, let

$$\mathbf{A} = \begin{bmatrix} a_{11} & a_{12} \\ a_{21} & a_{22} \\ a_{31} & a_{32} \end{bmatrix} \qquad (2.58)$$

Then,

$$\mathbf{AI} = \begin{bmatrix} a_{11} & a_{12} \\ a_{21} & a_{22} \\ a_{31} & a_{32} \end{bmatrix} \begin{bmatrix} 1 & 0 \\ 0 & 1 \end{bmatrix} = \begin{bmatrix} a_{11} & a_{12} \\ a_{21} & a_{22} \\ a_{31} & a_{32} \end{bmatrix} = \mathbf{A} \qquad (2.59)$$

Also,

$$\mathbf{IA} = \begin{bmatrix} 1 & 0 & 0 \\ 0 & 1 & 0 \\ 0 & 0 & 1 \end{bmatrix} \begin{bmatrix} a_{11} & a_{12} \\ a_{21} & a_{22} \\ a_{31} & a_{32} \end{bmatrix} = \begin{bmatrix} a_{11} & a_{12} \\ a_{21} & a_{22} \\ a_{31} & a_{32} \end{bmatrix} = \mathbf{A} \qquad (2.60)$$

In Equation 2.59, a 2×2 identity matrix is required for conformity and in Equation 2.60, a 3×3 identity matrix is used.

The *inverse of a (square) matrix* is another important matrix operation. It is written as

$$\mathbf{A}^{-1} = \frac{\text{adj } \mathbf{A}}{\det \mathbf{A}} \qquad (2.61)$$

That is, $\mathbf{A}^{-1}$ is the adjoint matrix of $\mathbf{A}$, each element of which is divided by the determinant of $\mathbf{A}$. Consider the example of the 3×3 matrix $\mathbf{A}$

$$\mathbf{A} = \begin{bmatrix} 2 & 1 & 1 \\ 1 & 2 & 3 \\ 3 & 2 & 1 \end{bmatrix} \qquad (2.62)$$

It is easy to find the determinant of $\mathbf{A}$:

$$\det \mathbf{A} = 2(2 \times 1 - 2 \times 3) - 1(1 \times 1 - 2 \times 1) + 3(1 \times 3 - 2 \times 1) = -8 + 1 + 3 = -4 \qquad (2.63)$$

Now the elements of the adjoint of $\mathbf{A}$ can be found: $A_{11} = (2 \times 1 - 2 \times 3) = -4$, $A_{12} = -(1 \times 1 - 3 \times 3) = +8$, $A_{13} = (1 \times 2 - 3 \times 2) = -4$. The other six adj $\mathbf{A}$ elements are found similarly from the cofactors of $\mathbf{A}$: $A_{21} = 1$, $A_{22} = -1$, $A_{23} = -1$, $A_{31} = 1$, $A_{32} = -1$, $A_{33} = 3$. So finally we can write the inverse of $\mathbf{A}$ as

$$\mathbf{A}^{-1} = -\frac{1}{4} \begin{bmatrix} -4 & 1 & 1 \\ 8 & -1 & -5 \\ -4 & -1 & 3 \end{bmatrix} \qquad (2.64)$$

Other properties of *inverse matrices* are (Lathi 1974)

$$\mathbf{A}^{-1}\mathbf{A} = \mathbf{A}\mathbf{A}^{-1} = \mathbf{I} \tag{2.65}$$

and

$$(\mathbf{A}^{-1})^{-1} = \mathbf{A} \tag{2.66}$$

Inverting a large matrix is tedious pencil-and-paper work, it is best done by a computer using MATLAB. You will see that inverse matrices are used when solving sets of simultaneous equations. For example,

$$y_1 = a_{11}x_1 + a_{12}x_2 + a_{13}x_3 \tag{2.67A}$$

$$y_2 = a_{21}x_1 + a_{22}x_2 + a_{23}x_3 \tag{2.67B}$$

$$y_3 = a_{31}x_1 + a_{32}x_2 + a_{33}x_3 \tag{2.67C}$$

where $\mathbf{x}$ are the unknowns, $\mathbf{y}$ and $\mathbf{A}$ are given. In matrix notation, $\mathbf{y} = \mathbf{A}\mathbf{x}$, and the unknowns are found from (Lathi 1974)

$$\mathbf{x} = \mathbf{A}^{-1}\mathbf{y} \tag{2.68}$$

When a matrix is *time variable*, i.e., describes a *nonstationary system*, some or all of its elements are functions of time. Such a matrix is generally written as $\mathbf{A}(t)$. The time derivative of $\mathbf{A}(t)$ is a matrix formed from the time derivatives of each of its elements. For example,

$$\mathbf{A}(t) = \begin{bmatrix} a_{11}e^{-bt} & a_{12}U(t) & a_{13}\sin(\omega t) \\ a_{21} & a_{22}[1+\cos(ct)] & a_{23} \\ a_{31} & a_{32}t & a_{33}e^{vt} \end{bmatrix} \tag{2.69}$$

$$\dot{\mathbf{A}}(t) = \begin{bmatrix} -ba_{11}e^{-bt} & a_{12}\delta(t) & a_{13}\omega\cos(\omega t) \\ 0 & -a_{22}c\sin(ct) & 0 \\ 0 & a_{32} & a_{33}ve^{vt} \end{bmatrix} \tag{2.70}$$

There are many other theorems and identities in matrix algebra. For example, see the venerable texts by Guillemin (1949), Kuo (1967), and Lathi (1974).

2.4.2.3 Introduction to State Variables

The SV formalism provides a powerful, compact mathematical means for characterizing large, dynamic, linear systems. There are many physiological and PK systems which can be linearized and therefore analyzed using the state equation approach. Many others, however, are frankly nonlinear and can be solved easily and directly by simulation by specialized computer programs such as Simnon or MATLAB. (Linear SV systems are easily simulated with MATLAB, but can also be solved by Simnon.)

Any linear system describable by one or more linked, high-order ODEs can be reduced to SV form as we did for Equation 2.46. A general notation for an nth-order, linear, stationary, dynamic SV system is given below:

$$\dot{\mathbf{x}} = \mathbf{A}\mathbf{x} + \mathbf{B}\mathbf{u} \tag{2.71A}$$

$$\mathbf{y} = \mathbf{C}\mathbf{x} + \mathbf{D}\mathbf{u} \tag{2.71B}$$

where

$\mathbf{x}$ and $\dot{\mathbf{x}}$ are n-element column matrices (x_j is the jth system state)

$\mathbf{A}$ is the $n \times n$ *system matrix*

$\mathbf{B}$ is the $n \times b$ *input matrix*

$\mathbf{u}$ is a column vector of b inputs

$\mathbf{y}$ is a p-element column matrix of *system outputs* derived from the n states and the b inputs

$\mathbf{C}$ is a $p \times n$ element matrix

$\mathbf{D}$ is a $p \times b$ element matrix

Note that in the simplest case, the system can be SISO, so $\mathbf{y} = c_1 x_1$, $\mathbf{u} = u_n$, and

$$\mathbf{D} = \begin{bmatrix} 0 \\ 0 \\ 1 \end{bmatrix} \tag{2.72}$$

for $n = 3$.

In the first example, consider an LS described by a third-order, linear ODE. Following the approach we used in Equations 2.45 and 2.46, we write

$$\frac{d^3x}{dt^3} + 4\frac{d^2x}{dt^2} + 3\frac{dx}{dt} + 2x = u(t) \tag{2.73}$$

As in the example of Equation 2.46, we let $x_1 = x$, $x_2 = \dot{x}_1$, and $x_3 = \dot{x}_2$. Thus the matrices are found to be

$$\mathbf{x} = \begin{bmatrix} x_1 \\ x_2 \\ x_3 \end{bmatrix}, \quad \mathbf{A} = \begin{bmatrix} 0 & 1 & 0 \\ 0 & 0 & 1 \\ -2 & -3 & -4 \end{bmatrix}, \quad \mathbf{B} = \begin{bmatrix} 0 \\ 0 \\ 1 \end{bmatrix}, \quad \mathbf{u} = u(t) \tag{2.74}$$

If the output is $y = x_1 + 7u$, then the output state equation is written as

$$y = Cx + Du \tag{2.75}$$

where

$$C = \begin{bmatrix} 1 & 0 & 0 \end{bmatrix}, \quad D = \begin{bmatrix} 0 & 0 & 7 \end{bmatrix}, \quad \text{and} \quad u = \begin{bmatrix} 0 \\ 0 \\ u(t) \end{bmatrix} \tag{2.76}$$

A very important case is when the right-hand side of the ODE includes the derivatives of the input, u. In general, the order of the highest derivative on the right-hand side will be ≤n, the order of the highest derivative of x. Starting with a second-order SISO system example,

$$\frac{d^2x}{dt^2} + a_1 \frac{dx}{dt} + a_2 x = b_0 \frac{d^2u}{dt^2} + b_1 \frac{du}{dt} + b_2 u \tag{2.77}$$

The goal is to eliminate the input derivative terms from the ODE (Kuo 1967). One way to do this is to first define state $x_1 \equiv (x - b_0 u)$. Now $x = x_1 + b_0 u$. When this latter equation is substituted into Equation 2.82, we can write

$$\frac{d^2x_1}{dt^2} + b_0 \frac{d^2u}{dt^2} + a_1 \frac{dx_1}{dt} + a_1 b_0 \frac{du}{dt} + a_2 x_1 + a_2 b_0 u = b_0 \frac{d^2u}{dt^2} + b_1 \frac{du}{dt} + b_2 u$$

$$\downarrow \tag{2.78}$$

$$\frac{d^2x_1}{dt^2} + a_1 \frac{dx_1}{dt} + a_2 x_1 = \frac{du}{dt}(b_1 - a_1 b_0) + u(b_2 - a_2 b_0)$$

We next define state $x_2 \equiv \dot{x}_1 - (b_1 - a_1 b_0)u$, or

$$\dot{x}_1 = x_2 + (b_1 - a_1 b_0)u \tag{2.79}$$

Equation 2.84 is substituted into Equation 2.83. After simplifying, we obtain the state equation,

$$\dot{x}_2 + (b_1 - a_1 b_0)\dot{u} + a_1[x_2 + (b_1 - a_1 b_0)u] + a_2 x_1 = \dot{u}(b_1 - a_1 b_0) + u(b_2 - a_2 b_0)$$

$$\downarrow \tag{2.80}$$

$$\dot{x}_2 = -a_2 x_1 - a_1 x_2 + [(b_2 - a_2 b_0) - a_1(b_1 - a_1 b_0)]u$$

To summarize, the state equations are

$$\dot{x}_1 = x_2 + (b_1 - a_1 b_0)u \tag{2.81A}$$

$$\dot{x}_2 = -a_2 x_1 - a_1 x_2 + [(b_2 - a_2 b_0) - a_1(b_1 - a_1 b_0)]u \tag{2.81B}$$

In SV form,

$$\dot{x} = \overset{\mathbf{A}}{\begin{bmatrix} 0 & 1 \\ -a_2 & -a_1 \end{bmatrix}} x + \overset{\mathbf{B}}{\begin{bmatrix} (b_1 - a_1 b_0) \\ (b_2 - a_2 b_0) - a_1(b_1 - a_1 b_0) \end{bmatrix}} u(t) \tag{2.82}$$

Kuo (1967) gave a general form for resolving nth-order ODEs with input derivatives to standard SV form. The nth-order ODE is

$$\dot{x}^n + a_1 \dot{x}^{(n-1)} + a_2 \dot{x}^{(n-2)} + \cdots + a_{n-1}\dot{x} + a_n x = b_0 \dot{u}^n + b_1 \dot{u}^{(n-1)} + \cdots + b_{n-1}\dot{u} + b_n u \tag{2.83}$$

Kuo shows that the general form of the $\mathbf{A}$ matrix is, starting with $x = (x_1 + b_0 u)$,

$$\mathbf{A} = \begin{bmatrix} 0 & 1 & 0 & 0 & \cdots & 0 \\ 0 & 0 & 1 & 0 & \cdots & 0 \\ 0 & 0 & 0 & 1 & \cdots & 0 \\ . & . & . & . & . & . \\ 0 & 0 & 0 & \cdots & 0 & 1 \\ -a_n & -a_{n-1} & -a_{n-2} & \cdots & -a_2 & -a_1 \end{bmatrix} \tag{2.84}$$

The n elements of the $\mathbf{B}$ (column) matrix are then

$$\beta_1 = (b_1 - a_1 b_0)$$

$$\beta_2 = (b_2 - a_2 b_0) - a_1 \beta_1$$

$$\beta_3 = (b_3 - a_3 b_0) - a_2 \beta_1 - a_1 \beta_2 \tag{2.85}$$

$$\vdots$$

$$\beta_n = (b_n - a_n b_0) - a_{n-1}\beta_1 - a_{n-2}\beta_2 - \cdots - a_2\beta_{n-1} - a_1\beta_n$$

Another way to eliminate the input derivative terms in Equation 2.83 is to use SFGs. (Laplace transforms are described in chapter 3 in Northrop (2003), and SFGs and Mason's rule are covered in Section 2.6.)

Solution of a set of state equations in the time domain is first approached by considering the general solution of Equation 2.71A when u=0, i.e., the solution of the *homogeneous state equation*, $\dot{\mathbf{x}} = \mathbf{A}\mathbf{x}$, with initial conditions. A simple ODE of the form

$$\dot{x} = ax(t) \qquad (2.86)$$

can be rewritten as

$$\frac{dx(t)}{dx} = a\,dt \qquad (2.87)$$

which is easily integrated to

$$\ln[x(t)] = at + C_1 \qquad (2.88)$$

Let $C_1 = \ln[k]$, now Equation 2.88 can be rewritten as

$$\ln[x(t)] = \ln[e^{at}] + \ln[k] \rightarrow x(t) = ke^{at} \qquad (2.89)$$

The constant k is simply x(0), the initial value of x. For any initial time $t = t_0$, the initial value of x by definition is $x(t_0)$. From Equation 2.89, we can write

$$x(0) = e^{-at}x(t_0) \qquad (2.90)$$

Now substituting Equation 2.90 into Equation 2.89 we find

$$x(t) = e^{a(t-t_0)}x(t_0) \qquad (2.91)$$

which is the solution of the ODE Equation 2.86 for any initial condition (IC), $x(t_0)$.

By inference from the above development, it is reasonable to assume that the solution of the homogeneous state equations, $\dot{\mathbf{x}} = \mathbf{A}\mathbf{x}$, is of the form

$$\mathbf{x}(t) = e^{\mathbf{A}(t-t_0)}\mathbf{x}(t_0) \qquad (2.92)$$

Note that the exponential function of matrix $\mathbf{A}$ is given by

$$e^{\mathbf{A}(t-t_0)} = \exp[\mathbf{A}(t-t_0)] = \mathbf{I} + \mathbf{A}(t-t_0) + \frac{\mathbf{A}^2(t-t_0)^2}{2!} + \cdots + \frac{\mathbf{A}^k(t-t_0)^k}{k!} + \cdots$$

$$(2.93A)$$

or

$$e^{\mathbf{A}(t-t_0)} = \sum_{n=0}^{\infty} \mathbf{A}^n \frac{(t-t_0)^n}{n!} \qquad (2.93B)$$

The time-domain *state-transition matrix or fundamental matrix* $\mathbf{A}$ is defined as

$$\mathbf{\Phi}(t) \equiv e^{\mathbf{A}t} \tag{2.94}$$

Thus the solution of the homogeneous equation can finally be written as

$$\mathbf{x}(t) = \mathbf{\Phi}(t - t_o)\mathbf{x}(t_o) \tag{2.95}$$

Kuo (1967) gives the following properties of $\mathbf{\Phi}(t)$ with proofs

$$\mathbf{\Phi}(0) = \mathbf{I} \text{ (identity matrix)} \tag{2.96A}$$

$$\mathbf{\Phi}(t_2 - t_1)\mathbf{\Phi}(t_1 - t_o) = \mathbf{\Phi}(t_2 - t_o) \tag{2.96B}$$

$$\mathbf{\Phi}(t_1 - t_o) = [\mathbf{\Phi}(t_o - t_1)]^{-1} \quad \text{or} \quad [\mathbf{\Phi}(t)^{-1} = \mathbf{\Phi}(-t)] \tag{2.96C}$$

Solution of the homogeneous set of state equations in the time domain is formidable as a pencil-and-paper project. Our best advice is to avoid it at whatever cost. The same comment is emphatically underscored when solving state equations with many inputs $\mathbf{u}$ (forcing). That is,

$$\dot{\mathbf{x}} = \mathbf{A}\mathbf{x} + \mathbf{B}\mathbf{u} \tag{2.97}$$

Kuo (1967) shows that the general time-domain solution can be written as

$$\mathbf{x}(t) = \mathbf{\Phi}(t - t_o)\mathbf{x}(t_o) + \int_{t_o}^{t} \mathbf{\Phi}(t - \tau)\mathbf{B}\mathbf{u}(\tau)\,d\tau, \quad t \geq t_o \geq 0 \tag{2.98}$$

Thus the complete solution for the states, $\mathbf{x}(t)$, consists of the homogeneous solution independent of the solution due to specific inputs, $\mathbf{u}$. Although the use of IC time, $t_o > 0$, is more general, in most cases, we assume the solution starts at $t = 0$, and we take the ICs at $t_o = 0$.

Thus Equation 2.98 becomes

$$\mathbf{x}(t) = \mathbf{\Phi}(t)\mathbf{x}(0) + \int_{0}^{t} \mathbf{\Phi}(t - \tau)\mathbf{B}\mathbf{u}(\tau)\,d\tau, \quad t \geq 0 \tag{2.99}$$

Frequency-domain solutions are in general a much easier approach to the solution of SV systems. They are considered in detail in chapter 3 in Northrop (2003).

2.5 STABILITY OF LINEAR SYSTEMS

The stability of a closed-loop, feedback system is not guaranteed by the use of negative feedback. Under suitable conditions, both positive and negative feedback LTI systems can be stable or become unstable. Much has been written on the design of compensation filters required to preserve closed-loop system stability, and several tests have been developed to predict single-loop, LTI system instability from their loop gain, before the loop is closed.

Intuitively, we know that if we turn on a public address (PA) system, speak into the microphone, and hear a loud howl or screech that overwhelms the amplified speech, there is audio feedback that renders the PA system useless until we either reorient the microphone with respect to the loudspeakers or turn down the system's amplification. The howl is a form of system instability in which the system oscillates at a high audio frequency which is determined by a number of factors, such as the amplifier's gain and frequency response, and room acoustics including the sound propagation delay time from the loudspeakers to the microphone. The feedback itself *can be positive or negative*, and the instability can be the result of one or more of the factors cited above. The oscillations generally begin as soon as the speaker begins to talk into the microphone and grow exponentially in amplitude until some part of the system (loudspeaker, microphone, and amplifier) saturates. Then they persist at this obnoxious level until the microphone is moved to a better location.

Another form of instability in feedback systems occurs when the system output grows exponentially without oscillation to a level where the system is saturated and useless.

Fortunately, there are many tests that one can apply to the complex *loop gain transfer function*, $A_L(s)$, of a single-loop, linear feedback system that will predict whether it will be unstable in the closed-loop configuration, and if so, will it simply saturate its output, or will it oscillate and at what frequency? Some tests for nonlinear systems are practical, based on experimental frequency response data, such as the *Nyquist stability criterion*, the *describing function method*, and the *Popov criterion*. Other stability tests are mathematical, focusing on the poles, zeros and gain of an LTI system's loop gain, $A_L(s)$ (e.g., *The Routh–Hurwitz test*, *root locus analysis*, and the *Lyupanov* method).

Figure 2.4 illustrates a SISO, negative feedback system defining $A_L(s)$. In our treatment, the feedback is considered to be negative if there is a net minus sign preceding $A_L(s)$. In this chapter, we will examine only the venerable *root locus* test. The reader interested in pursuing the general topic of the prediction of system stability and compensation for system stability should consult one of the many venerable texts on control systems, such as Ogata (1990), Nise (1995), or Rugh (1996).

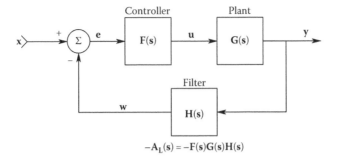

FIGURE 2.4   Block diagram of a SISO, single-loop, negative feedback system. The system's loop gain is negative: $-A_L(s) = -F(s)G(s)H(s)$. The minus sign in this case comes from the error-generating summer.

In general, a closed-loop, causal, continuous, LTI system is stable if all its *poles* lie in the left-hand **s**-plane. If one pole of a continuous system's closed-loop transfer function lies in the right-half **s**-plane, the output will theoretically grow exponentially without bound. Practically, some element in such a system always saturates (or burns out, etc.). If a linear system has a pair of complex-conjugate poles in the right-half **s**-plane, the system's output will be an exponentially growing oscillation. In either case, the unstable behavior is undesirable, and robust system design generally will compensate the closed-loop system so that unstable behavior cannot occur under any set of input conditions.

Note that system poles on the jω axis in the **s**-plane present a special problem. Two or more poles at the origin yield an unbounded impulse response. One pole yields a constant component to the impulse response (the impulse response of an integrator). A pair of conjugate poles on the jω axis will yield an oscillation of fixed peak amplitude and frequency dependent on initial conditions and the input. As we remarked above, a pair of complex-conjugate poles must lie in the right-half **s**-plane to produce exponentially growing oscillations.

It is important to realize that instability is deliberately designed into certain classes of man-made, *nonlinear feedback systems*. Such systems are intended to oscillate in *bounded limit cycles* around a desired *set point*. They generally use a simple, ON/OFF type of controller, or an ON/OFF controller with hysteresis. One example of limit-cycling control systems is the ubiquitous home heating (or cooling) system; another example is a closed-loop, drug infusion control system (Northrop 2000).

2.6 SIGNAL FLOW GRAPHS AND MASON'S RULE

2.6.1 INTRODUCTION

We start this section with a brief description of the use of linear SFGs as a means of introducing *graph theory* and the fact that many large, complex, nonlinear systems can be modeled by graphs characterized by *nodes* (aka *vertices*) and *branches* (aka *edges*). Such a graph can be analyzed for features including its *topology, node degree, degree distribution, cyclomatic number, signal processing ability, sensitivities*, and other metrics such as its *Shannon index* and *Weiner index*.

SFGs, a specialized subset of *mathematical graphs*, were developed specifically to do pencil-and-paper analysis of large *linear* dynamic systems characterized by sets of coupled, linear ODEs (Mason 1953, 1956, Northrop 2003). In the vocabulary of *graph theory*, SFGs are graphs with *signal nodes* (*vertices*) and *directed branches* (*edges*). (Note that in "graphspeak," an *arc* is a *directed* (from-to) edge or branch.) Furthermore, SFGs belong to the class of *directed graphs* or *digraphs*, because their *edges* (branches or arcs) denote *unidirectional information* (*signal*) *flow*. In graph theory, the *"order" of a graph is the number of (independent) vertices (nodes)* and its *"size" is the number of edges* (*branches*). Both increasing order and size of a graph are rough indications of its complexity. Below we consider linear SFGs as a special case of mathematical graphs.

Note that SFGs were initially developed for easy analysis of *linear* dynamic systems. However, in modeling CNLSs with digraphs, we lose the ability to use Mason's gain formula to calculate linear I/O transfer functions between certain nodes. An SFG with nonlinear branch gains still offers us insight into the CNLS's complex organization and topology, however.

2.6.2 Signal Flow Graphs

Signal flow graphs (SFG) is an LTI systems tool that was developed by S.J. Mason at MIT in 1953 for easy, pencil-and-paper analysis of large LTI systems. Mason published his SFG gain formula in 1956. SFGs enable one to take a large set of simultaneous, linear ODEs and linear algebraic equations describing a dynamic, LTI system and put them in graphical form having *nodes* and *directed branches* (signal conditioning paths). Then using *Mason's gain formula*, one can easily find the I/O, signal transfer function(s) for the system. The SFG approach is ideally suited for pencil-and-paper work because it replaces tedious, error-prone matrix inversion algebra as a necessary step to finding the state system's transfer function. Note that SFGs were developed more than a quarter century before PCs running linear system analysis applications such as MATLAB, Simulink, and Simnon appeared.

SFGs have found application in describing and analyzing linear analog electronic circuits, linear control systems (both discrete and analog), linear dynamic mechanical systems, linear PK systems, and linearized physiological and biochemical systems (fortunately, some physiological and biochemical systems can be approximated by linear ODEs). All signals in an SFG are assumed to be in the frequency domain (i.e., are represented by Laplace, Fourier, or z-transform variables). In this chapter, we use SFG digraph architecture to describe CNLSs;

however, the nonlinearities prevent the realization of transfer functions by Mason's formula. The nonlinear SFG architecture does give us insight into feedback paths and nodal connectivity, however.

An SFG has two components: *unidirectional branches* that condition signals and *signal nodes. Signals from incoming branches sum algebraically at a node.* The input signals summed at a node $\mathbf{p}$, $\mathbf{x_p}$, are multiplied by the *transmission* or *gain* of a branch leaving that node $\mathbf{p}$. The input to another node $\mathbf{q}$ from a branch $\mathbf{pq}$ is the product of the branch's transmission $\mathbf{T_{pq}}$ times the source ($\mathbf{p}$) node's signal, $\mathbf{x_p}$, i.e., $\mathbf{x_pT_{pq}}$. $\mathbf{x_pT_{pq}}$ does not directly affect the signal at source node $\mathbf{p}$ (i.e., it is not subtracted from $\mathbf{x_p}$, only added to $\mathbf{x_q}$). Some definitions relevant to SFGs are as follows.

Branch: A directed gain function that operates on the signal at a source node and adds the result to the signal at the sink node onto which the branch terminates. (A branch is also called an *edge*.) In SFGs, branch gains are *linear* operations or functions; in nonlinear SFGs, branch gains can be nonlinear operations. In nonlinear SFGs, Mason's gain formula does not apply.

Path: Any collection of a succession or concatenation of *branches* between nodes traversed in the same direction. The definition of path is general because it does not prevent any node to be traversed more than once (Kuo 1982).

Forward path: A path that starts at an *input* or *source node* and ends at an *output node*, along which no node in the SFG is traversed more than once. Many complex systems' SFGs have more than one forward path.

Forward path gain is defined as the net gain of a forward path.

Loop: A path that originates from a certain node $\mathbf{i}$ and also terminates on node $\mathbf{i}$. (This is also known as a *cycle* in graphspeak.) The loop path encounters no other node more than once. Loops with parallel paths can be reduced by adding their gains, forming an *independent loop*. An independent loop cannot be made up by simple addition from other loops (Truxal 1955).

Loop gain: The net path gain of a loop.

Node: Summing points where signals from all input paths are added algebraically. Paths leaving a node do not affect the node's signal. (Nodes are also called *vertices*.)

Nontouching loops share no nodes in common.

2.6.3 EXAMPLES OF LINEAR SFG REDUCTION

As an introductory example of signal manipulation in SFGs, consider the simple SFG of Figure 2.5A. Here, the signal $y_1 = T_1x_1$, $y_2 = T_2x_1$, and $y_3 = T_3x_1$. In Figure 2.5B, $y_1 = x_1T_1 + x_2T_2 + x_3T_3$. When several branches are in series forming a path as in Figure 2.5C, $y_4 = T_1T_2T_3x_1$. When they are in parallel as in Figure 2.5D, $y_1 = x_1(T_1 + T_2 + T_3)$.

The *systematic gain formula* for SFGs developed by Mason in 1956 is deceptively simple. It does require some interpretation, however. Mason's gain formula is written as

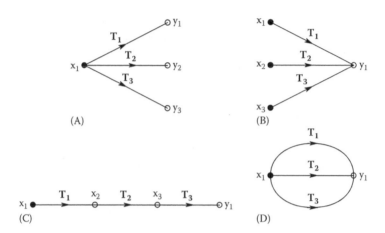

FIGURE 2.5 Four simple, directed SFGs. See text for description. (From Northrop, R.B., *Signals and Systems in Biomedical Engineering*, 2nd edn., CRC Press, Boca Raton, FL, 2010.)

$$\frac{V_{ok}}{V_{ij}} = H_{jk} = \frac{\sum_{n=1}^{N} F_n \Delta_n}{\Delta_D} \qquad (2.100)$$

where

V_{ok} is the frequency-domain (output) signal at node **k**

V_{ij} is the frequency-domain input signal at node **j**

H_{jk} is the net transmission (transfer function) from the **j**th (input) node to the **k**th (output) node

F_n is the transmission of the **n**th *forward path*. The **n**th forward path is a connected path of branches beginning on the **j**th (input) node and ending on the **k**th (output) node along which no node is passed through more than once. An SFG can have several forward paths which can share common nodes

Δ_D is the SFG gain *denominator*, or *determinant*. $\Delta_D \equiv 1 - [\text{sum of all indi-vidual loop gains}] + [\text{sum of products of pairs of all nontouching loop gains}] - [\text{sum of products of nontouching loop gains taken three at a time}] + \cdots$

Δ_n is the *cofactor* for the **n**th forward path. $\Delta_n \equiv \Delta_D$ evaluated for nodes *that do not touch* the **n**th forward path (see examples below). $\mathbf{n} = 1, 2, 3, \ldots.$

N is the total number of forward paths in the SFG

The best way to learn how to use Mason's formula on systems' SFGs is by example. In the first example, Figure 2.6 illustrates a simple SISO system. There is only one forward path. The loop touches two of its nodes, so $\Delta_1 = 1$. $\mathbf{F}_1$ is seen to be $1 \times K_v/(s+a)$, and

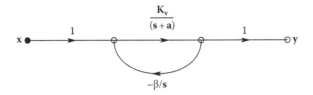

FIGURE 2.6 A simple, single-loop SFG used in example 1. See text for derivation of its transfer function, $\mathbf{Y}(s)/\mathbf{X}(s)$. (From Northrop, R.B., *Signals and Systems in Biomedical Engineering*, 2nd edn., CRC Press, Boca Raton, FL, 2010.)

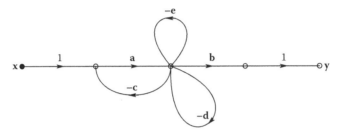

FIGURE 2.7 An SFG with three feedback loops and one forward path for the second example. See text for derivation of its transfer function, $\mathbf{Y}(s)/\mathbf{X}(s)$. (From Northrop, R.B., *Signals and Systems in Biomedical Engineering*, 2nd edn., CRC Press, Boca Raton, FL, 2010.)

$$\Delta_\mathbf{D} = 1 - \left[\frac{-\mathbf{K}_v}{(s+a)s} \right] \tag{2.101}$$

The overall Laplace transfer function is put in *time-constant form*. Note that the gain's denominator is second order.

$$\frac{\mathbf{Y}}{\mathbf{X}} = \frac{s/\beta}{s^2/\mathbf{K}_v\beta + sa/\mathbf{K}_v\beta + 1} \tag{2.102}$$

In a second example, the SFG is shown in Figure 2.7. Now $n=1$, $\mathbf{F}_1 = \mathbf{ab}$, $\Delta_\mathbf{D} = 1 - [-\mathbf{ac} - \mathbf{e} - \mathbf{d}] + 0$, and $\Delta_1 = 1$. From this we have the SISO gain

$$\frac{\mathbf{Y}}{\mathbf{X}} = \frac{\mathbf{ab}}{1 + \mathbf{ac} + \mathbf{e} + \mathbf{d}} \tag{2.103}$$

In a third example, shown in Figure 2.8, the SFG is more complicated; there are three forward paths, so: $n=3$, $\mathbf{F}_1 = \mathbf{abc}$, $\mathbf{F}_2 = \mathbf{def}$, $\mathbf{F}_3 = -\mathbf{akf}$, $\Delta_\mathbf{D} = 1 - [-\mathbf{bh} - \mathbf{eg}] + [(-\mathbf{bh})(-\mathbf{eg})] - 0$, $\Delta_1 = 1 + \mathbf{eg}$, $\Delta_2 = 1 + \mathbf{bh}$, $\Delta_3 = 1$. The transfer function is thus

$$\frac{\mathbf{Y}}{\mathbf{X}} = \frac{\mathbf{abc}(1 + \mathbf{ge}) + \mathbf{def}(1 + \mathbf{bh}) - \mathbf{akf}}{1 + \mathbf{bh} + \mathbf{ge} + \mathbf{bhge}} \tag{2.104}$$

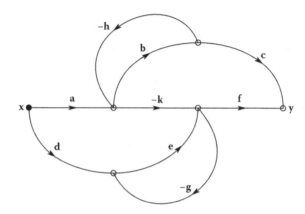

FIGURE 2.8 In the third example, the SFG has two feedback loops and three forward paths. See text for derivation of its transfer function, $\mathbf{Y}(s)/\mathbf{X}(s)$. (From Northrop, R.B., *Signals and Systems in Biomedical Engineering*, 2nd edn., CRC Press, Boca Raton, FL, 2010.)

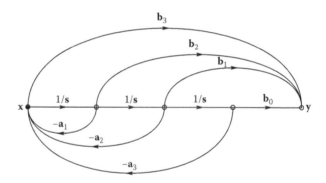

FIGURE 2.9 An SV format SFG of the fourth example. See text for derivation of its transfer function, $Y(s)/X(s)$. (From Northrop, R.B., *Signals and Systems in Biomedical Engineering*, 2nd edn., CRC Press, Boca Raton, FL, 2010.)

In a fourth example, Figure 2.9 illustrates an SV form SFG: Here $n=4$, $\mathbf{F}_1=\mathbf{b}_3$, $\mathbf{F}_2=\mathbf{b}_2/\mathbf{s}$, $\mathbf{F}_3=\mathbf{b}_1/\mathbf{s}^2$, $\mathbf{F}_4=\mathbf{b}_0/\mathbf{s}^3$, all $\Delta_k=1$, $\Delta_D=1-[-\mathbf{a}_1/\mathbf{s}-\mathbf{a}_2/\mathbf{s}^2-\mathbf{a}_3/\mathbf{s}^3]+0$. The SFG's Laplace transfer function is easily seen to have a cubic polynomial in the numerator and denominator:

$$\frac{\mathbf{X}}{\mathbf{Y}} = \frac{\mathbf{b}_3\mathbf{s}^3+\mathbf{b}_2\mathbf{s}^2+\mathbf{b}_1\mathbf{s}^1+\mathbf{b}_0}{\mathbf{s}^3+\mathbf{a}_1\mathbf{s}^2+\mathbf{a}_2\mathbf{s}^1+\mathbf{a}_3} \qquad (2.105)$$

In a fifth example, a molecule, $\mathbf{C}$, is synthesized in the mitochondria of a cell at a rate $\dot{\mathbf{Q}}_C$. Its concentration is $\mathbf{C}_m$ µg/L in the mitochondria. It diffuses out of the mitochondria into the cytoplasm where its concentration is C_c. It next diffuses through the cell membrane to what is basically zero concentration outside of the cell. The two compartments are: (1) the mitochondria and (2) the cytoplasm around the mitochondria. The compartmental state equations are based on simple diffusion (Fick's first law). V_m and V_c are compartment volumes.

$$V_m \dot{C}_m = \dot{Q}_0 - K_{12}(C_m - C_c) \quad \mu g/\text{min} \tag{2.106A}$$

$$V_c \dot{C}_c = K_{12}(C_m - C_c) - K_2 C_c \quad \mu g/\text{min} \tag{2.106B}$$

Written in state form, we have

$$\dot{C}_m = -C_m\left(\frac{K_{12}}{V_m}\right) + C_c\left(\frac{K_{12}}{V_m}\right) + \frac{\dot{Q}_0}{V_m} \quad \mu g/(\text{L min}) \tag{2.107A}$$

$$\dot{C}_c = -C_m\left(\frac{K_{12}}{V_c}\right) - C_c\frac{K_2 + K_{12}}{V_c} \quad \mu g/(\text{L min}) \tag{2.107B}$$

Note that mass diffusion rates depend on *concentrations* or mass/volume, so the diffusion rate constants, K_{12} and K_2, must have the dimensions of L/min.

From the ODEs, we see that the system is linear and can be described by an SFG as shown in Figure 2.10. The SFG can easily be reduced by Mason's

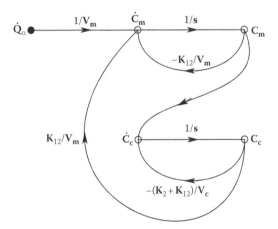

FIGURE 2.10 A two-state SFG representing the ODEs, Equations 2.107A and 2.107B in the fifth example. See text for description. (From Northrop, R.B., *Signals and Systems in Biomedical Engineering*, 2nd edn., CRC Press, Boca Raton, FL, 2010.)

rule to find the transfer function, $C_c/\dot{Q}_o(s)$. In this example, $n=1$, $F_1=(1/V_m)(1/s)$
$(K_{12}/V_c)(1/s), \Delta_1=1, \text{and} \Delta_D =1-\left[(-K_{12}/sV_m)+(-(K_2+K_{12})/sV_c)+\left(K_{12}^2/s^2V_mV_c\right)\right]+$
$\left[(-K_{12}/sV_m)(-(K_2+K_{12})/sV_c)\right]-0$.

This somewhat involved denominator turns out to be only a quadratic in **s**. After some algebra, we can write:

$$\frac{C_c}{\dot{Q}_C}=\frac{K_{12}/V_mV_c}{s^2+s[K_{12}/V_m+(K_{12}+K_2)/V_c+K_2K_{12}/V_mV_c]}=\frac{K}{(s+a)(s+b)} \qquad (2.108)$$

Thus, the two-compartment system governed by diffusion is seen to have linear, second-order dynamics with two real poles after factoring. MATLAB's *roots* utility can be used to numerically factor the denominator to find the **a** and **b** natural frequency values.

In our sixth and final example, we consider the SFG a *nonlinear* compartmental system described by Godfrey (1983). This is a two-compartment, PK system having six nodes in its SFG. It also has three loops and two nontouching loops. The ODEs describing the system can be written as

$$\dot{x}_1=-k_1x_1^2-\frac{V_mx_1}{K_m+x_1}+k_{12}x_2+u_1 \qquad (2.109A)$$

$$\dot{x}_2=\frac{V_mx_1}{K_m+x_1}-k_{12}x_2-k_{02} \qquad (2.109B)$$

Now we define: $f_1(x_1)=V_mx_1/(K_m+x_1)$, and $f_2(x_1)=k_1x_1^2$.

Sadly, because the system is nonlinear, Mason's formula cannot be used to find the gain or transfer function, x_2/u_1. The system's SFG is used only to illustrate the system's signal architecture (where positive and negative feedback loops are located). This nonlinear SFG has six nodes, two integrators, one squarer, a nonlinear function block, and a total of 10 directed branches and is shown in Figure 2.11. We can use the SFG to illustrate some properties of directed graphs, however. First, let us compile a table of the system's *vertex (nodal) degrees*:

The *degree sum formula* in graph theory states that for a given graph, $G=(V, E)$,

$$\sum_{v\in V}\deg(v)=2|E| \qquad (2.110)$$

since each edge (path) is incident on two vertices (nodes). This relation implies that in any graph, the number of nodes with odd degree is even. For the example system, $G=(6, 9)$ and $\sum\deg(v)=2(9)=18$. In Table 2.1, note that $\sum\text{Indegrees}+\sum\text{Outdegrees}=18$.

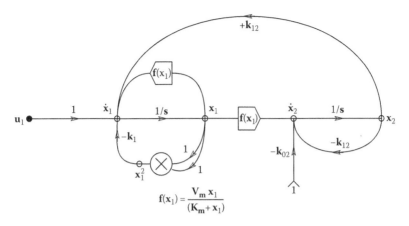

FIGURE 2.11   An SFG of the nonlinear compartmental model of the sixth example. See text for description.

TABLE 2.1
Tabulation of Vertex (Nodal) Degrees for the Nonlinear System of Equations 2.109A and 2.109B

Vertex (Node)	Indegree	Outdegree
u_1	0	1
$\dot{x}_1$	4	1
x_1	1	3
$\dot{x}_2$	3	1
x_2	1	2
k_{02}	0	1

2.6.4   MEASURES OF SFG COMPLEXITY

The complexity of a linear SFG generated from a set of linear, simultaneous algebraic equations or ODEs is largely in the eye of the beholder. Certainly we can agree on some objective criteria, such as the total number of nodes, **N**, the total number of directed branches, **B**, and the total number of independent feedback loops, **L**. If we create some monotonically increasing complexity index, $C_I = f(N, B, L)$, then we are still faced with establishing a criterion for the numerical functional, C_I, above which the system/SFG is considered complex. If we go back to Mason's Equation 2.100 for the SISO gain of a linear system, we see that generally it will yield a transfer function which is a rational polynomial in the Laplace variable, **s**. The denominator will be of order **n**, the numerator is of an order **m**, $0 \le m \le n$. **m** and **n** and the system's polynomial coefficients determine the number of the system's poles and zeros and its pole and zero positions in the complex s-plane. The pole and zero positions determine the system's dynamic, I/O behavior for an arbitrary input, as well as its steady-state, sinusoidal frequency response. The poles also determine the linear system's stability.

As we have seen above, an SFG can also be used to describe complex non-linear systems. However, Mason's rule can no longer be used to determine I/O gains for the nonlinear system. The SFG, in this case, "two-dimensionalizes" an abstract set of nonlinear ODEs and aids in visualizing functional relationships. Such a nonlinear SFG will in general have linear gains in some of the branches, and some branches will be nonlinear functions of the signal value at the origi-nating node (e.g., Hill functions, algebraic functions, e.g., $x_2 = a + bx_1 + cx_1^3$ or $x_2 = k \tanh(ax_1)$), or be parametric gain functions determined by another node variable, x_j. Obviously, such large sets of nonlinear ODEs are solved on com-puters. The index, C_I, will no longer be a valid measure of system complexity because large, internally interconnected nonlinear systems often have unexpected and surprising behavior (chaos), dependent not only on their initial conditions but also on the amplitude and rate of change of their inputs. See sections 1.4.4 and 1.5 in Northrop (2000), and also Chapter 9.

A comprehensive treatment of complex networks and graph theory may be found in the extensive review by Boccaletti et al. (2006). Besides reviewing graph theory, this paper considers social network phenomena, including dynamic effects (modeling cascading failures and congestion in communication networks); spreading processes (epidemics and rumors); the synchronization of coupled oscillators; as well as applications to social networks, the Internet and the World Wide Web, metabolic, protein, and genetic networks, CNS networks, etc.

2.7   SUMMARY

The purpose of this chapter has been to review some of the basics of LTI system theory. We began by considering the system properties of *linearity*, *causality*, and *stationarity*. Closed-loop system architecture was classified as simple-to-analyze SISO or the more complicated MIMO forms generally found in biochemical and physiological systems. LTI analog systems were shown to be described by one or more LTI ODEs. We examined the basics of solving first- and second-order LTI ODEs. System ODEs were shown to be reformatable in *SV form* in which all ODEs are first order and which can be solved in the time or frequency domains by matrix algebra. Matrix algebra was reviewed.

Characterization methods for LTI systems were next examined. The impulse response of an LTI system was defined, and real convolution was derived and its properties made clear. The usefulness of other transient inputs was described. The steady-state, sinusoidal frequency response of a continuous LTI system was shown to be an effective, frequency-domain system descriptor. LTI system fre-quency response is typically displayed as a Bode plot or a Nyquist (polar) diagram.

As we summarized above, Section 2.5 has dealt with discrete systems and signals, the z-transform, and discrete state equations. In Section 2.6, we reviewed the basis for various stability tests on continuous and discrete LTI systems. For continuous systems, we employ various tests to predict and/or detect closed-loop system poles in the right-half s-plane; for discrete systems, we test for closed-loop poles on or outside the unit circle in the z-plane. Note that long before a system becomes frankly unstable with an exponentially growing output, it can be useless for the purpose it was intended because its output is too oscillatory (underdamped). Good system design includes more than a theoretical test for instability; simulation is necessary.

PROBLEMS

2.1 When a certain drug is injected intravenously into a patient to build up its concentration in the blood, and then stopped, the drug's concentration is found to decrease at a rate proportional to its concentration. In terms of a simple, first-order, linear ODE, this can be written as

$$\dot{C} = -K_L C + \frac{R_{in}}{V_B} \quad (\mu g/L)/min$$

where

C is the drug concentration in the blood in $\mu g/L$
K_L is the loss rate constant in $1/min$
R_{in} is the drug input rate in $\mu g/min$
V_B is the patient's volume of blood in L

(A) Draw the simple, linear SFG for this system.
(B) Plot and dimension C(t) when $R_{in}(t) = R_o \delta(t)$ μg (a bolus injection at $t = 0$).
(C) Plot and dimension C(t) when $R_{in}(t) = R_D U(t)$ (a steady IV drip, modeled by a step function). Give the steady state C.

2.2 An LTI system is described by the second-order ODE below; x(t) is the system input, y(t) is the output.

$$\ddot{y} + 7\dot{y} + 10y = x(t)$$

(A) Use the **p** operator notation to find the roots of the characteristic equation, p_1 and p_2.
(B) Assume zero initial conditions. Find an expression for y(t) given $x(t) = \delta(t)$.
(C) Now let $x(t) = 10U(t)$. Find the steady-state value of y.

2.3 In this problem, there are two compartments where a drug can exist in a patient: In the blood volume (V_B) and in the extracellular fluid (ECF) volume (V_E). A drug is injected intravenously into the blood compartment at a rate R_{in} $\mu g/min$. The drug can diffuse from the blood compartment to the ECF compartment at a rate $K_{DBE}(C_B - C_E)$ $\mu g/min$, assuming $C_B > C_E$. Otherwise, the direction of diffusion is reversed if $C_B < C_E$. Drug is lost physiologically only from the blood compartment. The ECF compartment acts only for drug storage. Two linear, first-order ODEs describe the system:

$$\dot{C}_B V_B = -K_{BE}(C_B - C_E) - K_{LB}C_B + R_{in} \quad \mu g/min$$

$$\dot{C}_E V_E = K_{BE}(C_B - C_E)$$

(A) Draw the SFG for the system.
(B) Note that in the steady state, for $t \to \infty$, both $\dot{C}_B$ and $\dot{C}_E = 0$. Find expressions for $C_B(\infty)$ and $C_E(\infty)$, given a step infusion, $R_{in}(t) = R_D U(t) \mu g/min$.
(C) Plot and dimension $C_B(t)$ and $C_E(t)$, given zero initial conditions and a bolus injection of drug, $R_{in}(t) = R_o \delta(t) \mu g$. It is best if you solve this problem by simulation (e.g., use MATLAB). Let $V_E = 15\,L$, $V_B = 3\,L$, $K_{BE} = K_{LB} = 1\,L/min$.

2.4 An underdamped quadratic low-pass filter is described by the second-order ODE:

$$\ddot{y} + (2\xi\omega_n)\dot{y} + \omega_n^2 y = x(t)$$

(A) The characteristic equation for this ODE in terms of the **p** operator has complex-conjugate roots. Find an algebraic expression for the position of the roots, p_1 and p_2.
(B) Let the input, $x(t) = U(t)$ (a unit step). Find the steady-state output, y_{ss}.
(C) Let $x(t) = \delta(t)$. Find, sketch, and dimension $y(t)$ for $0 \le t \le \infty$, for $\xi = 0.5$, and $\omega_n = 1$ rev/s.

2.5 An alternate way of writing the ODE for an underdamped, quadratic, low-pass system is

$$\ddot{y} + 2a\dot{y} + (b^2 + a^2)y = x(t)$$

Let $a = 0.5$, $b = 0.86603$. Repeat (A), (B), and (C) in the preceding problem.

2.6 Consider a capacitor that is charged to V_{co} at $t = 0-$, and then allowed to discharge through a nonlinear conductance for which $i_{nl} = \beta V_c^2$, as shown in Figure P2.6. Thus for $t \ge 0$, system behavior is given by the nonlinear node ODE:

$$C\dot{V}_c + \beta V_c^2 = 0$$

We are interested in finding $V_c(t)$ for this circuit for $t \ge 0$. The ODE is a form of a nonlinear ODE called *Bernoulli's equation*, which can be linearized by a change in its dependent variable. In its general form, Bernoulli's equation is

$$\frac{dy}{dx} + P(x)y = Q(x)y^n$$

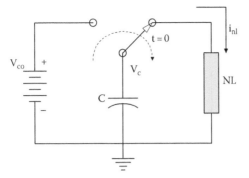

FIGURE P2.6

This equation can be linearized for solution by making the substitution, $u \equiv y^{1-n}$. Since $du/dx = (1-n)\, y^{-n}(dy/dx)$, this leads to the ODE linear in u. Sic:

$$\frac{1}{1-n}\frac{du}{dx} + P(x)u = Q(x)$$

Substitute $n=2$, $P=0$, $y=V_c$, $x=t$, and $Q=-\beta/C$ into the linearized ODE above and solve for $V_c(t)$. Plot and dimension $V_c(t)$ for $-C/(\beta V_{co}) \le t \le \infty$. (*Hint*: Note that the capacitor discharge through this nonlinearity generates an *hyperbola*, rather than an exponential decay.)

2.7 Find the nonlinear element, $i_{nl} = f(V_c)$, which when placed in parallel with the capacitor of Problem 2.6 will give a $V_c(t)$ of the form

$$V_c(t) = K \log_{10}\left(\frac{1}{t + T_o}\right), \quad t \ge 0$$

2.8 Find the determinants of the matrices below:

$$(A)\begin{bmatrix} 2 & 3 \\ -1 & 4 \end{bmatrix} \quad (B)\begin{bmatrix} 1 & 0 & 0 \\ 2 & 3 & 5 \\ 4 & 1 & 3 \end{bmatrix} \quad (C)\begin{bmatrix} 1 & 0 & 0 \\ 3 & 4 & 15 \\ 5 & 6 & 21 \end{bmatrix}$$

2.9 Find the adjoint matrices of the square **A** matrices below:

$$(A)\begin{bmatrix} a & b \\ c & d \end{bmatrix} \quad (B)\begin{bmatrix} 1 & 2 & 3 \\ 2 & 3 & 2 \\ 3 & 3 & 4 \end{bmatrix} \quad (C)\begin{bmatrix} 1 & 2 & 3 \\ 1 & 3 & 4 \\ 1 & 4 & 3 \end{bmatrix}$$

(D) Find **A** × adj **A** for the matrix of (B).

2.10 We have seen that the inverse of a square matrix can be found from $\mathbf{A}^{-1} = \text{adj } \mathbf{A}/\det \mathbf{A}$. Find the inverse matrices for

$$(A)\begin{bmatrix} 2 & 3 \\ 1 & 4 \end{bmatrix} \qquad (B)\begin{bmatrix} 1 & 2 & 3 \\ 1 & 3 & 4 \\ 1 & 4 & 3 \end{bmatrix} \qquad (C)\begin{bmatrix} 2 & 3 & 1 \\ 1 & 2 & 3 \\ 3 & 1 & 2 \end{bmatrix}$$

2.11 Solve the simultaneous equations using Cramer's rule:

$$(A)\, 3x_1 - 5x_2 = 0 \qquad\qquad (B)\, 2x_1 + x_2 + 5x_3 + x_4 = 5$$
$$x_1 + x_2 = 2 \qquad\qquad\qquad x_1 + x_2 - 3x_3 - 4x_4 = -1$$
$$3x_1 + 6x_2 - 2x_3 + x_4 = 8$$
$$2x_1 + 2x_2 + 2x_3 - 3x_4 = 2$$

2.12 A SISO, LTI system is described by the second-order ODE:

$$\ddot{y} + 5\dot{y} + 2y = r(t)$$

Let $y = x_1$, $x_2 = \dot{x}_1$. Find the $\mathbf{A}$ and $\mathbf{B}$ matrices in the SV form of this system:

$$\dot{x} = \mathbf{A}x + \mathbf{B}r(t)$$

2.13 A SISO LTI system is described by

$$2\ddot{y} + 3\dot{y} + y = \dot{r} + 2r$$

where $y = x_1$, $\dot{x}_1 = x_2$. Find the $\mathbf{A}$ and $\mathbf{B}$ matrices for this system.

2.14 A SISO LTI system is described by:

$$\ddot{y} + 3\dot{y} + 2y = 3\ddot{r} + 5\dot{r} + r$$

where $y = x_1$, $\dot{x}_1 = x_2$. Find the $\mathbf{A}$ and $\mathbf{B}$ matrices for this system, and the $\mathbf{C}$ and $\mathbf{D}$ matrices in the equation

$$y(t) = \mathbf{C}x + \mathbf{D}r(t)$$

2.15 Which of the LTI system transfer functions below are unstable? Plot their poles and zeros in the s-plane.

(A) $H(s) = \dfrac{s-8}{s^2+3s+2}$

(B) $H(s) = \dfrac{s+1}{s^2+3s+3}$

(C) $H(s) = \dfrac{s+1}{s^2+3s-3}$

(D) $H(s) = \dfrac{s+1}{s^2-3s-3}$

(E) $H(s) = \dfrac{s-1}{s^2+3s-4}$

2.16 A SISO LTI negative feedback system is shown in Figure P2.16. Note that the plant is unstable.
(A) Find the K value above which the closed-loop system is stable.
(B) Can the closed-loop system be stable for any negative value of K?

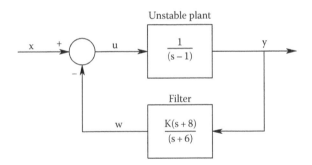

FIGURE P2.16

2.17 Find the transfer function for the "rabbit ears" SFG shown in Figure P2.17.

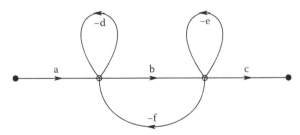

FIGURE P2.17

2.18 A two-input, two-output (2I2O) complex system with cross-coupling can be treated as linear. See Figure P2.18. It can be described by the two linear ODEs:

$$\dot{x}_1 = -a_{11}x_1 + a_{12}x_2 + b_1u_1$$

$$\dot{x}_2 = -a_{22}x_2 + a_{21}x_1 + b_2u_2$$

(A) Draw the SFG for the system.
(B) Use Mason's gain formula to find the direct transfer function, $X_1(s)/U_1(s)$.
(C) Use Mason's gain formula to find the direct transfer function, $X_2(s)/U_2(s)$.
(D) Use Mason's gain formula to find the cross-transfer function, $X_1(s)/U_2(s)$.
(E) Use Mason's gain formula to find the cross-transfer function, $X_2(s)/U_1(s)$.

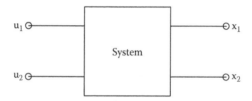

FIGURE P2.18

2.19 Consider the 2I2O system of Problem 2.18 to be in the dc steady state. Use Cramer's rule on the two ODEs to find expressions for X_{1SS} and X_{2SS}.

2.20 A *linear decoupling controller* is used to minimize x_2/u_1 and x_1/u_2. It is shown in Figure P2.20 by the dashed SFG branches attached to the system of Problem 2.18. The controller parameters a_{12}^*, a_{21}^*, a_{11}^*, and a_{22}^* are estimates of the 2I2O system's parameters. Ideally, $a_{12}^* = a_{12}$, etc.

(A) Assume ideal conditions. Use Mason's formula to show $(X_2(s)/U_1(s)) \rightarrow 0$. [Note the controlled system of Figure P2.20 has seven loops and four products each of two nontouching loops.] The two ODEs for the ideally decoupled system are

$$\dot{x}_1 = c_1v_1 - x_1P_1$$

$$\dot{x}_2 = c_2v_2 - x_2P_2$$

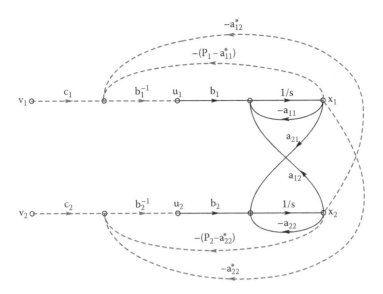

FIGURE P2.20

(B) Now assume there are errors in the decoupling controller's parameter estimates, so

$$(a_{11}^* - a_{11}) = \delta_{11}, (a_{22}^* - a_{22}) = \delta_{22}, (a_{12}^* - a_{12}) = \delta_{12}, \text{ and } (a_{21}^* - a_{21}) = \delta_{21}$$

Write the decoupled system's two ODEs in terms of c_1, c_2, P_1, P_2, and the δ_{xy}s.

(C) Draw the SFG for the two ODEs of part **B**.

2.21 Godfrey (1983) proposed a model for a nonlinear, two-compartment, cross-coupled, pharmacokinetic (2CPK) system. The PK system can be described by the two nonlinear ODEs below. Note that a saturating Hill function is used to describe the rate of increase of drug concentration x_2 from the x_1 drug concentration compartment.

$$\dot{x}_1 = -K_{01}x_1^2 + K_{12}x_2 - \frac{V_m x_1}{K_m + x_1} + b_1 u_1$$

$$\dot{x}_2 = -K_{12}x_2 - K_{02} + \frac{V_m x_1}{K_m + x_1} + b_2 u_2$$

(A) Draw the nonlinear SFG for the 2CPK system.

(B) A nonlinear decoupling controller is used to permit independent manipulation of the concentrations, x_1 and x_2. The decoupling controller's inputs are v_1 and v_2; its outputs are u_1 and u_2. u_1 and u_2 are now given by:

$$u_1 = b_1^{*-1} \left[-P_1 + a_1 v_1 + K_{01}^* x_1^2 - K_{12}^* x_2 + \frac{V_m^* x_1}{K_m^* + x_1} \right]$$

$$u_2 = b_2^{*-1} \left[-P_2 + a_2 v_2 + K_{12}^* x_2 - K_{02} - \frac{V_m^* x_1}{K_m^* + x_1} \right]$$

The starred controller parameters are estimates of the 2CPK system's parameters. P_1 is the designed natural frequency (rev/s) of the x_1 compartment; P_2 is the natural frequency of the x_2 compartment.

Draw the nonlinear SFG for the decoupler-regulated 2CPK system.

(C) Find the two ODEs for the case of the perfectly tuned decoupler-regulated system.

(D) While the strategy of using decoupling regulation appears to have utility, the fact that concentrations x_1 and x_2 are non-negative presents a problem. Discuss the consequences of using a decoupling controller on a 2CPK system. What is the significance of negative u_1 and u_2 values? (See Northrop 2000.)

3 Introduction to Biochemical Oscillators and Complex, Nonlinear Biochemical Systems

3.1 INTRODUCTION: SOME GENERAL PROPERTIES OF NONLINEAR SYSTEMS

First, note that many nonlinear systems are not complex; however, many complex systems are nonlinear. The first and most salient feature of a nonlinear system (NLS) is that it does not obey superposition, and thus cannot be fully characterized by its impulse responses, or by convolution, linear state-variable algebra, or transfer functions.

Large, nonlinear (and time-variable) systems are generally complex systems, that is, they exhibit unpredictable, emergent behavior, including chaos. They are also challenging to model, even approximately.

We know that if a sinusoidal input is given to a dynamic, SISO, linear system, then in the steady state (SS), the output is a sinusoid of the same frequency but with (in general) different phase and amplitude. When a stable CNLS is given a sinusoidal input, the SS output will generally be periodic, but not purely sinusoidal. The SS periodic output can be written as a Fourier series, showing the existence of the fundamental frequency and higher-order harmonics. The amplitude distribution of the harmonics will, in general, depend on the amplitude and frequency of the sinusoidal input signal. A nonlinear system can also generate intermodulation distortion terms at its output, given an input, which is the sum of two or more sinusoids of different frequencies. The intermodulation process can be illustrated by the use of a simple, static, power-law nonlinearity given by Equation 3.1:

$$\mathbf{y} = \mathbf{a}_0 + \mathbf{a}_1 \mathbf{x}^1 + \mathbf{a}_2 \mathbf{x}^2 + \mathbf{a}_3 \mathbf{x}^3 \tag{3.1}$$

If we let the input be $\mathbf{x}(t) = [b_1 \sin(\omega_1 t) + b_2 \sin(\omega_2 t)]$, then the output can be written as

$$y(t) = a_0 + a_1 \left[b_1 \sin(\omega_1 t) + b_2 \sin(\omega_2 t) \right]$$

$$+ a_2 \left[b_1^2 \sin^2(\omega_1 t) + 2 b_1 b_2 \sin(\omega_1 t) \sin(\omega_2 t) + b_2^2 \sin^2(\omega_2 t) \right]$$

$$+ a_3 \left[b_1^3 \sin^3(\omega_1 t) + b_2^3 \sin^3(\omega_2 t) + 3 b_1^2 b_2 \sin^2(\omega_1 t) \sin(\omega_2 t) \right.$$

$$\left. + 3 b_1 b_2^2 \sin(\omega_1 t) \sin^2(\omega_2 t) \right] \tag{3.2}$$

From trigonometric identities, we see that the output, $y(t)$, not only contains terms at the input frequencies, but also dc terms, plus terms with frequencies of $(\omega_1 - \omega_2)$, $(\omega_1 + \omega_2)$, $2\omega_1$, $2\omega_2$, $3\omega_1$, $3\omega_2$, $(2\omega_1 - \omega_2)$, $(2\omega_1 + \omega_2)$, $(2\omega_2 - \omega_1)$, and $(2\omega_2 + \omega_1)$. (The situation becomes even more algebraically complicated if an x^4 term is present in the nonlinearity.)

Certain simple nonlinear ODEs can also generate subharmonics at their outputs. Tomovic (1966) gave an example of a dynamic NLS, given a sinusoidal input with a frequency of 6 rad/s:

$$\ddot{y} = -y - 0.2 y^3 + A \cos(6t) \tag{3.3}$$

This system is easily simulated with Simnon™; the state equations are

$$\dot{x}_1 = -x_2 - 0.2 x_2^3 + A \cos(6t) \tag{3.4A}$$

$$\dot{x}_2 = x_1, \quad y = x_2 \tag{3.4B}$$

A Simnon program to simulate Equations 3.4A and B is given in Appendix A.2. The simulation shows that the SS output, $y(t)$, contains terms [B cos(6t) + C cos(2t)] (see Figure 3.1). (The cos(2t) term is the subharmonic term.)

The behavior of a nonlinear system is generally dependent on its input's amplitude and frequency, or rate of change, as well as its initial conditions. As you will see below, certain nonlinear systems can exhibit *limit cycles*, which are bounded, periodic, steady-state oscillations of the output and other states in the absence of a periodic forcing input. NLSs can also exhibit input-amplitude-dependent damping, where the system's output in response to an input step shows less and less damping as the input step amplitude is increased. In some cases, the damping can go to zero or a negative value, causing an unbounded output. Initial conditions can also determine whether a NLS's response is stable, oscillatory, or unbounded for a given input.

The nonlinearity in a nonlinear system can often be modeled by a functional nonlinearity in series with otherwise linear dynamics. The functional nonlinearity can be half-wave rectification, otherwise known as nonnegativity, common

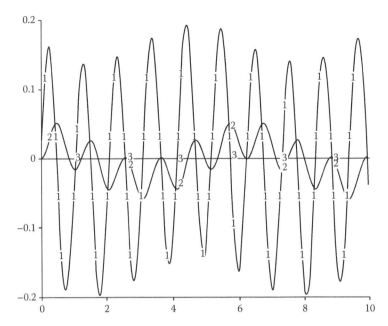

FIGURE 3.1   Steady-state oscillations in the states of the nonlinear system, **tomovic1**, given 6 rad/s sinusoidal excitation. Results of a Simnon™ simulation. Trace 1 is state x1; trace 2 is state x2. (*cf.* Equations 3.4A and B). (From Northrop, R.B., *Endogenous and Exogenous Regulation and Control of Physiological Systems*, Chapman & Hall/CRC Press, Boca Raton, FL, 2000.)

to all physiological concentrations and densities or it can be an odd saturating nonlinearity on **x**, such as $y(\mathbf{x}) = A \tanh(b\mathbf{x}) = A(1 - e^{-2b\mathbf{x}})/(1 + e^{-2b\mathbf{x}})$. Many physiological systems exhibit rate saturation. For example, the steady-state secretion rate of a gland in response to a hormone concentration, [**H**], can be written as a first-order (hyperbolic) Hill function. In general, a saturating Hill function (in [**H**]) has the form

$$\dot{Q}_G = \frac{[\mathbf{H}]\dot{Q}_{GMAX}}{(1/K + [\mathbf{H}])}, \quad \text{for } [\mathbf{H}] \geq 0; \quad \dot{Q}_G \rightarrow \dot{Q}_{GMAX} \text{ when } [\mathbf{H}] \gg \frac{1}{K} \quad (3.5)$$

To be nonlinear, an NLS need not have specific functional nonlinearities (e.g., Hill functions, tanh(x) functions) in a gain pathway. The nonlinear behavior can be due to one or more nonlinear ODEs having terms that are the products of two or more variables, or a variable raised to a power. This is generally the case in modeling physiological systems involving chemical kinetics. Such state equations are derived from the laws of chemical mass action, they can contain algebraic terms that are products of states, and states raised to integral powers. Often these ODEs are "stiff," which means that during numerical solution, certain terms can $\rightarrow 0$ while other terms become very large. There are often problems in

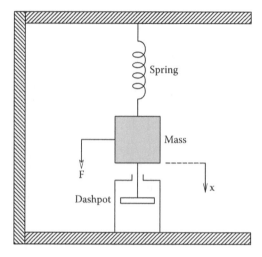

FIGURE 3.2 Nonlinear mechanical system consisting of a series spring, mass, and dashpot (Duffing system). The spring is made nonlinear to investigate jump resonance phenomena.

the numerical solutions of sets of stiff ODEs; special integration routines must be used to prevent numerical over- and underflows (Hultquist 1988), and oscillating solutions.

Under sinusoidal excitation, certain simple algebraic nonlinear systems can exhibit the phenomena of multivalued responses and jump resonances. These interesting nonlinear characteristics may be demonstrated on a simple, mechanical, second-order nonlinear system consisting of a mass in series with a linear dashpot and a nonlinear spring, as shown in Figure 3.2. The ODE governing the mass position, x, can be written as

$$M\ddot{x} + D\dot{x} + kx + k'x^3 = P\cos(\omega t) \tag{3.6}$$

Equation 3.6 is known as *Duffing's equation*; it can be rewritten as (Tomovic 1966)

$$\ddot{x} + k_2\dot{x} + \omega_0^2 x + k_1 x^3 = A\cos(\omega t) \tag{3.7}$$

The second-order, NL ODE, Equation 3.7, can be put in nonlinear state form and solved using Simnon:

$$\dot{x}_1 = -k_2 x_1 - k_1 x_2^3 - \omega_0^2 x_2 + A\cos(\omega t) \tag{3.8A}$$

$$\dot{x}_2 = x_1 \tag{3.8B}$$

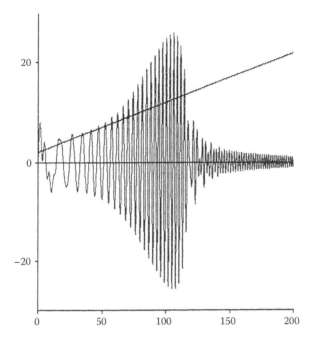

FIGURE 3.3   Response of the stiff-spring Duffing system (see Equations 3.6 and 3.7) to a linearly increasing force driving frequency. Straight line denotes frequency in hertz. Note the sudden decrease in the amplitude of the mass displacement, x(t). (From Northrop, R.B., *Endogenous and Exogenous Regulation and Control of Physiological Systems*, Chapman & Hall/CRC Press, Boca Raton, FL, 2000.)

ω is swept by setting $\omega = (\omega_s + \delta_\omega t)$. The amplitude jump phenomenon with increasing frequency for a stiff spring is shown in Figure 3.3. The straight line is a plot of $10\omega(t)$. Note that in the stiff-spring system, $k_1 > 0$, and the amplitude of **x** abruptly decreases at about $t = 110$ s, where $\omega \cong 1.3$ rad/s. The sharpness of the jump in amplitude and the frequency at which it occurs depend on the driving cosine's amplitude, A; A = 15 in this case. In Figure 3.4, the frequency of the source is swept down in the stiff-spring system from 2.5 to 1.875 rad/s. Now, the amplitude of **x** increases to a peak at $t = 100$ ($\omega = 2.25$), and then linearly decreases with frequency.

Figure 3.5 shows a more pronounced amplitude jump behavior with slowly decreasing frequency. In this case, the system has a soft spring ($k_1 < 0$). The jump in this case occurs when ω reaches 1.45 rad/s. A transitional behavior on increasing frequency for the soft-spring system is illustrated in Figure 3.6. At $\omega \geq 0.96$ rad/s, the peak amplitude of **x** begins to decrease linearly in the soft-spring system.

In conclusion, we wish to stress that nature gives us simple, nonlinear, physical and chemical systems. Such systems can show unexpected behavior. Modeling such systems is relatively easy using MATLAB® and Simulink®, or Simnon.

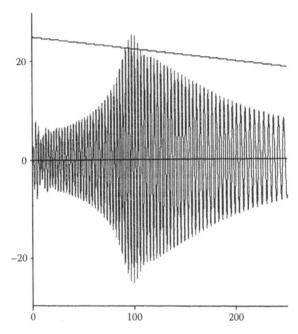

FIGURE 3.4 Mass position x(t) of Duffing system with a stiff spring to a linearly decreasing forcing frequency. (From Northrop, R.B., *Endogenous and Exogenous Regulation and Control of Physiological Systems*, Chapman & Hall/CRC Press, Boca Raton, FL, 2000.)

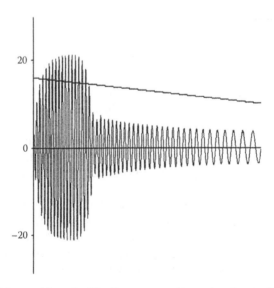

FIGURE 3.5 Mass position x(t) of Duffing system with a soft spring to a linearly decreasing forcing frequency. (From Northrop, R.B., *Endogenous and Exogenous Regulation and Control of Physiological Systems*, Chapman & Hall/CRC Press, Boca Raton, FL, 2000.)

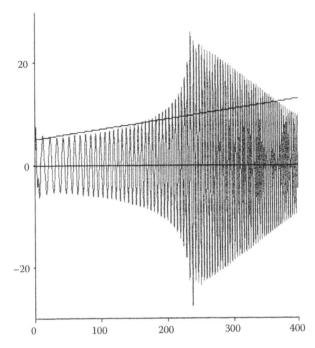

FIGURE 3.6   Mass position x(t) of Duffing system with a soft spring to a linearly increasing forcing frequency. (From Northrop, R.B., *Endogenous and Exogenous Regulation and Control of Physiological Systems*, Chapman & Hall/CRC Press, Boca Raton, FL, 2000.)

3.2   ALL LIVING SYSTEMS ARE NONLINEAR

3.2.1   INTRODUCTION

No matter at what level of organization or scale we examine biological systems, we come to the inescapable conclusion that they are all nonlinear. This is one reason why large biological systems are complex. One source of their nonlinearity stems from the fact that molecular concentrations are necessarily all nonnegative. Also, there is no such thing as a negative transcription rate, secretion rate, biosynthesis rate, or injection rate. This nonnegativity causes all biochemical mass-action ODEs to be nonlinear, regardless what other algebraic or functional nonlinearities they may contain. (On the other hand, simple diffusion can be treated as a linear process; its rate depends only on concentration differences, which can be negative.)

We will see below that yet another nonlinearity can enter into mass-action models of biochemical regulatory systems, that is, parametric feedback. Parametric feedback is implemented by parametric modification of mass-action equation rate constants.

3.2.2 Mass-Action Kinetics and Nonlinearity in Chemical Systems

Biochemical reactions, such as the regulated synthesis of hormones and proteins, underlie nearly all physiological regulatory and control processes. All chemical reactions proceed at rates governed by the concentrations of the reactants, the temperature of the reaction, and by the number of molecules of reactants required to produce a molecule of product. One underlying assumption in formulating mass-action descriptions of chemical kinetics is that the reactants are uniformly distributed (in a compartment) and are free to move around and collide (and react) with a probability that is proportional to their concentration(s). Another assumption is that mass-action formulations are based on very large numbers of molecules interacting; mass action is therefore a bulk formulation of chemical behavior.

It will be seen that describing mass-action systems' behavior generally involves solving nonlinear ODEs. Thus dynamic solutions are best done by computer simulation. We will give some simple examples here that illustrate the procedures to be used and also some typical biochemical reaction architectures.

3.2.3 Examples of Mass-Action Kinetics

1. In the first example, let us first consider a bimolecular reaction in which one molecule of **A** combines irreversibly with one molecule of **B** to form one molecule of product, **C**, that is, in chemical notation,

$$\mathbf{A} + \mathbf{B} \xrightarrow{\;k1\;} \mathbf{C}$$

   In the first kinetic formulation, we start at t=0 with a moles of **A**, b moles of **B**, and no **C**. Let x=moles of **C** made at time t. By the mass-action formalism, we can write the nonlinear ODE:

$$\dot{x} = k_1(a-x)(b-x) = k_1[ab - x(a+b) + x^2] \tag{3.9}$$

   where k_1 is the reaction rate constant, generally an increasing function of Kelvin temperature.

   An alternative mass-action ODE uses the running concentration (RC) of $\mathbf{A} = y = (a-x)$, the RC of $\mathbf{B} = z = (b-x)$, and the RC of $\mathbf{C} = x$. Initial conditions are used. Thus, the rate of appearance of **C** is

$$\dot{x} = k_1 yz \tag{3.10}$$

and the rate of disappearance of A is

$$\dot{y} = -k_1 yz \qquad (3.11)$$

2. A second example of mass-action kinetics is given for the reversible oxidation of nitrogen oxide: The chemistry is

$$2NO + O_2 \overset{k_1}{\underset{k_2}{\rightleftharpoons}} 2NO_2 \qquad (3.12)$$

Let
  $k_1 =$ forward rate constant
  $k_2 =$ reverse rate constant
  $a =$ initial conc. of NO
  $b =$ initial conc. of O_2
  $x =$ amount of O_2 reacted at t

The forward-reaction rate is $\dot{x}_f = k_1(a-2x)^2(b-x)$.
The reverse reaction rate is $\dot{x}_r = k_2(2x)^2$.
The net reaction rate is $\dot{x}_{net} = \dot{x}_f - \dot{x}_r = k_1(a-2x)^2(b-x) - k_2(2x)^2$.
(Note that when two molecules react, the concentration is squared; when three molecules react, it is cubed.)

3. For the third example, we consider the physiologically important formation and decomposition of carbonic acid. This reaction is important in vertebrate metabolism; the blood carries carbonic acid to the lungs or gills where it is converted to CO_2 gas and lost:

$$H_2CO_3 \overset{k_F}{\underset{k_R}{\rightleftharpoons}} H_2O + CO_2 \qquad (3.13)$$

Let
  $x =$ concentration of carbonic acid
  $w =$ concentration of water
  $g =$ concentration of dissolved CO_2

Thus, we can write for the rate of appearance of carbonic acid:

$$x = -k_F x + k_R \left(wg \right) \qquad (3.14)$$

The rate of appearance of CO_2 is simply

$$\dot{g} = k_F x - k_R(wg) = -\dot{x} \qquad (3.15)$$

If water is present in excess, then the w factor can be eliminated from the ODEs above and its effect incorporated in k_R.

4. In the fourth example, we examine a typical biochemical "two-step" reaction, in which two reactants combine reversibly to form a complex, and then the complex is rapidly converted to the end product with the release of the **E** reactant unchanged. The product **P** decays at rate $k_4 P$. This reaction form is known as the Michaelis–Menten architecture (Northrop 2000):

$$X + E \underset{k_2}{\overset{k_1}{\rightleftharpoons}} E * X \xrightarrow{\ k_3\ } E + P \xrightarrow{\ k_4\ } * \tag{3.16}$$

where

$x_1 =$ running conc. of substrate **X**
$u =$ running conc. of free enzyme, **E**
$x_2 =$ running conc. of complex
$x_3 =$ running conc. of product **P**
$k_1, k_2, k_3 =$ reaction rate constants

The reaction is assumed to take place in a closed vessel, so enzyme is conserved, that is, $u_o = u + x_2$, or $u = u_o - x_2$. The nonlinear, mass-action ODEs can be written as

$$\dot{x}_1 = -k_1(u_o - x_2)x_1 + k_2 x_2 \tag{3.17}$$

$$\dot{x}_2 = k_1(u_o - x_2)x_1 - \left(k_2 + k_3\right)x_2 \tag{3.18}$$

$$\dot{x}_3 = k_3 x_2 - k_4 x_3 \tag{3.19}$$

From enzyme conservation, we see that $\dot{x}_2 = -\dot{u}$. If we assume steady-state conditions, $x_1 = x_2 = x_3 = 0$, and we have

$$x_{2SS} = \frac{k_1 x_1 u_o}{(k_2 + k_3) + k_1 x_1} = \frac{x_1 u_o}{x_1 + K_M} \tag{3.20}$$

where $K_M \equiv (k_2 + k_3)/k_1$ is the well-known (to biochemists) Michaelis constant.

If we set $x_1 \equiv x_{10}$ (constant) and $u \equiv u_o$ (constant) and again assume steady-state conditions where all derivative terms $\rightarrow 0$, the equilibrium concentration of P is easily shown to be

$$x_{3SS} = \frac{k_3 x_2}{k_4} = \frac{k_3}{k_4} \frac{k_1 u_o x_{10}}{(k_2 + k_3)} = \frac{k_3 u_o x_{10}}{k_4 K_M} \tag{3.21}$$

It is clear that dynamic solutions of the three Michaelis–Menton equations are best done by computer simulation.

5. As a fifth example, consider a compartment surrounded by a diffusion membrane. (Perhaps this is a cell.) In the compartment is an enzyme in excess concentration that catalyzes the reversible transformation of A into B. Furthermore, A must diffuse into the compartment, and B diffuses out. The concentration of A outside the membrane is a_o; inside the compartment it is a. Likewise, the concentration of B outside the compartment is b_o, and it is b inside. Refer to Figure 3.7. The rate of increase of A and B inside the compartment is thus governed by both mass action and diffusion. The ODEs are

$$\dot{a} = K_{da}(a_o - a) + k_{-1}b - k_1 a \quad \text{(kg/s)} \qquad \text{(3.22A)}$$

$$\dot{b} = -K_{db}(b - b_o) + k_1 a - k_{-1}b \quad \text{(kg/s)} \qquad \text{(3.22B)}$$

K_{da} and K_{db} are diffusion rate constants. The k_1 and k_{-1} terms relate to simple mass action. Note that in this special case, the ODEs are linear and can be solved for the states, a and b, using conventional linear algebraic methods and Laplace transforms. Also note that concentrations, by definition, are nonnegative, making the overall system implicitly nonlinear.

Diffusion and mass-action dynamics are often necessarily used together to describe intracellular biochemical events, such as the synthesis of a hormone in response to a signal from a control hormone. The control hormone must enter the cell, and the synthesized hormone

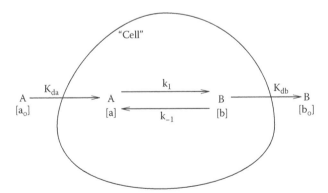

FIGURE 3.7 Diffusion/reaction in a cell, relevant to the fifth example (cf. Equations 3.22A and B). See text for analysis. (From Northrop, R.B., *Endogenous and Exogenous Regulation and Control of Physiological Systems*, Chapman & Hall/CRC Press, Boca Raton, FL, 2000.)

must diffuse out through the cell's membrane to enter the bloodstream in order to reach its intended, remote target cells. In addition, reaction substrates, such as glucose and oxygen, must diffuse into the cell. The mass-action dynamics governing the hormone's synthesis are generally nonlinear and high order; no simple Laplace solution is possible, as in Example 4.

6. A very important class of reactions in evolutionary biochemistry is the autocatalytic reaction (ACR) considered in this sixth example (Kauffman 1995, Korthof 2009). In a simple ACR, 1 mol of **A** combines reversibly with 1 mol of **B** to form 2 mol of **B**. Sic:

$$\mathbf{A} + \mathbf{B} \underset{k_r}{\overset{k_f}{\rightleftharpoons}} 2\mathbf{B}$$

The MA ODE describing this autocatalytic reaction is $\dot{b} = k_f ab - k_r b^2$, where a and b are the running concentrations of molecules **A** and **B**, respectively. The misfolding of prion proteins in BSE, CJD, and CWD is an autocatalytic process.

7. In the seventh and final example, we consider the hypothetical situation where there is a finite number, N, of cell membrane receptors per unit volume. At any instant, there is a density, **F**, of free receptors, which can bind with molecules of an input hormone, $\mathbf{H}_1$, to form a complex. Once bound, the complex initiates the biosynthesis and release of a second messenger hormone, $\mathbf{H}_2$. The complex is broken down enzymatically to release an inactivated input hormone molecule, $\underline{\mathbf{H}}_1$, and a free membrane receptor, **F**. Intuitively, we see that this system will saturate for high concentration of $\mathbf{H}_1$, that is, there can be no further increase in the rate $\mathbf{H}_2$ is produced, because nearly all of the free receptors are complexed at any time. These processes can be represented by

$$\mathbf{H}_1 + \mathbf{F} \underset{k_2}{\overset{k_1}{\rightleftharpoons}} \mathbf{H}_1 * \mathbf{F} \xrightarrow{k_3} \mathbf{H}_1 + \mathbf{F} \qquad (3.23A)$$

$$\mathbf{H}_1 * \mathbf{F} \xrightarrow{k_4} \mathbf{H}_2 \xrightarrow{k_5} * \qquad (3.23B)$$

The total number of receptors $N = f + c$. f is the number of free receptors and c is the number of receptors complexed with $\mathbf{H}_1$. The system mass-action equations are

$$\dot{c} = -(k_2 + k_3)c + k_1 h_1 f = -(k_2 + k_3)c + k_1 h_1 (N - c) \qquad (3.24A)$$

$$\dot{h}_2 = -k_5 h_2 + k_4 c \qquad (3.24B)$$

In the steady state, we see from Equation 3.24A that as $h_1 \gg K_M$,

$$c_{SS} = \frac{Nh_1}{h_1 + K_M} \to N \qquad (3.25)$$

K_M is the Michaelis constant for the system. Also,

$$h_{2SS} = \frac{k_4 c_{SS}}{k_5} \qquad (3.26)$$

Thus, the concentration of H_2 exhibits saturation as a function of H_1 as well.

Remember that mass-action kinetics depend on the assumptions that large numbers of molecules of a given species are uniformly distributed in a volume where the reaction is taking place. Many biochemical reactions take place on membrane surfaces where concentration gradients of reactants and products can exist. Such gradients invoke diffusion dynamics in addition to mass-action ODEs. This latter situation is often neglected in modeling because of the inhomogeneity of tissues surrounding the cell membranes in question.

3.3 PARAMETRIC REGULATION IN COMPLEX BIOLOGICAL SYSTEMS

Biochemical, genomic, hormonal, and physiological regulatory systems generally use parametric regulation (PR) or control. Other CNLSs can also use PR. Because PR is a nonlinear process, it makes analysis and modeling of systems with PR much more challenging (Northrop 2000).

A SISO parametric regulator is one in which a function of the difference between a set point (or desired output state or goal) and an actual output state is used to manipulate one or more system gains, rate constants, or edge parameters, in order to force that output toward the desired, set value. In a biochemical PR system, especially in "wet" systems, output variables are regulated by the alteration of diffusion and/or chemical reaction rate constants (i.e., enzyme activities) and loss rates by regulatory molecules such as metabolic pathway products, transcription factors, microRNAs, enzymes, hormones, and other autacoids (Hardman and Limbird 1996). Hormones also affect the active "pumping" of certain ions and molecules across cell membranes. A hormone's release rate, hence its concentration, is generally a function of regulated variables. Neurotransmitters also alter specific ion permeabilities or conductances.

To illustrate the effects of parametric regulation, we will consider the following two simple examples: (1) a theoretical cellular system and (2) an example of endpoint inhibition in a Michaelis chemical reaction.

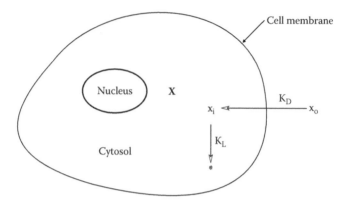

FIGURE 3.8 First example of unregulated diffusion of a molecule **X** with external concentration x_o into a cell. K_D is the diffusion constant. The internal concentration of the molecule, x_i, is lost at a rate K_L times x_i. See Equation 3.27. (From Northrop, R.B., *Signals and Systems in Biomedical Engineering*, 2nd edn., CRC Press, Boca Raton, FL, 2010.)

1. In the first example, we first consider a theoretical, unregulated, cellular diffusion system illustrated in Figure 3.8. We assume that x_i is the concentration of a certain substance, **X**, inside a cell. The substance **X** is also present outside the cell in concentration $x_o > x_i$. (Concentrations are in µg/L.) The substance **X** diffuses passively into the cell according to Fick's first law, and is metabolized inside the cell at a rate proportional to its internal concentration. These processes are described by a simple, linear, first-order ODE, Equation 3.27. K_D is the diffusion rate constant, K_L is the loss rate constant, and V is the cell's volume. Thus,

$$\dot{x}_i = \left(\frac{K_D}{V} \right)(x_o - x_i) - \left(\frac{K_L}{V} \right)x_i \quad \text{µg/(L} \times \text{min)} \tag{3.27}$$

This linear ODE can be rearranged and Laplace-transformed to yield the (linear) Laplace transfer function:

$$\frac{X_i}{X_o}(s) = \frac{K_D/V}{s + (K_D + K_L)/V} = H(s) \tag{3.28}$$

From the transfer function, given a constant input, x_o, the steady-state internal concentration, x_i, can be written as

$$x_{iSS} = \frac{x_o}{1 + K_I/K_D} \tag{3.29}$$

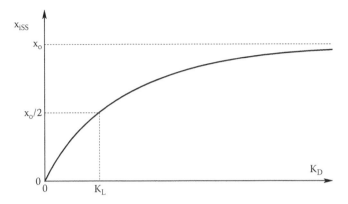

FIGURE 3.9   Plot of the unregulated, steady-state, internal concentration of the molecule, x_{iss}, vs. K_D. (From Northrop, R.B., *Signals and Systems in Biomedical Engineering*, 2nd edn., CRC Press, Boca Raton, FL, 2010.)

Figure 3.9 shows a plot of x_{iss} vs. K_D. It is clear that manipulation of the diffusion rate constant, K_D, will control the value of x_{iss}.

Now, let us investigate a simple, hypothetical, parametric feedback system that can regulate the intracellular concentration of X. We will assume that a biochemical feedback mechanism exists in which the diffusion constant, K_D, is made to decrease linearly as x_i increases, that is,

$$K_D = K_{D0} - \rho x_i, \quad 0 \le x_i \le \frac{K_{D0}}{\rho}$$

$$= 0, \quad x_i > \frac{K_{D0}}{\rho} \tag{3.30}$$

This relation is plotted in Figure 3.10. Note that K_D vs. x_i is linear over $0 \le x_i \le K_{D0}/\rho$. Assuming the system is operating in the linear range of Equation 3.30, the ODE describing the system can now be written as

$$\dot{x}_i = \left[\frac{K_{D0} - \rho x_i}{V}\right](x_0 - x_i) - \left(\frac{K_L}{V}\right) x_i \quad \mu g/(L \times min) \tag{3.31}$$

In the steady state, $\dot{x}_i \equiv 0$. Where x_i lies in the linear range for K_D, we can write

$$x_{iss} = \frac{x_0}{1 + K_L/[K_{D0} - \rho x_{iss}]} \tag{3.32}$$

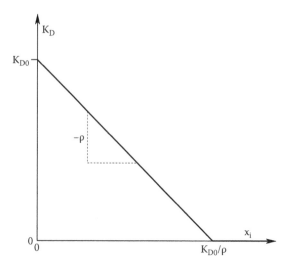

FIGURE 3.10 The parametric regulator law for $K_D(x_i)$. See Equation 3.30. (From Northrop, R.B., *Signals and Systems in Biomedical Engineering*, 2nd edn., CRC Press, Boca Raton, FL, 2010.)

Equation 3.32 can be put in the standard quadratic form:

$$x_{iss}^2 - x_{iss}\left[\frac{K_{D0} + K_L}{\rho} + x_o\right] + x_o\left(\frac{K_{D0}}{\rho}\right) = 0 \qquad (3.33)$$

The solution of this quadratic equation has the familiar form

$$x_{iss} = \frac{(K_{D0} + K_L)/\rho + x_o}{2}$$

$$\pm \left(\frac{1}{2}\right)\sqrt{\left[\frac{(K_{D0} + K_L)}{\rho} + x_o\right]^2 - 4x_o\left(\frac{K_{D0}}{\rho}\right)} \qquad (3.34)$$

If we factor out the $[(K_{D0} + K_L)/\rho + x_o]$ term from the square root, and we consider the negative root, we can write

$$x_{iss} \cong \frac{x_o(K_{D0})/(K_{D0} + K_L)}{1 + (\rho x_o)/(K_{D0} + K_L)} = \frac{x_o\left[1/(1 + K_L/K_{D0})\right]}{1 + (\rho x_o/K_{D0})/(1 + K_L/K_{D0})} \qquad (3.35)$$

If the term, $\rho x_o/K_{D0}$, is $\gg 1$, then Equation 3.35 reduces to

$$x_{iss} \approx \frac{K_{D0}}{\rho} \qquad (3.36)$$

which is independent of the external concentration, x_o. If the desired set point for $x_i = x_{iset}$, then make $\rho = K_{D0}/x_{iset} \gg (K_{D0} + K_L)/x_o$. Under this condition, the sensitivity, $S_{xi} = (dx_{iSS}/dx_o)(x_o/x_{iss}) \rightarrow 0$.

2. For a second example of a parametric regulatory system, consider the well-known Michaelis–Menton chemical reaction architecture:

$$\mathbf{X + E} \underset{k_2}{\overset{k_1}{\Longleftrightarrow}} \mathbf{E * X} \xrightarrow{k_3} \mathbf{E + P} \xrightarrow{k_4} * \qquad (3.37)$$

where

$x_1 =$ running concentration (RC) of the reactant, X
$x_2 =$ RC of the complex, $\mathbf{E*X}$
$x_3 =$ RC of the product, $\mathbf{P}$
$e_o =$ conc. of the total enzyme present

Thus, $\mathbf{E}$ is conserved so the RC of the enzyme $e = e_o - x_2$. k_1, k_2, and k_3 are reaction rate constants; k_4 is the loss rate constant of $\mathbf{P}$. We first assume the $\{\mathbf{k}\}$ are fixed at a constant temperature.

With no parametric regulation, we can express the reaction kinetics through three coupled, nonlinear ODEs based on chemical mass action (Northrop 2000):

$$\dot{x}_1 = -k_1(e_o - x_2)x_1 + k_2 x_2 \qquad (3.38A)$$

$$\dot{x}_2 = k_1(e_o - x_2)x_1 - (k_2 + k_3)x_2 \qquad (3.38B)$$

From enzyme conservation, we see that $\dot{x}_2 = -\dot{e}$. To find a steady-state solution for the reaction, we set $\dot{x}_1 = \dot{x}_2 = \dot{x}_3 \equiv 0$, and solve for x_{2SS} and x_{3SS}. Thus, from Equation 3.38B we can write

$$x_{2SS} = \frac{x_1 k_1 e_o}{(k_2 + k_3) + x_1 k_1} = \frac{x_1 e_o}{x_1 + \mathbf{K_M}} \quad \text{(enzyme * reactant complex)} \quad (3.39)$$

where $\mathbf{K_M} \equiv (k_2 + k_3)/k_1$ is the well-known Michaelis constant.

Now, assume reactant concentration x_1 is held constant ($x_1 = x_{1o}$). Equation 3.39 for x_{2SS} can now be substituted into Equation 3.38C and we find

$$x_{3SS} = \frac{k_3 k_1 x_{1o} e_o}{k_4[(k_2 + k_3) + k_1 x_{1o}]}$$

$$= \frac{k_3 e_o x_{1o}}{k_4(x_{1o} + \mathbf{K_M})} \quad \text{(the SS product concentration depends on } x_{1o}) \quad (3.40)$$

Now let us assume the Michaelis reaction is regulated parametrically by end-product inhibition of the forward-reaction rate constant, k_1. We shall approximate this inhibition by the hyperbolic relation: $k_1 = f(x_3) \cong \beta/x_3$, $x_3 > 0$. Thus, we find

$$x_{3SS} = \frac{k_3(\beta/x_{3SS})x_{1o}e_o}{k_4[(k_2 + k_3) + x_{1o}(\beta/x_{3SS})]} \tag{3.41}$$

Equation 3.41 for the steady-state product concentration when there is feedback inhibition of the enzyme action is seen to reduce to

$$x_{3SS} \cong \frac{e_o k_3}{k_4} \tag{3.42}$$

when $(\beta x_{1o})/x_{3SS} \gg (k_2 + k_3)$. In other words, under this inequality, the steady-state product concentration is set by the total enzyme E used, e_o, and is substantially independent of x_{1o}.

3.3.1 MODELING PARAMETRIC CONTROL OF ENZYME FUNCTION

We have seen that parametric regulation or control can result from the direct action of a regulatory molecule X on an enzyme molecule, either slowing or speeding up a reaction with a rate constant, K. A regulatory molecule can also signal for the transcription of an enzyme protein from its gene, inhibit its transcription, or prevent its assembly post-transcriptionally (by microRNAs). And as we have remarked above, a regulatory molecule can also affect the rate of trans-membrane import or export of other molecules or ions.

In the examples above, we have used both linear and hyperbolic approximations to the PR of reaction rate constants. Namely, $K \cong (K_o - \beta x)$, and $K \cong \rho/x$. Other models for PR also exist. Often, the parametric action of the regulatory molecular concentration, x, is graded and can be approximated by an nth order Hill function (Alon 2007a). A decreasing (inhibitory) regulation by x can be modeled by

$$K_{(-)} = \frac{K_0 \beta^n}{\beta^n + x^n}, \quad K_{(-)}(0) = K_0, \quad K_{(-)}(\infty) = 0, \quad 0 \le x \le \infty, \quad n \ge 1 \tag{3.43}$$

Increasing (stimulatory) regulation by x can be modeled by the increasing (saturating) Hill function:

$$K_{(+)} = \frac{K_0 x^n}{x^n + \beta^n}, \quad K_{(+)}(0) = 0, \quad K_{(+)}(\infty) = K_0, \quad n \ge 1 \tag{3.44}$$

As the positive exponent, n, gets larger and larger, the Hill functions (decreasing and increasing) approach step functions, that is, the control molecule concentration x behaves as an on–off switch (i.e., 0, K_0) around $x = \beta$.

Another sigmoid function sometimes used to model enzyme activation by a regulatory molecule, **x**, makes use of the hyperbolic tangent function:

$$K = K_0(\tfrac{1}{2})\{1 + \tanh[(\mathbf{x} - a)\beta]\} \qquad (3.45)$$

where a, $\beta > 0$.

In Equation 3.45, $K = K_0(\tfrac{1}{2})\{1 + \tanh[(-a)\beta]\} > 0$ for $\mathbf{x} = 0$, $K = \tfrac{1}{2}K_0$ for $\mathbf{x} = a$, and $K \to K_0$ for $\mathbf{x} \gg a$. A decreasing hyperbolic model is obtained by simply changing the sign from $+$ to $-$ in front of the $\tanh(^*)$ function.

3.4 STABILITY, LIMIT-CYCLE OSCILLATIONS, AND CHAOS IN NONLINEAR BIOLOGICAL SYSTEMS

3.4.1 Introduction

Quantitative (systems) biologists, physiologists, biomedical engineers, biophysicists, and mathematicians have been fascinated for many years with the molecular organization and modeling of homogeneous chemical oscillators and biochemical oscillators (BCOs) (Chance et al. 1964, Goldbeter 1996, Sagués and Epstein 2003, Alon 2007a). In this section, we review some examples of homogeneous chemical oscillators, intracellular biochemical oscillators, and putative models for biochemical oscillators, and show that they generally can be modeled with relatively simple sets of coupled, nonlinear ODEs. You will see that one exception to the use of algebraically and topologically simple models for oscillators lies in the more complex glycolysis metabolic pathway. The relatively simple models for single, intracellular biochemical oscillators morph into a complex scenario when the synchronization of large groups of autonomous, intracellular, biochemical oscillators is treated at the model level.

Synchronization of BCOs can be modeled by assuming a density of a common autoinducer (AI) molecule in the extracellular compartment surrounding the many cells containing the individual oscillators. The AI molecules can be of exogenous origin (from another master oscillator) or can be produced by the intracellular BCOs themselves.

Note that the eukaryote BCOs with 24 h periods (circadian clocks) have the ability to entrain or synchronize with external stimuli (e.g., photoperiod) as well as exhibit robustness in their cyclic operation in the presence of biochemical and genetic "noise." Such robustness ensures that a clock runs accurately and stimulates (or suppresses) the expression of clock-dependent genes at the correct time of day.

Every engineering textbook on linear control systems has a detailed section on predicting closed-loop system stability. Very simply, predicting instability in an LTI system boils down to finding whether the closed-loop system's linear transfer function, $H(s) = Y(s)/X(s)$, has any finite poles (polynomial denominator roots) in the right-half **s**-plane. (When the complex variable $\mathbf{s} = \sigma + j\omega$ is equal to a root value, the denominator polynomial, $\mathbf{Q(s)} \to 0$, and $|\mathbf{H(s)}| \to \infty$.) A pair of complex-conjugate poles in the right-half s-plane means that the system's transient response will contain an exponentially growing sinusoidal oscillation. In the real world, the oscillations of electronic oscillators and BCOs are bounded. This is because these systems are in fact nonlinear—they contain saturation nonlinearities that limit the growth of oscillations to a practical level. In the phase plane, when we plot $\dot{x}(t)$ vs. $x(t)$, we see that the SS bounded oscillations create a simple, closed, limit cycle (LC) in the phase plane. This closed LC path is called the system's attractor.

Some workers have called certain biological and genetic oscillators relaxation oscillators by analogy to the waveforms they produce, as well as their generating mechanisms (McMillen et al. 2002). In electronic systems, a relaxation oscillator generally involves a device with a switching threshold, such as a neon bulb, analog comparator, silicon-controlled rectifier (SCR), or a unijunction transistor (UJT) (Nanavati 1975), etc., which has an electronic on–off condition. (See relaxation oscillator in the Glossary.) However, the molecular control of gene expression by transcription factors (TFs) can be graded, or resemble an all-or-none (binary) process making a conceptual bridge to the switching analogy.

Below, we first introduce the concept of feedback system stability, and how to describe unstable system behavior using the phase-plane and trajectory attractors. We consider simple mathematical models for oscillation and also how time delays can destabilize even simple, linear, SISO feedback systems.

3.4.2 STABILITY IN SIMPLE LINEAR AND NONLINEAR SYSTEMS: INTRODUCTION TO PHASE-PLANE ANALYSIS

A widely used method of characterizing the behavior of dynamic nonlinear systems, especially for IVPs, views their behavior in the phase plane. The phase plane is generally a 2-D, (or in certain cases, 3-D) Cartesian plot of the transient behavior (impulse or step inputs, or an IVP) of a nonlinear system. The output, $x(t)$, and its derivative, $\dot{x}(t)$, are plotted for various t values beginning at $t = 0$, and ending when the system reaches a steady-state behavior. This plot of $\dot{x} = f(x)$ is called a phase-plane trajectory. A family of trajectories (generated using different initial conditions [ICs]) or input amplitudes constitutes a phase portrait.

Figure 3.11A through F summarizes the phase-plane behavior of a linear second-order ($n = 2$) system characterized by the second-order, linear, ODE, $\ddot{x} + a\dot{x} + bx = c$, where a and b are real constants. A systems engineering format for the classical second-order ODE is $\ddot{x} + (2\zeta\omega_n)\dot{x} + \omega_n^2 x = c(t)$. Here ω_n is the system's undamped natural (radian) frequency, and ζ is the system's damping factor ($\zeta \geq 1$ overdamped, $1 > \zeta > 0$ underdamped, $\zeta = 0$ oscillator, $\zeta < 0$ unstable).

We find that for $c(t) \equiv 0$, and certain initial conditions on $x(t)$ and $\dot{x}(t)$, x_0 and $\dot{x}_0$, respectively, and certain real values of a and b, the system's phase-plane trajectories (plots of $\dot{x}(t)$ vs. $x(t)$) behave differently.

In Figure 3.11A, the poles (complex roots of the ODE's characteristic equation) are complex-conjugate and lie in the left-hand s-plane (i.e., they have negative real parts, corresponding to $1 > \zeta > 0$). This pole location gives rise to stable

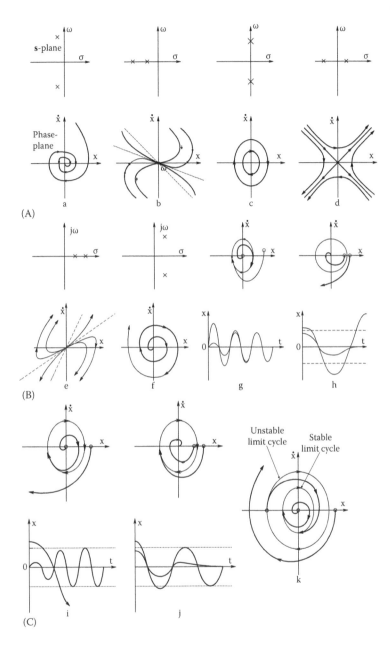

FIGURE 3.11   (A) Examples of phase-plane trajectories and attractors for a simple linear, second-order system. Characteristic equation s-plane roots are shown above each phase-plane plot. (B) e and f: Phase-plane plots for two unstable, quadratic, linear systems. g and h: Phase-plane plots for nonlinear systems showing representative trajectories and attractors. (C) i, j, and k: More examples of trajectories and attractors for nonlinear system. See text for discussion.

focal (underdamped) behavior. Below each pole-zero plot is a typical phase-plane "portrait" of the system, or simply, its phase portrait. Note that in Figure 3.11a, as time increases, the damped, oscillatory trajectory of $(x, \dot{x})$ spirals into the origin (0 amplitude, 0 velocity) as $t \to \infty$; this behavior is because the transient oscillations in $\dot{x}(t)$ and $x(t)$ die out.

In Figure 3.11b, the system is also stable; both poles lie on the left-hand real axis in the s-plane ($\zeta > 1$). This creates phase trajectories that have a stable nodal behavior. They all converge on the origin, wherever they start in the phase plane. There are no oscillations.

In Figure 3.11c, a and b are such that the conjugate roots lie exactly on the $j\omega$ axis of the s-plane, and thus have zero damping ($\zeta = 0$). This center behavior generates an ideal oscillator whose phase-plane trajectories are closed, elliptical orbits. Once started at a certain amplitude, this system oscillates sinusoidally forever at that amplitude. Note that the frequency of the oscillation can be obtained by examining the maximum $\dot{x}$ (when $r = 0$). Because the output is sinusoidal, the oscillation frequency is $\omega_o = \dot{x}_{max}/x_{max}$ rad/s.

Figure 3.11d and e illustrates two types of unstable, non-oscillatory behavior that is produced when one or two system poles lie on the right-hand real axis in the s-plane. With one pole in the right-half s-plane, the trajectories have saddle point behavior in which a trajectory approaches zero, then shoots off to infinity. Two poles in the right-half s-plane yield a family of trajectories that increase monotonically toward infinity. This is called *unstable nodal behavior*.

Note that a linear quadratic system can only produce stable oscillations in the improbable event that both conjugate poles lie exactly on the $j\omega$ axis and the initial conditions are $x_0 = R$, $\dot{x}_0 = 0$, and no noise influences the system's behavior.

In Figure 3.11f, the complex-conjugate poles lie in the right-hand s-plane, and thus have negative damping ($\zeta < 0$). This means that once started, the peak amplitude of the output oscillations grows exponentially toward infinity, as e^{+kt}. This is called unstable focal behavior. Of course in the real world, some component would saturate of failure in this unstable oscillating system.

Figure 3.11g through k illustrates representative phase portraits of general nonlinear systems for five cases of limit-cycle (LC) oscillations. In Figure 3.11g, the LC is stable. Regardless of the initial conditions on $\dot{x}$ and x, all trajectories lead to the attractor's closed pathway in the steady state. In Figure 3.11h, the attractor is unstable. All trajectories starting inside or outside of the closed attractor diverge from it; only ICs lying exactly on the LC will follow the attractor giving stable oscillations. The inside of this attractor is called a basin of attraction for the point 0,0.

Figure 3.11i illustrates semistable LC behavior in which any trajectory originating outside of the attractor diverges from it, going to ∞. Any trajectory originating inside the attractor will enter the LC. Figure 3.11j also illustrates semistable behavior: A trajectory originating outside the attractor will converge on it, while a trajectory originating inside the attractor will decay to the origin. The interior of this attractor is also a basin. In Figure 3.11k, we have the phase portrait of a nonlinear system with two concentric attractors. The inner attractor is stable, the outer is semistable. Trajectory ICs outside the outer attractor give rise to trajectories that converge on ∞. ICs between the two attractors will converge on the stable, inner attractor. Any ICs inside the inner attractor will generate a trajectory that follows the inner attractor.

Often a system's phase portrait shows symmetry around the x-axis, the $\dot{x}$-axis, or both axes. Ogata shows that given the system, $\ddot{x} + f(x,\dot{x}) = 0$, x-axis symmetry will occur if $f(x,\dot{x}) = f(x,-\dot{x})$. Symmetry of the portrait about the $\dot{x}$-axis occurs when $f(x,\dot{x}) = -f(-x,\dot{x})$, and x- and $\dot{x}$-axis symmetry is seen when $f(-\dot{x},x) = -f(\dot{x},-x)$.

Phase-plane plots can also be done in 3-D, (x, y, z), as with the well-known Lorenz equations (see below). The reader interested in the functionality of phase portraits should see Chapter 12 in the venerable control systems text by Ogata (1970). Ogata gives an exhaustive treatment of phase-plane analysis applied to nonlinear dynamic systems. More examples of phase-plane descriptions of NL systems are found below.

In one design architecture of practical, electronic, sinusoidal oscillators using feedback, a quadratic (or higher-order) feedback system is deliberately made unstable (given unstable focal behavior, having a pair of complex-conjugate [CC] poles in the right-half s-plane). The feedback is passed through a soft saturation nonlinearity that effectively increases the system's damping as the output oscillations grow, that is, it pushes the two CC poles to the left in the s-plane (Northrop 2004). At some critical output amplitude, the overall oscillator has an output amplitude-sensitive loop gain that effectively places its conjugate poles on the jω axis in the s-plane and thus, the system exhibits stable oscillations. If some disturbance (e.g., a load) causes a transient decrease in output amplitude, this causes a simultaneous increase in the effective loop gain of the oscillator; the poles shift into the right-hand s-plane and the output oscillations grow to their stable amplitude again. This simple nonlinear oscillator thus exhibits a stable limit cycle (Figure 3.11g). The oscillator's loop gain is thus regulated to be the critical value for instability by the nonlinearity.

3.4.3 EXAMPLES OF SIMPLE NONLINEAR ODES WITH OSCILLATORY BEHAVIOR

One mathematical model for this simple type of oscillator is the well-known, second-order, nonlinear, van der Pol equation (Cunningham 1958):

$$\ddot{x} + \mu(x^2 - 1)\dot{x} + x = 0, \quad \text{given initial conditions on } \dot{x} \text{ and } x \qquad (3.46)$$

The van der Pol equation can be written in an alternate state-variable form (Wang 1999):

$$\dot{x} = \mu[y - f(x)] \qquad (3.47A)$$

$$\dot{y} = -\frac{x}{\mu} \qquad (3.47B)$$

where

$$f(x) = \left(\frac{x^3}{3} - x\right) \qquad (3.47C)$$

When we plot system trajectories in the **x, y** plane for steady-state oscillations in the case where $\mu \gg 1$, we see that the q–r and s–p trajectory transitions are fast in the time domain, and the r–s and p–q transitions are relatively slower. A typical result is shown in Figure 3.12. Note that the x(t) waveform grows to a constant amplitude where it is said to be in a stable limit cycle.

When the van der Pol equation is simulated, we see that for $\mu = +1$, and various initial conditions on x(0) and $\dot{x}$(0), stable limit-cycle oscillations occur.

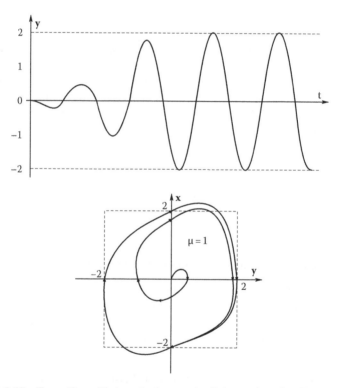

FIGURE 3.12 Top: y(t) oscillations in the van der Pol equation. See Equations 3.47A through C. Bottom: **x, y** phase plane for the van der Pol equation, showing growing oscillations converging on the closed attractor. (From Northrop, R.B., *Signals and Systems in Biomedical Engineering*, 2nd edn., CRC Press, Boca Raton, FL, 2010.)

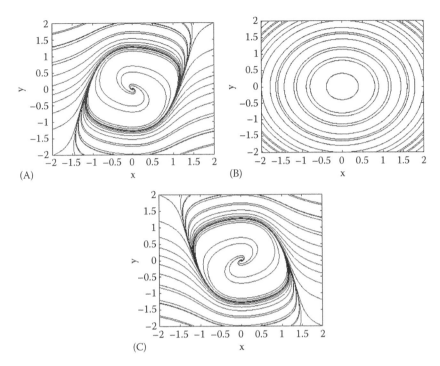

FIGURE 3.13   (A) A family of stable van der Pol phase-plane trajectories for the case of $\mu=+1$. Trajectories originating inside and outside the attractor converge on the attractor. The system oscillates with a bounded limit cycle. (B) Trajectories for $\mu=0$. The system is a sinusoidal oscillator. (C) Unstable van der Pol system: $\mu=-1$. Trajectories outside the attractor go to ∞, inside the attractor, they $\rightarrow 0,0$. (From Northrop, R.B., *Signals and Systems in Biomedical Engineering*, 2nd edn., CRC Press, Boca Raton, FL, 2010.)

Regardless of the initial conditions, all trajectories converge to a simple, closed attractor in the $(x,\dot{x})$ phase plane. An example of these stable trajectories is shown in Figure 3.13A. Note that as a result of the nonlinear damping (the middle term) in Equation 3.46, the output oscillations are not purely sinusoidal; there is a harmonic distortion—a major drawback to this type of oscillator architecture. Phase-plane trajectories for $\mu=0$ and $\mu=-1$ are shown in Figure 3.13B and C, respectively. The oscillations for $\mu=0$ are sinusoidal.

The Belousov–Zhabotinski reaction (BZR) is often cited as an example of a complicated, homogeneous chemical oscillator. It was first proposed as an inorganic analog to the Krebs cycle in the 1950s by Belousov. The BZR is a concatenated system of multiple chemical reactions. One feature of this system is the color changes that accompany its chemical cycles that spread in 3-D waves in the reaction vessel. In the BZR, three solutions are initially mixed in a glass vessel to form a green color, which then turns blue $\rightarrow$ purple $\rightarrow$ red $\rightarrow$ green $\rightarrow$ etc. The oscillations eventually die out as mass is lost from the reaction vessel and chemical, and chemical and thermodynamic equilibria are reached.

The three solutions used are: (1) 0.23 M KBrO, (2) 0.31 M Malonic acid + 0.059 M KBr, (3) 0.019 M Cerium(4+) ammonium nitrate + 2.7 M H_2SO_4. The overall BZ reaction is the cerium-catalyzed oxidation of malonic acid by bromate ions in dilute sulfuric acid. It can be written as

$$3CH_2(CO_2H)_2 + 4BrO_3^- \rightarrow 4Br^- + 9CO_2 \uparrow + 6H_2O \tag{3.48}$$

This reaction can be broken into two processes: Process I is dominant when $[Br^-]$ is high; Process II dominates when $[Br^-]$ is low. Process I involves the reduction of bromate ions by bromide in two electron transfers. Three steps are involved, which are shown in Equations 3.49, 3.50, and 3.51:

$$BrO_3^- + Br^- + 2H^+ \rightarrow HBrO_2 + HOBr \tag{3.49}$$

$$HBrO_2 + Br^- + H^+ \rightarrow 2HOBr \tag{3.50}$$

$$3HOBr + 3Br^- + 3H^+ \rightarrow 3Br_2 + 3H_2O \tag{3.51}$$

The net reaction in Process I is thus

$$BrO_3^- + 5Br^- + 6H^+ \rightarrow 3Br_2 + 3H_2O \tag{3.52}$$

The free bromine released in Process I reacts with the malonic acid according to the reaction

$$Br_2 + CH_2(CO_2H)_2 \rightarrow BrCH(CO_2H)_2 + Br^- + H^+ \tag{3.53}$$

Process II becomes dominant as the concentration of Br^- is reduced in Process I. The overall reaction of Process II is

$$2BrO_3^- + 12H^+ + 10Ce^{+++} \rightarrow Br_2 + 6H_2O + 10Ce^{4+} \tag{3.54}$$

Process II may be decomposed into the following five steps:

$$BrO_3^- + HBrO_2 + H^+ \rightarrow 2BrO_2\bullet + H_2O \tag{3.55}$$

$$BrO_2\bullet + Ce^{+++} + H^+ \rightarrow HBrO_2 + Ce^{4+} \tag{3.56}$$

$$2HBrO_2 \rightarrow HOBr + BrO_3^- + H^+ \tag{3.57}$$

$$2HOBr \rightarrow HBrO_2 + Br^- + H^+ \tag{3.58}$$

$$HOBr + Br^- + H^+ \rightarrow Br_2 + H_2O \tag{3.59}$$

Key elements in this sequence are found by adding $2 \times$ Equation 3.56 to Equation 3.55, giving

$$2Ce^{+++} + BrO_3^- + HBrO_2 + 3H^+ \rightarrow 2Ce^{4+} + H_2O + 2HBrO_2 \tag{3.60}$$

Equation 3.60 illustrates that hydrobromous acid is produced autocatalytically; this is a positive feedback step in this complicated series of reactions. The autocatalysis does not begin until the reagents are depleted because there is a second-order destruction of $HBrO_2$ (see reaction 3.57). Reactions 3.58 and 3.59 represent the disproportionation of hyperbromous acid to bromous acid and Br_2. Ce^{4+} ions and Br_2 oxidize malonic acid to form bromide [Br⁻] ions. This, in turn, causes an increase in [Br⁻] concentration, reactivating process I. The shifting colors in the BZ reaction are due to the redox products forming and disappearing.

Chemists who have investigated the BZ oscillator have found that manganese ions could be used instead of cerium, and a number of carboxylic acids, including citric acid, could replace malonic acid.

Originally, the BZ reaction was carried out in a well-mixed volume (flask), then, as workers discovered the spatiotemporal waves of color accompanying the reaction processes, they tried dimensional constrains on reaction geometry. For example, BZ reactions were examined in a thin, 2-D volume between glass plates, a 1-D volume in a capillary tube, and a point volume in a microdrop (Sagués and Epstein 2003).

In conclusion, we note that the BZ chemical equations above can be used to forge a large, complicated set of nonlinear, mass-action ODEs with which we can model the complex dynamic behavior of the BZ system.

Because of the complicatedness of the BZ dynamic model, Field and Noyes (1974), and Field et al. (1972) proposed a simpler model of BZ dynamics they called the *Oregonator*. The Oregonator reduction of the BZ reactions can be written as five irreversible chemical reactions:

$$\mathbf{A} + \mathbf{Y} \xrightarrow{\quad k_3 \quad} \mathbf{X} + \mathbf{P} \tag{3.61A}$$

$$\mathbf{X} + \mathbf{Y} \xrightarrow{\quad k_2 \quad} 2\mathbf{P} \tag{3.61B}$$

$$\mathbf{A} + \mathbf{X} \xrightarrow{\quad k_5 \quad} 2\mathbf{X} + 2\mathbf{Z} \quad \text{(autocatalytic step)} \tag{3.61C}$$

$$2X \xrightarrow{k_4} A + P \tag{3.61D}$$

$$B + Z \xrightarrow{k_c} \frac{1}{2} fY \tag{3.61E}$$

where

$A = BrO_3^-$

$B =$ organic species such as malonic acid or bromomalonic acid

$P = HOBr$

$Y = Br^-$

$Z =$ oxidized catalyst (Ce^{4+})

$f =$ a stoichiometric factor that serves as an "adjustable parameter"

In the Oregonator mass-action ODEs, let us assume that A and B are kept at fixed concentrations (a_o and b_o) and examine the dynamics of X, Y, and Z. The corresponding three MA equations are (Sagués and Epstein 2003)

$$\dot{x} = k_3 a_o y + k_5 a_o x - k_2 xy - 2k_4 x^2 \tag{3.62A}$$

$$\dot{y} = \frac{1}{2} f k_c b_o z - k_3 a_o y - k_2 xy \tag{3.62B}$$

$$\dot{z} = 2k_5 a_o x - k_c b_o z \tag{3.62C}$$

Figure 3.14 illustrates our nonlinear signal flow graph representation of the three state equations of the Oregonator system of Sagués and Epstein. Note that there are three integrators, three multipliers, and one squarer in the SFG. However, because a and b are assumed to be constant, only one multiplier is really required to form the xy product, and also one squarer to make x^2. Sagués and Epstein simplified the Oregonator state model by introducing the dimensionless parameters

$$x' \equiv \frac{2k_4 x}{k_5 a_o} \quad y' \equiv \frac{k_2 y}{k_5 a_o} \quad z' \equiv \frac{k_c k_4 b_o z}{(k_5 a_o)^2} \quad t' \equiv k_c b_o T \quad \text{(time scaling)}$$

They then wrote the simplified equations

$$\dot{x}' = \frac{qy' - x'y' + x'(1 - x')}{\varepsilon}, \quad \text{where } \dot{x}' \equiv \frac{dx'}{dt'} \tag{3.63A}$$

$$\dot{y}' = \frac{qy' - x'y' + fz'}{\gamma} \tag{3.63B}$$

$$\dot{z}' = x' - z' \tag{3.63C}$$

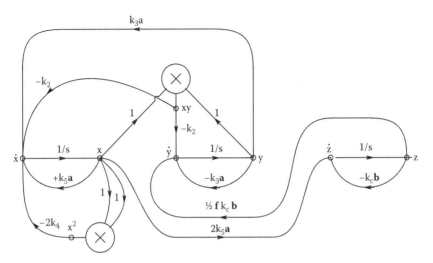

FIGURE 3.14 Nonlinear signal flow graph constructed from the three oregonator ODEs, Equations 3.62A through C. Because this SFG is nonlinear, Mason's rule is not applicable. See text for discussion. (From Northrop, R.B., *Signals and Systems in Biomedical Engineering*, 2nd edn., CRC Press, Boca Raton, FL, 2010.)

Sagués and Epstein's simulations have shown that for certain ranges of parameters and ICs, this system can have a limit-cycle attractor (i.e., it will oscillate in the SS). They used parameters in the range of, $\varepsilon \approx 10^{-2}$, $\gamma \approx 10^{-5}$, and $q \approx 10^{-4}$, in their simulations. (Note that these three ODEs can easily be simulated using MATLAB or Simnon.)

The Brusselator is a nonlinear, two-ODE system that exhibits a Hopf bifurcation in its phase plane behavior. The Brusselator is a hypothetical chemical system. The history of the Brusselator was given by Noyes and Field (1974) and Prigogine (1980). The four chemical reactions below form the basis for the Brusselator oscillator. Molecules **A**, **B**, **D**, **E**, **X**, and **Y** participate in the reactions

$$\mathbf{A} \overset{k_1}{\Rightarrow} \mathbf{X} \tag{3.64A}$$

$$2\mathbf{X} + \mathbf{Y} \overset{k_2}{\Rightarrow} 3\mathbf{X} \quad \text{(autocatalytic reaction)} \tag{3.64B}$$

$$\mathbf{B} + \mathbf{X} \overset{k_3}{\Rightarrow} \mathbf{Y} + \mathbf{D} \tag{3.64C}$$

$$\mathbf{X} \overset{k_4}{\Rightarrow} \mathbf{E} \tag{3.64D}$$

The two nonlinear Brusselator ODEs are found by applying mass-action kinetics to the chemical reactions above:

$$\dot{x} = ak_1 + x^2yk_2 - xk_4 - bxk_3 \qquad (3.65A)$$

$$\dot{y} = bxk_3 - x^2yk_2 \qquad (3.65B)$$

where
the $\{k_j\}$ are rate constants
a, b, x, and **y** are running concentrations
a and **b** are assumed to be constant ($\mathbf{a_0 = A, \ b_0 = B}$)

We sometimes see the Brusselator equations written under the assumption that the rate constants, $k_1–k_4$, are unity, giving

$$\dot{x} = A + x^2y - (B+1)x \qquad (3.66A)$$

$$\dot{y} = Bx - x^2y \qquad (3.66B)$$

Figure 3.15 illustrates the nonlinear SFG topology of the two Brusselator equations, Equations 3.65A and B. Note that the system has 10 nodes, only 2 integrators, 3 multipliers, and has positive as well as negative feedback loop gains, albeit nonlinear ones. Is this a complex system? It certainly is complicated.

Noyes and Field (1974) were among the first investigators to study the Brusselator system numerically. Using the simplified Equations 3.66A and B, they observed bounded limit-cycle behavior for $\mathbf{b_0} > (\mathbf{a_0^2} + 1)$. They found that the

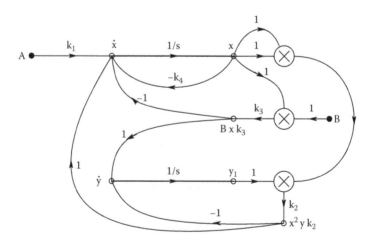

FIGURE 3.15 Nonlinear SFG for the Brusselator system. See Equations 3.65A and B. (From Northrop, R.B., *Signals and Systems in Biomedical Engineering*, 2nd edn., CRC Press, Boca Raton, FL, 2010.)

LC trajectories of the Brusselator are uniquely defined by the reactant concentrations, a_o and b_o, and the four rate constants, $\{k_j\}$. When b_o is only a little greater than $(a_o^2 + 1)$, the intermediate concentrations follow smooth, nearly sinusoidal oscillations around their mean values. As b_o/a_o^2 becomes very large, the oscillations morph into sharp pulses similar to those produced by an electronic relaxation oscillator.

Still another example of a simple two-ODE, nonlinear dynamic system that can have stable limit-cycle behavior is the well-known, Lotka–Volterra (L–V) basic ecological model for predator–prey interaction. Note that the L–V system formulation is based on mass-action principles (Lotka 1920, Volterra 1926). It is the basis for many models in population biology/ecology. The two L–V equations are commonly written as

$$\dot{\mathbf{x}} = k_1 \mathbf{x} - k_2 \mathbf{xy} \tag{3.67A}$$

$$\dot{\mathbf{y}} = -k_3 \mathbf{y} + k_4 \mathbf{xy} \tag{3.67B}$$

As written, this is a closed system, which requires initial conditions on $\mathbf{x}$ and $\mathbf{y}$, $\mathbf{x}_o$ and $\mathbf{y}_o$, respectively. $\mathbf{x}$ represents the number of prey animals, for example, hares. $\mathbf{y}$ is the number of predators, for example, lynxes. The (autocatalytic) rate of reproduction of hares, $+k_1\mathbf{x}$, is proportional to the number of hares, $\mathbf{x}$. The rate of loss of hares to predation is proportional to the product of the hare population and the predator population, $-k_2\mathbf{xy}$. The lynxes die off at a rate proportional to their population, $-k_3\mathbf{y}$, and their population grows at a rate proportional to their predation on hares, $+k_4\mathbf{xy}$, that is, the rate at which predators and prey meet. Figure 3.16 shows the simple, nonlinear SFG architecture of the L–V system. Note that without predation (k_2, $k_4 = 0$), the hare population, $\mathbf{x}$, will grow exponentially, and the lynx population will collapse.

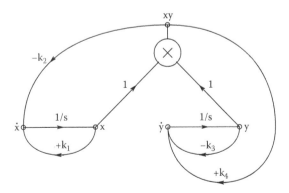

FIGURE 3.16 Nonlinear SFG for the two Lotka–Voterra equations. The system can exhibit limit-cycle oscillations. See Equations 3.76A and B. (From Northrop, R.B., *Signals and Systems in Biomedical Engineering*, 2nd edn., CRC Press, Boca Raton, FL, 2010.)

A phase-plane plot of **y** vs. **x** under certain conditions on $\{k_1, k_2, k_3, k_4\}$ can show stable limit-cycle oscillations of **y** and **x**, or exhibit oscillations that damp out as $t \rightarrow \infty$. When $k_1 = 0.6$, $k_2 = 0.05$, $k_3 = 0.4$, and $k_4 = 0.005$, we observe oscillations in **x** and **y** and a stable limit cycle in the **y**, **x** phase plane, shown in Figure 3.17A and B. Maron (2003) did data fitting of the L–V model (the set of constants $\{k_j\}$) to match the observed ecological data on hares and lynxes from Hudson's Bay, Canada.

We note in summary that few ecosystems are really closed. Populations are also affected by hunting and trapping (human predation), weather, disease, loss of environment, prey food supply, etc. Some workers have introduced delay operations in the L–V equations to account for growth to sexual maturity and gestation. (As described in Section 3.4.4 otherwise, stable feedback systems can

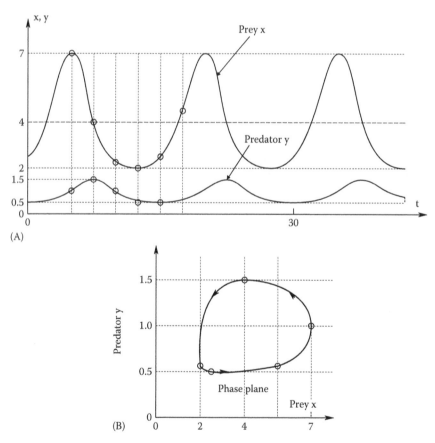

FIGURE 3.17 (A) Limit-cycle oscillations in predator–prey populations modeled by the LV equations. (B) Steady-state attractor in the predator–prey phase plane. (From Northrop, R.B., *Signals and Systems in Biomedical Engineering*, 2nd edn., CRC Press, Boca Raton, FL, 2010.)

be destabilized by transport delay operations.) Thus, we can conclude that the LV equations are a gross oversimplification of a more complex ecological system.

The Lorenz equation system is a complicated, three-ODE, nonlinear dynamic system that has a variable, chaotic, oscillatory, limit-cycle behavior. In 1962, meteorologist B. Saltzman developed a set of partial differential equations to model convection in a fluid heated from below using the Navier–Stokes equations for Bènard fluid flow. Another meteorologist, Lorenz (1963), showed that the 12 Navier–Stokes equations could be simplified to the 3 ODEs given below that bear his name. This "simplified" system exhibited tipping-point behavior. That is, a slight perturbation in an initial condition (ca. 0.025%) caused a marked change in the system's dynamic behavior—the so-called *butterfly effect* (analogous to the slight disturbance of the air by a butterfly causing a global weather change in the future at a remote location). The global behavior of the Lorenz system was shown to depend on the numerical value of a parameter, ρ.

The three, coupled Lorenz equations are

$$\dot{x} = \sigma(y - x) \tag{3.68A}$$

$$\dot{y} = x(\rho - z) - y \tag{3.68B}$$

$$\dot{z} = xy - \beta z \tag{3.68C}$$

where
σ = the Prandtl number
ρ = the Rayleigh number

All σ, ρ, and $\beta > 0$.

Typically $\sigma = 10$, $\beta = 8/3$, and ρ is varied between 1 and 400 to study the system's behavior. For $\rho \approx 24.74$, the system shows chaotic behavior.

Figure 3.18 illustrates the complicated, tangled, signal flow graph topology of this chaotic, nonlinear system using a nonlinear SFG diagram; there are eight nodes, two multipliers, and three integrators. (The product terms in the Lorenz equations require two multiplications.) Clearly, this is a nonlinear system whose behavior is best understood by simulation. The chaotic behavior of the Lorenz system is generally illustrated by 2-D or 3-D phase-plane plots (x vs. y vs. z in rectangular coordinates). Figure 3.19 illustrates the 3-D, complex, chaotic, steady-state trajectories for the Lorenz system for $\sigma = 10$, $\beta = 8/2$, and $\rho = 28$. Note that part of the time, the x(t) oscillations are > 0 (the X^+ region) and at other times they are < 0 (the X^- region in the 3-D phase plane). In this plot, z(t) is always > 0.

An excellent description of how the Lorenz equations are related to the Navier–Stokes equations can be found in the honors thesis by Danforth (2001). Danforth did an ingenious analog simulation using a thermosiphon, yielding Lorenz-system-type results, which were startlingly similar to his simulation results of the Lorenz equations.

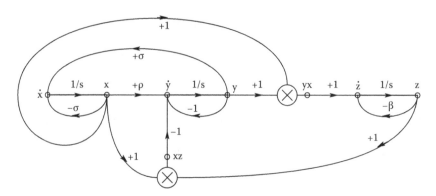

FIGURE 3.18 Nonlinear SFG for the Lorenz equations, Equations 3.68A through C. (From Northrop, R.B., *Signals and Systems in Biomedical Engineering*, 2nd edn., CRC Press, Boca Raton, FL, 2010.)

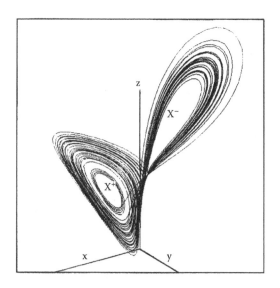

FIGURE 3.19 A typical 3-D (x, y, z) phase-plane plot of the chaotic oscillations obtained from the Lorenz equation system. (From Northrop, R.B., *Signals and Systems in Biomedical Engineering*, 2nd edn., CRC Press, Boca Raton, FL, 2010.)

Schaffer (2001) explored the Lorenz-system behavior by simulation with $\sigma \equiv 10$ and $\beta \equiv 8/3$, and ρ is varied—He found that for $\rho \geq 24.74$, the strange attractor shown in Figure 3.19 can assume various forms, the topology of which changes continuously as ρ is increased. Finally, at $\rho = 313$, there is a single stable periodic orbit. Thus the chaotic Lorenz system can morph into a simple oscillator.

A cautionary word to those readers interested in simulating the BZ, Oregonator, Lotka–Volterra, Brusselator, or Lorenz equations: Choose your integration routine wisely. Lack of integrator numerical precision can introduce noise into the system, exacerbating chaotic behavior (attractor-jumping, tipping-points) from within.

We have demonstrated above that simple, second-, and third-order, nonlinear ODEs can exhibit stable limit-cycle oscillations, as well as complex, nine-state systems. The system nonlinearities cause the chaotic behavior as well as cause the limit cycles to be bounded. Such small oscillators can be embedded in (as modules) and drive larger complex nonlinear systems with their oscillating product concentrations.

3.4.4 INSTABILITY IN SYSTEMS WITH TRANSPORT LAGS

Another well-known source of instability in both linear and nonlinear feedback systems of all sizes is the incorporation of one or more transport lags (TLs) in the feedback loop gains (Northrop 2000). One or more TLs (aka *dead times* or *time delays*) can destabilize simple linear as well as complex nonlinear feedback systems, and under certain conditions, lead to bounded limit-cycle oscillations. TLs can occur in many CNLSs, including but not limited to process control, molecular biological, physiological, ecological, economic, and social systems. What can be delayed is a signal, a reactant (mass or concentration) (in biochemical and physiological systems), or information (in ecological, economic, and social systems). Genomic and molecular biological control pathways are rife with TLs. TLs are implicated in the import and export of molecules across nuclear membranes, the transcription process itself, the diffusion of mRNA to ribosomes, the final folding of a polypeptide to the form of an operational protein, etc. Some of these lags may be significant in modeling systems such as biological clocks, but others may have little effect.

A TL can be expressed mathematically in the time domain: An input $x(t)$ enters the TL element, and emerges unchanged as the output, y, τ seconds later. Mathematically, we can write: $y(t) = x(t - \tau)$. A TL is a linear operation, thus in the frequency (Laplace) domain we can write its transfer function: $TL(s) = Y(s)/X(s) = e^{-\tau s}$. If we are considering steady-state sinusoidal signals in the system, then $\mathbf{s} = j\omega$, and $\mathbf{TL}(j\omega) = \mathbf{Y}(j\omega)/\mathbf{X}(j\omega) = e^{-j\omega\tau}$, that is the radian phase lag of a TL element increases linearly with the radian frequency of $\mathbf{x}(t)$. Hence in polar notation, $|\mathbf{TL}(j\omega)| = 1.0$ and $\angle\mathbf{TL}(j\omega) = -\omega\tau$ rad.

There are four, venerable, easy-to-use, linear systems tools we can apply to demonstrate the instability that may occur in a linear feedback system with a TL: One tool is the root-locus plot [see section 3.2 in Northrop 2000], which shows the positions of the poles of a linear, SISO, closed-loop system in the s-plane as a function of its scalar loop gain. Another tool is the Nyquist stability criterion (Kuo 1982, section 9.2) applied to a polar plot of the frequency response of the loop gain function, $\mathbf{A_L}(j\omega)$ (Ogata 1970). The third approach, the Popov stability criterion, also makes use of a polar plot of $\mathbf{A_L}(j\omega)$ (Northrop 2000). It assumes a single-loop, SISO system architecture in which system error, $\mathbf{e} = \mathbf{c} - \mathbf{r}$, is the input to a nonlinear controller whose output, $\mathbf{u}$, is the input to a plant with linear dynamics and output, **r**. The fourth technique is the Describing Function (DF) method [see chapter 11 in Ogata (1970) and Northrop (2000)], which can also be applied to SISO, single-loop systems in which there is a nonlinearity in series with linear dynamics in the feedback loop. In the DF method, the intersection(s) of $\mathbf{A_L}(j\omega)$ and the negative reciprocal describing function in the polar plane are examined to determine stability. The intersection(s) tell us the gain required for stable oscillations, and the frequency of oscillation.

We will consider the root-locus approach below, because of its ease of application to linear, single-loop feedback systems with delays. (See Appendix B for a primer on root-locus plotting.)

As an example of instability caused by a TL, consider the very simple SISO negative feedback system shown in the block diagram of Figure 3.20. This system has the loop gain, $A_L(s) = -Ke^{-\tau s}/(s+1)$, from which we can construct the root-locus diagram shown in Figure 3.21. The root-locus criterion tells us that for a pair of closed-loop system poles on the $j\omega$ axis, the vector equation 3.69 must be satisfied:

$$\mathbf{A_L}(j\omega_o) = \frac{-Ke^{-j\omega_o\tau}}{j\omega_o + 1} = +1\angle[-\pi(2k+1)], \quad k = 0, 1, 2, \ldots \qquad (3.69)$$

This equation is solved for the gain K required, and the frequency of $j\omega$ axis crossing, ω_o, by equating the magnitude and angle parts of $\mathbf{A_L}(j\omega_o)$ in Equation 3.69. Thus,

$$\frac{K}{\sqrt{(\omega_o^2+1)}} = 1 \quad \text{(magnitude)} \qquad (3.70)$$

$$-\tan^{-1}(\omega_o) - \omega_o t = -\pi \quad \text{(angle in radians)} \qquad (3.71)$$

Equation 3.71 is first solved by trial and error and ω_o is found to be 2.029 rad/s for $\tau = 1.0$ s. for the first pair of closed-loop poles to cross the $j\omega$ axis as the gain K is increased >0. This ω_o value is substituted into Equation 3.70, and the critical

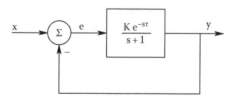

FIGURE 3.20 Block diagram of a simple, linear, SISO, negative feedback system incorporating a transport lag. (From Northrop, R.B., *Signals and Systems in Biomedical Engineering*, 2nd edn., CRC Press, Boca Raton, FL, 2010.)

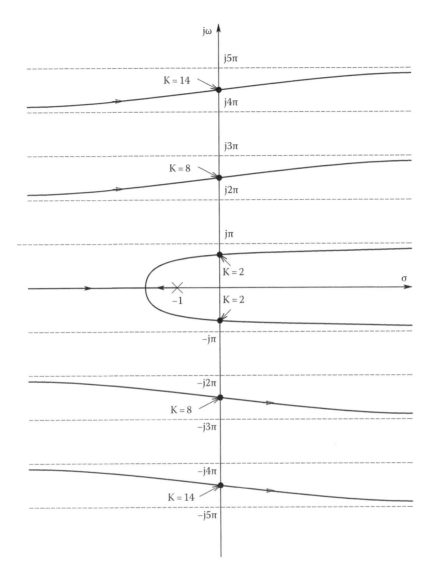

FIGURE 3.21   Root-locus plot for the feedback system of Figure 3.20. At $K=1.871$, the first pair of complex-conjugate poles of the closed-loop system crosses the $j\omega$ axis at $\omega=\pm 1.582\,\text{rad/s}$ into the right-half s-plane. (From Northrop, R.B., *Signals and Systems in Biomedical Engineering*, 2nd edn., CRC Press, Boca Raton, FL, 2010.)

value of K for instability is found to be 2.262. Thus, a pair of complex-conjugate, closed-loop system's poles lie in the right-half s-plane for $2.26 < K < 8$. Two more closed-loop poles enter the right-half s-plane at $K > 8$, and two more for $K > 14$, etc. So for $2.26 < K < 8$, the TL system will oscillate near $\omega_0 = 2\,\text{rad/s}$ with an exponentially growing sinusoidal waveform. For a stable limit cycle to exist, we must

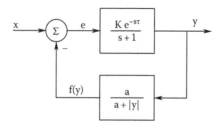

FIGURE 3.22 The delay feedback system of Figure 3.20 is modified to include a soft-saturating function in the feedback path. This causes stable limit-cycle oscillations to exist for K>2. (From Northrop, R.B., *Signals and Systems in Biomedical Engineering*, 2nd edn., CRC Press, Boca Raton, FL, 2010.)

place a soft-saturating, Hill-type nonlinearity of the form, $a/(a+|y|)$, in the feedback path, as shown in Figure 3.22.

Note that single-loop, negative-feedback, linear, SISO systems with delays in their loop gains generally have multiple pairs of conjugate pole loci that originate at $s=-\infty$ and cross the $j\omega$ axis into the right-half s-plane at regularly spaced (conjugate) points as the scalar loop gain K is increased. It is the lowest K to put a single, closed-loop system pole pair just into the right-half s-plane that is of interest to us here.

Komarova (2006) presented a simple model for oscillations in population sizes, in which the system can be described by one, single-time-lag, logistical ODE of the form

$$\dot{x}(t) = rx(t) - \frac{rx(t)x(t-\tau)}{K} \tag{3.72}$$

where
  x is the population size
  $x(t-\tau)$ is the population size delayed τ time units
  r is the growth rate of the population
  K is the carrying capacity
  τ is the delay time

Komarova observed that this nonlinear ODE, which has both positive and delayed negative feedback, can have steady-state, limit-cycle oscillations (closed trajectories in $\dot{x}(t)$ vs. $x(t)$). In an ecological context, the population growth rate is diminished by competitive interaction between adults and recently matured offspring.

Oscillatory biochemical systems are far more complicated than the oscillator models described above because they generally involve large sets of coupled, nonlinear ODEs, and involve parametric feedback, as well as delays. We examine some biochemical oscillators, both theoretical and living, in the section below.

3.4.5 Some Models for Complicated and Complex Biochemical Oscillators

A biochemical oscillator (BCO) consists of a nonlinear biochemical system in which the concentrations of certain reactants can vary periodically in the steady state in the absence of any periodic inputs. In this section, we will consider dynamic models for several intracellular BCOs. These models are instructive because they are complicated, not complex, and they illustrate the art of modeling small, isolated, homogeneous chemical reaction systems having feedback. In nature, such small-state BCOs are not "stand-alone"; they are generally embedded in larger metabolic systems from which they derive substrates and interact. Also, each cell in a like group of cells can contain a BCO, and nature has devised ways to synchronize them both in frequency and phase. They are, in effect, arrays of natural phase-locked loops (see Northrop 2004, section 12.5).

Biochemical oscillators are the basis for natural, circadian (diurnal) behaviors in eukaryote organisms ranging from fungi to humans. Their bounded, steady-state oscillations affect the concentrations of hormones and signaling substances that affect behavior, as well as non-circadian events such as menstrual periods in humans (Foster and Kreitzman 2005), estrus in other mammals, and crustacean locomotor activity linked to tides.

Biochemical circadian behavior occurs almost universally throughout the biosphere in eukaryotes in Plantae, Animalia, and Fungi (Dunlap 1999, Hastings et al. 2007). Some examples of behavior linked to internal "clocks" in vertebrates are sleep cycles; body temperature cycles; and metabolic activity, including the cyclical production of certain hormones. Also, the times of song production in birds and insects, locomotor activity, and feeding behavior are all affected by diurnal "clocks" and their biochemical products whose concentrations cycle periodically.

It is believed that the mammalian central circadian oscillator is located in the suprachiasmatic nucleus (SCN) of the hypothalamus. Interacting feedback loops affecting gene transcription and translation interact to regulate this mammalian neural master clock. Its synchronization is also affected by the protein hormone, VIP. The SCN circadian oscillator acts to entrain or synchronize other intracellular biochemical oscillators located in peripheral tissues such as the adrenal cortex, liver, and pancreas. Physiological processes such as cell division, tissue repair, and metabolism are all influenced by the SCN circadian clock.

The intracellular cyclical concentration rhythms are generated by stable limit-cycle oscillations that involve coupled biochemical reactions. In one scenario, interacting transcription/translation-based feedback loops involving "clock" or "period" genes periodically express critical proteins that affect behavior. For example, in mice, three period genes have been found (**mPer-1**, **-2**, and **-3**), for which their respective mRNA concentrations are expressed cyclically in vivo (Matsuo et al. 2003). More recently, Fujimoto et al. (2006) have shown evidence that oscillation in the **mPer-2** gene product may not be accompanied by oscillations in its coding mRNA. This implies that the endogenous clock system has an ability to modify the **mPer-2** gene, post-transcriptionally.

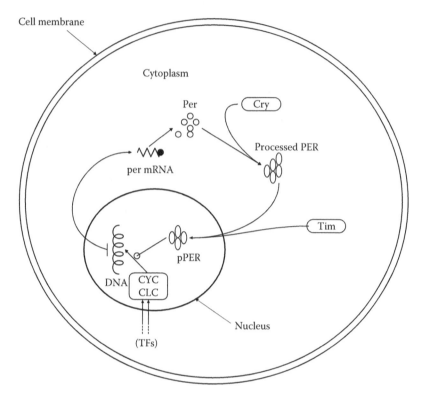

FIGURE 3.23 Schematic of the intracellular processes that form the Per oscillator. See text for description. (From Northrop, R.B., *Signals and Systems in Biomedical Engineering*, 2nd edn., CRC Press, Boca Raton, FL, 2010.)

A simplified model for the oscillations seen in the expression of the Per protein in the fruit fly, *Drosophila*, is shown in Figure 3.23. This is a genetic oscillator. We have adapted it from a book chapter by Tyson (2002), and it involves the end-product, feedback regulation of the expression of a single, generic *Per* gene. In this simple model (Per is a protein that regulates some diurnal activity, such as looking for food.) Figure 3.23 illustrates the schematic of one cell in which the Per oscillator operates. We assume that Per-mRNA is made in the nucleus from the *Per* gene at a rate proportional to the concentrations of two TF proteins, CLC and CYC, times a parametrically controlled rate constant, K_T, which is a decreasing function of the nuclear concentration of "processed" Per protein. Once made, nuclear Per-mRNA is transported to the cytoplasm where it acts with ribosomes to make Per protein, which is post-transcriptionally processed by phosphorylation and binding to two other proteins, TIM and CRY. This processed protein, pPer, then enters the nucleus where it disrupts the binding of the TFs, CLC and CYC to the promoter region of the *Per* gene, and thus inhibits the synthesis of Per-mRNA. This inhibition can be modeled mathematically by making the rate constant, K_T, a decreasing function of the nuclear concentration of pPer.

We have expressed these events mathematically by assuming a mass action–diffusion model in which we use the following parameters:

x_1 = nuclear concentration of Per-mRNA

[CLC] and [CYC] = nuclear concentrations of the TF proteins, CLC and CYC

[TIM] and [CRY] = cytoplasmic concentrations of the proteins, TIM and CRY

x_2 = cytoplasmic conc. of Per-mRNA

x_3 = cytoplasmic conc. of unmodified Per protein (made by ribosomes)

x_4 = cytoplasmic conc. of processed Per protein, pPer

x_5 = nuclear conc. of pPer

The five (linear) system ODEs are

$$x_1 = -[CLC][CYC]x_5 K_T(x_5) - x_1 K_{D1} \quad (\text{Per}-\text{mRNA in nucleus}) \qquad (3.73A)$$

$$\dot{x}_2 = x_1 K_{D1} - x_2 K_{L2} \quad (\text{Per}-\text{mRNA in cytoplasm}) \qquad (3.73B)$$

$$\dot{x}_3 = x_2 K_C - x_3 [TIM][CRY] K_{P3} \quad (\text{newly made Per protein}) \qquad (3.73C)$$

$$\dot{x}_4 = x_3 [TIM][CRY] K_{P3} - x_4 K_{D4} \quad (\text{"processed" Per in cytoplasm [cpPer]}) \quad (3.73D)$$

$$\dot{x}_5 = x_4 K_{D4} - x_5 K_{L5} \quad (\text{nuclear conc. of pPer, npPer}) \qquad (3.73E)$$

The loop is closed using the decreasing parametric Hill function, Equation 3.73F. (K_{T0}, a, and **n** are nonnegative constants.)

$$\mathbf{K_T}(x_5) = \frac{K_{T0}}{1 + (x_5/a)^{\mathbf{n}}} \qquad (3.73F)$$

In the above model, for simplification, we have assumed that the concentrations of the proteins, CLC, CYC, TIM, and CRY are fixed. (However, it is certain that, in vivo, these TF proteins are themselves regulated and their concentrations may be affected by one or more model states.) The instability of the single-loop, negative-feedback Per system, as modeled above, results from the linear, fifth-order part of the loop gain. (No unstabilizing transport delays of the form, y(t) = x(t − τ) were included.) Application of the root-locus technique (Northrop 2004, chapter 5) shows that for a large enough loop gain, the system will have a pair of complex-conjugate poles in the right-half s**-**plane, and thus exhibit growing oscillations in the cytoplasmic Per concentration. As the oscillations grow, the effective loop gain is reduced parametrically by $\mathbf{K_T}(x_5)$, causing a stable limit cycle to be entered.

There is ample evidence that "master" circadian "clocks" require exogenous entrainment (synchronizing) signals, that is, zeitgebers, such as natural photoperiod or tidal changes, to maintain a precise, phase-locked period over many cycles. A clock without entrainment signals tends to oscillate with a longer or shorter period, but it still is periodic. Below, we examine some simple mathematical models of biochemical oscillators based on chemical kinetics (see Goldbeter 1996, Yamaguchi et al. 2003, Garcia-Ojalvo et al. 2004, Sharma and Chandrashekaran 2005, Wang and Chen 2005, Li et al. 2007).

Many coupled biochemical reaction systems have parametric feedback inherent in their architectures. The effectiveness of critical enzyme catalysts is often modulated by molecular concentrations from within or without the system, altering reaction rates. The modulation can be from other enzymes, product concentration, or substrate concentrations. Control of transcription/translation of a protein enzyme is one means of affecting its concentration. Regulation of vital chemical reactions by parametric control is a fundamental property of life.

The kinetics of biochemical systems with endogenous feedback loops can be approximated by systems of nonlinear ODEs based on mass action. Large, nonlinear systems are capable of bizarre and counterintuitive behaviors, including sustained, oscillatory, stable limit-cycle behavior in which the concentrations of reactants and products vary periodically within bounds around average values. A stable limit cycle is a bounded, steady-state oscillation in the amplitudes of the states of a nonlinear system that has constant (or zero) input(s). In such a system, a stable limit cycle is entered regardless of initial conditions (Ogata 1970, chapter 12). The concentration waveforms in a nonlinear, biochemical oscillator are generally not sinusoidal. In fact, they can be pulsatile (see the Hodgkin–Huxley model for nerve impulse generation), and the observed oscillations in insulin and glucagon secretion by isolated pancreatic beta cells looks like distorted sine waves (Stagner et al. 1980, Bergsten 2000, Gilon et al. 2002, Simon and Brandenberger 2002).

Some biochemical and physiological systems that exhibit limit-cycle behavior are listed in Table 3.1.

The fact that a coupled biochemical system can exhibit stable limit-cycle behavior was demonstrated by Chance et al. in 1964. Working with anaerobic yeast cultures, in vitro, chance observed that there were periodic fluctuations in the concentration of NADH, a reactant in the glycolysis pathway, whereby sugars are converted to alcohol. The fluctuations in [NADH] were measured as nearly sinusoidal variations in the UV-induced fluorescence of NADH that had a ca. 5 min period. More recent studies on glycolytic oscillations have shown that the period of glycolytic oscillations in yeast can be modulated by the [NAD+]/[NADH] ratio without affecting the instability of the system (Madsen et al. 2005).

Oscillations in the glycolysis metabolic pathway in yeast have been described and analyzed by a number of workers (Chance et al. 1964, Goldbeter 1996, chapter 2, Brusch et al. 2004, Madsen et al. 2005). The complex glycolysis/fermentation pathway is shown in Figure 3.24. The net reactions of glycolysis are

TABLE 3.1
Some Typical Endogenous Biochemical and Physiological
Oscillators and Their Approximate Periods

System	Period (Approximate)
Human ovarian cycle	ca. 28 days
Circadian rhythms (behavioral)	24 h
Cell cycle	ca. 30 min to > 24 h
Gonadotropic hormone secretion	ca. hours
Insulin/glucagon/Ca++/glucose oscillations	ca. 10 min, also 50–120 min
cAMP oscillations	ca. 10 min
Glycolytic NADH oscillations (in yeast)	1 min–1 h
Protoplasmic streaming	1 min
Calcium oscillations	Seconds–minutes
Smooth muscle contractions	Seconds–hours
Cardiac pacemakers (human)	0.4–2 s
Insect fibrillar flight muscle	0.005–0.1 s
Neural membrane potential oscillations. (modeled by the Hodgkin–Huxley equations)	0.002–1 s

$$\text{Glucose} + 2\text{ADP} + 2\text{NAD}^+ + 2\text{Pi} \Rightarrow 2\text{Pyruvate} + 2\text{NADH} + 2\text{H}^+ + 2\text{H}_2\text{O} + 2\text{ATP}$$

$$(3.74\text{A})$$

$$2\text{Pyruvate} + 2\text{NADH} + 2\text{H}^+ \Leftrightarrow 2\text{Lactate} + 2\text{NAD}^+ \qquad (3.74\text{B})$$

(Fermentation is seen to use up 1 NADH to produce 1 NAD^+ [NADH is oxidized], CO_2, and of course 1 ethanol.)

According to Madsen et al., the NADH concentration [NADH] oscillations in yeast extracts appear to be so-called relaxation oscillations [see Glossary], in which the [NADH] increases rapidly, then slowly decays over the remainder of the cycle, then increases rapidly again, etc. However other [NADH](t) waveforms have been measured that are more sinusoidal. Periods are about 5–7 min. Various mechanisms have been proposed for the steady-state oscillations observed in [NADH] in yeast glycolysis, and some attempts have been made to model the mass-action ODEs that describe the glycolysis pathway in yeast (Wolf et al. 2000, Hynne et al. 2001). And yes, the models have shown limit-cycle oscillations in [NADH] and other chaotic behavior.

In their 2005 paper, Madsen et al. concluded from the results of their in vivo experiments and their modeling that the yeast extract glycolytic system instability is largely due to the "on–off switching" (allosteric regulation) of phosphofructokinase (PFK) enzyme expression, and also depends less critically on the ATP-ADP-AMP system. However, they remark: "The implication is that the oscillations are a property of the entire network, and that one cannot dissect the

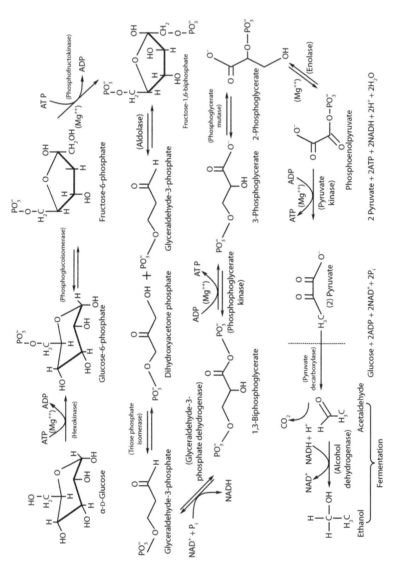

FIGURE 3.24 Schematic of the reactions in the glycolysis metabolic pathway, with one type of alcoholic fermentation. (From Northrop, R.B., *Signals and Systems in Biomedical Engineering*, 2nd edn., CRC Press, Boca Raton, FL, 2010.)

network and identify the mechanisms responsible for the oscillations.... Hence it may not be that surprising that all components of these models are important for the dynamics." [Our emphasis—the whole is greater than the sum of its parts; it is the system that oscillates.]

As we mentioned above, another fascinating intracellular BCO that has recently been given attention is the cyclical secretion of the hormones insulin and glucagon by "resting" pancreatic beta cells. These oscillations have been studied in vivo in monkeys and in man, and in vitro with single beta cells, clusters of cells, and entire islets using dog and mouse pancreases. The slow, in vivo, oscillations of plasma insulin and glucose concentrations have periods of 50–110 min (Stagner et al. 1980, Simon and Brandenburger 2002, Schaffer et al. 2003). The research of Schafer et al. has demonstrated that the long-period insulin oscillations in man are correlated with the nearly simultaneous extracellular release of the amino acid, L-arginine along with insulin. Exogenous L-arginine is known to induce insulin release through a poorly understood mechanism, as well as have other metabolic effects. It acts as a substrate for the formation of the neurotransmitter, nitric oxide (NO), and has other anti-asthenic effects on the body including the restoration of endothelial function in patients with cardiovascular problems. But most significantly in this discussion, the extracellular L-arginine surrounding the pancreatic beta cells that secrete it may act as a pool of synchronizing molecules for the biochemical oscillators that secrete insulin, synchronizing the net insulin secretion rate oscillations. (Synchronizing molecules are considered in more detail in models of *Escherichia coli* repressilator oscillators in Section 3.4.5.)

More rapid insulin and intracellular Ca^{++} concentration ($[Ca^{++}]_i$) oscillations have also been measured in mouse beta cells. The $[Ca^{++}]_i$ oscillations are in phase with the insulin concentration oscillations; their periods are on the order of 5–10 min. Gilon et al. (2002) hypothesized: "Because a rise in β-cell $[Ca^{++}]_i$ is required for glucose to stimulate insulin secretion, and because glucose-induced $[Ca^{++}]_i$ oscillations occur synchronously within all β-cells of an islet, it has been proposed that oscillations of insulin secretion are driven by $[Ca^{++}]_i$ oscillations."

Thus, the bounded oscillations in $[Ca^{++}]_i$ may drive the insulin secretion mechanism by parametrically modulating the sensitivity of β-cells to glucose. But what causes the oscillations in $[Ca^{++}]_i$ to occur? L-Arginine? It may turn out that the lower frequency component of insulin and glucose oscillations in vivo is related to a different mechanism involving time delays, and perhaps the glucagon control system. Perhaps there are two, distinct, cross-coupled oscillator mechanisms leading to the two, superimposed limit cycles in beta cells. Two questions for future research: Are the two cycles phase locked? Are they significantly cross-correlated?

A possible mechanism to describe certain intracellular biological clocks was proposed in 1961 by Spangler and Snell, using a simple, theoretical, chemical-kinetic model architecture. Their model was based on the concept of cross-competitive inhibition. (It did not involve gene expression or transport delays.) The four hypothetical chemical reactions shown below are based on mass-action and diffusion

(Northrop 2000). Eight system state equations, given in Equations 3.76A through H, are based on the following four coupled chemical reactions:

$$(A) \xrightarrow{J_a} A + E \underset{k_{-1}}{\overset{k_1}{\rightleftharpoons}} B \xrightarrow{k_2} E + P \xrightarrow{k_3} * \tag{3.75A}$$

$$nP + E' \underset{k'_{-4}}{\overset{K'_4}{\rightleftharpoons}} I' \tag{3.75B}$$

$$nP' + E \underset{k_{-4}}{\overset{k_4}{\rightleftharpoons}} I \tag{3.75C}$$

$$(A') \xrightarrow{J'_a} A' + E' \underset{k'_{-1}}{\overset{k'_1}{\rightleftharpoons}} B' \xrightarrow{k'_2} E' + P' \xrightarrow{k'_3} * \tag{3.75D}$$

The hypothetical oscillator works in the following manner: Substrates A and A' diffuse into the reaction volume at constant rates J_a and J'_a, respectively. There, A and A' combine with enzyme/catalysts E and E', respectively, and are transformed reversibly into complexes B and B'. B and B' next dissociate to release active enzymes and products P and P'. P and P' next diffuse out of the compartment to effective zero concentrations with rates k_3 and k'_3, respectively. The two, main, Michaelis–Menton-type reactions are entirely independent. However, the system is cross-coupled; therefore, dependent and possibly unstable because of the reversible combination of the products with the complementary free enzyme.

The entire system is described by eight state equations (mass-action ODEs), in which a=running concentration (RC) of A, b=RC of B, p=RC of P, i=RC of I, a'=RC of A', b'=RC of B', p'=RC of P', i'=RC of I'. There are no ODEs for e and e'. We assume the enzymes are conserved in the compartment, that is, enzymes are either free or tied up as complexes; enzymes are not destroyed or created. Thus, in a closed reaction volume: $e_0 = e + b + i$, and $e'_0 = e' + b' + i'$, where e_0 is the initial (total) conc. of enzyme E, and e'_0 is the initial (total) conc. of E'. Note that n molecules of P in the compartment combine with one molecule of E' to reversibly form one molecule of complex I', etc. Thus when I' breaks down, n molecules of P are released with one molecule of E', etc.

Using the notation detailed above and mass-action kinetics, we wrote the eight ODEs and two algebraic equations (for enzyme conservation) describing the coupled, Spangler and Snell system:

$$\dot{a} = J_a - k_1 ae + k_{-1} b \tag{3.76A}$$

$$\dot{b} = k_1 ae - b(k_{-1} + k_2) \tag{3.76B}$$

$$\dot{p} = k_2 b - k_3 p - k'_4 p^n e' + k'_{-4} n i' \tag{3.76C}$$

$$\dot{i} = k_4 p'^n e - k_{-4} i \tag{3.76D}$$

$$\dot{a}' = J'_a - k'_1 a' e' + k'_{-1} b' \tag{3.76E}$$

$$\dot{b}' = k_1'a'e' - b'(k_2' + k_{-1}')$$ (3.76F)

$$\dot{i}' = k_4'p^n e' - k_{-4}'i'$$ (3.76G)

$$\dot{p}' = k_2'b' - k_3'p' - k_4p'^n e + k_{-4}ni$$ (3.76H)

$$e_0 = e + b + i$$ (3.76I)

$$e_0' = e' + b' + i'$$ (3.76J)

To verify that this system is capable of sustained, limit-cycle oscillations, we simulated the system's behavior at "turn-on" using the Simnon nonlinear system modeling software. Note that the initial conditions used for states **b** and **b'** were 4 units each. Other states had zero initial conditions. The initial amounts of free enzymes, e_0 and $e_0' = 5$ units. We used the system parameters listed in the program, **snell2.t**, given in Appendix A.1.

Oscillatory behavior did indeed occur in the model system, as shown in Figure 3.25. To verify that the model system has a SS limit cycle, we plotted **dp/dt** vs.

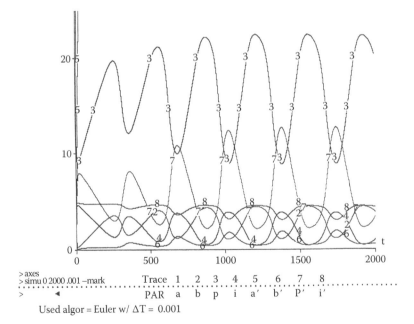

FIGURE 3.25 Plot of Simnon simulation results for the Spangler and Snell oscillating system (cf. Reactions 3.75A through D; ODEs and Equations 3.76A through J). See text for details. Trace IDs: 1=a, 2=b, 3=p, 4=i, 5=a', 6=b', 7=p', 8=i'. (From Northrop, R.B., *Signals and Systems in Biomedical Engineering*, 2nd edn., CRC Press, Boca Raton, FL, 2010.)

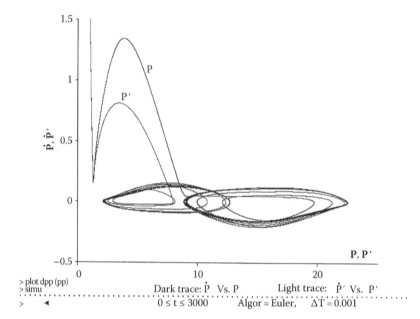

FIGURE 3.26 Simnon phase-plane plots of p, $\dot{p}$; and $\dot{p}'$, p'. The closed attractors indicate steady-state oscillations. (From Northrop, R.B., *Signals and Systems in Biomedical Engineering*, 2nd edn., CRC Press, Boca Raton, FL, 2010.)

$p(t)$ and dp'/dt vs. $p'(t)$ (phase-plane trajectory plots) in Figure 3.26. Note that a stable, closed trajectory in the phase plane indicates a stable, finite-amplitude, steady-state, limit-cycle attractor. That is, continuous, bounded oscillations occur. Note that double-letter variables ending in "p" in the **snell2.t** program are the primed variables in the ODEs above. Also note that in the simulation, we have "half-wave rectified" the concentration states. (In nature, obviously there are no negative concentrations; if the system ICs or constants were ill-chosen in the model, it is possible that a concentration could go negative, creating a paradox. In a well-posed model, all the unrectified states remain nonnegative, eliminating the need for rectification.) Note that in the Simnon program **snell2.t**, the ODEs are written as simple text; no graphic interface is required.

Note that the Spangler and Snell model biochemical oscillator is based entirely on coupled, chemical-kinetic equations. Figure 3.27 illustrates a nonlinear SFG representation of the Spangler and Snell system derived from the state equations of Equations 3.76. In the model, there are eight integrators (eight states), four multipliers, and two exponentiations (p^n and p'^n). Although there are only eight states, the nonlinearities and many feedback loops make computer simulation essential to understand the system's behavior. The system is certainly complicated. In the simulation, the rate constants and initial enzyme levels (e_0 and e_0') were held constant. In the repressilator examples described below, you will see that oscillations can be associated with gene expression controlling protein synthesis.

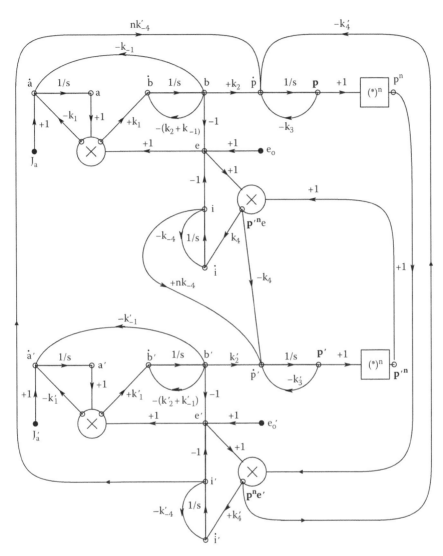

FIGURE 3.27 Nonlinear SFG for the Spangler and Snell system. See text for discussion. (From Northrop, R.B., *Signals and Systems in Biomedical Engineering*, 2nd edn., CRC Press, Boca Raton, FL, 2010.)

In an innovative project encompassing both genetic engineering and mathematical biology, Elowitz and Leibler (2000) designed and installed in vivo a synthetic biochemical network called the repressilator in the *E. coli* lac⁻ strain MC4100 genome. (The lac operon is described in section 6.4.2 in Northrop and Connor 2008.) This genetically modified (GM) bacterium exhibited steady-state, limit-cycle oscillations in three protein concentrations. It used three genes in a sequential, (nonlinear) negative feedback topology illustrated in Figure 3.28.

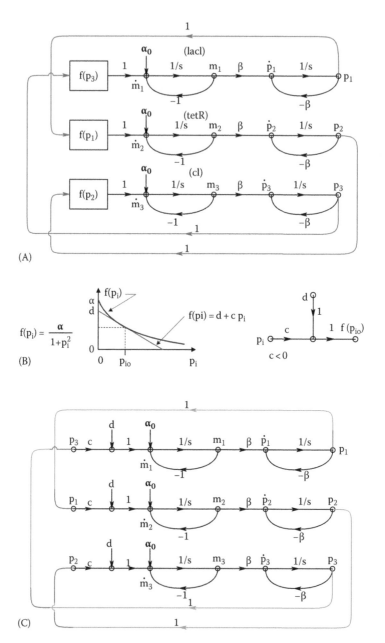

FIGURE 3.28 (A) Nonlinear SFG for the repressilator oscillator. (See Equations 3.77A and B.) The system has a concatenated, sixth-order architecture. (B) Hill function model for decreasing gain nonlinearity that stabilizes repressilator oscillation amplitudes. (C) Linearized sixth-order repressilator system. System loop gain has six poles on the negative real axis in the s-plane. (From Northrop, R.B., *Signals and Systems in Biomedical Engineering*, 2nd edn., CRC Press, Boca Raton, FL, 2010.)

Each gene produces a protein, each of which inhibits the expression of the next gene in the cycle. As the concentration of one of the three produced proteins increases, the expression of the next gene in the cycle is inhibited, and the concentration of the second protein falls, permitting increased expression of the third gene. As the third protein increases in concentration, the first gene now becomes inhibited, allowing expression of the second gene, etc. Note that there are probably transport lags or delays in the living system due to protein synthesis mechanisms, etc., which even though not used in the original Elowitz and Leibler (E & L) model, could certainly contribute to system instability.

The E & L oscillating genomic network periodically induced the synthesis of a green fluorescent protein (GFP) marker in individual GM bacteria as a sign of the oscillation. The GFP oscillations have periods of ca. 150 min, which are slower than the cell division cycle (ca. 1 h), so the oscillator (with its initial conditions) has to be transmitted from generation to generation of GM *E. coli*. How this transfer of oscillator initial conditions occurs through bacterial generations is not known. Perhaps some sort of epigenetic transfer is involved.

A plasmid containing the GM repressilator oscillator and another coupled "reporter" plasmid that makes the GFP were introduced into target bacteria. The repressilator itself is a cyclic negative feedback loop composed of three repressor genes and their corresponding promoters. Molecular details of the repressilator plasmid design and the reporter can be found in the Elowitz & Leibler paper. These authors noted that their GM bacterial "clock" displayed noisy behavior, possibly because of stochastic fluctuations of its components.

Elowitz and Leibler (2000) constructed a simplified, continuous, kinetic, mathematical model of the in vivo repressilator system that exhibited limit-cycle oscillations in protein concentrations and mRNAs under certain boundary conditions and initial conditions. Using their notation, p_i = repressor protein concentration and m_i = the corresponding mRNA concentration. (Ribosomal action was not specifically simulated.) To simplify their model, they considered only the symmetrical case in which all three repressors are identical except for their DNA-binding specificities. Thus, the kinetics of the system were represented by six, coupled, first-order, nonlinear ODEs. (Details of their simulation are found in their paper.)

$$\dot{m}_i = -m_i + \frac{\alpha}{(1+P_j^{n_i})} + \alpha_0 \quad i = 1(\text{lacI}), 2(\text{tetR}) \text{ or } 3(\text{cI}) : \ j = 1(\text{cI}), 2(\text{lacI}) \text{ or } 3(\text{tetR})$$

$$(3.77A)$$

$$\dot{p}_i = -\beta(p_i - m_i) \qquad (3.77B)$$

where the number of protein copies per cell per minute produced from a given promoter type during continuous growth is α_0 in the presence of saturating amounts of repressor (owing to the "leakiness" of the promoter), and $(\alpha + \alpha_0)$ in its absence. β is the ratio of the protein decay rate to the mRNA decay rate. The exponent n_i is a Hill coefficient (the fraction is the familiar Hill hyperbola, which models gene expression inhibition by the preceding gene product).

See Figure 3.28A for a signal flow graph representation of the repressilator ODE set of Equations 3.77A and B. Note that the linear part of this system's loop gain has six real poles. All $n_i = 2$ in the figure. The larger the p_k value, the smaller will be the output of the hyperbolic Hill nonlinearities.

Figure 3.28B illustrates a linearization of the Hill hyperbolic functions. We assume an average operating point, p_{io}, and draw a tangent to the $f(p_i)$ curve at $p = p_{io}$. The tangent defines a linear transfer function, $f(p_i) = cp_i + d$ based on the assumption that $p_i \approx p_{io}$. The linear gain, $c = d[f(p_{io})]/dp_i = -2\alpha p_{io}/(1 + p_{io}^2)^2$. Thus, for small changes of p_i around the operating point, p_{io}, the output of the linearization function is

$$f(p_i) = -2\alpha p_i \left[\frac{p_{io}}{(1 + p_{io}^2)^2} \right] + d \qquad (3.78)$$

Inspection of the linearized repressilator system in Figure 3.28C shows that its (linear part) loop gain in Laplace transform format is

$$A_L(s) = \frac{-|c^3\beta^3|}{(s+1)^3(s+\beta)^3} \qquad (3.79)$$

Thus, we see that the linearized system approximation has a negative loop gain and six real poles. Application of root-locus theory shows that the system will exhibit unstable oscillations for $|c^3\beta^3|$ above some critical value (Northrop 2000). Note that the nonlinear gain, $f(p_i)$, is a decreasing function of p_i, acting to create stable limit-cycle oscillations. A number of other workers have elaborated on the repressilator model set forth by Elowitz and Leibler; for example, see Drennan and Beer (2006) and Alon (2007a).

Not all models of biological oscillators that have been proposed have the relatively simple formats described above. Vilar et al. (2002) described a study of a theoretical genetic oscillator system that used a repressor protein, **R**, and an activator protein, **A**, that form an inactive complex **C**. Figure 3.29 summarizes the reactions in this oscillating system, which include the formation of activated genes for **R** and **A**. The genes for **R** and **A** make **mRNAs**, then the proteins are formed. Positive feedback from the product **C** increases the rate of formation of **R**. The Vilar model is not simple; in fact, it is tempting to say it is complex. Using mass action, we can write the nine nonlinear state equations for Vilar's genetic oscillator:

$$\dot{C} = K_{AR}AR - K_{LC}C \quad \text{(conc. of protein complex formed from A \& R)} \quad (3.80A)$$

$$\dot{R} = K_{MR}M_R + K_{LC}C - K_{LR}R - K_{AR}AR \quad \text{(conc. of repressor protein)} \quad (3.80B)$$

$$\dot{A} = K_{MA}M_A + K_{GR}G'_R + K_{GA}G'_A - K_{LA}A - K_{AR}AR - K_{GA}AG_A$$
$$- K_{GR}AG_R \quad \text{(activator protein concentration)} \quad (3.80C)$$

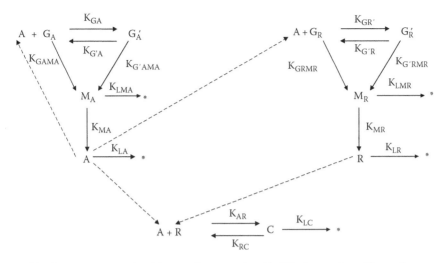

FIGURE 3.29 Schematic of the reactions postulated in Vilar's genetic oscillator. See text for description. (From Northrop, R.B., *Signals and Systems in Biomedical Engineering*, 2nd edn., CRC Press, Boca Raton, FL, 2010.)

$$\dot{\mathbf{M}}_\mathbf{A} = -K_{LMA}\mathbf{M}_\mathbf{A} + K_{GM'A}\mathbf{G}'_\mathbf{A} + K_{GMA}\mathbf{G}_\mathbf{A}$$
$$\text{(mRNA for Activator (Promoter) Protein conc.)} \qquad (3.80D)$$

$$\dot{\mathbf{M}}_\mathbf{R} = -K_{LMR}\mathbf{M}_\mathbf{R} + K_{GM'R}\mathbf{G}'_\mathbf{R} + K_{GMR}\mathbf{G}_\mathbf{R}$$
$$\text{(mRNA for Repressor (Promoter) Protein conc.)} \qquad (3.80E)$$

$$\dot{\mathbf{G}}_\mathbf{A} = -K_{GA}\mathbf{A}\mathbf{G}_\mathbf{A} + K_{GA'}\mathbf{G}'_\mathbf{A} \quad \text{(number of activator genes without A)} \quad (3.80F)$$

$$\mathbf{G}'_\mathbf{A} = -K_{GA}\mathbf{A}\mathbf{G}_\mathbf{A} + K_{GA'}\mathbf{G}'_\mathbf{A} \quad \text{(number of activator genes with A bound to promoter)}$$
$$(3.80G)$$

$$\dot{\mathbf{G}}_\mathbf{R} = -K_{GR}\mathbf{A}\mathbf{G}_\mathbf{R} + K_{GR'}\mathbf{G}'_\mathbf{R} \quad \text{(number of repressor genes without A)} \quad (3.80H)$$

$$\dot{\mathbf{G}}_{\mathbf{R}'} = -K_{GR'}\mathbf{G}'_\mathbf{R} + K_{GR}\mathbf{A}\mathbf{G}_\mathbf{R} \quad \text{(number of repressor genes}$$
$$\text{with A bound to promoter)} \qquad (3.80I)$$

Figure 3.30 shows the complicated nonlinear SFG derived from the nine ODEs above. There are nine integrators and three multipliers. Note that there are both positive and negative loop gains in this SFG. Vilar et al. simulated this system and demonstrated its circadian limit-cycle behavior. Initial conditions and parameter values were given in their paper. We observe that no delay operations were used

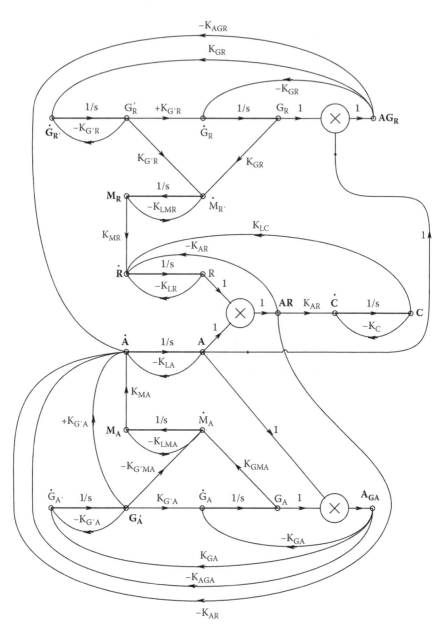

FIGURE 3.30 Nine-state, nonlinear SFG describing the Vilar genetic oscillator based on the nine ODEs, Equations 3.80A through I.

in their model, and the proteins **A** and **R** appear at a rate proportional to their mRNA concentrations; no dynamics were invoked to describe protein synthesis by mRNA, ribosomes, and tRNAs. Proteins **A** and **R** combine to form a product protein, **C**. **C** causes an increase in the rate of appearance of **R**.

For review of some other synthetic (bioengineered) genetic oscillators, see the review paper by Mukherji and Oudenaarden (2009).

In summary, biochemical oscillators are found throughout living systems. They presumably form the basis for endogenous, "biological clocks," which modulate hormone secretion, diurnal behavior, behavior in response to tides, monthly behavior, etc. Circadian biological clocks are generally entrainable.That is, they use an exogenous synchronizing signal such as sunrise, the time of low tide, etc., to insure their oscillations are phase locked to environmental conditions even for several cycles in the short-term absence of synchronizing signals.

The fact that coupled chemical systems have parametric feedback, parametric feed-forward (autocatalytic behavior), transport lags (phase shifts), and are nonlinear often leads to bounded limit-cycle behavior, even for systems with a relatively low numbers of states. If a closed-loop biochemical system were to become frankly unstable, the organism with that system in its genome would quickly be eliminated from existence by natural selection. Small, bounded limit-cycle oscillations are not harmful, in fact they can be useful; overly large ones can be lethal.

3.4.6 SYNCHRONIZATION OF GROUPS OF BIOLOGICAL OSCILLATORS: QUORUM SENSING

The theoretical problem of synchronizing a large array of intracellular oscillators has been addressed by Garcia-Ojalvo et al. (2004), and McMillen et al. (2002). Garcia-Ojalvo et al. postulated a common, extracellular, spatiotemporally averaged, autoinducer (AI) molecule that affects all *E. coli* cells containing repressilator oscillators. In their model, Garcia-Ojalvo et al. assumed that the AI molecule was freely diffusible in and out of all oscillator cells, and its net intracellular concentration acted as a TF that stimulates expression of the *lacL* gene in a repressilator, thus synchronizing it with other repressilators bathed in the external AI solution. Garcia-Ojalvo et al. have called this process *quorum sensing*. Quorum sensing in yeast biochemical oscillators was also described in a review paper by Klevecz et al. (2008). Synchronization of cell division in *S. cerevisiae* cultures is thought to involve the secretion of acetaldehyde and H_2S gas by the cells. Quorum sensing causes many individual, endogenous cellular oscillators to become phase locked.

Wang and Chen (2005) developed a mathematical model based on the quorum-sensing synchronization model described by Garcia-Ojalvo et al. (2004). They further investigated how individual repressilator oscillators in neighboring *E. coli* cells can become phase locked (both in frequency and phase). In the Wang and Chen synchronization model, there are **N** *E. coli* cells that have repressilator oscillators (first described by Elowitz and Liebler 2000). These model cells are clustered together in a common compartment.

In the Wang and Chen and Garcia-Ojalvo models, a small, exported, synchronizing, autoinducer (AI) molecule is made intracellularly at a rate determined by the intracellular concentration of a protein, LuxI. (The kinetics of LuxI protein production were assumed to be identical to that of the *lac* operon protein, TetR, so both Garcia-Ojalvo and Wang and Chen used the intracellular concentration of TetR = A instead of a new variable for LuxI concentration.) The AI diffuses freely through the cell membrane into a common extracellular volume (compartment). The intracellular AI molecule can also combine with a second protein, LuxR, and this complex acts as a stimulating TF that up-regulates the production of the *lac* operon protein, LacI.

The Wang and Chen model uses the six, original, Elowitz and Liebler oscillator equations describing the repressilator oscillations in each of $i = 1, 2, ..., N$ bacterial *lac* operon oscillators (one per cell).

For each of N bacteria, the three mRNA equations are

$$\dot{a}_i = -d_{1i}a_i + \frac{\alpha C}{\mu C + c_i^n} \quad (a_i = \text{mRNA transcribed from the } \textbf{\textit{tetR}} \text{ gene}) \quad (3.81A)$$

$$\dot{b}_i = -d_{2i}b_i + \frac{\alpha_A}{\mu_A + a_i^n} \quad (b_i = \text{mRNA transcribed from the } \textbf{\textit{cI}} \text{ gene}) \quad (3.81B)$$

$$\dot{c}_i = -d_{3i}c_i + \frac{\alpha_B}{\mu_B + b_i^n} + \frac{\alpha_S S_i}{\mu_S + S_i} \quad (c_i = \text{mRNA transcribed from the } \textbf{\textit{lacI}} \text{ gene})$$

$$(3.81C)$$

where n is the Hill coefficient ($n = 4$ in the Wang and Chen model). α_j and μ_j are nonnegative Hill function constants ($j = C, A, B, S$).

A_i, B_i, and C_i are concentrations of the proteins, TetR, CI, and LacI, respectively. The synchronizing autoinducer up-regulates the production of mRNA from the *lacI* gene. (The Hill function in Equation 3.81C describes the stimulation [and synchronization] of *lacI* mRNA production expression by intracellular AI.) α_S and μ_S are Hill function constants, and S_i is the ith cell's intracellular concentration of AI protein.

For each bacterium, the ODEs describing the dynamics of the proteins, TetR, CI, and LacI are, respectively,

$$\dot{A}_i = -d_{Ai}A_i + \beta_a a_i \quad (\text{TetR protein in the ith cell}, \quad i = 1, 2, ..., N) \quad (3.82A)$$

$$\dot{B}_i = -d_{Bi}B_i + \beta_b b_i \quad (\text{CI protein in the ith cell}) \quad (3.82B)$$

$$\dot{C}_i = -d_{Ci}C_i + \beta_c c_i \quad (\text{LacI protein in the ith cell}) \quad (3.82C)$$

where

$\{d_{Xi}\}$ are the loss rate constants for the three proteins (for each bacterium, $i = 1, \ldots, N$)

$\{\beta_k\}$ are the translation rate constants of the proteins from the mRNAs (the same in each cell)

The kinetics for the ith cell's intracellular, autoinducer concentration, $\mathbf{S_i}$, are given by the $\mathbf{N}$ ODEs:

$$\dot{\mathbf{S}}_i = -d_S \mathbf{S}_i + \beta_S \mathbf{A}_i - \eta_S[\mathbf{S}_i - \mathbf{S}_e], \quad 1 \leq i \leq N \tag{3.83}$$

where d_S, β_S, and η_S are the loss rate, translation rate, and diffusion rate constant for intracellular and extracellular AI, respectively. The average, extracellular concentration of AI is modeled by the ODE

$$\dot{\mathbf{S}}_e = -d_e \mathbf{S}_e + (\eta_e N^{-1}) \sum_{j=1}^{N} [\mathbf{S}_j - \mathbf{S}_e] = -d_e \mathbf{S}_e + (\eta_e N^{-1}) \sum_{j=1}^{N} [\mathbf{S}_j - N\mathbf{S}_e] \tag{3.84}$$

where

d_e and η_e are the loss rate constant and diffusion constant for the (homogeneous) extracellular autoinducer concentration, $\mathbf{S}_e$

$N^{-1} \sum_{j=1}^{N} \mathbf{S}_j = \overline{\mathbf{AI}_i}$ is the average $\mathbf{AI}$ concentration inside the $\mathbf{N}$ cells

To simplify their model, Wang and Chen approximated the average extracellular $\mathbf{AI}$ concentration $\mathbf{S}_e$ by

$$\mathbf{S}_e \cong Q_e \left(N^{-1} \sum_{j=1}^{N} \mathbf{S}_j \right) = Q_e \overline{\mathbf{AI}_i} \tag{3.85}$$

where $Q_e \equiv \eta_e/(d_e + \eta_e)$, $0 \leq Q_e \leq 1$, and $\mathbf{S}_j = \mathbf{AI}$ concentration inside the jth cell.

In their paper, Wang and Chen (2005) used simulations of the equations above to illustrate that their model indeed entrains the $\mathbf{N}$ cells to synchronized oscillations, starting with $\mathbf{N}$ different initial conditions on the $\mathbf{N}$ intracellular repressilator oscillators. They commented: "The change from a non-oscillatory state to an oscillatory one is induced by appropriate coupling, which also entrains all cells to synchronization." Wang and Chen concluded: "...we not only provided a general model of cellular synchronization but also derived a sufficient condition to ensure global convergence of the collective dynamics." The interested reader should see their paper for simulation details and graphical results. In Figure 3.31, we illustrate the repressilator synchronization system in the form of a nonlinear SFG in order to stress its complexity. The state equations above were used in its formulation. Only one of $\mathbf{N}$ intracellular oscillators and its accompanying AI molecular dynamics are shown.

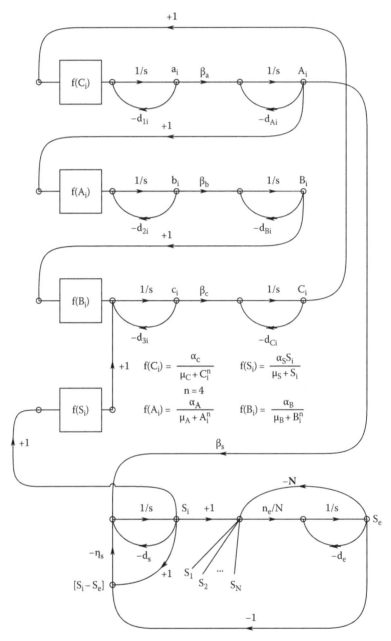

FIGURE 3.31 Nonlinear SFG describing the Wang and Chen repressilator synchronization system. Model is for only one cell (out of **N**). Hill-type nonlinearities are used. See text for details. (From Northrop, R.B., *Signals and Systems in Biomedical Engineering*, 2nd edn., CRC Press, Boca Raton, FL, 2010.)

In summary, we note that repressilator synchronization has been a popular topic in systems biology. For example, Li et al. (2007) have considered the problem of how the **N** repressilator genetic oscillators with AI synchronization respond to noise, that is, they investigated the robustness of the synchronized system.

The suprachiasmatic nucleus (SCN) of mammals is known to contain the master clock affecting a number of physiological systems: locomotor activity, feeding, drinking, hormone secretion (corticosterone, growth hormone, serotonin, melatonin), the sleep–wake cycle, and N-acetyltransferase concentration. There may be a direct synchronizing action of the hormone melatonin on the component cellular oscillators in the SCN, giving it a role as a zeitgeber (Sharma and Chandrashekaran 2005). In this respect, extracellular melatonin evidently plays the role of the synchronizing, extracellular, autoinducer substance, AI, in the Wang and Chen (2005) model described above. The fact that melatonin secretion is regulated by external photoperiod allows it to play, among other functions, the role of the mammalian, molecular zeitgeber.

3.5   SUMMARY

In this chapter, we have introduced some important properties of nonlinear systems and several ways we can analyze and model them. Small arrays of nonlinear ODEs were shown, given appropriate parameters and initial conditions, to have the property creating of bounded, limit-cycle oscillations. Chemical mass-action kinetics were shown to be useful in describing coupled biochemical reactions, and in particular, biochemical oscillators (BCOs) and biological clocks (zeitgebers). Transport lags in both linear and nonlinear systems were shown to cause limit-cycle oscillations under appropriate circumstances.

The phase plane was introduced as a means of visualizing the behavior of BCOs and other nonlinear coupled systems in initial value problems (IVPs). Typical phase-plane trajectories and attractors were shown for nonlinear systems.

Models for small and large homogeneous biochemical oscillators were described, including the Per protein oscillator, oscillations in the glycolysis system, the *lac* operon repressilator, the Spangler and Snell (1961) theoretical biochemical oscillator, and Vilar's (2002) theoretical genetic oscillator.

Models for the synchronization of groups of autonomous intracellular BCOs through a common pool of common, secreted autoinducer (AI) molecules were described. We introduced the use of signal flow graphs to examine the feedback topology of BCOs and other coupled nonlinear systems.

PROBLEMS

3.1   Make a nonlinear SFG for the two, coupled Tomovic equations, text Equations 3.4A and B.

3.2   Make a nonlinear SFG for the two, coupled Duffing equations, text Equations 3.8A and B.

3.3 Consider the **n**th-order Hill function:

$$\dot{Q}_G = \frac{[H]^n \dot{Q}_{GMAX} K}{(1+[H]^n K)}, \quad \text{for } [H] \geq 0; \quad \dot{Q}_G \rightarrow \dot{Q}_{GMAX} \text{ as } [H]^n K \gg 1$$

(A) Find an expression for $(d\dot{Q}_G/d[H])$ for $[H]^n \equiv 1/K$, $n = 1, 2, 3, \ldots$
(B) Plot $(d\dot{Q}_G/d[H])$ vs. n for $n = 1, 2, 3, \ldots$. Let $K = \dot{Q}_{GMAX} = 1$.

3.4 The circuit uses parametric control to charge a capacitor to $v_c = V_D$. See Figure P3.4. As v_c approaches V_D, the conductance $G(v_c)$ decreases linearly to 0 at V_D. That is, $G(v_c) = G_o - kv_c$ Siemens for $0 \leq v_c \leq V_D$, and 0 for $v_c > V_D$. $k = G_o/V_D$. Simulate the behavior of the circuit as the capacitor charges. Let $V_S = 10\,V$, $V_D = 1\,V$, $G_o = 10^{-3}$ s, and $C = 10^{-3}$ F. $v_c(0) = 0$. Plot $v_c(t)$, $i_c(t)$ mA, and $\tau(t) = C/G(v_c)$. Use a vertical axis = 0, 10; time axis = 0, 1.5 s.

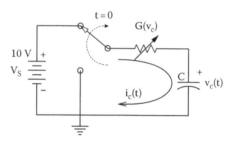

FIGURE P3.4

3.5 Consider the circuit of Figure P3.4: Now let the conductance be determined by $G(v_c) = G_o \exp(-v_c/2V_D)$ s, $0 \leq v_c \leq V_D$. When $v_c > V_D$, $G \equiv 0$. Simulate and plot $v_c(t)$, $i_c(t)$, and $\tau(t)$. Use the same parameters as in Problem 3.4.

3.6 This problem is about parametric regulation of the eye's intra-ocular pressure (IOP). (In this steady-state model, we will neglect the fluid capacitances of the posterior and anterior chambers of the eye.) Assume that aqueous humor (AH) is made at a constant rate, $\dot{Q}_{AH}$ mL/min, in the posterior chamber. It flows past the lens with a relatively small hydraulic resistance, R_{IL}, into the anterior chamber, thence through the trabecular network into the canal of Schlemm, thence out of the eyeball to the episcleral veins. The hydraulic resistance of the outflow pathway is R_{TS}. Hydraulic resistance to slow flow is analogous to Ohm's law: $R = V/I \rightarrow R_H = \Delta P/\dot{Q}$. P_{EV} is the hydraulic pressure in the episcleral veins (the outflow of AH). From analogy to Ohm's law, the (unregulated) pressure in the anterior chamber (the IOP) is: $P_{AC} = \dot{Q}_{AH} R_{TS} + P_{EV}$ mmHg. It is desirable to regulate P_{AC} at a nominal value of ca. 15 mmHg to permit adequate blood circulation in the retina. We assume that increased P_{AC} stretches the walls of the trabeculae and the canal of Schlemm, thereby facilitating the outflow of aqueous humor; that is, it decreases R_{TS}. Assume this parametric variation of R_{TS} can be modeled by

$$R_{TS} = R_{TS_0} - \rho P_{AC}, R_{TS} > 0$$

A second postulated mechanism for regulating P_{AC} is the action of P_{PC} on the AH inflow rate, $\dot{Q}_{AH}$. Assume this law is: $\dot{Q}_{AH} = \dot{Q}_{AH_0} - \gamma P_{PC}$, $\dot{Q}_{AH} = 0$ for $P_{PC} > \dot{Q}_{AH_0}/\gamma$.

(A) Find an expression for $P_{AC} = f(P_{EV}, \gamma, \rho, \dot{Q}_{AH_0}, R_{TS_0}, \text{etc.})$.

(B) Calculate the sensitivities for the P_{AC} regulator: $[dP_{AC}/d\dot{Q}_{AH_0}]$ $(\dot{Q}_{AH_0}/P_{AC})$, $[dP_{AC}/dP_{EV}](P_{EV}/P_{AC})$, and $[dP_{AC}/dR_{TS}](R_{TS}/P_{AC})$.

3.7 This problem is about managed, sustainable fishing. A proposed, Lotka–Volterra-type model is given below consisting of two nonlinear ODEs describing the growth, death, and predatory interaction of two fish species:

$$\dot{x}_1 = fd\, x_1(t - D_1) - \alpha_1 x_1 x_2 - \beta x_1^2 \quad \text{Rate of change of "Herring pd"}$$

$$\dot{x}_2 = \alpha_2(x_1 x_2)(t - D_2) - L_c x_2 - b_0 x_2 \quad \text{Rate of change of "Codfish pd"}$$

$$\dot{c}c = L_c x_2 \eta_c \quad \text{Cumulative cod catch} = cc$$

where
  x_1 is the herring density (a prey species) in thousands of fish per km^2
  $x_1(t - D_1)$ is $x_1(t)$ delayed D_1 months
  x_2 is the cod (the predatory species) density
  cc is the cumulative (integrated) cod catch over time (a quantity to be maximized)
  η_c is the catch-to-market efficiency for cod ($\eta_c \leq 1$)
  fd is herrings' food (plankton), which varies cyclically with season, sic:
    $fd = f_0 + f_1 \sin(2\pi t/P)$, $P = 12$, a 12-month period
  D_1 is the delay in months for herring growth to maturity
  D_2 is the delay for cod growth after feeding on herring (that is, the product $(x_1\, x_2)$ is delayed by D_2 months)
  L_c is the cod-fishing law: $L_c = $ if $x_2 > \varphi_c$ then b1, else 0

(That is, cod fishing is prohibited until the cod density exceeds a threshold density, φ_c.)

(A) Make a nonlinear SFG describing this system.

(B) Simulate the system's behavior using MATLAB or Simnon. The ODEs are stiff, so use an appropriate integration routine. Use the following system parameters: Initial values: $x_1(0) = 3$, $x_2(0) = 3$. Parameters: $\alpha_1 = 0.5$, $\alpha_2 = 1$, $\beta = 0.1$, $b_1 = 8$, $D_1 = 3$ months, $D_2 = 5$ months, $\varphi = 2$, $f_0 = 7$, $f_1 = 7$, $P = 12$ months, time t is in months.

Plot x_1, x_2, c, and fd vs. time for 72 months (6 years). Also do a phase-plane plot of y vs. x. Does the system have a stable attractor?

(C) Using simulations, find the φ_c value ≥ 0 that yields the largest sustainable cumulative cod catch over 72 months. Plot cc(72) vs φ_c.

(D) Examine the model's behavior when herring are also fished according to a threshold law: $L_h = $ if $x_1 > \varphi_h$ then b_2, else 0. (Thus the herring ODE is now $\dot{x}_1 = f\, x_1 (t - D_1) - \alpha_1 x_1 x_2 - \beta x_1^2 - L_h x_1$.) Define a cumulative herring catch rate, $\dot{c}h = L_h x_1 h_h$. Examine the effect of regulated herring fishing on the cod yield. Start with the parameters above and also let: $\varphi_h = 4$, $b_2 = 5$, $\eta_h = 1$. Can the thresholds φ_c and φ_h be adjusted to maximize both cc and ch at $t = 72$ months? (Remember, an increased herring catch means less food for cod.)

3.8 Consider the three coupled equations of Rössler (1976):

$$\dot{x} = -(y + z) \tag{11.66A}$$

$$\dot{y} = x + ay \tag{11.66B}$$

$$\dot{z} = b + xz - cz \tag{11.66C}$$

(A) Make a nonlinear SFG for the Rössler system. Compare it to that of the Lorenz system.

(B) Use an appropriate simulation language to examine this system's behavior in the x, y, z phase plane. Begin by using the parameters in the text. Now let $a = 0.343$, $b = 1.82$, and $c = 9.75$. Define what is meant by "screw chaos."

3.9 (A) Simulate the Lorenz system as an IVP. Use a precise integration routine suitable for stiff ODEs. Use
$\sigma = 10$, $\beta = 8/3$, $\rho = 26$. Make a 3-D, x, y, z phase-plane plot.

(B) Examine the "butterfly effect" for ρ values around 24.74.

3.10 Simulate the simple delay system shown in text Figure 3.22. Let $a = 4$, $\tau = 2$, $y(0) = 0$. Use a unit step input at $t = 0$ [i.e., $x(t) = U(t)$]. Find the K value required for sustained oscillations. Plot the oscillations in y in time, and also in the $\dot{y}$, y phase plane. Try K values between 1 and 5.

3.11 (A) Make a nonlinear SFG for the Komarova equation, $\dot{x} = ax(t) - ax(t)\, x(t - \tau)/K$.

(B) Examine the stability of this system in the $\dot{x}$, x phase plane for different ICs on $\dot{x}$ and x. Examine the effect of different values of the delay τ, the feedback gain a, and the constant K.

3.12 (A) Make a nonlinear SFG of the Tyson (2002) per protein oscillator model of text Equations 3.73A through E. Make the TF concentrations, [TIM], [CRY], [CLC], and [CYC], constant. Note that the constant system input to $\dot{x}_1$ decreases with increasing x_5 because of the Hill function, $K_T(x_5)$. This is, in effect, parametric negative feedback.

(B) Simulate the per system. Find conditions for oscillation. Treat the system as an IVP. Start with the parameters: $[CLC][CYC]K_{T0} = 0.76$, $\mu M/h$, $A = 1.0$, $K_C = 3.2$, $K_{DI} = 0.5$, $K_{D4} = 1.3$, $K_I = 1$, $K_{L2} = 3.2$, $K_{L5} = 1.3$, [TIM] $[CRY]K_{P3} = 3$. Assume unity IVs for the states. Plot x_1, ..., x_5 vs. t. Examine the effect of raising the loop gain by increasing K_C or K_{D4}.

3.13 Tyson (2002) simplified the five Per equations to two nonlinear ODEs:

$$\dot{M} = \frac{v_m}{1 + (P_T/A)^2} - k_m M \quad \text{Per mRNA}$$

$$\dot{P}_T = v_p M - k_3 P_T - \frac{k_1 P_T}{J + P_T} \quad \text{Total per protein}$$

where $v_m = 1$, $v_p = 0.5$, $A = 0.1$, $J = 0.05$, $k_1 = 10$, $k_2 = 0.03$, $k_m = k_3 = 0.1$, $J = 0.05$.

(A) Make a nonlinear SFG for the system.

(B) Simulate the system. Plot the system's behavior in the M,$\dot{M}$ and P_T, $\dot{P}_T$ phase planes. Examine the effect of varying v_m. What parameters of this simple system will most effectively control the SS period of oscillation?

3.14 [This problem was adapted from Tyson (2002).] It is a cellular biochemical oscillator called a *substrate-inhibition oscillator*. Two extracellular reactants with concentrations A_0 and B_0 diffuse passively into a cell where their concentrations are A and B, respectively. A+B react, catalyzed by an enzyme E, to form a product C. C is lost with rate K_{LC}. The effectiveness of enzyme E is down-regulated parametrically by the intracellular concentration of A according to the rule shown below. The system's diffusion/mass-action reactions are

$$\dot{A} = K_{DA}(A_0 - A) - \frac{k_e AB}{K_{m1} + A + A^2/K_{m2}}$$

$$\dot{B} = K_{DB}(B_0 - B) - \frac{k_e AB}{K_{m1} + A + A^2/K_{m2}}$$

$$\dot{C} = \frac{k_e AB}{K_{m1} + A + A^2/K_{m2}} - K_{LC}C$$

(A) Draw a nonlinear SFG describing the system. (It will have three integrators, two multipliers, and a divider.)

(B) Simulate the system. Plot A(t), B(t), and C(t), and A($\dot{A}$) and B($\dot{B}$) [phase plane]. Initially use: $K_{DA} = 1$, $K_{DB} = 0.15$, $k = 30$, $K_{m1} = 1$, $K_{m2} = 0.005$, $K_{LC} = 1$, $A_0 = 1$, $B_0 = 6$. Assume A(0) = B(0) = 0.

(C) For $K_{DB} = 1$, $A_0 = 1.5$, and $B_0 = 4$, plot $\dot{A}$(A), $\dot{B}$(B) and A vs. B. (Other parameters as in (B) above.)

(D) Assume the system has no substrate inhibition from A, that is,

$$\dot{A} = K_{DA}(A_0 - A) - k_e AB$$

$$\dot{B} = K_{DB}(B_0 - B) - k_e AB$$

$$\dot{C} = k_e AB - K_{LC}C$$

Draw a nonlinear SFG for the system of (D). Compare this SFG to that for the Lotka–Volterra system.

3.15 A Goldbeter (1996) model for calcium-induced [Ca⁺⁺] release oscillations is given below:

$$\dot{z} = v_o + v_1\beta - v2 + v3 - kz + k_1 y$$

$$\dot{y} = v2 - v3 - k_f Y$$

where $v2 \equiv V_{m2}z^2/(K_2^2 + z^2)$, $v3 \equiv V_{m3}y^2/(K_R^2 + y^2) + z^4/(K_A^2 + x^4)$, $z =$ [Ca⁺⁺] in the cytosol, $y =$ vesicular [Ca⁺⁺], $v_o =$ slow leak rate of Ca⁺⁺ into the cytosol from extracellular fluid, $v_1\beta = IP_3$-induced release of Ca⁺⁺ into the cytosol from intracellular stores, $v2 =$ ATP-dependent Ca⁺⁺ pumping rate, $v3 =$ [Ca⁺⁺]-induced Ca⁺⁺ release rate from storage vesicles, $k =$ Ca⁺⁺ elimination rate through the plasma membrane, and $k_1 =$ Ca⁺⁺ leak rate from storage vesicles.

Estimated parameter values are: $v_o = 1\,\mu M/s$, $v_1 = 7.3\,\mu M/s$, $\beta = 0$, $V_{m2} = 65\,\mu M/s$, $K_2 = 1\,\mu M$, $V_{m3} = 500\,\mu M/s$, $K_R = 2\,\mu M$, $K_A = 0.9\,\mu M$, $k = 10\,s^{-1}$, and $k_f = 1\,s^{-1}$.

(A) Make an NLSFG for this system.

(B) Simulate the system with the parameter values given above. Plot $z(t)$, $y(t)$, and $y(x)$ (phase plane).

(C) To simulate an input pulse of IP3, let $\beta(t) = b_o e^{-t/\tau}$, with $\tau = 10\,s$. Try $b_o = 10$. Plot $z(t)$ and $y(t)$ for both cases.

3.16 The general mathematical model for enzymatic system oscillations first proposed by Goodwin in 1968 has been tinkered with by several workers. In its first version, no LC oscillations occurred. Tyson and Othmer (1978) [The dynamics of feedback control circuits in biochemical pathways. *Prog. Theor. Biol.*, 5, 1–62.] tinkered with Goodwin's model and rewrote the m, MA ODEs as

$$\dot{x}_1 = \frac{k_1}{\left(K_I^n + x_m^n\right)} - b_1 x_1$$

$$\dot{x}_2 = a_1 x_1 - b_2 x_2$$

$$\ldots$$

$$\dot{x}_m = a_{m-1} x_{m-1} - b_m x_m$$

Draw the nonlinear SFG for this system of m equations.

3.17 The well-known Hodgkin–Huxley (1952) model for nerve spike generation is given by the nonlinear ODEs and functions below. For a unit area of active membrane, we can write the node equation (Northrop 2001):

$$C_m \dot{v} = J_{in} - (J_K + J_{Na} + J_L) \text{ amps/cm}^2$$

where

J_L is the transmembrane leakage current density $= g_{Lo}(v - V_L) \; \mu A/cm^2$

J_K is the potassium ion transmembrane current density $= g_{Ko} n^4 \; (v - V_K)$ $\mu A/cm^2$

J_{Na} is the sodium ion transmembrane current density $= g_{Nao} \; m^3 \; h \; (v - V_{Na})$ $\mu A/cm^2$

J_{in} is an input current density used to trigger nerve impulses

V_L, V_K, and V_{Na} are dc Nernst equilibrium potentials

C_m is the electrical capacitance per unit area of axon membrane

The auxiliary dynamic parameters, n, m, and h, are given by the three nonlinear ODEs

$$\dot{n} = -n(\alpha_n + \beta_n) + \alpha_n$$

$$\dot{m} = -m(\alpha_m + \beta_m) + \alpha_m$$

$$\dot{h} = -h(\alpha_h + \beta_h) + \alpha_h$$

Although the three ODEs above appear linear, they are in fact nonlinear functions of v, the instantaneous millivolt hyperpolarization potential of the nerve membrane patch. v is measured as a change from the neuron's dc resting potential. If $v < 0$, the actual transmembrane potential, V_m, is depolarizing (i.e., going positive from the dc resting potential, $V_{mr} = -70 \, mV$). In fact, $v \equiv (V_{mr} - V_m)$. (This rather baroque sign convention follows faithfully from the original 1952 H-H paper.) The six nonlinear coefficients of the three ODEs above are

$$\alpha_n(v) = \frac{0.01(v + 10)}{[\exp(0.1v + 1) - 1]} \qquad \alpha_m(v) = \frac{0.01(v + 25)}{[\exp(0.1v + 2.5) - 1]}$$

$$\alpha_h(v) = 0.07 \exp\left(\frac{v}{20}\right)$$

$$\beta_n(v) = 0.125 \exp\left(\frac{v}{80}\right) \quad \beta_m(v) = 4 \exp\left(\frac{v}{18}\right) \quad \beta_h(v) = \frac{1}{[\exp(0.1v + 3) + 1]}$$

Draw a nonlinear SFG describing the H-H system. It is interesting to note that the H-H system is also current-to-frequency oscillator for constant J_{in} (see section 1.4.2 in Northrop 2001).

3.18 In examining the frequency response of the Duffing system (Equations 3.8A and B) by simulation, we swept the input frequency up or down in time according to the rule: $\omega = (\omega_s + \delta_\omega t)$. Comment on why the frequency response observed is not a true steady-state frequency response. What modifications on the input frequency function could you make to give a true SS frequency response?

3.19 Balagaddé et al. (2008) devised a synthetic "predator–prey" ecosystem using two *E. coli* genotypes. These cells effectively engage in a Red Queen survival contest that exhibits oscillatory population densities. Predator cells inhibit the proliferation of prey, and prey cells derepress the proliferation of predators. Two representative cells from the mixed predator and prey populations are shown in Figure P3.19. In their paper, Balagaddé et al. do not show a mathematical model for their synthetic predator–prey ecosystem. (Note that the predator cells in this model do not eat the prey, they use an AI molecule, Las*l*, to suppress prey reproduction. In turn, a high prey population leads to more rapid proliferation of predator cells.)

(A) In this problem you will use mass-action principles to write the nine nonlinear ODEs that model the system. For example, the predator cell density, P_r, is given by the ODE

$$\dot{P}_r = \frac{a_{pr}}{(1 + CcdB/k_1)} - Pr * K_{LPr} \text{ cells/unit volume}$$

The $(1 + CcdB/k_1)$ denominator in the first term models the inhibition of predator cell proliferation by CcdB intracellular concentration. The second term models predator cell death. The intracellular inhibitor protein concentration, CcdB, is given by the ODE

$$\dot{CcdB} = \frac{a_{CB}}{(1 + CcdA/k_2)} - CcdB * K_{LCB} \quad \text{(molecules/unit volume)}$$

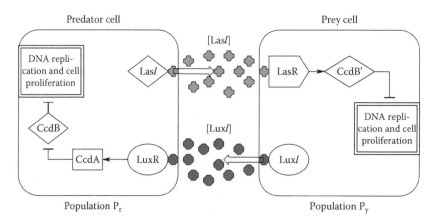

FIGURE P3.19

The inhibitory autoinducer (iAI) protein, LasI, is made constantly inside one predator cell, released into the medium, and is given by the ODE

$$\dot{Lasl} = a_{LI} - Lasl * K_{LLI}$$

(molecules/unit volume inside each predator cell)

The steady-state concentration of iAI in the medium [LasI] is assumed to be proportional to the density of predator cells, Pr. [LasI] ultimately inhibits reproduction of the prey cells. Thus,

$$[Lasl] = Lasl * P_r * KP_r$$

(molecules/unit volume in extracellular space)

Expression of the inhibitor molecules, CcdA, is stimulated by the promoter, LuxR. Thus, the ODE for CcdA is

$$\dot{CcdB} = \frac{a_{CA} * LuxR}{(1 + LuxR/k_3)} - CcdA * K_{LLR} \quad \text{(molecules/unit volume)}$$

Now all you have to do is write the remaining five ODEs that describe the system.

(B) Using the nine ODEs from (A), make a nonlinear signal flow graph describing the predator–prey system.

4 Modularity, Redundancy, Degeneracy, Pleiotropy, and Robustness in Complex Biological Systems

4.1 INTRODUCTION

Often when viewing a 2-D or 3-D plot of a complex system's graph, the eye can pick out clusters of nodes that apparently "talk more amongst themselves" than to other peripheral nodes. The local density of the edges is the give-away in this subjective identification of a structural module, or subsystem. Modularity is an abstract concept that can refer to patterns of physiological and molecular interactions, the organization of graphs for modeling living processes, or in genetics to the distribution of mutational effects on the phenotype (Wagner et al. 2007). However, there are objective, mathematical criteria based on topology and thresholds for identifying what collections of nodes comprise a module in a large graphical network. Some of these metrics are described below.

We are familiar with modules that occur in man-made systems. Modules in the form of software subroutines can be linked into different, complex computer programs. A modular design of interchangeable, replaceable parts in mechanical systems leads to easier design and maintenance, as well as robust operation. In analog electronic systems, the operational amplifier (op amp) often serves as a vital part of a modular gain element or signal filter. Other integrated circuits serve as analog RF and power amplifiers, digital ICs store data (RAMs and ROMs), and programmable logic array (PLA) modules are used in digital controllers.

Modularity is widely encountered in all biological systems, regardless of scale. Alon (2003) reminds us that "Biology displays the same [modular] principle, using key wiring patterns again and again throughout a network." He remarked that many metabolic networks use biochemical regulatory architectures such as feedback inhibition. "The [gene] transcriptional network of *E. coli* has been shown to display a small set of recurring circuit elements termed 'network motifs.'" The network motifs can perform tasks such as filtering out random input fluctuations, generating oscillating gene expression patterns, and up-regulating gene expression. The same motifs were also found in the transcriptional networks of yeast. Alon is of the opinion that nature appears to converge on these functional circuit patterns over and over again in different, nonhomologous systems. They are evidently part of a universal, functional architecture found in all biochemical systems. Wang and Zhang (2007) reported on the modular structures in protein networks.

The significance of modularity in molecular systems biology, genetics, and evolutionary biology has also been discussed in an insightful review paper by Wagner et al. (2007). These authors point out that two areas of interest in modularity research that need further attention are as follows: (1) how it affects evolution, evolvability, and adaptive radiations; and (2) the origins of modularity.

It appears that modularity allows the creation of robust complex systems from smaller, complex subsystems. Lipson et al. (2002) stated that "It has long been recognized that architectures that exhibit functional separation into modules are more robust and amenable to design and adaptation." These authors pointed out that modularity creates a functional separation (or isolation) that reduces the amount of coupling between internal (i.e., inside a module) and external changes. Thus, evolutionary pressures may more easily rearrange the inputs to modules without altering their intrinsic behaviors. Evidence for this preservation of modular function is seen in certain developmental experiments with *Drosophila* larvae. Using targeted misexpressions of the *eyeless* gene's cDNA, functional compound eyes were caused to grow on appendages such as wings, antennae, and legs (Halder et al. 1995). This induced, anomalous growth suggested that the whole compound eye is represented by a mobile, modular, genomic regulatory unit. (See Section 4.4.2.)

Albert (2005) observed that "...modules should not be understood as disconnected components but rather as components that have dense intracomponent connectivity but sparse intercomponent connectivity." The challenge in identifying functional modules is that a module is not always a clear-cut subnetwork linked in a clearly defined manner, "...but there is a high degree of overlap and crosstalk between modules."

In some biological systems, modularity may be hierarchical, a kind of nested architecture. Albert also remarked on graph properties related to modularity: "...a heterogeneous degree distribution, inverse correlation between degree and clustering coefficients (as seen in metabolic and protein interaction networks) and modularity taken together suggest hierarchical modularity, in which modules are made up of smaller and more cohesive modules, which themselves are made up of smaller and more cohesive modules, etc."

We see modularity allowing an evolutionary change in one module causing minimum disturbances in the functioning of other modules in the complex nonlinear system (CNLS), as long as the paths between modules are held relatively constant.

4.2 MEASURES OF MODULARITY

A challenge exists in finding objective algorithms for the identification of modules in a large graph or network. Hallinan (2003, 2004a), Hallinan and Smith

(2002), and Halinan and Wiles (2004) have reviewed some of the approaches taken for module detection and have done research on quantifying modularity in graphs. Hallinan developed a criterion called the *iterative vector diffusion algorithm* (IVDA) and applied the IVDA to randomly generated networks, social networks derived from the internet relay chat, and protein regulatory networks in *Saccharomyces cerevisiae* (brewer's yeast).

Hallinan (2004b) defined an average cluster coefficient **C** as a measure of the extent to which the neighbors of nodes are linked to each other in a graph:

$$\mathbf{C} = \left(\frac{1}{N}\right) \sum_{i=1}^{N} \frac{c_i}{n_i(n_i - 1)/2} \tag{4.1}$$

where
   N is the total number of nodes in the network
   c_i is the number of connections between neighbors of root node i
   n_i is the number of neighbors of root node i

(A neighbor node k of a root node i is connected to it by an i ↔ k edge.)

Hallinan also devised a modular coherence algorithm that is applied to the nodes of a previously identified module and measures the relative proportions of intra- and inter-module links (edges). It assigns a cluster coherence value, χ, given by

$$\chi = \frac{2k_i}{N(N-1)} - \left(\frac{1}{N}\right) \sum_{j=1}^{N} \frac{k_{ji}}{(k_{jo} + k_{ji})} \tag{4.2}$$

where
   N is the total number of nodes in the entire network
   k_i is the total number of edges between all nodes in the module
   k_{ji} is the number of edges between node j and other nodes within the module
   k_{jo} is the number of edges between node j and other nodes outside the module

Cluster coherence is a measure of the relative proportion of edges between nodes within and outside of a previously identified module. χ ranges from −1 (no coherence) to +1 (a fully connected, stand-alone module or subgraph). The first term in the cluster coherence equation is simply the proportion of possible links between the nodes of the graph. The summation term is the average proportion of edges per node that are internal to the module. Thus a highly connected module with few external edges will have a lower χ value than a highly connected node with many external edges. Note that the formal condition for a strongly connected component (SCC) (or module) in a directed graph is where there is a path from each node in the SCC to every other node in the SCC. In particular, this means paths in each direction: a path from node **p** to **q**, and also a path from **q** to **p**, generating simple feedback.

In applying this coherence measure to the yeast protein network, Hallinan found that it has a strongly hierarchical modular organization. On the other hand, when a computer-generated random network was analyzed, no significant level of modularity was found at any level, validating her analysis algorithm.

4.3 MODULARITY IN PHYSIOLOGICAL ORGAN SYSTEMS

4.3.1 INTRODUCTION

As we have seen, physiological systems can be studied at various levels of scale: intracellular (molecular biological) metabolic pathways, cells, segments, organ systems, and whole organisms. Physiological modularity is present at all levels of scale, ranging from intracellular metabolic pathways, to segments or organs, to developmental units, to entire organisms. In this section, we will examine the properties of physiological modularity at the organ level. The first organ we will examine for modularity is the mammalian pancreas.

4.3.2 MODULARITY IN ORGAN SYSTEMS

4.3.2.1 Pancreas

The pancreas is composed of two major types of tissues: (1) The *acinar cells* that secrete digestive enzymes directly into the pancreatic–duodenal duct, which empties into the duodenum through the sphincter of Oddi. The pancreatic juice is a mixture of proteolytic enzymes (trypsin, chymotyrpsin, carboxy-polypeptidase, and several elastases and nucleases); pancreatic amylase (an enzyme to digest carbohydrates); and pancreatic lipase, phospholipase, and cholesterol esterase (enzymes to digest fats). (2) Surrounded by the acinar cells, the *islets of Langerhans* contain three different endocrine islet cell types: α-, β-, and δ-cells. There are ca. 1–2 million islets, each about $300\,\mu m$ in diameter, organized around a bed of capillaries into which they secrete their endocrine products (Guyton 1991). The α-cells secrete the protein hormone, glucagon (GLC). The β-cells secrete insulin (I), and the δ-cells secrete the hormone, somatostatin (ST). About 60% of the islet cells are β-cells, 25% are α-cells, and 10% are δ-cells. The functions of these hormones are described in Section 6.4.2.

Note that Type I diabetes has an autoimmune pathology. For some reason, the human immune system (hIS) attacks and damages β-cells destroying their function, causing insufficient circulating insulin for normal glucoregulation. The patient must take exogenous insulin to correct the intrinsic insulin deficit. In spite of the damage to the β-cell module, the rest of the pancreas generally can perform normally, given exogenous insulin therapy.

4.3.2.2 Modularity in the hIS

It is tempting to consider the complex hIS as a three-module system. Module I is the biochemical innate IS (complement system [CS] and antimicrobial peptides [AMPs]). See Figure 4.1 for a summary of the linked reactions of the CS.

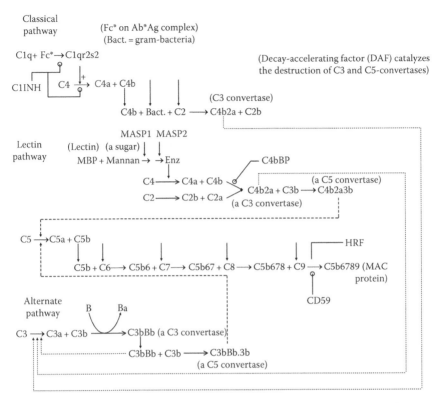

FIGURE 4.1   Schematic of the three complement system pathways. Parametric reaction inhibition is shown by small circles, and activation by arrows. The MAC protein (C5b6789) bores holes in cell membranes and kills them.

There are over 30 interacting protein molecules (most of them soluble enzymes and reactants) in the CS whose main function is to attack and destroy invading pathogens (e.g., viruses, bacteria, fungi, and parasites). The CS acts in a complicated, concatenated cascade of reactions to provide humoral defense against these pathogens. There are three pathways by which the CS can be activated, and there is active inhibition of the membrane attack complex (MAC) (C5b6789) of the CS by CD59 protein. (We describe the CS in more detail in Section 6.4.5.) In addition to the CS, there are various AMPs that attack pathogens directly without the complicated steps found in the CS. AMPs are described in Sections 5.4 and 6.4.6.

Module II of the hIS is the innate, cell-mediated system in which hIS leukocytes known as *natural killer cells* (NK), *cytotoxic T-cells* (CTL), *dendritic cells* (DCs), *macrophages* (Mϕ), and *neutrophils* (Np) directly attack pathogens, or somatic cells bearing external molecular evidence of internal pathogen infection. This module is connected to the adaptive immune system by the mechanism of antigen presentation (AP).

Module III of the hIS is called the adaptive immune system (AIS). This module creates antibodies (Abs) for specific pathogen molecules and has a "molecular memory" that "remembers" the specific Ab used to fight a pathogen in the past. Note that Abs can activate the complement system (Module I) that can also join in the battle against the pathogens. The function of the AIS involves the process known as *antigen presentation* of specific molecular epitopes (fragments) from pathogens by infected somatic cells, B-cells, macrophages, neutrophils, and dendritic cells to helper T-cells (Th), which then stimulate clonal expansion of the B-cells that make the specific Abs that fight the pathogens. (See Sections 5.5.2, 6.4.3, and 6.4.4.) Through a molecular Monte Carlo system, each Th cell is provided with a surface T-cell receptor (TCR) molecule combined with one of hundreds of thousands of different paratope molecules that may have an affinity to the antigenic epitope molecular fragments being presented. The epitope is generally a part of the antigen (Ag) molecule. A molecular (logical) triple AND operation must occur in order for AP to occur: The TCR must dock with the major histocompatibility complex (MHC) molecule on the AP cell, and the Th's paratope molecule must have an affinity to the presented Ag epitope, and lastly, the Th's surface CD4 (or CD8) protein must dock with a receptor on the AP cell's MHC complex. We close by remarking that all three hIS modules operate cooperatively, that is, they send C^3 signals back and forth.

4.3.2.3 Modularity and Plasticity in the hCNS

Still another complex physiological system that is clearly organized into both anatomical and functional modules is the human brain. Turner (2007) makes a clear case for this modularity in his interesting book, *The Tinkerer's Accomplice*. As it turns out, the boundaries between these functional modules are fuzzy (Zadeh et al. 1996), even plastic. The brain has the ability to change its functional organization within limits; it is a complex adaptive system (CAS).

An important instrument used for quantifying the brain's functional modularity is *functional magnetic resonance imaging* (fMRI) (Cohen and Bookheimer 1994). fMRI allows tomographic reconstruction of thin slices of a living patient's brain while the patient is performing cognitive tasks or certain types of mentation. Blood oxygen level-dependent (BOLD) fMRI senses local variations in hemoglobin oxygen saturation (HbO/ΣHb) in capillaries carrying blood to metabolically active regions of neural tissue. In a volume of brain tissue where neurons are more active (firing), the CNS's hemodynamic response causes the local blood volume flow to increase in capillaries intimate with the active neural tissue. The brain, in fact generally overcompensates, bringing in more oxyhemoglobin (HbO) than is converted to deoxyhemoglobin (HbR) by loss of its O_2 to the metabolically active neural tissue. Thus, the HbO content of venous blood (paradoxically) increases during brain modular activation. The fMRI senses the differential magnetic susceptibility between HbO and HbR. Hb is paramagnetic when HbO, but diamagnetic when in the HbR form. (Paramagnetic materials have stronger induced magnetic fields than do diamagnetic substances; hence, the fMRI signal is stronger from neuronally active brain regions.)

The temporal dynamics of the BOLD fMRI response to brain O_2 utilization are relatively slow compared to neural dynamics. The buildup of excess HbO has a lag of 1–5 s, rises to a peak over 4–5 s, and then falls back to the baseline in ca. 4 s upon cessation of activity. Spatial resolution of fMRI is blurred because of the random distribution of capillaries in neural tissue. BOLD fMRI voxels are about 3 mm on a side; however, resolution of standard MRI of the brain can approach voxels of 0.1 mm/side.

Volumes of the brain tissue involved in a specific task appear denser in BOLD fMRI scans because of increased, localized HbO in capillaries. BOLD fMRI has been used to localize the area of the brain involved with facial recognition, and can plot in real time how activity moves around in the brain's volume as a subject thinks about typing a sentence, plans the finger movements, executes them, feels them on the keyboard, then sees the typed words and verifies them. Clearly, the brain's modules can communicate neuronally within it, but their functions are localized.

Other examples of application of BOLD fMRI studies include, but are not limited to, the following: (1) Gaillard et al. (2000) studied the localization of brain activity associated with the development of verbal fluency in children and adults. (2) Büchel et al. (1998) examined the functional anatomy of attention to visual motion. (3) Schlosser et al. (1998) examined regions of the hCNS in a verbal fluency task. (4) Zarei et al. (2006) examined the functional anatomy of interhemispheric cortical connections in the hCNS.

The modularization of hCNS operations allows for evolutionary pressures to operate on the modules that need to adapt, while leaving other successful modules relatively unchanged. Think what an evolutionary breakthrough occurred when the human speech centers developed. This may have been over 2×10^5 ya. First came a module to enable spoken language (the coordination of breath, tongue, lips, and jaw producing word sounds instead of grunts, hoots, etc.); at the same time, a hCNS module must have emerged to process speech sounds (recognize and associate word sounds, i.e., vocabulary storage). Finally, written language (symbols = sounds) emerged ca. 3400–5000 yBCE (First 2010a,b). Appropriate modules in the brain were modified to link the modules for vocabulary, word association (a symbol lookup table) in speech, writing, and reading, and the hand–eye coordination needed for writing.

Just when we think we have seen firm evidence for modularity in hCNS function, along comes the observation of neuroplasticity or cortical remapping. fMRI activity associated with a particular brain function can move to a different brain location as the result of normal experience, or CNS injury (from trauma or stroke). Cortical remapping can be seen if a brain volume associated with a particular sensory input is abruptly deprived of that input. Some very important, yet-to-be-discovered, brain management system eventually allocates the input-deprived neural volume to another (usually adjacent) sensory input (Wall et al. 2002). This functional plasticity can occur in the lowest neocortical processing areas as well as the cortex, and these changes can significantly alter the patterns of neuronal activation in response to experience.

This behavior underscores the fact that the brain is indeed a CAS that can "learn" to change its "wiring," as well as its behavior. The ongoing studies in physiological psychology using BOLD fMRI are providing us with more evidence of CNS functional modules, their interconnections, and functional plasticity.

4.4 MODULARITY IN GENE REGULATION AND DEVELOPMENT

4.4.1 OPERONS IN PROKARYOTES AND EUKARYOTES

Some prokaryote genes needed for "housekeeping" (cellular homeostasis) are expressed constitutively, that is, they are always "turned on" to some degree, continuously making proteins required for metabolism, homeostasis, etc. Other genes are only expressed when the cell needs their products, so they are normally inactive ("turned off").

The *operon* is a form of modularized gene transcription regulator found in prokaryote cells (bacteria and archaea), as well as in some eukaryotes. An operon can contain 2 to more than 10 structural genes that code for 2 to over 10 specific proteins as a group. It has a common promoter region of DNA that binds to RNA polymerase (RNAp), an operator region to which a segment of a regulatory protein (RP) binds, and the regulatory elements needed to regulate expression of the structural genes. The most widely described operon in the literature is probably the *lac* operon of the bacterium *E. coli*, first reported by Jacob and Monod in 1961. Many operons have also been found in the genome of the eukaryote worm, *Caenorhabtitis elegans* (Blumenthal 2005, Blumenthal and Gleason 2003).

Now that so many bacterial genomes have been sequenced, it is possible to analyze them statistically for pairs of genes belonging to the same operon. Ermolaeva et al. (2001) analyzed 34 bacterial and archaeal genomes and found more than 7600 pairs of genes, each pair of which is highly likely ($P \geq .98$) to have its own operon. Similar analysis is being applied to eukaryotic genomes.

The genes of eukaryotes are mostly monocistronic, each with its own promoter at the 5′ end and a stop codon at the 3′ end. However, cases of polycistronic (operon) transcription have now been found in eukaryotes, including protists, nematode worms (e.g., *C. elegans*), and certain chordates (Blumenthal 2004, Price et al. 2006). Initial studies reported by Blumenthal (2005) and Blumenthal and Gleason (2003) on the number and types of operons in the *C. elegans* worm's genome have indicated that there are at least 1000 operons, 2–8 genes long. They stated that these operons contain ca. 15% of all *C. elegans* genes. Eukaryotic operons, like bacterial operons, generally cause the co-expression of functionally related proteins. *C. elegans* operons encode proteins involved in the basic mechanism of gene expression and energy metabolism. It is thought that *C. elegans* operons contain genes that are mostly regulated at the level of mRNA stability or translation (Blumenthal 2004).

Lawrence (2003) theorized that the high incidence of operons in *C. elegans* is due to the fact that the nematode has the ability to process polycistronic mRNAs

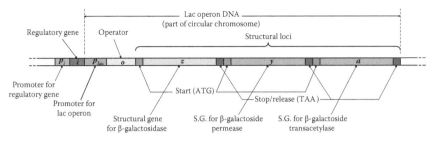

FIGURE 4.2   Schematic of the *lac* operon. See text for description. (From Northrop, R.B. and Connor, A.N., *Introduction to Molecular Biology, Genomics and Proteonomics for Biomedical Engineers*, CRC Press, 2009.)

so that ribosome reentry is not required for the translation of all the downstream genes in the operon: "The precursor mRNAs are trans-spliced onto special leader sequences that allow for independent translation of the separate processed mRNAs made from the original polycistronic message." Lawrence further observed that "Nematode operons differ notably from bacterial operons in that they do not contain all the genes necessary to confer for a single, selectable function." One might say that they are "quasi-modular," compared with bacterial operons.

The best way to introduce an operon is by example. Figure 4.2 illustrates a schematic of the well-studied *E. coli lac* operon. First in the DNA is the promoter region for the regulatory gene (p_i), then the regulatory gene (**i**) followed by the *lac* operon promoter sequence (p_{lac}), then the operator sequence (**o**), then a group of three structural genes, z, y, and a, coding certain enzymes for lactose metabolism. The regulatory gene p_i codes for a repressor protein that obstructs the *lac* promoter, hence inhibits transcription of the three structural genes (p_i is the master "OFF" switch; *lac* is normally OFF, i.e., not being expressed).

The structural genes are separated by initiation and termination signals (stop/start base sequences), and are expressed by ribosomes as a common group (from a single, triple gene section of mRNA) because they work synergistically as a biochemical module. The *lac z, y*, and a structural genes code for β-*galactosidase*, β-*galactoside permease*, and β-*galactoside transacetylase*, respectively.

Note that the *lac* genes are inducible; the presence of lactose sugar around and in a bacterium turns them on (de-represses them). Thus, lactose acts like a hormone, transcription factor, or signal substance. In the absence of lactose, the *lac* repressor protein (LRP) binds to the *E. coli lac* DNA near the *lacZ* gene promoter. This prevents RNA polymerase (RNAP) from binding, inhibiting the transcription process, thus saving energy and substrate molecular resources. When lactose is present inside the bacterium, it binds to the LRP, inducing a conformational change that causes the repressor to dissociate itself from the DNA, allowing transcription to proceed. Then a single mRNA for the concatenated z, y, and a structural gene exons is made and used by ribosomes to manufacture the three enzymes required for *E. coli* to metabolize lactose. The *lac* operon is an example of positive regulation (the inhibitor is itself inhibited).

But wait. There are actually two regulatory inputs to the *lac* operon. One, described above, is the presence of lactose molecules; the other uses the presence of glucose molecules. When glucose is abundant, even in the presence of lactose, it is not efficient for the cell to metabolize lactose (more energy is required than for glucose), so *lac* gene expression is repressed. (Only D-glucose drives the cell's metabolism.) Absence of the LRP is necessary but not sufficient for *lac* gene expression. RNAP activity and gene expression also depends on the presence of another DNA-binding protein called *catabolite-activating protein* (CAP) (also called cAMP receptor protein [CRP]). The binding of the cAMP–CRP complex to the *lac* operon stimulates RNAP activity ca. 20–50-fold. Similar to the LRP, CAP has two binding sites: one is to free cyclic AMP (cAMP) molecules, the other to a sequence of 16 base pairs upstream of the *lac* promoter. The presence of intracellular glucose (even with lactose) inhibits the activity of the enzyme adenyl cyclase, thus reducing the production and concentration of cAMP. But CAP can bind to DNA only when cAMP is bound to CAP, so when cAMP levels in the bacterium are low, CAP cannot bind to DNA and thus the RNAP cannot begin its work of making mRNA, and the operon is repressed (Kimball 2009j).

Catabolite repression is a positive regulatory mechanism while the repressor system is a negative regulator. In Boolean terms, the activation of the *lac* operon occurs for lactose and not glucose. It follows the PRO logic illustrated in Figure 4.3 where CRP=glucose and AP=lactose. Thus the combined action of the two regulatory mechanisms ensures that the three lactose metabolic enzymes will be produced only when lactose concentration is high and glucose concentration is low, but not when both lactose and glucose concentrations are both high.

van Hoek and Hogeweg (2006) developed a continuous, dynamic, mass-action-based computer model for the *lac* operon using 10 nonlinear ODEs applied to a population of a few hundred model *E. coli* cells. Each of the cells was programmed to respond to external lactose and glucose, to divide with possible mutation, to

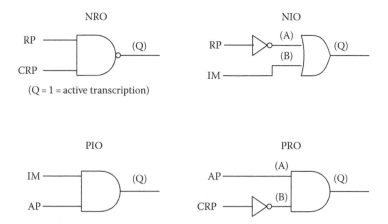

FIGURE 4.3 Logic gate representations of the four basic types of controls for gene expression. See text for description.

move in a random walk across a 25×25 grid "culture dish," and to "die"' in a density-dependent way or when its modeled energy ≤ 0. The in silico evolved *lac* operons were found to exhibit bistability in response to artificial inducers, but not for lactose. The authors argued that their modeling results suggest that the bistability observed in the in vivo *lac* operon is in fact an artifact of using artificial inducers and has not, in fact, evolved for lactose.

E. *coli* bacteria also need the AA tryptophan (Trp) for protein synthesis. They can extract it from the proteins the cell consumes, or if unavailable, they can manufacture it, de novo, using the *trp* operon products. The synthesis of the AA Trp from precursor molecules within the bacterium requires five enzymes, TrpA– TrpE. The genes encoding these five enzymes are grouped together on the E. *coli* DNA plasmid, along with its own repressor, promoter, and operator. When Trp is present within the cell, it shuts down the *trp* operon. Two molecules of Trp bind to identical sites on the two identical halves of the dimer Trp repressor molecule. This Trp binding enables the repressor to bind to the operator, stopping transcription. In the absence of Trp, the repressor dimer leaves its operator, and transcription of the five structural genes can begin.

The *trp* operon is an example of negative regulation. The end-product of the *trp* operon activation is the AA tryptophan. Tryptophan inhibits activation of the *trp* operon, regulating its own rate of synthesis. This is a classical example of a molecular negative feedback motif.

To generalize, regulation of operon gene expression can be either negative or positive, by induction or repression (i.e., four classes). Positive control of an operon is effected by the binding of an activator protein (AcP) to the operon's DNA at a site other than the operator. There are two classes of positively controlled operons as follows. (1) Positive inducible operons (PIOs): In PIOs, APs are normally unable to bind to the pertinent DNA. When an inducer molecule (IM) reacts with the AP, it undergoes a conformational change, permitting it to bind to the DNA and initiate transcription. (2) Positive repressible operons (PROs): The activator proteins are normally bound to the pertinent DNA segment. However, when a corepressor protein (CRP) binds to the AP, it is prevented from binding to the DNA, which stops activation of transcription.

In addition, there are two classes of negatively controlled operons as follows. (1) Negative inducible operons (NIOs): A regulatory repressor protein (RrP) is normally bound to the operator and prevents operon gene transcription. When an inducer molecule is present, it binds to the RrP and changes its conformation so that it is unable to bind to the operator; this allows the operon's genes to be expressed. (2) Negative repressible operons (NROs): In NROs transcription of the operon genes normally takes place continuously. Repressor proteins are produced by a regulator gene, but cannot bind to the operator in their normal configuration. However, the repressor protein can bind to corepressor molecules that change its configuration so that the RrP can bind to the operator site and block operon gene transcription.

Assuming the four classes of operon have all-or-nothing regulation, we can represent their regulation using simple Boolean operations of logic gates. Figure 4.3

illustrates these four simple logic circuits. In the PIO, operon transcription is determined by a simple AND operation between an activator protein (AcP) and an inducer molecule (IM). (AND gate HI Q output = transcription.) In the PRO, the control AND gate output is normally HI (transcription enabled). The AcP is present at one AND gate input (A), and the corepressor protein (CRP) acts through an inverter to the AND gate's other (B) input. When the corepressor is present, the inverter output (= B input to the AND gate) is LO, and transcription is turned off. In the NIO, the regulating logic is an OR gate. In the presence of a regulatory RrP, the A input to the OR gate from the inverter is 0, the input from the inducer is also 0, and the OR gate output is 0, hence transcription is blocked. If an inducer molecule (IM) is present, the B input to the OR is 1, its Q output is 1, and the operon's genes are expressed. Finally, consider the Boolean NRO: the transcription continues until a repressor protein binds to a corepressor molecule, stopping transcription. The logic for the NRO is a NAND gate with inputs RrP and CRP. When both RrP and CRP are present, the A and B inputs to the NAND gate are 1s, and its Q output is 0, turning off the operon's gene expression. Otherwise the NAND's Q output is 1, and the operon is transcribed. Be aware that operon expression can be graded, as well as all-or-none.

Why are there operons? How did they evolve? Theories abound; Price et al. (2006) noted that operon modules can, in theory, reduce stochastic differences between protein levels. We note further that by modularizing gene expression, the protein enzymes produced can be made in nearly constant (1:1:1: …) ratios, that is, the stoichiometry for the end reaction is ensured, making an efficient use of reaction substrates (no excesses, no shortfalls). Since all operon genes are expressed as a bundle, simpler genomic regulatory architecture is needed.

Price et al. observed that operons have a "life cycle." They point out that operon creation, modification, and destruction may be driven by selection on gene expression patterns.

4.4.2 *Homeobox* and *Hox* Genes

One of the more common (eukaryote) animal body plans is that of the bilaterally symmetrical tetrapod (four-legged) vertebrate, which includes all mammals, birds, reptiles, and amphibians. Some animal groups, such as the snakes, cetaceans, bats, and most birds do not outwardly appear to be four-legged, but they are modified tetrapods, with wings replacing forelegs, and flippers being derived from both pairs of legs (UCMP 2007). Other metazoan body plans include the bilaterally symmetrical, segmented, exoskeletal, hexapods (insects); the symmetrical, segmented, exoskeletal, octopods (spiders, ticks); the decapod crustaceans, and segmented worms. However, echinoderms have radial symmetry, and soft-bodied animals including molluscs have bilateral symmetry, although this may not be so obvious in molluscs with shells.

The genes that regulate development of the body plan in bilateral eukaryote embryogenesis are known as *Homeobox* genes. *Homeobox* genes were first discovered in the fruit fly, *Drosophila* sp. (McGinnis et al. 1984, Scott and Weiner

1984). A homeobox is a special DNA sequence found within a general class of selector genes that control development; it is only ca. 180 base pairs long (Bürglin 2005). (*Selector genes* are genes whose products regulate the expression of other genes, acting as "master switches" in programmed development.) Thus, the homeobox codes a protein with ca. 60 AAs, called a *homeodomain* that can bind to DNA and act as a transcription factor (TF) that can switch on cascades of other genes. Homeodomain proteins generally act in the promoter region of their specific target genes along with other TFs, forming a genomic, molecular "AND gate" with high specificity. *Homeobox* genes have been found in cnidarians, fungi, unicellular yeasts, and in plants, as well as in metazoans. They are ubiquitous.

Hox genes are an evolutionarily ancient subclass of homeodomain genes. They direct segment identity, that is, whether a segment of an embryo forms part of the head, thorax, or abdomen. It is clear from an examination of a dendrogram illustrating the evolution of *Hox* clusters, that the number of *Hox* genes has increased with organismic complexity (Lappin et al. 2006). For example, plants and sponges each have 1, hydroids 2, nematodes 5, *Drosophila* 8, *Amphioxus* 10, the pufferfish 31, mice 39, and humans 39. Oddly, the zebrafish has 52 *Hox* genes.

In vertebrate animals, *Hox* genes are found in modules or clusters on chromosomes. Humans have a total of 39 *Hox* genes organized in four modules: hHOXA1-7 and hHOXA9-13 are on chromosome 7; hHOXB1-9 and hHOXB13 are on chromosome 17; hHOXC4-6 and hHOXC8-13 are on chromosome 12; and hHOXD1, hHOXD3-4, and hHOXD8-13 are located together on chromosome 2. Mice also have 39 *Hox* genes: their HoxA cluster has HoxA1-7, HoxA9-11, and HoxA13; the mouse HoxB cluster contains HoxB1-9 and HoxB13; their C cluster contains HoxC4-6 and HoxC8-13; and their D cluster has HoxD1, HoxD3-4, and HoxD8-13. Mammalian *Hox* genes contain only two exons and a single intron that varies from less than 200 bps to several kilo-bps. The homeobox is always present in the second exon in *Hox* genes.

An amazing property of all vertebrate *Hox* genes is that they are spatiotemporally collinear, that is, they act in sequential anatomical zones of the developing embryo in the same order that they occur on a chromosome. As the embryo's body develops in an anterior-to-posterior direction, there is a sequential expression of its homeotic gene complex.

Kimball (2009i) observed that all the genes in the mammalian Hox clusters show some base sequence homology to each other (especially in their homeobox) but very strong sequence homology to the equivalent genes in *Drosophila*. Human HOXB7 differs from fly *antp* at only two AAs. He noted that when the mouse *mHoxB6* gene is inserted in *Drosophila*, it can substitute for *Antennapedia* (*antp*), and produce legs in place of antennae just as mutant *antp* genes do. Kimball observed further that this interspecies *Hox* gene substitution indicates clearly that "these selector genes have retained, through millions of years of evolution, their function of assigning particular positions in the embryo, … and: …the structures actually built depend on a different set of genes specific for a particular species." Another interspecies *Homeobox* gene swap described by Kimball is also amazing. Mice have a gene, *small eyes* (aka *Sey*, aka *Pax6*), that is similar in sequence to the *Drosophila eyeless* gene. The bp sequences of *Sey* and *eyeless* are so similar that the mouse gene can be substituted for *eyeless* in *Drosophila* with the same effect as *eyeless*.

So what we have is a modular, master control mechanism for embryonic development that must have appeared very early in evolutionary history. *Hox* genes typically switch on cascades of other genes, such as all the ones needed to make a leg or a thorax. This is similar to a computer program "calling" subroutines. *Homeobox* and *Hox* genes exhibit hierarchical functional modularity in their expression. *Hox* gene modularity leads to organ system modularity. For example, mouse *mHoxa3*, *mHoxb3*, and *mHoxd3* have been found to coordinate and direct the development of some of the nerves that control eyeball and facial muscle movements, directing the functional "wiring" of the correct parts of the hindbrain.

The saying that "ontogeny recapitulates phylogeny" really must have its basis in *Hox* genes. In mammals, for instance, all early embryos are segmented, for example, mice, rabbits, and horses. The segments (somites) of a human embryo develop into ribs, vertebrae, and back muscles under the guidance of sequentially expressed *Hox* genes. Severe morphological mutations occur when there is a defective (mutated) *Hox* gene. High doses of vitamin A (retinoic acid) act as a teratogen. In mouse embryos exposed to high levels of vitamin A, *Hox* genes *1–4* are expressed in groups of cells that usually do not express them, leading to severe skeletal malformations, that is, the carefully regulated sequential expression of *Hox* genes is altered by the teratogen.

"Normal" mutational changes of the homeotic complex can lead to conditions that permit a relatively rapid evolution of the body plan. Genomic and fossil evidence suggests that a distant, common ancestor of protostomes and deuterostomes was a prototypical metazoan called *Urbilateria* that had *Hox* genes, dorso-ventral polarity, anterio-posterior polarity, genes regulating the formation of elementary photoreceptors, a contractile blood vessel (acting as a single-chambered heart), and segments. It is hypothesized that all contemporary *Bilateria* evolved through modifications (additions, deletions) of *Hox* genes found in *Urbilateria*, which lived ca. 550 mya.

How could this evolution of the Hox system of developmental control occur? First, there could be variations in the number of homeotic genes by duplication or deletion. Second, there could be increases in the number of Hox complexes. Third, there could be mutations affecting the timing, position, or level of homeotic gene activation that may generate small adaptive changes (think Darwin's finch beaks, on the Galapagos). Fourth, there could be alterations in the regulatory interactions between Hox TF proteins and their targets caused by mutations of *Hox* gene coding sequences. It appears that small changes in TF modules can have large leverage in altering animal phenotype, more than a small mutation to a non-homeotic structural gene might have.

A great challenge in understanding the complex regulation of development is describing the mechanisms that turn *Hox* genes on and off at the appropriate times. *Hox* gene clusters are controlled by a global enhancer, which acts on the specific gene promoters. MicroRNA (miRNA) strands located in *Hox* clusters may inhibit the more anterior *Hox* genes post-transcriptionally (Lempradl and Ringrose 2008). Noncoding RNA has been found in *Hox* clusters; perhaps 231 varieties of noncoding RNAs are present in humans. One of these, HOTAIR, is transcribed from the hHOXC cluster and has been found to inhibit the late *hHOXD* genes by binding to polycomb-group proteins (Rinn et al. 2007).

4.4.3 *Dlx* Genes

The *Dlx* gene family (also known as *Distal-less* homeobox genes) encodes home-odomain proteins (TFs) required for vertebrate craniofacial and forebrain development, as well as aspects of hematopoiesis. The six human and mouse *Dlx* genes are primarily expressed in ectodermal derivatives, for example, the nervous system and the surface ectoderm (Panganiban and Rubenstein 2002). *Dlx* genes are found in all chordate phyla; there are six known *Dlx* genes in humans and mice, and eight in zebrafish (*Danio rerio*). The six human and mouse *Dlx* genes are transcribed in pairs, each of which is linked to a *Hox* gene cluster (see above). Specifically, *Dlx1* and *2* are linked to *HoxD*, *Dlx3* and *4* are linked to *HoxB*, and *Dlx5* and *6* are linked to *HoxA* (Panganiban and Rubenstein 2002). Strangely, *Dlx4, 7, 8,* and *9* are the same gene in vertebrates; they were apparently named before they were sequenced.

Dlx genes are required for the tangential migration of interneurons from the subpallium to the pallium during vertebrate brain development. They act to promote the outgrowth of axons and dendrites. Knockout mice lacking *Dlx1* exhibit epilepsy and histological evidence of loss of interneurons (Cobos et al. 2005). *Dlx2* has been associated with development in the prethalamus. *Dlx4* is associated with development of bone marrow (Shimamoto et al. 1997). Development of normal lower jaw morphology depends on expression of *Dlx5* and *6* (Depew et al. 2002). Using *Dlx5/6* knockout mice, Beverdam et al. (2002) found that the simultaneous inactivation of the mouse *Dlx5* and *6* genes resulted in severe malformations in the skull vault and base, and in all derivatives of the branchial arches. They found that, in utero, the lower jaws were "...gradually transformed into upper jaws depending on the gene dosage, resulting in a symmetrical snout in mutants lacking all four alleles. Their mandibular process gives rise to a structure, which is the mirror image of that derived from the maxillary portion..."

Clearly, there are complex interactions between the 39 *Hox*, 6 *Dlx*, and the 3 *Msx homeobox* genes that regulate cellular proliferation, differentiation, and apoptosis in the development of mammals. One reason for the study of the *Dlx* genes is because autism is associated with chromosomes 2q and 7q that contain *Dlx1* and *2* and *Dlx5* and *6*, respectively. It is known from experiments with *Dlx* knockout mice that *Dlx* genes control differentiation of a subset of GABA-ergic neurons of the basal ganglia and cerebral cortex (GABA = gamma amino butyric acid, a neurotransmitter) (Panganiban and Rubenstein 2002). The pieces of the development puzzle are slowly being assembled.

4.4.4 Msx Genes

Vertebrate muscle segment homeobox (*Msx*) genes code homeodomain protein
TFs that generally act as transcriptional repressors. Mice have three *mMsx* genes:
1, *2*, and *3*. According to Ramos and Robert (2005), "*mMsx1* and *mMsx2* are
expressed during embryogenesis, in overlapping patterns, at many sites of epi-
thelial-mesenchymal inductive interactions, such as limb and tooth buds, heart,
branchial arches and craniofacial processes, but also in the roof plate and adjacent
cells in the dorsal neural tube and neural crests." Ramos and Robert point out that
mMsx3 is expressed exclusively in the mouse dorsal neural tube.

In mice, *mMsx1* gene expression maintains cyclin D1 expression and thus prevents exit from the cell cycle, thus inhibiting terminal differentiation of progenitor (stem) cells (Hu et al. 2001).

In humans, the *hMsx2* product is implicated in determining the survival
and apoptosis of neural crest-derived cells required for normal craniofacial
development.

Regulation of development is very complex; we have just grazed the surface of
the many regulatory genes and the roles of their products. For an in-depth view of
gene regulation, I recommend the text by Latchman (2010).

4.5 NETWORK MOTIFS AND MODULARITY

4.5.1 Introduction

Network motifs (NMs) are a form of functional modularity seen in complex
genomic systems. NMs were first identified in *E. coli* bacterial metabolism, in
which they appear as patterns in the regulatory architecture in protein transcription networks. The same motifs have now been found in other bacteria, as well as
in eukaryotes (yeast, plants, and animals) (Alon 2007a,b). In the sections below,
we will describe several network motifs in the transcriptional regulation used by
E. coli (Shen-Orr et al. 2002).

4.5.2 Motifs in E. coli Transcription Regulation

Figure 4.4 illustrates schematically the eight types of three-node, feed-forward
loop (FFL) motif architecture described by Alon (2007b). (From graph theory,
these are actually feed-forward paths [FFPs], not loops. However, in deference to
established use, we will continue to refer to them as FFLs.) In the directed paths,
arrows represent excitatory directed edges and edges with balls are directed inhibitory edges. The protein **X** is a regulatory TF. Its concentration can act directly
on a gene for **Z**, and also on a gene for **Y**. The **Y** product acts directly on **Z**. The
actions can be excitatory or inhibitory; because there are three paths, and they
can be either excitatory or inhibitory on their targets, thus there are $2^3 = 8$ types of
three-node FFLs. Alon subdivided the eight FFL types into what he calls *coherent* and *incoherent* architectures. In a coherent FFL, the sign of the direct $X \rightarrow Z$
path is the same as the overall sign of the $X \rightarrow Y \rightarrow Z$ path. Incoherent FFLs have

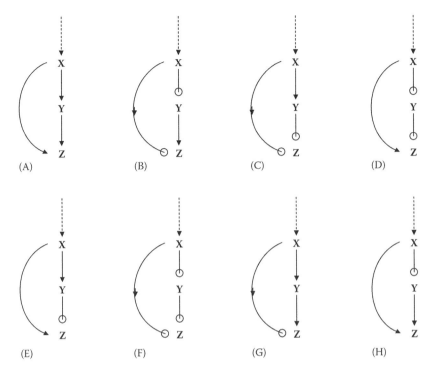

FIGURE 4.4   The eight types of feed-forward, regulatory motifs described by Alon (2007b). Top row: (A) coherent type 1, (B) coherent type 2, (C) coherent type 3, and (D) coherent type 4. Bottom row: (E) incoherent type 1, (F) incoherent type 2, (G) incoherent type 3, and (H) incoherent type 4. Arrows indicate activation and circles indicate repression in reactions.

opposite signs for the two paths. The sign of a path refers to whether its action is excitatory (+) or inhibitory (−).

Alon (2007b) extended his three-node, FFL motif architecture to include threshold logic and an AND gate. Figure 4.5A and B illustrate a coherent type-1 FFL (C1-FFL), and its relevant concentration waveforms. Alon described how the C1-FFL motif/module operates. Both **X** and **Y** are transcriptional activators. A signal molecule, S_X, activates **X**, which then binds to its downstream promoters, activating **Y** expression, hence **Y** accumulation. **Z** production starts only when the concentration of **Y** crosses the activation threshold, φ, for the **Z** promoter. This causes a delay in **Z** expression following the (step) appearance of S_X. (This delay is not a true transport lag, but one caused by the time for [**Y**] to reach φ.) When the signal, S_X, is removed, the **X** concentration [**X**] rapidly decays. As a result, **Z** production stops because the AND action requires both **Y** and **X** to be present at the **Z** promoter. Alon pointed out that there is no delay in the deactivation of the **Z** promoter when the S_X signal →0. He called this dynamic response *sign-sensitive delay*; there is a delay in the production of **Z** to an input step of S_X, but no delay for an OFF step of S_X.

Figure 4.6A illustrates the schematic of what Alon (2007b) calls an *incoherent type-1 feed-forward loop* (I1-FFL). To stimulate the product **Z**, the AND gate

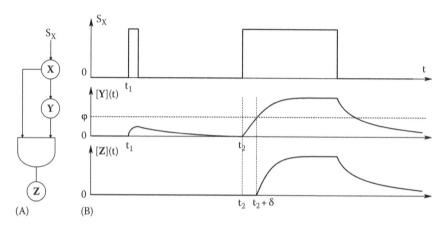

FIGURE 4.5 (A) Logic gate schematic of a coherent type-1 feed-forward loop (C1-FFL) (data from Alon 2007b). Note that the AND gate has a threshold φ for [**Y**]. (B) Waveforms in the C1-FFL gene regulator.

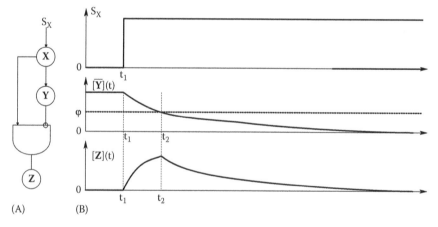

FIGURE 4.6 (A) Logic gate schematic of an incoherent type-1 feed-forward loop (I1-FFL) (data from Alon 2007b). Note that the AND gate has a threshold φ for [**Y**]. (B) Waveforms for the I1-FFL.

input must be **X** and not **Y** (**Y** inhibits **Z** production). The concentration waveforms in Figure 4.6B illustrate that the I1-FFL generates a "pulse" of **Z** when a step of S_X triggers a slow rise in **Y**. When [**Y**] reaches the inhibitory threshold, φ_z, for **Z**, **Z** expression is inhibited and the concentration of **Z** decays, forming an exponential "pulse" of **Z** in time. (The pulse decays to zero if the inhibition by **Y** is complete.)

Yet another motif architecture proposed by Alon (2007b) is shown in the logic schematic and plots of Figure 4.7. Alon calls this a C1-FFL gene regulatory motif. Control of **Z** expression is mediated by a molecular OR logic. The presence of

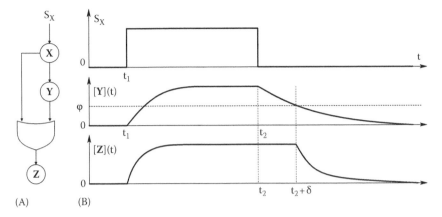

FIGURE 4.7 (A) Logic gate schematic of a C1-FFL gene regulatory motif. Note that the OR gate has a threshold φ for [**Y**]. (B) Waveforms for the C1-FFL gene regulatory motif.

X or **Y** molecules will signal **Z** production. This motif architecture is associated with the motor genes of the flagella system of *E. coli*. The waveforms shown in Figure 4.7B illustrate the waveforms of the C1-FFL-OR motif, and the threshold logic associated with the molecular OR process signaling **Z** expression. This system shows a delay in [**Z**] decay after signal removal, but not at onset. When the signal S_X appears, **X** alone through the OR logic, instantly activates expression of the **Z** flagella motor operon. When $S_X \to 0$, activation of the **Z** system and flagella expression persists because of stored **Y**, and then dies out gradually as **Y** $\to 0$.

A single input (SIM) motif is illustrated in Figure 4.8A and B. (This is a single-input, multiple-output (SIMO) module.) Here, a single **X** molecule activates a number of genes in parallel. Alon makes the point that given a monotonic increase in **X**, the various Z_k genes have different activation thresholds, φ_{zk}, and hence the various Z_k concentrations begin to rise with different delays relative to the beginning of the rise in [**X**]. Fall in [Z_k] is also delayed because of thresholds acting on [**X**].

Yet another regulatory motif described in *E. coli* is called the *dense overlapping regulon* (DOR) (Shen-Orr et al. 2002, Alon 2007b). Figure 4.9 illustrates this MIMO system architecture, also found in yeast. Many other variations of gene transcription network motifs (modules) are described in Alon (2007b).

In the emerging field of synthetic biology, An and Chin (2009) describe the use of two layers of effective biochemical AND-gate logic (Alon 2007a) in the design of orthogonal ribosomes (ORs). This network architecture is used in the biomolecular engineering design of ORs used to make novel, bioengineered proteins. Wang et al. (2006) commented on the methodology required to introduce an unnatural AA into a desired site in an "engineered" protein: "... one requires a unique tRNA-codon pair, a corresponding aminoacyl-tRNA synthetase, and significant intracellular levels of the unnatural amino acid." They further observed that to make sure that the specified unnatural AA is incorporated in the engineered polypeptide at the site specified by its codon, "...the

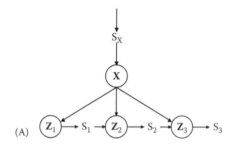

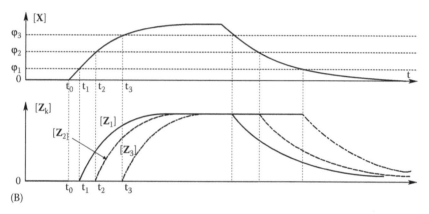

FIGURE 4.8 (A) Schematic of a single-input motif (SIM). Different thresholds exist for the $\mathbf{Z}_k$ functions. (B) Waveforms for the SIM.

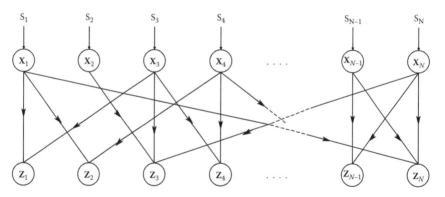

FIGURE 4.9 Schematic of a dense overlapping regulon (DOR) found in *E. coli.*

tRNA must be constructed such that it is not recognized by the endogenous aminoacyl-tRNA synthetases of the host, but functions efficiently in translation (an orthogonal tRNA)."

For further reading on synthetic biology, *o*-tRNA and ORs, the reader is recommended to read the review papers Lu et al. (2009), Mukherji and van

Oudenaarden (2009), Purnick and Weiss (2009), Win et al. (2009), de Lorenzo and Danchin (2008), Wang et al. (2007), Cropp and Chin (2006), Heinemann and Panke (2006), Chin (2006), Rackham and Chin (2005a), and Rackham and Chin (2005b). Workers in the field of synthetic biology are truly biochemical engineers; they are tackling the design and regulation of novel genomic networks head-on. Their molecular networks work around and alongside the complex, natural, genomic C^3 systems in cells.

The trend started by Alon (2007a) using binary logic gates to describe genomic regulatory motifs has gathered momentum. Although formal binary logic gates are based on inputs and outputs with just two states (0, 1), existing biological systems exhibit continuous values for their input/output functions (e.g., concentrations and percentage of receptors active). Viewed at large, however, a gene may be viewed as either being expressed or not. The regulatory molecules that determine the gene's state include TFs that act on TF sites as either repressors or activators, and on each other. RNAP also figures in gene activation. Silva-Rocha and de Lorenzo (2008) have demonstrated theoretically how various Boolean operations can be related to the regulation of gene expression; these include amplification, NOT, AND, OR, NAND, and ANDN. (For example, the gene is expressed with AND logic when two activating TFs are present at the same time.) They also proposed hypothetical gene regulatory models using NOR, ORN, XOR, and XNOR operations with interacting TFs and RNAP. These hypothetical "binary" scenarios may eventually be bioengineered into synthetic biological systems as gene expression regulatory mechanisms.

4.5.3 ROLE OF MicroRNAs in REGULATORY MOTIFS

MicroRNAs (miRNAs) are endogenous, ca. 22 Nt RNAs that selectively bind to the 3′-noncoding region of specific mRNAs by base pairing, and thus suppress gene expression posttranscriptionally. miRNAs evidently interact functionally with various components of many cellular networks (Cui et al. 2006). Alon (2007b) stated, "...an I1-FFL in mammalian cells involves MYC as activator **X**, E2F1 as the target gene **Z**, and a miRNA in the role of the repressor **Y**. Diverse FFL motifs with miRNAs have been found in [the worm] *Caenorhabditis elegans*."

Because miRNAs can specifically and directly down-regulate protein expression, it is quite probable that they play an important role in the regulation of the strength and specificity of animal cellular signaling networks by directly regulating the concentrations of their component proteins at posttranscriptional and translational levels. Cui et al. stated that "The diversity and abundance of miRNA targets offer an enormous level of combinatorial possibilities and suggest that miRNAs and their targets appear to form a complex regulatory network intertwined with other cellular networks such as signal transduction networks."

(miRNAs are treated in more detail in section 6.2.8 of Northrop and Connor 2008.)

4.5.4 SUMMARY

In conclusion, we observe that nature has modularized genetic regulation in bacteria. Not only are certain bacterial genes expressed in small groups called operons, but also the regulatory architecture makes use of a small number of reaction sequences called motifs. Several motifs have evolved in *E. coli*. Motif architectures are topologically simple, but can exhibit complicated dynamic response patterns (e.g., the FFLs). Motifs can be modeled by using threshold functions (such as Hill power functions, Alon 2007a) in their graphs, and AND, ANDnot, NOR, and OR functions to represent the action of certain TFs on end-products. Workers in synthetic biology are developing engineered motifs to regulate designer genes that lead to orthogonal mRNA, *o*-ribosomes, and ultimately custom proteins.

There is recent evidence (Cui et al. 2006) that microRNAs have an important role in down-regulating eukaryote genomic networks (metabolic, development, growth).

4.6 ROBUSTNESS, DEGENERACY, PLEIOTROPY, AND REDUNDANCY

4.6.1 INTRODUCTION

Robustness (adj.) is a lot like complexity in that there are many context- and system-dependent definitions of robustness. One can find robustness in various systems, for example, cellular homeostasis, organismic homeostasis, physiological homeostasis, computer software, aircraft avionics, and automobile braking. Engineers strive to incorporate robustness into their designs; nature has been doing it for eons. (For an overview of robustness, modularity, and redundancy, see the concept paper by Hammerstein et al. 2006.) Like complexity, robustness is also graded.

Jen (2003) called robustness a measure of "feature persistence" for systems. She continued, "...robustness is a measure of feature persistence in systems where the perturbations to be considered are not fluctuations in external inputs or internal system parameters, but instead represent changes in system composition, system topology, or in the fundamental assumptions regarding the environment in which the system operates." Here, we favor a broader view, in which a robust system not only can function normally in the face of changes in system internal composition, but it is also resistant to particular parametric changes on branch gains, small changes in system architecture, and to disturbance inputs. Feature persistence implies nominal, normal system functioning in the face of internal and external changes. In biological organisms, robustness generally refers to the organism's ability to survive to reproduce in the face of severe environmental stressors (in terms of temperature, starvation, loss of water, habitat damage, salinity changes, radiation, chemicals, etc.). In other words, the robust system has low sensitivities, as defined below, to these stressors.

Examples of robust biological systems can be found at all levels of scale, from biochemical to ecological. Such robustness may arise from the nonlinear, dynamic feedback networks between the elements, or reflect the properties of individual elements. For example, the expression of a certain metabolic functions may be robust in the face of temperature changes (i.e., have low temperature sensitivities [Q_{10}s]) because critical enzymes maintain their shapes and functionalities over a broad range of temperatures. Low Q_{10}s may also be due to an interconnected network of reactions that sustains the rate of supply of the product (output), even when some enzyme fails (parallel redundancy), or be due to functional degeneracy.

Note that cancer is a highly robust disease in which tumor cells grow rapidly, and develop their own blood supply (by angiogenesis). Cancers can spread by metastasizing. Many types of cancer cells have developed resistance to apoptosis therapeutically induced by radiation and chemotherapy agents. They are also robust against attack by certain cells of the hIS. In addition to resistance to programmed cell death (PCD), cancer cells often exhibit chromosome instability, which generates a high degree of genetic heterogeneity. This heterogeneity allows some few malignant cells to tolerate therapy and continue growing (Kitano 2004).

A genome may be robust because it encodes DNA proofreading and repair systems that reduce replication errors and errors from external sources (radiation, free radicals, etc.). An ecosystem can be called robust if it resists the extinction of some pivotal species, because the surviving species can compensate over physiological, demographic, or evolutionary timescales (Lenski et al. 2006).

A basic question that we will address is as follows: Does modularity in complex systems lead to their increased robustness over non-modular complex systems of equivalent size (numbers of nodes and edges, degree of nonlinearity)?

4.6.2 SENSITIVITIES

Sensitivities are a useful quantitative measure by which we can measure a CNLS's robustness. Engineers often evaluate robustness in the context of regulatory systems by using signal flow graph (SFG) models of CNLSs to simulate, measure, or calculate system sensitivities as a function of an edge parameter (gain) change. Mathematically, this type of *parameter sensitivity* is defined by

$$S_{P_j}^{Y_k} \equiv \frac{\text{Fractional change in the function (node parameter, or state)}, Y_k}{\text{Fractional change in the ith system parameter}, P_j}$$

$$= \lim_{\Delta P \to 0} \frac{\Delta Y_k / Y_k}{\Delta P_j / P_j} \to \frac{\partial Y_k}{\partial P_j} \frac{P_j}{Y_k} \qquad (4.3)$$

where all inputs and other graph parameters are held constant.

From this definition, an $S_{P_j}^{Y_k} \to 0$ is a quantitative illustration of robustness, that is, robustness can be viewed as the ability of a system to withstand perturbations in a gain parameter with a minimum change in output function, all other variables remaining constant. One type of complex adaptive system (CAS) can express robustness by dynamically "tuning" its parameters to maintain its effectiveness (minimize its sensitivity set) without changing its basic architecture.

Another way of looking at parameter sensitivity is to consider the changes in a node-to-node gain, $\mathbf{G}_{ki}$, due to a parameter change in the system. The *gain sensitivity*, $\mathbf{S}_{P_j}^{G_{ki}}$, can be written as

$$\mathbf{S}_{P_j}^{G_{ki}} \equiv \frac{\text{Fractional change in the gain from node } \mathbf{i} \text{ to node } \mathbf{k}, \mathbf{G}_{ki}}{\text{Fractional change in the parameter}, \mathbf{P}_j}$$

$$= \lim_{\Delta P \to 0} \frac{\Delta G_{ki}/G_{ki}}{\Delta P_j/P_j} \to \frac{\partial G_{ki}}{\partial P_j} \frac{P_j}{G_{ki}} \tag{4.4}$$

Robustness in the context of a regulatory system can also be expressed as a cross-parameter sensitivity, $\mathbf{S}_{xi}^{y_k}$, where $\mathbf{y}_k$ is the desired (**k**th) output, and $\mathbf{x}_i$ is the *i*th input ($i \neq k$), a disturbance. Ideally, a robust system should not respond to disturbance inputs, $\{\mathbf{x}_i\}$, so $\mathbf{S}_{xi}^{y_k} \to 0$ ($i \neq k$). The *cross-parameter sensitivity* (CPS) is

$$\mathbf{S}_{xi}^{y_k} = \frac{\partial \mathbf{y}_k}{\partial \mathbf{x}_i}\left(\frac{\mathbf{x}_i}{\mathbf{y}_k}\right), \quad i \neq k \tag{4.5}$$

This CPS can be approximated as

$$\mathbf{S}_{xi}^{y_k} = \frac{\Delta \mathbf{y}_k}{\Delta \mathbf{x}_i}\left(\frac{\mathbf{x}_i}{\mathbf{y}_k}\right)(\text{at an operating point}) \tag{4.6}$$

4.6.3 Noise Sensitivity

A well-engineered or designed robust system must be fairly immune to internal noise, in particular, random fluctuations in edge gains around a mean value, $\underline{G}_{jk}$. Assume this gain varies in magnitude described by a probability function, $p(G_{jk})$, having a mean $\underline{G}_{jk}$ and a standard deviation, $\sigma_{G_{jk}}$. Also assume this noise has a one-sided power density spectrum (Northrop 2004), $S_{G_{jk}}(f)$, that characterizes its behavior in the frequency domain. We define the *noise sensitivity* (NS) of the *i*th system node parameter to noise in the j–*k*th edge gain as

$$\mathbf{NS}_i \equiv \frac{\langle \mathbf{x}_i^2 \rangle / \overline{\mathbf{x}_i^2}}{\langle \mathbf{G}_{jk}^2 \rangle / \underline{\mathbf{G}}_{jk}^2} \tag{4.7}$$

where
$\langle \mathbf{x}_i^2 \rangle$ is the mean squared variation of x_i around its mean, $\overline{x}_i$
$\langle \mathbf{G}_{jk}^2 \rangle$ is the MS variation of the j–kth edge gain around its mean, $\underline{G}_{jk}$

NS should be as small as possible in a robust system, that is, the variation in G_{jk} should not affect x_i. (One way to minimize **NS** is by selective, linear band-pass filtering (see Northrop 2003, section 8.4.3).

Output noise factor (Northrop 2004), given a noise input, n_i, at node i, is another well-known measure of system noise performance:

$$F_{ki} = \frac{\text{msSNR at ith system input}}{\text{msSNR at kth (output) node}} \tag{4.8}$$

In an ideal system, $F \rightarrow 1$. (By using signal averaging, it is possible to make $F < 1$.)

4.6.4 TEMPERATURE SENSITIVITY

Q_{10} (pronounced "Q-ten"), is a venerable measure of temperature sensitivity in physical–biochemical reaction systems. Broadly, Q_{10} is the fractional increase in the reaction rate ΔR of an isothermal biochemical or physiological process produced by a temperature rise of $\Delta T = +10°C$. It can also be described as the ratio of the velocity of a reaction at some temperature T_2 to that of the same reaction at $T_1 = T_2 - 10°C$. Mathematically, we can write

$$Q_{10} \equiv \left(\frac{R_2}{R_1}\right)^{[10/(T_2 - T_1)]} \tag{4.9}$$

where
  R_1 is the reaction rate at T_1
  R_2 is the reaction rate at $T_2 > T_1$

$Q_{10} \approx 2$ for most biochemical and inorganic reactions. At high temperatures, critical enzymes begin to denature, and R_2 decreases. Thus there is generally a temperature T_2 at which Q_{10} peaks.

Another, more general way of looking at the temperature sensitivity (S_T^P) of any parameter P is

$$S_T^R \cong \frac{\Delta P}{\Delta T}\left(\frac{T_1}{P_1}\right) \tag{4.10}$$

where $\Delta P = P_2 - P_1$, $\Delta T = T_2 - T_1$, $T_2 > T_1$, $P_1 = P$ at T_1, etc.

4.6.5 DEGENERACY, PLEIOTROPY, AND REDUNDANCY

The noun, *degeneracy*, is normally associated with the deterioration or a loss in quality. In biology; however, it has an entirely different range of meanings, especially when applied to the hIS, genomics, and other molecular biological systems. For example, in immunology, degeneracy refers to the ability of one recognition structure (a molecular immune receptor) to bind to a number of different ligands. (Do not confuse degeneracy with pleiotropy, where a single intercellular signaling molecule can evoke different physiological effects on different kinds of cells that have receptors that bind to that signaling molecule, or to different epitopes on that one molecule.)

One reason robustness occurs in biological systems is because all life employs degeneracy and redundancy (as well as modularity) in their designs. *Degeneracy is a ubiquitous condition in biological systems at all levels of scale* (Edelman and Gally 2001, Whitacre and Bender 2009). We define degeneracy in a physiological context broadly as the ability of elements that are structurally different in a system to perform the same functions, or yield the same outputs. ("Elements" can be biochemical pathways, cell organelles, cells, or organs.) For example, the DNA protein code is said to be degenerate because, in most cases, one amino acid can be specified by more than one three-base codon (Northrop and Connor 2008). Figure 4.10 illustrates the protein coding "wheel"; 22 codon-coded amino

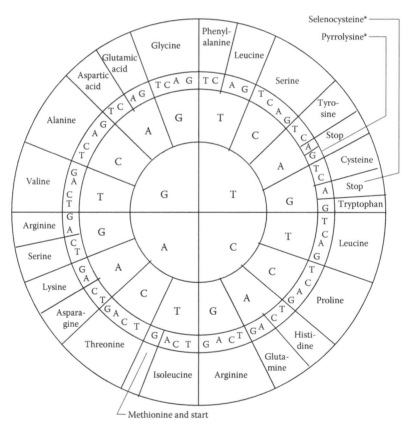

FIGURE 4.10 Amino acid coding wheel; three-base DNA codons specify which of 22 AAs will be assembled into a polypeptide/protein molecule. The AAs pyrrolysine and selenocysteine are found in bacteria.

acids (AAs) are shown (Proteinogenic 2010). The AA selenocysteine is found in some bacterial and eukaryotic proteins, and pyrrolysine is found in bacteria and archaea. The **TGA** codon (normally a stop codon) is made to code for selenocysteine when a gene's mRNA also contains an SECIS element (**SE**leno**C**ysteine **I**nsertion **S**equence). The AA pyrrolysine is found in methanogenic archaea; it is coded by the "amber" DNA stop codon, **TAG**, in the presence of the *pylT* gene that codes an unusual tRNA with a **CUA** anticodon, and the *pylS* gene that encodes a class II aminoacyl-tRNA synthetase that attaches the *pylT*-derived tRNA to the pyrrolysine AA. Pyrrolysine is incorporated into enzymes that are part of the methane-producing metabolism. (The only other AAs, tryptophan and methionine/start, are coded by single DNA codons-**TGG** and **ATG**, respectively.) Thus there are 22 AAs coded by codons, but only 20 generally appear in vertebrate proteins.

Sercarz and Maverakis (2004) cited a number of examples of degeneracy at different levels of biological organization: the genetic code, protein folding, units of transcription, gene regulatory sequences, gene control elements, post-transcriptional processing, metabolic pathways, nutrition from different diets, intracellular localization of cell products, intra- and extracellular signaling, immune responses, connectivity in neural networks, sensory modalities, and inter-animal communications. (These examples were attributed to Edelman and Galley 2001.)

Using computer models, Krakauer et al. (2002) explored… "adaptive theories for the diversity of protein translation based on the genetic code viewed as a primitive immune system… degeneracy and redundancy in the translational machinery of the genetic code has been explored in relation to infection with parasites utilizing host protein synthesis pathways". They speculated that a Red Queen scenario exists where mechanisms that facilitate epigenetic modification of anticodons that modify codon–anticodon binding specificity are likely to have evolved to bring the rate of host evolution in line with the rapid evolution of parasites.

A physiological example of degenerate regulation involves cardiac output (CO): Inherent in cardiac muscle is the Frank–Starling mechanism in which the heart pumps harder in response to ventricles stretched by increased filling blood volume. CO output can also be increased by circulating epinephrine released by the adrenal glands as also by neural stimulation by sympathetic nerves innervating heart muscle (Guyton 1991).

Mammalian glucoregulation is also a degenerate process. Glucose, required for ATP synthesis, can be obtained through diet, for example, through direct input of sugars and hydrolysis of starches. Glucose can also be generated internally by the process of gluconeogenesis, regulated principally by the pancreatic hormone glucagon. Eighteen of the 20 essential AAs can be deaminated and biochemically converted to glucose in the gluconeogenesis process. For example, deaminated alanine is pyruvic acid, which can be converted to glucose or glycogen. Or it can be converted into acetyl-CoA, which then can be polymerized into fatty acids. These AAs can be obtained by breaking down proteins (Guyton 1991). Fats can also be converted to glucose using other biochemical pathways.

In an interesting review paper, Leonardo (2005) considered in detail the evidence for degenerate coding of information in three neural systems: (1) the pyloric network (PN) of the lobster (controls muscle activity in the pyloric filter), (2) the song control system of the zebra finch (causes song), and (3) the olfactory system of the locust (governs feeding behavior). The lobster pyloric network is a ca. 14 neuron subset of the stomatogastric ganglion. The PN produces a triphasic spike pattern that controls the muscles of the lobster's pyloric filter. According to Leonardo, birdsong is a learned behavior and is generated by a set of brain nuclei. The song is learned by juveniles and is sung throughout their lives. Auditory feedback is required both to learn the song and to maintain it. The locust olfactory system is a classic example of a degenerate sensory system. Only a few types of receptors are used to resolve many odorant molecules. Data processing takes place in the locust's mushroom body, the output of which is the firing of Kenyon cells, which respond to specific, multicomponent odors (e.g., fermenting fruits). Leonardo offered the hypothesis that two of the degenerate systems he describes (the locust olfactory system and the finch song control system) use the neural equivalents of hash functions to gain their specificities.

The hIS and AP, long viewed from the perspective of the clonal selection theory (CST), is viewed by Cohen et al. (2004) to be a degenerate process. Immune system degeneracy is seen to be the capacity of any single antigen receptor molecule on a naive helper T-cell (Th) to bind and respond to ("recognize") many different ligands (epitopes), albeit with a limited range of affinities. According to Cohen et al., the main consequence of degeneracy in antigen receptors is *poly-recognition*, in which a single Th clone can bind to (recognize) different antigenic epitopes. Andrews and Timmis (2006), in considering Cohen et al.'s paper, remarked that "This causes a problem for the traditional clonal selection theory view of immunology that relies on the strict [narrow] specificity of lymphocyte clones."

Perhaps the perceived narrow specificity of the hIS to an invading pathogen is the result of some kind of cytokine postprocessing in which the naive Th cell that binds most strongly to the presented Ag epitope sends out cytokine signals that suppress the clonal expansion of Th cells having weaker (less specific) related responses. The AP cells probably also have a role in the dynamics of this hypothetical postprocessing.

Andrews and Timmis (2006) reported on their innovative computer modeling study of AP to degenerate naive Th cells in a lymph node. In one scenario, they used 10 naive Th, each with a random 8-bit "receptor molecule." Twenty different random 16-bit Ags were presented. Chemokines were included in their model to induce Th cell clustering. (A chemokine is an attractant to other Th cells in their model.) In the results presented, only one Ag activated two different Ths. All other Ags activated from three to six different Ths. Clearly, degeneracy and postprocessing were demonstrated. Another modeling study on degeneracy in the hIS was reported by Mendao et al. (2007). They modeled clonal selection of B-cells interacting with antigens, ignoring the role of Th cells. Their model is far too complicated to describe here; the interested reader should study their paper. The salient result of their simulations was that the time for Ag recognition (out of 45,360 Ags) was far shorter using a clonal selection model of B-cells with degenerate receptors than for Ag recognition requiring a perfect Ag–Ab match. In the degenerate B-cell model, two B-cells had to bind to the Ag for recognition; in the single-paratope B-cell scenario, only one B-cell had to bind to the target Ag. There were differences between the receptors on the two classes of B-cells modeled.

Hopefully, future in silico models of AP will make use of Th- and B-cell combined scenarios, as well as inhibitory cytokines as well as chemokines. Remember, the hIS is very complex, and AP and clonal expansion of B-cells and Abs is just one piece of the system.

Tononi et al. (1999), commenting on degeneracy stated, "The ability of natural selection to give rise to a large number of non-identical structures capable of producing similar functions appears to increase both the robustness of biological networks and their adaptability to unforeseen environments by providing them with a large repertoire of alternative functional interactions." ("Structures" can refer to biochemical pathways.) From the results of their stochastical mathematical modeling studies on small nodal systems (eight input and inter-nodes, plus four output nodes), Tononi et al. (1999) also commented, "The relationship between degeneracy and redundancy is therefore the following: to be degenerate, a system must have a certain degree of functional redundancy. However, a completely redundant system will not be degenerate, because the functional differences between distinct elements and thus their ability to contribute independently to a set of outputs will be lost."

Sussman (2007), and Edelman and Gally (2001) argued that degeneracy is a product of evolution, and that it enables evolution. Perhaps degeneracy is selected for in evolution because only life-forms that have a threshold amount of degeneracy are sufficiently adaptable (robust) to allow survival to reproduce as their environment changes.

Whitacre and Bender (2009) treat degeneracy and its relationship to the presence of biological robustness and evolvability. These authors concluded from their extensive model simulation experiments "…that purely redundant systems have remarkably low evolvability while degenerate, i.e., partially redundant systems tend to be orders of magnitude more evolvable." "This suggests that degeneracy, a ubiquitous characteristic in biological systems, may be an important enabler of natural evolution."

Redundancy is found in all living systems, at all levels of scale. It is clear that a proper measure of redundancy creates cellular and organismic robustness and fitness. Redundancy is not without cost, however. There is an energy (metabolic) cost in maintaining many identical, redundant systems. They also must be integrated into the organism's C^3 system.

On the macroscale, we have some multiple organs/features (tentacles, teeth), some paired organs (kidneys, lungs, eyes, arms, legs, ears), and at the organ level we have many copies of cells and modules having the same function(s) (e.g., hepatocytes, kidney collecting-duct cells, RBCs, pancreatic α-, β-, and δ-cells, T-lymphocytes, macrophages, dendritic cells, and mast cells). At the subcellular level, we have multiple, parallel, metabolic pathways (many mitochondria, ribosomes, copies of transmembrane proteins, etc.). Also, DNA occurs in complementary pairs of bases. (Special enzymes can repair damages to DNA by making use of dsDNA's complementarity.) This redundancy is required for robustness because cells and cell organelles have half-lives, that is, cells constantly turn over ribosomes, mitochondria, and transmembrane proteins and replace them with new ones. At any instant, some intracellular organelles are functioning, others are being destroyed, and others are being assembled.

One can speculate that in an intracellular environment, there might be some kind of chemical reciprocal inhibition (negative feedback) between redundant organelles and cells. Thus, the failure of a group of like cells or organelles (modules) removes inhibition on the functioning of the remainder cells or organelles, and their increased output replaces that lost by the failed modules, and perhaps stimulates their replacement by specific, adult stem cells.

Going up a level of scale, we see that in most organs, their cells have finite lives; they undergo programmed cell death (PCD = apoptosis) and are continually replaced by new like ones (e.g., skin cells, intestinal epithelia, erythrocytes, white cells, and olfactory neurons). Of course when cells are damaged (by physical injury or pathogens), they also can undergo PCD and in most cases, will be replaced by the body. Stem cells are involved in these replacements; there are multiple copies of stem cells, and they can reproduce themselves (the ultimate redundancy).

In conclusion, think of redundancy as insurance against component failure, either by PCD (apoptosis) or accidentally (by physical injury or infection). The cells in animal bodies are continually dying and being replaced, fortunately, asynchronously, or there would be no point in redundancy. Epithelial cells in the GI tract and respiratory system and in the integument are continually undergoing PCD. In a digraph, redundancy appears as parallel branches, all with the same input node and output node. Although we have only one heart, there are millions of parallel pathways for peripheral blood flow. There are billions of RBCs, platelets, and leukocytes having similar functions.

A point to ponder: Does Darwinian evolution (survival of the fittest) favor redundancy? Is there a trade-off between the cost of maintaining redundant systems and their size (e.g., the number of pancreatic β-cells)? Why do we have two eyes when spiders have eight eyes?

Pleiotropy is another important regulatory property of hormonal regulatory systems. Physiological pleiotropy is where a single hormone or cytokine can have different physiological effects on different kinds of cells having surface receptors for that hormone. The different effects can be either directed toward a common purpose or result or not. An example of pleiotropy is seen in the hormone, epinephrine. Epinephrine from the adrenal medullae causes increased cardiac output by general stimulation of heart muscle cells; it also generally increases cell metabolism in the body, by as much as 100% above normal, increases blood glucose concentration, and acts as a powerful bronchodilator (Guyton 1991).

A second example of a pleiotropic hormone is leptin, which is secreted by fat cells. Leptin affects different types of cells in different ways. For example, it suppresses insulin release by β-cells, it decreases the rate of production of the neurohormone, neuropeptide **Y** (NPY), and it downregulates lipogenesis. Behaving like a cytokine, leptin is also important to the normal functioning of the hIS. Faggioni et al. (2001) have shown that leptin plays a role in innate and adaptive immune responses. Leptin levels increase sharply during infection and inflammation. A leptin deficiency increases susceptibility to infections and inflammation, and is associated with dysregulation of cytokine production, and causes a defect in hematopoiesis. Elevated leptin affects T-cell phenotype, producing a higher Th1/Th2 cell ratio.

The rate of leptin release from white adipose tissue (WAT) (fat) cells is raised by insulin, NPY, triiodothyronine, epinephrine, cortisol, and the adipose tissue mass (Remesar et al. 1996). Raised leptin concentration in the pancreas, in turn, lowers or suppresses the rate of release of insulin by pancreatic β-cells (at a given plasma glucose concentration), forming a negative feedback loop. A lower insulin secretion rate means lower portal and plasma insulin concentrations than normal. Lower insulin concentrations mean less glucose enters the liver cells and other insulin-sensitive cells. Thus less glycogen is stored, and less fatty acids are made. Lower insulin concentration leads to increased secretion of the hormone glucagon. Higher glucagon means that more fatty acids are released from adipocytes into the blood. Elevated glucagon also inhibits the storage of triglycerides in the liver.

A third example of a pleiotropic hormone is the neuro-hormone, 5-*hydroxytryptamine* (5HT). 5HT has a number of effects in the gastrointestinal tract: It causes smooth muscle contraction in the intestines, esophagus, and fundus of the stomach. 5HT stored in platelets is released by an initial platelet aggregation, it then stimulates clot formation by causing accelerated platelet aggregation, and also causes vasoconstriction (these actions stop wound bleeding). In the CNS, 5HT acts through four different receptors and affects the sleep–wakefulness cycle, aggression and impulsivity, and anxiety and depression (Hardman and Limbird 1996).

The digraphs of Figure 4.11 illustrate the distinction between pleiotropy, degeneracy, and redundancy in biochemical/physiological systems using basic SFG notation. Branch transmissions are **a**, **b**, **c**, and the node variables are **w**, **x**, **y**, and **z**.

4.6.6 OPERONS AND ROBUSTNESS

As we have seen in Section 4.4.1, an *operon* is a module of two or more structural genes connected to a common promoter and operator region. Its expression is regulated as a unit to produce mRNA, leading to the synthesis of its coded proteins. Some operons are normally turned on, others are normally off.

Operons pose an interesting problem when considering genetic robustness. Clearly, the operon is an efficient way to manage biochemical regulation; all the

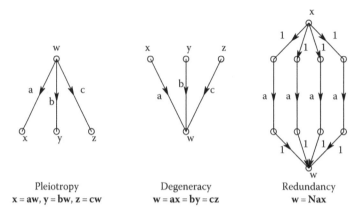

Pleiotropy	Degeneracy	Redundancy
$x = aw$, $y = bw$, $z = cw$	$w = ax = by = cz$	$w = Nax$

FIGURE 4.11 Three digraphs illustrating the differences between pleiotropy, degeneracy, and redundancy. A pleiotropic hormone can have different effects on different kinds of cells. In degeneracy, several different hormones (x, y, and z) can have the same effect on a cell. In redundancy, an input x can act through multiple pathways to produce the same effect.

protein enzymes required for a required metabolic operation are expressed at the same time, presumably with the correct stoichiometry. However, if the expression control mechanism of an operon is damaged, all of its gene products are disabled at the same time. It appears that nature has designed bacteria, trading-off robustness for efficiency and rapid growth. In eukaryotes, there are few operons; individual genes are generally regulated in a vast, complex, network.

4.7 SUMMARY

In this chapter, we have examined certain properties of CNLSs: modularity, redundancy, degeneracy, and pleiotropy, and shown how they may affect the robustness of the system, and if the CNLS is living, its fitness. We considered operons in prokaryotes and eukaryotes, *Homeobox* and *Hox* genes, *Dlx* genes, and *Msx* genes.

Network motifs and modularity were also explored. Sensitivities were introduced as a measure of a CNLS's robustness.

PROBLEMS

4.1 Consider the graph shown in Figure P4.1. This graph has two modules (M_1 and M_2), and a total of $N = 25$ nodes.
  (A) Calculate the Hallinan Average Cluster coefficient C for the graph. (See Equation 4.1.) (Take any indeterminate terms (0/0) as 0.)
  (B) Calculate the cluster coherence value, χ_1, for module M_1. (See Equation 4.2.)
  (C) Calculate the cluster coherence value, χ_2, for module M_2.

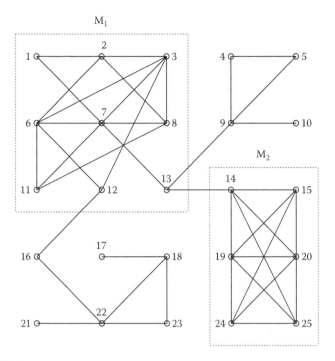

FIGURE P4.1

4.2 Make a list of the anatomical and functional modules found in the human brain.

4.3 In Section 4.6.5, we cited the hormones epinephrine, leptin, and 5-*hydroxy-tryptamine* as being pleiotropic. Give three more hormones, cytokines, or autacoids that are pleiotropic; list the systems each hormone affects, and the effects each hormone has on its several targets.

4.4 What human diseases with genetic etiologies have defects in *Hox* homeobox genes?

4.5 One proposed form of degeneracy in cell biology is the property of certain types of cell surface receptors to bind to and be signaled by two or more cytokines, hormones, autacoids, or epitopes that produce the same effect on the cell. (The effect may be graded.) List some cells that have degenerate receptors, the ligands, and how the cells respond.

4.6 Give examples of five pleiotropic hormones (cytokines, autacoids) (other than epinephrine, leptin, and 5-HT) that act on mammalian biochemical and physiological systems.

4.7 Nature has given us many important organs and organ systems that are paired, a design that can be viewed as a form of redundancy (e.g., kidneys, lungs, eyes, ears, testicles, ovaries, and adrenal glands). Why do we have only one heart, liver, pancreas, etc.?

4.8 MicroRNAs have been shown to have many roles in regulating gene products and in the incidence of certain diseases (cf. Chapter 7). Describe the regulation of miRNAs themselves.

4.9 Consider the cross-coupled linear system of Figure P4.9.

(A) Write the system's two-state ODEs.

(B) Find the system's dc gain sensitivity, $S_{a11}^{x1} = (\partial x_1/\partial a_{11})(a_{11}/x_1)$. (*Hint:* Substitute the gain $x_1/(u_1, u_2)$ for x_1.)

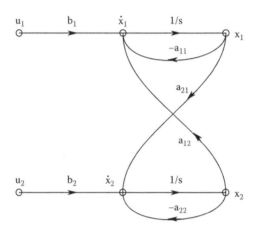

FIGURE P4.9

4.10 Use the system of Problem 4.9. Find the system's dc gain sensitivity: $S_{a12}^{x2} = (\partial x_2/\partial a_{12})(a_{12}/x_2)$

4.11 Discuss the relationship between apoptosis and redundancy.

4.12 Human growth hormone (hGH) is released by cells in the anterior pituitary gland.

(A) What are its functions in young and mature humans? Is it pleiotropic?

(B) Describe the regulation of hGH secretion. What other hormones are involved? What regulates their concentrations? Make a block diagram.

(C) It has been suggested that exogenous hGH may act as an anti-aging factor (anti-agathic) in elderly people.

What do you think the unexpected consequences of its use would be?

4.13 What do you think the effect will be on the fitness of surviving species in ecosystems in the eastern United States if, in the next 5 years, populations of bats were to be brought near extinction by White Nose Syndrome, and also populations of frogs and toads are brought near extinction by *Chytrid* fungus infections?

5 Evolution of Biological Complexity: Invertebrate Immune Systems

5.1 INTRODUCTION

In this chapter, we treat two apparently unrelated topics: evolution and invertebrate immune systems. They are, in fact, closely related. Invertebrate immunity evolved with the organisms; it was required to fight the Red Queen battles for survival of the fittest against other infective and/or competing species. One dimension of an invertebrate animal's fitness is seen to be a robust, competent immune system.

It is evident that living "higher" animals and plants have more complex functional structures and systems than do single-celled forms. Their biochemistry, metabolisms, cells, organs, organ systems, and overall physiology are all more complex. This increase of biological complexity evidently has accompanied the evolution of advanced life-forms over the millennia. Once a basic life "scaffold" was laid down billions of years ago (a DNA genome, ribosomes for protein synthesis, photosynthesis, glycolysis, and the Krebs cycle making ATP for energy used in biosynthesis), the way was open for biochemical evolution. A complex hierarchy of biochemical regulatory pathways evolved to govern reproduction, development, and adaptation to the organism's environmental niche, and to permit evolution to occur.

A major evolutionary leap forward was the appearance of eukaryotic cells, and then metazoan plants and animals appeared as coordinated associations of eukaryotic cells. The evolution of metazoan architectures required intercellular communication, hence the evolution of endocrine hormones, the electrotonic coupling of cells, and then a nerve net, which evolutionarily morphed into a CNS with motor control and sense organs. Sex evolved as a means of shuffling genes, seen as an aid to natural selection (NS).

The fossil record has demonstrated that these evolutionary trends toward increasing complexity and size of organisms have not occurred gradually, but in surges. The reason for the salient character of evolution is closely linked to abrupt changes in the habitat: factors such as temperature, oxygen, and CO_2 concentrations in the atmosphere; available photon energy (sunlight); and ocean levels, ocean temperature, and salinity all figure in NS. Events such as massive volcanic eruptions, and Earth strikes by asteroids and comets possess factors responsible for sudden climate changes, global glaciation, or hot periods.

Bonner (1988), in his insightful book, *The Evolution of Complexity*, commented, "By becoming multicellular, an organism can preserve all the advantages for efficient metabolism and proper gene distribution and at the same time become very large." Multicellularity requires efficient command, communication, and control (C^3) systems. It permits robustness through modularity and redundancy. Exoskeletons (found in Cnidaria, Mollusca, Bryozoa, Brachiopoda, Echinodermata, and Arthropoda) offer physical protection from predation and environmental factors, as well as expediting mobility on mobile forms. Endoskeletons (found in vertebrates and chordates) offer increased mobility and limb dexterity; the integument provides external protection. Turtles have endoskeletons that evolved to be partly exoskeletal as well. The integumentary system is continually renewed, and possesses the properties of self-repair and protection from microbial invasion by secreted antimicrobial proteins (AMPs). Both endo- and exoskeletal animals have been evolutionarily successful.

Yet, we still have with us bacteria (prokaryotes) that have evolved over billions of years. Their fitness has been tuned to survive in many niches. In evolution, they have evidently traded off complexity for reproductive speed and fitness in their niches. Once eukaryotes appeared and became metazoan, the path was open for complexity in evolution.

Mukherji and van Oudenaarden (2009) made the observation that the human genome contains about the same number of genes as the fruit fly, *Drosophila melanogaster*. They note that one way you get one organism to look like a fly and the other like a human using the same number of genes is through the evolution of a rich repertoire of nonprotein-coding sequences (coding for miRNAs, TFs, promoters, and other nonprotein regulatory molecules). They are of the opinion that organismic complexity "...arises from novel combinations of preexisting proteins, and the ability to evolve new phenotypes rests on the modularity of biological parts."

In the following sections, we examine evolution, speciation, complex physiological systems, and how they may have evolved to their present states.

5.2 DARWINIAN EVOLUTION: A BRIEF SUMMARY

5.2.1 INTRODUCTION

Darwinian theory is remarkable because of what it explained at the time; the fact that its precepts have withstood the criticisms of critics over time (both scientific and faith-based) and new evidence from the fossil record, geology, and from comparative genomics. Darwin's *On the Origin of Species* was first published in 1859. The two cornerstones of Darwinian theory are the theory of NS and the fact that evolution does in fact occur. NS provides an easy-to-understand mechanism for the evolution of plants and animals. We know now that the pressures of NS interact with phenotypes determined by genes and genetic variation, that is, the probability of survival of those individuals with more favorable genes will be higher, and their rate of reproduction will also be higher. This way, the more favorable genes will be passed on to their descendants. Conversely, genetic mutations leading to phenotypes that have low probability of survival to breed (low fitness) are gradually eliminated from their population. Such disadvantaged individuals can have difficulty in acquiring food, difficulty in finding a mate, competing with other species for food or habitat, coping with changes in their environment, etc.

In summary, evolution of a fit species requires three conditions: variation, selection, and replication. Variation is also seen as phenotypic plasticity, that is, the ability of an organism to express different phenotypes. An organism having a single genotype can alter its chemistry, physiology, development, morphology, or behavior in response to environmental changes. Such plasticity leads to fitness because it permits organisms to adjust to widely different, novel, environmental challenges within its own lifetime, and breed. There has been selective pressure for organisms to develop the ability to adapt, that is, "Over evolutionary time, the tools to generate innovation have been built into successful organisms" (Johnson 2009). Evolution also selects for those genomic regulatory motifs that generate the phenotypic plasticity required for fitness.

Of course, Darwin had no knowledge of heredity, genes, and genetics; this came later, beginning with the work of the monk Gregor Mendel, who, in 1866, was the first person to describe the rules governing the phenotypic variation of blossom colors with his experiments cross-breeding sweet peas (Field Museum 2007, Northrop and Connor 2008).

Bonner (1988) commented, "Therefore there is a great spectrum of levels in which one can consider the effects of NS: the level of the gene, the cell, the multicellular individual, the level of interaction between individuals of the same species, and finally between individuals of different species." Bonner goes on to note that the trends toward increased complexity and size are evident at all these levels, and they are undoubtedly evolutionary trends.

Some difficulty in reconciling Darwinian evolution and NS comes from the fossil record. Paleontologists using isotopic dating procedures have observed abrupt, not gradual changes, in the order of the types of fossil species discovered versus time. It is now suspected that these "quantum changes" in extinctions and evolution were caused by corresponding, sudden, adverse changes in the animals' ecosystems with which they could not cope. There is evidence that some of these changes were caused by collisions of asteroids or comets of considerable size with the Earth, causing sudden climate cooling and/or massive tsunamis. Other sudden environmental challenges to ancient life-forms can be attributed to massive volcanic eruptions that put cubic kilometers of volcanic ash and poison gasses into the atmosphere, also causing cooling. An example of such extreme volcanism was the "Siberian traps" eruptions ca. 251 mya. Not only ash, but large quantities of poisonous volcanic gasses such as CO_2, SO_2, and H_2S were emitted. Ecosystems were suddenly changed in terms of temperature, weather, pH, food supplies, etc.

The creatures best able to cope with these sudden coolings, warmings, droughts, floods, etc., were the ones that survived.

The evolution of the complex, adaptive CNS in primates was no doubt driven in part by the challenges posed by sudden ecological catastrophes. The ability to remember dangers, the location of edible foods, recognize individuals, and communicate this information between like-species contributed to primate survival, that is, there was evolution of interspecies communication capability. *Homo sapiens* use speech (and writing); cetaceans, birds, and insects use sound; and many animals use pheromones (odor molecules) to signal each other (e.g., mark territory—"stay out"; and ready to mate [estrus]).

Most evolutionary scenarios are viewed over billions, millions, or hundreds of thousand years. However, one sees rapid evolution today, albeit in bacteria and viruses. Bacteria reproduce very rapidly, and so NS is accelerated (compared to vertebrates) when they are exposed to toxic (to them) antibiotics or an adverse environment. Thus, we have recently seen the rapid rise of antibiotic-resistant strains of bacteria, such as methicillin-resistant *Staphococcus aureus* (MRSA). In perfect hindsight, this evolution has been driven, in part, by the excessive use of antibiotics to treat human diseases actually caused by viruses.

Evolutionary biologists are fond of drawing evolutionary "trees" versus time based on evidence from the fossil record, and from modern genomic analysis, showing the appearance (and disappearance) of species. All such plots demonstrate that, in general, species have developed more complicated bodies over time, that is, there has been an increase in more complicated and specialized cells, organs, and physiological systems.

Clearly, there was a jump in organismic complexity when single-celled eukaryotes learned to associate with others of their kind, presumably for survival advantage. Bonner (1988) listed three main forces that direct the evolution of multicellular organisms: (1) A group of cells can harvest energy (feed) more effectively than separate, single cells. Bonner cited the example of slime molds, which can break down food more easily by virtue of being in a large aggregation of cells. The group can secrete a large, focused quantity of concentrated extracellular digestive enzymes in order to process more food/cell than is possible by a single cell where diffusion dissipates the enzyme. (2) The second advantage is a more effective dispersal of progeny throughout the organism's niche. The specialized structures (spores, seeds, and eggs) require metazoan morphology. (3) Very large, multicellular organisms make predation possible on smaller animals. When there is ample food (energy) in an animal's ecosystem, size increase also offers the advantage of some protection from predation.

Another important advantage of multicellularity is that it provides functional redundancy, and hence a robust organism, and also the ability to form efficient, specialized organ systems (e.g., pancreas, heart, liver, muscles, and immune system). Multicellularity only became possible when cells developed the ability to adhere to one another, and more importantly, to communicate with each other locally and at a distance. Intercellular communication is generally chemical (hormones, cytokines, AI molecules), but also can be by nerves (propagating action potentials release neurotransmitters or hormones), or cell-to-cell signals can be purely electrical (electrotonic communication).

5.3 WHY DOES A RISE IN COMPLEXITY ACCOMPANY BIOLOGICAL EVOLUTION?

5.3.1 INTRODUCTION

Bonner (1988) argued from contemporary ecological evidence, as well as from the fossil record, that evolution produces larger and larger species in a given ecosystem (the size "arms race"). Considering representative samples of invertebrates, mammals, reptiles, birds, and amphibians, Bonner showed that one can construct a scatter diagram in which the $\log_{10}$ (density) of animals in their ecosystem is a linearly decreasing function of the logarithm of an animals' body mass. We have examined the linear approximation fit to the data in Bonner (1988, fig. 16), and this approximation can be modeled by the power law equation:

$$\rho_A = \frac{K_o}{M_B K_M} \tag{5.1}$$

Logging both sides of Equation 5.1, we get the log-linear relation of Equation 5.2:

$$\log_{10}(\rho_A) = \log_{10}(K_o) - K_M \log_{10}(M_B) \tag{5.2}$$

where
ρ_A is the density of a given species in its ecosystem in no./km^2
$K_o = \rho_A$ when the body mass $M_B = 1$ kg
K_M is the slope constant

From Bonner's fig. 16, we estimate $\log (\rho_A) \cong 1.95 = K_o$, when $M_B = 1$ kg. If we let $M_B = 10^{-4}$ kg, then $\log (M_B) = -4$, and $\rho_A \cong 4 \times 10^5$ and $\log (\rho_A) = 4.60$.
Thus, we can solve for the exponent $K_M = 0.663$, and there is a good fit for

$$\log_{10}(\rho_A) = 1.95 - 0.663 \log_{10}(M_B) \tag{5.3}$$

This exercise underscores the observation that the density of larger animals in a given ecosystem (e.g., blue whales in the Antarctic) is far less than the density of the smaller ones (e.g., krill). Bonner argued that the general relation between population density and body mass of Equation 5.2 "…is rigidly and explicitly laid down by energy considerations." The big critters eat the smaller ones, etc.

Metazoan animal size is determined by NS, which can operate on many levels of scale and function: (1) On the genes and their regulated expression, especially homeobox genes. There is a relatively rapid evolution associated with microRNAs that posttranscriptionally down-regulate gene products. NS also operates on transcription factors (TFs), molecules that directly regulate gene expression. (2) On the cells and their functions. (3) On the multicellular organism. (4) On interactions of an organism between individuals of the same species, individuals of different species, and physical factors in its environment. Animal behavior is ultimately important in selection. Animals have intrinsic, "hard-wired" behaviors built in to their CNSs, and as brains have evolved complex adaptive neural circuits, animals have developed learning to cope with selection pressures. Plasticity in behavior has also evolved.

Bonner (1988) illustrated that animal body mass $\propto$ (brain mass)^b, where $0.2 \le b \le 0.4$. Bonner goes on to state: "… that brain size and body size are not linked in any rigid way." Yes, memory and complex behavior require more neural circuits, but so does the control of a large body with many muscles and autonomic functions (e.g., the octopus). Marine mammals such as the porpoise have large brains because in addition to the factors cited above, porpoises must generate and process signals from their complex echolocation sonar systems. So which came first? Did the survival requirements of a competent echo location system drive the selection of brain circuits for size and function in signal processing, or did an already complex adaptive CNS enable the animal to develop its sophisticated echo location system? Perhaps they happened together.

Many vertebrates use external sensory modalities other than vision, hearing, touch, and basic chemosensing: (1) As we remarked above, cetaceans must generate and process underwater sound signals for echolocation. (2) Bats generate and process air-borne ultrasonic signals to communicate, navigate, and capture prey. (3) Pit vipers and rattlesnakes use long-wave, infrared photons that are passively emitted from small mammals to guide their prey capture. (4) Sharks, eels, and certain fish sense and process underwater electric fields. Indeed, certain fish generate time-modulated electric fields to guide their navigation and prey capture in murky waters by sensing E-field distortions (Northrop 2001). (5) Fish can sense low-frequency water-pressure waves with their lateral line organs. (6) Many terrestrial vertebrates have acute olfaction to sense pheromones (e.g., bears, canines, and moose). (7) There is evidence that certain vertebrates (certain birds, sea turtles, spiny lobsters, and migratory fish) have magnetosensing organs with which they can orient themselves in the Earth's magnetic field and use it to navigate (Northrop 2001). Clearly, as animals evolved novel sensory modalities, their complex adaptive CNSs had to simultaneously develop circuitry to process and associate the incoming information.

For a clue as to how vertebrates evolved sound production, audition, and CNS sound processing, we note that certain fish, such as the toadfish, generate a limited repertoire of underwater sounds (croaking, buzzing) for mating and territorial marking. Perhaps in the distant past, toadfish ancestors made accidental sounds using their air bladders. The ancestors of one sex that could "hear" and locate the noisy fish of the other sex could breed. Thus, the ability to hear and produce directed behavior was reinforced, as was the ability to make louder sounds. The important principle in this hypothetical scenario is that the sound did come first and was accidental. Sound detection sensitivity was enhanced by evolution. Sound production became purposeful only when the early toadfish were able to associate making sounds with mating and territorial defense. While this "chicken or egg first" argument may sound plausible, one can imagine other situations where an animal evolved acute hearing to locate and capture prey, and the mating calls evolved later.

5.3.2 SPECIATION

Currently, we remain largely ignorant about the fundamental processes governing the origin of species. Ongoing research on what triggers speciation, the genetic changes involved, and the role of the organism's environment and niche promises to shed light on the multiple causes of speciation. A new interdisciplinary focus on speciation called *ecological genetics* (Via 2002) should help our understanding of this important, complex issue. Via stated, "If ecological genetics is the study of the evolution of ecologically important characters, then the ecological genetics of speciation is the study of the evolution of characters that cause reproductive isolation or the ways in which natural selection on the phenotype produces postzygotic isolation in geographically separated populations." Via maintained that ecological genetics is an approach to the study of evolution that has the premise that the process of evolution is ongoing and can be studied in contemporary populations. Coyne and Orr (1998) gave a comprehensive review of the role of genetics in speciation. They stressed that the study of speciation has increasingly grown to be based on genetics/genomics, and that no comprehensive view of speciation is possible until we can understand the genetic mechanisms of reproductive isolation.

One definition of a species has been given by evolutionary biologist Ernst Mayr (Kimball 2009g): "A species is an actually or potentially interbreeding population that does not interbreed with other such populations when there is opportunity to do so." (Clearly, this definition implies that sex is involved.) All definitions of "species" assume that an organism gets all its genes from one or two parents, which are very much like that organism. However, in prokaryotes and some single-cell eukaryotes, horizontal gene transfer (HGT) can occur in lieu of sex. Another common definition of species is "If two organisms cannot reproduce to produce fertile offspring, then they are of different species." (This is a definition of exclusion, rather than inclusion.) When a horse mare (*Equus caballus*) is bred with a jackass (*Equus asinus*), their non-fertile offspring (the mule) is evidence that horses and donkeys are separate species. However, boundaries between species are often fuzzy. There are examples where members of population 1 can produce fertile offspring when mated with population 2, and members of the population 2 can produce fertile offspring with members of population 3, but members of populations 1 and 3 cannot produce fertile offspring (Wiki 2009b). Presumably, the reason for this conundrum lies in the genomics of the three population groups, since mating evidently occurred. It appears that examination of a putative species' DNA relative to other organisms in its genera is a modern, useful tool for determining speciation, taken with other "obvious" criteria of phenotype, behavior, habitat preference, and F1 fertility.

By way of review, the hierarchy for complete biological classification begins with Domain, which contains Kingdom, then Phylum, Class, Order, Family, Genus, and finally Species, and then Subspecies, if any. Defined (universally agreed-on) species are generally given two or three names: the genus is listed first, then the species, then the subspecies. Sometimes the subspecies is followed by the name of the person who discovered and first described the organism (e.g., the conch (mollusc): *Strombus marginatus septimus* Linnaeus). However, in botany, species can be followed by Variety, Subvariety, and Form.

Clearly, the understanding of the origin of a species is a complex problem; it draws evidence from the complex areas of paleobiology, genomics, ecology, and behavior. Most species have heretofore been defined on the basis of morphology. Other important criteria involve observing behavior (e.g., mate selection and niche preference) and genomic analysis (chromosome number, dsDNA sequencing).

5.3.3 ORIGINS OF SPECIES

We shall address two problems in this section: One is why different species do not (or cannot) interbreed, the other is how species arise. The answer to the interbreeding problem can be as simple as allopatric (other country) distribution of closely related species, that is, the various subspecies (races) of animals almost never occupy the same territory, hence seldom interact, court, mate, and breed. Also, in the case of birds, differences in songs (behavior) and morphology (plumage, beak size, and shape) tend to isolate a breeding population. Other specific attractants for breeding include courtship behavior and specific pheromones. Plumage, beaks, and pheromones must be determined genetically, songs and behavior may be the result of genetic information, or be acquired (learned) behavior, or a bit of both. In some cases, interbreeding is prevented genomically; there may be an alteration in the chromosome number between two species (Kimball 2009g).

In the case of plants, differences in the times of flowering of different species can prevent cross-pollination, and if insects are involved in pollination, there may be specificity favoring only one species' flower. Consider what could happen if two allopatric species were to suddenly be able to interbreed. Hybrids would be generated generally lacking the adaptive features of either species, and these would be able to survive poorly in either niche. As we observed above, when two distinct but related species breed (horse and donkey), the F1 (mule) is sterile and thus eliminated from competing in either species' gene pool.

A classic example of speciation is seen in Darwin's 13 species of finches found on the Galapagos Islands off the coast of Ecuador. These birds are held as a case study of how a single breeding pair of finches reaching these islands several million years ago could give rise to 13 species sharing the same habitat. Notably, the birds occupy 13 different niches on the islands, each with a different energy source (seeds, nuts, insects, etc.) and microclimate. The Galapagos islands are a scattered volcanic archipelago, where occasional volcanism has provided extreme selection pressure. Climate changes have also provided survival challenges, and the separation of the islands gives a degree of allopatric isolation. Kimball (2009g) pointed out that a 14th species of finch can be found on Cocos Island, 500 mi NE of the Galapagos. He asked why is only this one species found on Cocos Island?

The complex genetic mutations that led to the speciation of Darwin's finches can be traced by detailed DNA sequencing of their genomes. The lability of beak size in the finches may involve the regulation of the homeobox genes (e.g., *Dlx* genes) that in turn, regulate craniofacial development (see Section 4.4). These genes control how and how much of a structure is built in embryonic development.

Kimball (2009g) cited another example of speciation on an isolated island, Madeira, off the west coast of Morocco. Some six different species of house mice, *Mus musculus domesticus*, have been observed living allopatrically in isolated valleys. Each species has a diploid chromosome number less than the "normal" *Mus* (2N = 40). He states that the distinct and uniform karyotype of each species probably rose from genetic drift rather than from NS. The six different populations are technically described as races because there is no opportunity for them to interbreed. Kimball asserted that these races surely meet the definition of true species.

In addition to the allopatric speciation model, two examples of which we described above (finches and mice), there are three other speciation models that population biologists and ecologists consider. These are peripatric, parapatric, and sympatric speciation models. In the peripatric model, the root species enters a new niche attached to the original habitat. The niche becomes isolated, preventing any mutations in the niche from exchanging genes with the original root species. Genetic drift may contribute to the rise of the new species in its isolated niche. An example cited (Wiki/Speciation 2008) is the London Underground mosquito; it lives, feeds, and breeds there.

In the parapatric model for speciation, the root population invades two new niches and evolves behaviors and morphologies that adapt the two new populations to their environments. However, there is not complete isolation of the two new species; they may come in contact but they differ in that they cannot breed fertile offspring.

In sympatric speciation, species diverge while inhabiting the same, common environment. Two or more descendant species from a single root species all inhabit the same location. The existence of sympatric speciation is controversial. If the same environment or niche affects both putative species, what are the mechanisms for NS? A chance mutation could affect sexual selection (song, behavior, pheromones in an F1 generation), but what can keep them from breeding back to F0?

Two examples of artificial, "forced" speciation involve populations of fruit flies, *Drosophila melanogaster*. (*Drosophila* reproduces rapidly and its genome has been sequenced.) In one experiment, Rice and Salt (1988) created two new fly species by using mazes with different binary choices of habitat including light/dark, up/down, and ethanol/acetaldehyde vapor. They stated, "Thus the flies made three binary decisions that divided them into one of eight spatial habitats."

"Each generation was placed into the maze, and the groups of flies that came out of two of the eight exits [4L & 5E] were set apart to breed with each other in their respective groups." After 25 generations of disruptive selection, these authors were able to isolate 99% brown-eyed flies from habitat 5E, and 98% of the flies from habitat 4L were yellow eyed (eye color was used as a fixed genetic marker). "Because the flies mated locally in their selected habitats, two reproductively isolated populations using different spatiotemporal resources were produced." "By generation 25, the experimental flies derived from habitats 4L and 5E differed in all components of habitat preference (for all χ^2 tests, P<0.01)," that is, because of their strong habitat preferences, the two groups and their offspring were found to be reproductively isolated; they mated only within the areas they preferred, and so did not mate with flies that preferred the other six areas.

In another clever experiment, Dodd (1989) was able to demonstrate allopatric speciation in only eight generations using fruit flies. She began with a single species (F0), divided them in half, then fed the two groups either on a starch-based food, or a maltose (sugar) based food, and bred them. At the end of eight generations on their diets, the F8s showed mating preferences according to their diets; the starch-fed F8s only bred together, as did the F8s fed maltose. Others have replicated Dodd's experimental speciation with other species of fruit flies and foods (Kirkpatric and Ravigné 2002). How do you think this rapid speciation could have happened?

An example of rapid, natural speciation in crickets was described by Mendelson and Shaw (2005). *Laupala*, a forest-dwelling Hawaiian cricket achieved reproductive isolation by developing differences in the male courtship song. The closely related species of *Laupala* are morphologically identical, and differ only in the pulse rate of the male song. Females prefer the pulse rates of their own species. A speciation rate of 4.17 species per million years was estimated from a monophyletic clade of *Laupala* from Hawaii Island.

Speciation is complex because it not only involves the interaction of an animal with its environment and niche, but includes complex behaviors (mating, territory defense, food, mating calls, pheromones, etc.), and the possible induction and suppression of regulatory genes by environmental factors and diet. Speciation is expedited by (allopatric) isolation of a breeding population. External factors such as chemicals, UV, ionizing radiations, etc., that cause rapid, unrepaired genetic mutations can also lead to new species if the mutations are nonlethal and selected for.

Further complications in speciation are seen in the presence of operon-like transcription of functionally related genes. Operons were first described in bacteria, but studies have revealed that many eukaryote genomes carry them, as well. For example, they are abundant in the genomes of nematode worms, but are found less frequently in other eukaryotes (Ben-Shahar et al. 2007). Specifically, they have been found in the genomes of the nematode, *C. elegans*, the mouse, humans, *Drosophila*, yeasts, mosquitoes, *Plasmodium falciparum*, and *Guillardia theta* (Trachtulec 2007). In more complex organisms, operons generally consist of pairs of colocated genes. For example, Ben-Shahar et al.

discovered an operon in *Drosophila* consisting of a gene they called *lounge lizard* (*llz*) that encodes a *degenerin/ENaC* transmembrane positive ion (cation) channel, and another gene of unknown function, *CheB42a*. Both genes are transcribed from a single promoter as one primary transcript and are further processed to generate two, individual mRNAs. (The jury is still out on the function of *CheB42a*; perhaps it codes a chaperonin protein for llz protein.) Ben-Shahar et al. go on to state that their preliminary analysis suggests that the [fruit] "...fly genome contains numerous gene pairs that may share a chromosomal arrangement like that of *CheB42a/llz*."

We note that environmental factors (e.g., diet and cold) can affect promoters that switch operons off and on. How all these factors contribute to the models of forced speciation in fruit flies described above is a complex puzzle yet to be solved.

5.3.4 Speciation by Horizontal Gene Transfer and Transposons

Another interesting genomic mechanism that can generate new varieties of organisms, and perhaps even new species is HGT. HGT was first described in bacteria (Ochiai et al. 1959, Akiba et al. 1960), and now there are cases found in eukaryotes, as well. HGT occurs when an organism transfers its genetic material to organisms that are not its offspring. Known HGTs include DNA exchanges between different species of bacteria, protists, different plant families, animals, between bacteria and plants, and bacteria and insects (Citizendium 2008a). One way HGT must first have been expressed is extracellularly by eukaryotic cells; mitochondria may have been derived from ingested alpha-proteobacterial cells, and chloroplasts probably came from ingested cyanobacteria. Then eukaryote mitochondria gradually lost certain genes to the nucleus in animal cells, and some chloroplast genes now reside in the nucleus of plant cells.

There are three main mechanisms of HGT in bacteria and archaeans: (1) Bacterial conjugation, in which bacterial DNA plasmids are exchanged when bacteria temporarily fuse membranes. (2) Transduction, a process in which a bacteriophage virus moves sections of bacterial DNA from one bacterial host to another. (The F1 phages pick up DNA from the initial host, then infect the second F2 host.) (3) Transformation, which is the genetic alteration of a bacterial cell resulting from the direct introduction, uptake, and expression of extracellular genetic material (DNA or RNA). Transformation is used in bioengineering/biotechnology applications.

HGT is relatively common in prokaryotes that acquire mosaic genes. A mosaic gene is an allele that has acquired a new DNA sequence from a different species of bacteria through transformation and subsequent integration. The new, mosaic gene is composed of sequence polymorphisms identical to the original allele in some parts, but polymorphisms derived from the integrated DNA in other parts. Often the integrated sequence contains a genetic marker, such as antibiotic resistance. HGT transfer in bacteria makes the creation of phylogenetic trees difficult. If two distantly related bacteria are found to have the same gene, one supposes them to be closely related, when in fact, they are not, and the gene came from transduction or transformation.

HGT has also been noted in eukaryotes: In the protists, analysis of the genome sequence of *Entamoeba histolytica* shows 96 cases of HGT from prokaryotes. The protist *Trichomonas vaginalis* also has acquired a gene for a biosynthetic enzyme from bacteria related to *Pasturella* sp. genomic analysis of the protist, *Cryptosporidium parvum* shows that it has acquired 24 different genomic candidates from bacteria. The point is made that in unicellular grazing organisms, foreign DNA is constantly entering the cell from its bacterial diet, and evidently, some of these fragments can enter the grazer's genome (Citizendium 2008a)!

Some "junk DNA" (intron base sequences) may be evidence of heritable HGT in humans (Pace and Fescotte 2007). An example in the human genome of possible HGT are the mariner family transposons (*Homo sapiens mariner*), *Hsmar1* and *Hsmar2*. The mariner transposon family has been widely studied in both invertebrate and vertebrate animals. Demattei et al. (2000) cite the number of single terminal inverted repeats associated with mammal **mar1** elements as 900 in the mouse, 7000 in humans, and 10^4 in sheep and cows. Ongoing studies of mariner transposons in the human genome promises to enhance our understanding of how molecular rearrangements involving repeated DNA sequences in the human genome can cause diseases, but also reveal the manner in which the human genome itself continues to change.

Robertson et al. (1996) commented, "*SETMAR* is a new primate chimeric gene resulting from fusion of a *SET* histone methyltransferase gene to the transposase gene of *Hsmar1* mobile DNA. The transposase gene is thought to have been recruited as part of *SETMAR* 40–58 mya in an anthropoid ape, after the insertion of an *Hsmar1* transposon downstream of a preexisting *SET* gene" (Citizendium 2008c). Close relatives of the mariner transposons have been found in the genomes of mites, flatworms, hydras, insects, nematodes, humans, and other mammals.

Retroviruses (RNA viruses) and retrotransposons are other examples of mobile, horizontally transferred genetic material found in animals (Citizendium 2008a). In humans, heritable transposons can cause disease by inserting itself into a critical gene and disabling it. A transposon can leave a gene and the break may not be repaired correctly. Multiple copies of the same sequence, such as ALU, can hinder chromosomal pairing during mitosis, resulting in unequal crossovers, hence mutations. In addition, many transposons contain promoters that drive transcription of their own transposases. These promoters can cause aberrant expression of linked genes, causing disease or mutant phenotypes. Some diseases that can be caused by transposons include: Hemophilia A & B, severe combined immunodeficiency, porphyria, predisposition to cancer, and possibly Duchenne muscular dystrophy (Citizendium 2008b).

The evolution of transposons in organisms' genomes is the subject of current genomic research. They are found in all major groups of organisms on earth. How this came to pass may be because (1) transposons were present in the last universal common ancestor (cell), (2) they have arisen a number of times independently (a property of genome biochemistry), and (3) they can be spread by viruses or other mechanisms of HGT.

HGT and self-replicating transposons have the ability to alter genes (cause mutations that alter protein structures, or silence genes), leading to changes in organismic fitness that opens the door to NS, but only if the changes are heritable. In rare cases, speciation may occur, also there is the possibility that a new variety or race of animal could emerge. We have certainly seen that transposons can lead to genetic diseases.

5.4 EXAMPLES OF BIOLOGICAL COMPLEXITY IN INVERTEBRATE IMMUNE SYSTEMS

5.4.1 Introduction

We have remarked earlier that all physiological systems are complex, in particular, the hIS and the hCNS. In seeking to better understand and model the workings of the hIS, studies of simpler, invertebrate ISs have shown us some general principles: in invertebrate ISs, like the hIS, there is non-self-recognition; there are also cytophagic and cytotoxic cells (hemocytes), a great number of antimicrobial peptides (AMPs), and various immune mediators. Research in invertebrate ISs has been driven not only by our desire to better understand human immune function, but to learn how to protect commercially important, farmed invertebrates (such as shrimp, clams, oysters, abalones, mussels) from bacterial, viral, and fungal infections, and importantly, to discover new drugs to fight infections in humans. Because present invertebrates have evolutionary lineages far more ancient than humans (hundreds of millions of years), NS has literally had eons to perfect their ISs. Their ISs have fought and won any Red Queen contests with pathogens, or they wouldn't be here today with some degree of fitness. We have much to learn to benefit modern medical pharmacology by characterizing invertebrate antibiotic proteins.

In metazoan animals, innate immunity and adaptive immunity are the two forms of immune defense against infectious pathogens (e.g., viruses, bacteria, parasites, and fungi). Adaptive or acquired immunity, found in all jawed vertebrates, involves antigen presentation, antigen specificity, and immunological memory, while the innate immune responses also require a level of molecular pattern recognition of pathogen surface molecules; Abs and immune memory are not involved, however. All animals have innate ISs, while only jawed vertebrates have adaptive ISs. In higher animals, the innate and adaptive IS modules signal each other, that is, they interact to a certain degree. The jawed vertebrates have also evolved MHC class I and II cell surface receptors to facilitate antigen presentation (AP) in their adaptive ISs, and to limit autoimmune responses.

To better understand how the complex hIS has evolved and works, and how invertebrate innate ISs have been so successful, there has been a recent revival in comparative immunology using the tools of modern genomics and proteomics. For example, it has recently been reported (Dong et al. 2006) that the *Anopheles gambiae* mosquitoes use alternative splicing of their immunoglobulin protein, Dscam, to produce a highly diverse set of over 31,000 alternative splice forms, which enables specific recognition and protection against bacteria and *Plasmodium* (malaria) parasites. Thus, mosquito Dscam proteins function similar to vertebrate Abs in selecting and binding to pathogens. You will also see that certain invertebrates also have blood cells that phagocytose pathogens similar to vertebrate macrophages and dendritic cells, however, there are no lymphocytes or functional Igs in invertebrate ISs, nor apparently do they do antigen presentation by their hemocytes.

Hemocytes (circulating blood cells) in various taxa of insects play varied roles in immunity; some are capable of phagocytosis of foreign targets, including bacteria, malaria sporozoites, and strangely, latex beads. Larger targets can elicit encapsulation responses, where layers of hemocytes cling to and encyst the target object. Insect hemocytes also produce some of the effector molecules for humoral immunity (Shi and Paskewitz 2006). Bangham et al. (2006) described three classes of *Drosophila* hemocytes: crystal cells are involved in melanization that occurs at wound sites or around microbes, plasmatocytes are specialized for phagocytosis, and lamellocytes are associated with the encapsulation of larger parasites.

The Red Queen is alive and well in insect immunity. Bangham et al. cited the case where a parasitic wasp injects poly-DNA viruses into the host (a moth, *Manduca sexta*) along with its parasitoid egg. The virus suppresses the host's IS and inhibits encapsulation/melanization of the egg (which would deactivate it).

In another dipteran Red Queen scenario, an infective fungus, *Beauveria bassiana*, somehow inhibits the immune system's production of phenol oxidase (PO), which is important in fighting fungus. Fortunately, there is degeneracy in the flies' ISs. The flies manage to use an alternate pathway to up-regulate their Toll systems and fight *B. bassania* infections.

Invertebrate immunity has been studied in a number of species and an extensive and growing literature exists on the topic. The ongoing challenge is to put together the "big picture" of the complex molecular C^3 structures involved. To complicate such research, immune C^3 systems differ from order to order. Studies have included a wide range of creatures. We cite the ones receiving the most attention: *Anopheles gambiae* (a mosquito), bees, *C. elegans* (a nematode), *Ciona intestinalis* (an ascidian), clams, cnidarians, cockroaches, diptera (including *Drosophila*, fly maggots, mosquitoes), lice, lobsters, mollusks (snails, mussels), moths, prawns, sea urchins (echinoderms), shrimps, sponges, ticks, trematodes, tunicates, the water flea (*Daphnia*), etc.

Functional immune adaptation in invertebrates can be defined as any case where an initial exposure to a pathogen generates a greater response for subsequent exposures to that pathogen. This adaptation can be as simple as the animal's IS remaining activated after the first challenge. This simple persistence of activation has been noted in *Drosophila* and in mealworms (Pham et al. 2007). However, as we have described above, more complicated models for invertebrate adaptive phenomena have been formulated. Strain-specific immunity has been observed in flour moths (*Galleria* sp.), *Daphnia*, cockroaches, *Drosophila*, and bumblebees. For example, Pham et al. found that fruit flies inoculated with dead *Streptococcus pneumoniae* bacteria were protected from a subsequent lethal challenge with these bacteria. The response was specific for *S. pneumoniae*, and lasted the life of the fly. They also demonstrated that the fly's specific immunity depends on phagocytic hemocytes and the Toll pathway. The fly's specific immunity is independent of the Imd pathway and AMPs. This behavior suggests that *Drosophila*'s IS is, at least in part, adaptive.

The ISs of many invertebrates have been studied. See, for example, shrimp (Chou et al. 2009), bees (Xu et al. 2009), fruit flies (Loker et al. 2004, Pham et al. 2007), mosquitoes (Kurtz and Armitage 2006, Fragkoudis et al. 2009), Copepods, waterfleas, prawns (Little et al. 2005), tunicates (Nair et al. 2005), fly maggots (Nigam et al. 2006a,b), various mollusks (Litman et al. 2005), *C. elegans* (Loker et al. 2004), *Ciona intestinalis*, a non-vertebrate chordate (Loker et al. 2004), *Tridacna* (giant clam) (Loker et al. 2004), cockroaches (Chiang et al. 1988), and sea urchins (Rast et al. 2006).

In the following sections (5.4.2, 5.4.3, and 5.4.4), we describe the structure and physiology of the innate ISs of some invertebrates.

5.4.2 SOME EXAMPLES OF INVERTEBRATE IMMUNITY

5.4.2.1 Sea Urchins

Rast et al. (2006) cited evidence that between 4% and 5% of genes in the sea urchin genome are directly involved in immune functions. There are ca. 222 *Toll-like receptor* (TLR) genes that have been shown to participate in the recognition of conserved pathogen-associated molecular patterns (PAMPs). Also, ca. 203 NACHT-domain-LRR (NLR) genes with similarity to vertebrate nucleotide-binding and oligomerization domain cytoplasmic receptors, and ca. 218 genes coding scavenger receptor, cysteine-rich (SRCR) proteins. Rast et al. stated, "Other classes of immune mediators, such as key components of the complement system, peptidoglycan-recognition proteins (PGRPs), and Gram-negative [bacteria] binding proteins (GNBPs) are equivalent in numbers to their homologs in protostomes and other deuterostomes."

Sea urchins do not appear to have an adaptive immune response, but they do have RAG genes: *SpRag1L* and *SpRag2L* are evolutionary homologs of vertebrate *RagI* and *RagII* genes (Fugmann et al. 2006). In vertebrates, the *Rag* genes rearrange DNA given appropriate sequence cues. They locate, shuffle, and splice codons in antibody genes, part of the adaptive hIS.

5.4.2.2 *Drosophila melanogaster* and Other Dipteran Insects

Innate immunity in insects is generally based on molecular recognition between AMPs and microbial surface molecules such as LPS, peptidoglycans, and β-1, 3-glucans. This molecular adhesion triggers the subsequent activation of various immune effector responses.

Like other invertebrates, fruit flies rely only on their innate ISs to fight invading pathogens. The *Dscam* gene in *Drosophila* codes a humoral immune protein,

Dscam. *Dscam* is operated on by genomic alternative splicing machinery (recombining exons by mutually exclusive excision) in the fly's genome to produce ca. 38,000 Dscam isoforms, any one of which may have an affinity to a portion of a pathogen coat protein. Binding of a Dscam molecule to a pathogen (e.g., *E. coli*) is the first step in its destruction. In the *Drosphila Dscam* gene, there are four variant arrays (exons 4, 6, 9, and 17) that can produce 12, 48, 33, and 2 variants, respectively. Assuming alternative exons are expressed independently, they can be combined to make $(12 \times 48 \times 33 \times 2) = 38,018$ Dscam isoforms (Chou et al. 2009). The 38,000 variant Dscam molecules behave like Ab analogs.

Chou et al. reported that they cloned and characterized the full-length cDNA sequence of the first Dscam from the shrimp, *Litopenaeus vannamei*, LvDscam. From examination of LvDscam, these workers concluded that there were 8970 possible isoforms of this protein, which has 1587 AAs.

There are a total of ca. 19,000 hemocyte-specific Dscam isoforms (Bowden et al. 2007). Hemocytes ingest pathogens and lyse them similarly to mammalian macrophages and dendritic cells. Curiously, Dscam is also used for neural wiring specificity in fruit flies. Not surprisingly, different Dscam isoforms are expressed in the brain compared to hemocytes.

Certain antimicrobial peptides (AMPs) of the fruit fly are synthesized in its fat body (the functional equivalent in insects of the vertebrate liver). Examples of invertebrate AMPs include, but are not limited to diptericins, drosomycins, metchnikowins, defensins, attacins, cecropins, and drosocins (Rowley and Powell 2007) (see Glossary). According to Bartlett et al. (2002), more than 170 AMPs have been found in insects alone. (These AMPs are in addition to the many isoforms of the Dscam protein.)

Dong et al. (2006) have shown that the mosquito, *Anopheles gambiae*, is protected by its Dscam protein system against both Gram-positive (G⁺) and Gram-negative (G⁻) bacteria, as well as two species of *Plasmodium*. (*Plasmodium* is the protozoan that causes malaria.) They claimed the *AgDscam* gene is capable of producing over 31,000 alternative splice forms (isoforms), forming a variable range of binding affinities to non-self cells.

Not surprisingly, there is some unique evidence that the humoral (AMP) and cellular (hemocyte) components of certain invertebrates interact. In *Drosophila*, Lemaitre et al. (1997) reported that flies challenged with a fungal infection biosynthesized antifungal AMPs, while infection by a Gram-negative bacterium was correlated with an increase of AMPs specific for that pathogen. Rowley and Powell (2007) commented that the findings of Lemaitre et al. do not appear to be universal for other invertebrates and other pathogens. Rowley and Powell cited more recent studies that either failed to observe an increase of expression of genes for AMPs following a microbial challenge (lobsters) or found that the nature of the challenge agent is uncorrelated with the AMPs expressed (in mussels).

Returning to *Drosophila*, Rowley and Powell commented that: "It has long been thought that there may be a link between phagocytic hemocytes and the fat body cells that are responsible for AMP biosynthesis. A recent report… identified a gene, *psidin*, that codes for a protein found in the lysosomes of the hemocytes of

Drosophila." They go on to comment that in *psidin*-deficient mutants, the expression of psidin is severely reduced. This suggests that hemocytes are important in controlling or stimulating AMP synthesis. Thus, hemocytes may act analogously to vertebrate APCs; they may produce cytokine signals that regulate AMP production by the fat body, or they phagoticise pathogens and then digest complex antigens in their lysosomes in such a way to present Ag components on their surfaces to fat body cells, a putative *Drosophila* AP scenario. It seems reasonable to speculate that dipteran insects, after millions of years of evolution and Red Queen challenges, have developed their own version of adaptive immunity.

In summary, there are four, major immune responses in *Drosophila* (Bangham et al. 2006): (1) Encapsulization and melanization of parasitoid eggs. (2) Destruction of G^+ bacteria and fungi by the fly's IS through activation of the Toll pathway. (3) Destruction of G^- bacteria by the fly's IS through activation of the Imd/protease cascade pathway (Tanji et al. 2007, Gupta 2008, Maillet et al. 2008). (4) Antiviral activity triggered by the Jak-STAT pathway and the expression of antiviral genes.

Drosophila is an ideal experimental animal used to study immunity. They have had their genome sequenced, they reproduce rapidly, and much is known about their genetics. Motivation for studying the *Drosphila* IS appears to be purely scientific; fruit flies as far as we know, are not disease vectors, a food source, nor do they get AIDS, autoimmune diseases, the flu, etc. (some factors that have driven in-depth research on the hIS). *Drosophila* do expedite the rotting of ripe fruit and vegetables, however, giving it some economic importance. Hopefully, using the now sharper tools of genomics and proteomics, we will soon have a more complete picture of how their complex, innate/adaptive IS works.

5.4.3 Maggot Therapy

Blow fly maggots have three principal actions when used in the therapy of chronic, infected lesions: (1) They debride wounds by dissolving (and eating) only necrotic, infected tissues. It takes only 1–3 days for them to clean up most wounds. (2) They disinfect wounds by killing bacteria, even MRSA. (3) They stimulate wound healing.

The written history of the medical infestation of fly larvae (maggots) in human wounds (myiasis) goes back to the Old Testament Bible (Job 7:5) (Whitaker et al. 2007). Whitaker et al. cited evidence that maggots have been used for wound healing in the last thousand years by various ancient cultures. A more recent use of maggots for wound healing was documented by Napoleon's head surgeon, Baron Dominique-Jean Larrey (1766–1842). During the French campaign in Syria, Larrey used maggots of the "blue fly" to treat war wounds, and observed that they not only removed necrotic tissues, but also had a positive effect on wound granulation and healing. Maggots were also used to treat festering wounds in the American Civil War, and in subsequent wars. However, their use was largely discontinued following the discovery and use of mold-based antibiotics and sulfonamides in the 1940s.

What has stimulated a renaissance of maggot therapy has been the evolution of antibiotic-resistant bacteria as a result of chronic overuse of broad-spectrum antibiotics to treat diseases with viral etiologies. As early as 1948, only 4 years after the widespread use of penicillin, over 50% of nosocomial *S. aureus* infections were penicillin resistant due to their production of penicillinase (β-lactamase), an enzyme that inactivates β-lactam antibiotics. Nigam et al. (2006a) reported that in 2006, ca. 80%–90% of *S. aureus* infections were penicillin resistant. The synthetic antibiotic methicillin was introduced in 1960, but *S. aureus* developed resistance to this drug, as well, hence MRSA has spread and now causes serious hospital and community infections all over the world. For a while, vancomycin was effective against MRSA, but vancomycin-resistant *S. aureus* (VRSA) strains have now been reported in Japan, and no doubt will spread, as did MRSA. Nigam et al. (2006a) and Bowling et al. (2007) reported on studies that maggots do possess the ability to kill clinical isolates of MRSA (in vitro and in vivo).

Sterile maggots of freshly emerged, sterile larvae of the common Green-Bottle Blowfly (*Phaenicia sericata*, aka *Lucilia sericata*), and the Northern Blowfly (*Protophormia terraenovae* aka *Lucilia terraenovae*) are now being actively used for the debridement of chronically infected and necrotic wounds. They are effective in healing diabetic foot ulcers, as well as pressure sores, pilonoidal ulcers, and various other infected, nonhealing abscesses (Sherman 2003).

First instar maggots (just hatched from eggs) are ca. 1–2 mm in length. As they feed and grow, they secrete proteolytic enzymes that liquefy only the host's necrotic tissues. They feed and grow for ca. 4–5 days, going through two molts, reaching 8–12 mm in their mature third instar. At maturity, the third instar maggots leave their food source and crawl to some dry place to pupate and then undergo metamorphosis to adult flies.

Maggots have been evaluated for their wound-healing properties, especially against MRSA (Nigam et al. 2006b). In addition to consuming and metabolizing necrotic infected tissues, maggot waste secretions contain ammonia that raises local pH in the wound, which inhibits bacterial growth. Also secreted are the antimicrobial substances phenylacetic acid and phenylacetaldehyde, both secreted by a commensal organism, *Proteus mirabilis*, in the maggot's midgut. Perhaps the main way maggots fight wound infections is by eating wound bacteria and killing them in their digestive system. In in vitro experiments, it was found that live maggots are particularly effective against *Staphococcus aureus* and *Group A and B Streptococci*. They show some activity against *Pseudomonas* sp., but none against *E. coli* or *Proteus* sp. (Whitaker et al. 2007).

Also found in the maggots' waste excretions is a small, yet-to-be identified, potent AMP of <500 Da (Nigam et al. 2006b). Nigam et al. cited Nibbering (2004), who incubated endothelial cells with maggot waste excretions. Nibbering found an increase in the immunocytokines IL-8, IL-10 and growth factor β-FGF. Nigam et al. concluded, "We are clear that that maggots do produce accelerated healing in wounds that have remained stationary and non-healing for a long time." Clearly, more work needs to be done to isolate the natural antibiotics (AMPs) in maggot excretions.

We view maggots as micromachines adapted for chronic wound healing. The ongoing identification and characterization of their internal antibiotic proteins offers us the promise of new weapons in the ongoing Red Queen war between antibiotic-resistant bacteria (e.g., MRSA and VRSA) and pharmacological medicine.

5.4.4 ANTICIPATORY RESPONSES OR IMMUNE MEMORY?—INOCULATION IN INVERTEBRATES

Shrimp are economically important as a human food. When raised in mariculture farms, they are susceptible to *Vibrio* infection as larvae, and viral disease as adults. Rowley and Powell (2007) observed that there is a commercially available "vaccine" for farmed shrimp larvae, AquaVac Vibromax, made by Schering-Plough Animal Health. This is a multivalent vaccine designed to protect the larvae from a range of pathogenic *Vibrios*. Rowley and Powell go on to comment that the vaccine appears to give a certain improvement in the health and survival of the larvae; however, its mode of action and specificity are unknown. The effectiveness of the Vibromax vaccine does suggest that the larval shrimp's IS is in some manner adaptive. We note that it appears that the status of shrimp "vaccines" is currently at the state of human vaccination in 1796 when done by Edward Jenner (Northrop and Connor 2008).

Little et al. (2005) cited "vaccination" studies of the prawn, *Penaeus monodon*, that showed different responses to conditioning with two different envelope proteins derived from white spot syndrome virus (WSSV). Primary exposure to the VP28 protein isolated from WSSV provided protection from subsequent viral challenge, while exposure to the VP19 protein did not (Witteveldt et al. 2004). Little et al. also cited evidence that both specific and general immunity can be passed from mother to offspring, providing the F1 generation... "of pathogen-exposed parents with improved defense against infection." They concluded, "We still cannot dismiss the possibility that at least some invertebrates have an immune system that is functionally equivalent to the acquired [adaptive immune] response of vertebrates."

Inoculation of the shrimp, *Litopenaeus vannamei* with dsRNA sequences taken from the WSSV genome were shown to protect the animal from subsequent WSSV infection (Robalino et al. 2005). This suggests that shrimp can use pathogen-specific RNAi systems to obtain highly specific protection against viral diseases.

A detailed discussion of antiviral RNAi as a defense mechanism against arboviruses in mosquitoes can be found in a review paper by Fragkoudis et al. (2009). These authors stated, "...the studies detailed above suggest that there is induction of antimicrobial immune pathways, including Toll, JAK/STAT and Imd/Jnk, in arbovirus-infected mosquitoes; the [arbovirus] activators of these systems remain unknown." Their final thought was, "We are probably only just beginning to comprehend how complex the interactions between arboviruses and their arthropod vectors really are."

Bowden et al. (2007) reviewed pathogen-specific host defense in invertebrates. They stated that *Drosophila* is capable of fine discrimination between pathogens. They cited the work of Pham et al. (2007), in which immune response against *Streptococcus pneumoniae* could be enhanced by prior inoculation of *Drosophila* with heat-killed *S. pneumoniae*. The inoculated flies did not have protection from other pathogens, even closely related species. In addition, prior exposure to other bacteria did not generate protection against *S. pneumoniae*. The same type of protection could be seen when flies were vaccinated with *Beauveria bassania*, a natural pathogen of fruit flies. The adaptive range of protection afforded by the *Drosophila* IS is apparently limited. Inoculation with *Salmonella typhimurium*, *Listeria monocytogenes*, and *Mycobacterium marinum* did not induce pathogen-specific protective responses. Bowden et al. commented that the enhanced immune responses in *Drosophila* appear to be mediated by Toll-like pathways and phagocytic responses.

Another example of pathogen-specific immunity among insects is seen in bumblebees (*Bombus terrestris*). Bowden et al. (2007) cited research (Sadd and Schmid-Hempel 2006) in which bees exhibited specific immune response against *Pseudomonas fluorescens*, *Paenibacillus alvei*, and *Paenibacillus larvae*. Curiously, it took 22 days for the bees to develop their highly specific immune responses. Bowden et al. go on to summarize that there is experimental evidence for acquired, pathogen-specific immunity in a wide range of invertebrates: flies, bees, certain crustaceans, and mollusks. Inoculation with certain bacteria, trematodes, tapeworms, and trypanosomes has been used in these studies. They commented, "...there is still insufficient evidence to tell whether these systems are widespread throughout the [invertebrate] metazoa. It is also unclear if they can generate targeted protection against many different pathogens, or whether they have been selected only to provide protection against pathogens that are the most relevant to the host species." So the jury is still out on invertebrate adaptive immune systems.

5.4.5 BACTERIAL AMPs

One of the ways bacteria have maintained fitness over the millennia is by producing AMPs called bacteriocins that enable them to compete for resources by killing other bacterial species (Red Queen wars among the prokaryotes) (Cotter et al. 2005). Bacteriocins are generally much more potent cell killers than AMPs from eukaryotes (Nissen-Meyer et al. 2009). G^+ lactic acid bacteria (LAB) produce two classes of bacteriocins: Class I is the lanthionine-containing AMPs. Class I is posttranslationally modified. Class II are the non-lanthionine-containing AMPs; they are not subject to extensive posttranslational modifications (lanthionine is a nonstandard AA; see Glossary). Nissen-Meyer et al. recognized four subclasses of Class II bacteriocins: Subclass-IIa are the anti-Listerial, one-peptide, pediocin-like AMPs. Subclass-IIb includes the two-peptide bacteriocins (see below). Subclass-IIc includes the cyclic bacteriocins whose N- and C-termini are covalently linked. Subclass-IId are linear, non-pediocin-like, one peptide bacteriocins with no peptide sequence similarity to the pediocin-like AMPs. Thirty-one class IId bacteriocins are listed by Nissen-Meyer et al. (2009).

Fimland et al. (2005) listed some 24 pediocin-like bacteriocins (subclass-IIa) secreted by a variety of lactic acid bacteria. These AMPs display anti-*Listeria* activity and kill the target cells by permeabilizing their cell membranes. The pediocin-like bacteriocins are cationic (have a net positive charge) that presumably allows interactions with the negatively charged bacterial phospholipid-containing membranes and acidic bacterial cell membranes. Pediocins differ markedly in their target cell specificities; they have been shown to be active in various degrees against strains of *Lactobaccillus, Pediococcus, Enterococcus, Carnobacterium, Leuconostoc, Lactococcus, Clostridium,* and *Listeria* (Fimland et al. 2005).

The bacteria that produce pediocin-like AMPs also produce cognate immunity proteins to protect themselves from their own AMPs. Fimland et al. state that the genes coding an immunity protein are generally located near or on the same operon that codes the pediocin-like AMP. Expression of the two genes is generally co-regulated, and bacteria may thus be sensitive to their own bacteriocin when in a non-AMP-producing state.

Enhanced production of many bacteriocins involves a quorum-sensing mode of regulation. This is mediated by an exogenous, autoinducer (AI) molecule (which has been called an induction peptide [IP], or a peptide pheromone), a histidine protein kinase (HPK), and a response regulator (RR). Genes of these regulatory elements are generally found on the same bacterial operon (Diep et al. 2009). At a low colony cell density, there is a low basal expression of the IP. The regulatory operon for the IP can be turned on by several cues: the near presence of some competing bacterial species, or by a critical threshold concentration of the secreted IP when the colony density, hence IP, reaches a certain level.

Morgan et al. (2005) have reported on a bacteriocin, Lacticin 3147, from *Lactococcus lactis lactis DPC3147.* This is an AMP (subclass-IIb) that consists of two synergistic peptides; it is lethal in nanomolar concentrations ($MIC_{50} = 7\,nM$) to *Lactococcus lactis cremoris HP.* Lacticin 3147 is similar to the bacteriocin nisin in terms of its biological activity; it also has a broad spectrum of activity against such G+ bacteria as *Staphococcus* sp. and *Streptococcus* sp. Cotter et al. (2005) stated that Lacticin 3147 has in vitro activity against: *S. aureus* including MRSA, enterococci, streptococci (including *S. pneumoniae, S. pyogenes, S. agalactiae, S. dysgalactiae, S. uberis,* and *S. mutans*), *Clostridium botulinum,* and *Propionibacterium acnes.*

It is clear that bacteria have had a long history in fighting Red Queen survival wars among themselves, and thus have evolved a potent array of bacteriocins, means of inducing their expression, and natural defenses in the form of cognate immunity proteins (Fimland et al. 2002, Johnson et al. 2004). Some questioned to be answered: Can bacteria "fight back" against hIS cells (e.g., macrophages, dendritic cells, and neutrophils) with certain bacteriocins? Do bacterial species B's cognate immunity proteins confer any protection against bacteria species A's bacteriocins? How do *Lactobacillus* sp. bacteriocins promote "intestinal health" (think yogurt) (see Diep et al. 2009)? Through genetic engineering and biotechnology, can bacteriocins be developed into effective antibiotics against bacteria such as MERSA (see Rossi et al. 2008)? Finally, do you think chronic use of a certain bacteriocin (e.g., Lacticin 3147) in a common food (e.g., milk) will lead to adventitious lacticin-resistant bacteria?

5.4.6 INVERTEBRATE AMPs: APPLICATIONS IN BIOMEDICINE

AMPs are important because they have the potential to fight many pathogens that have acquired resistance to modern antibiotic drugs. Insects and other invertebrates have been using them for millions of years successfully. Tincu and Taylor (2004) defined AMPs as proteins less than 10 kDa in mass that show antimicrobial properties, providing a rapid and non-delayed response to invading pathogens.

In a review paper, Bell and Gouyon (2003) described the properties of what they called ribosomally synthesized, antimicrobial peptides or RAMPS. (Note that AMPs can be directly ribosomally synthesized, or enzymatically modified from RAMPs.) They stated that the bacterial RAMP nisin is already widely used as a food preservative! However, the bacteria that produce nisin are self-immune. Bell and Gouyon stated that secretion of RAMPs is often stimulated by the host bacteria recognizing nearby G⁻ bacteria by the binding of host receptor proteins such as CD14 to lipid A of the G⁻ bacterial cell membrane. As might be expected, there is a Red Queen coevolutionary contest between the bacterial RAMP secretors and the evolution of RAMP immunity by competing bacterial species. Bell and Gouyon gave a simple mathematical model for the evolution of resistance to RAMPs. They warned that while the use of RAMPs to fight human bacterial infections shows great promise, extended use, as with "-mycin" antibiotics, will eventually result in RAMP-resistant strains and the problems they create. They observed that nisin resistance has already been seen in common food-spoiling organisms such as *Listeria*, *Clostridium*, *Baccillus*, and *Staphylococcus*. Beware the Law of Unintended Consequences.

Bulet and Stöcklin (2005) reported that a total of ca. 700 AMPs are listed in protein databases, including Swissprot and TrEMBL. They are found in plants, invertebrates, and vertebrates. With modern proteomic analysis tools, the exact structures of AMPs are being elaborated; now Bulet and Stöcklin estimate that their own compilation of AMPs will reach ca. 1500 from animals alone. These authors state that in holometabolic insects, AMPs are synthesized by both the fat body (an organ corresponding to the mammalian liver) and various epithelial cells. In hemimetabolic insects, AMPs are expressed in hemocytes and secreted into the hemolymph during an infection. Insect AMPs are cleaved in vivo from a large precursor protein that contains a signal domain and sometimes a pro-domain up- or downstream from the mature peptide.

Insect AMPs are usually cationic (positively charged molecules), and their primary structures can vary markedly. The *Cecropins* (including *sarcotoxins*, *hyphancin*, *enbocin*, and *spodopsin*) have α-helical domains. *Cecropins* occur in both *Lepidoptera* and *Diptera*.

All insect species investigated to-date (2005) were reported to have Defensins. Bulet and Stöcklin (2005) commented, "More than 60 defensins have been isolated from insects belonging to phylogenetically recent orders (*Diptera, Lepidoptera, Coleoptera, Hymenoptera*) and to the ancient order of Odonata (dragonflies)." They go on to comment that the dragonfly Defensin is closely related to the *defensins* of scorpions and mollusks. Antibacterial defensins are more effective against Gram-positive (G+) bacteria (including human pathogens). They are less effective against Gram-negative (G−) bacteria, yeast, and filamentous fungi. It has been observed that the bactericidal effectiveness of defensins in vitro decreases as the ionic strength of the culture medium increases.

Because the genome of the fruit fly, *Drosphila melanogaster*, has been sequenced, it is an ideal model system to study the biochemistry, genomics, and proteomics of its innate immune responses, including its seven inducible AMPs. These include defensin (against G+ bacteria); diptericin, drosocin, attacins, cecropins (against G− bacteria); and drosomycin and metchnikowin (against filamentous fungi). Following experimental infection, *Drosophila* synthesize these seven AMPs in its fat body; their concentrations in the flies' hemolymph reaches overall values of 0.5 mM, far higher than the LD50s for the microorganisms in vitro.

Curiously, *Drosophila's* IS can discriminate between an infection by a filamentous fungus and bacteria. Two different signal transduction pathways (acting in fat body cells) control the discrimination between bacteria and fungi; the Toll pathway and the immune deficiency (Imd) pathway. We cannot go into the biochemical details here; the interested reader should consult Bulet and Stöcklin (2005). In summary, the Toll pathway predominantly activates the expression of the gene encoding the antifungal peptide, drosomycin, through an extracellular proteolytic cascade of reactions in which serine proteases are important (Shi and Paskewitz 2006). Toll activation can also be observed during infection by G+ bacteria. On the other hand, the Imd pathway is involved with the resistance of *Drosophila* to G− bacterial infections, and the up-regulation of the genes coding the AMPs fighting G− bacterial infections.

Chou et al. (2009) constructed a phylogenetic tree of vertebrate and invertebrate Dscams. They stated that most invertebrate Dscams have a three-domain architecture: extracellular, transmembrane, and a cytoplasmic tail. From the evidence presented, Chou et al. concluded that (shrimp) LvDscam has a unique domain architecture; a standard extracellular domain structure, but no transmembrane domain or cytoplasmic tail. They state that the LvDscam they isolated only occurs in this tailless, secreted form. They suggested that the soluble Dscam binds to the pathogen and is then "recognized" by the phagocytic host cells via homophilic interaction with the membrane-bound Dscam on the phagocytic hemocyte cell surfaces. This putative model of Dscam's involvement in invertebrate adaptive immunity is analogous to the way Abs function in mammals with NK cells. Dong et al. (2006) found that AgDscam was required for the phagocytic response against pathogens in the mosquito.

TABLE 5.1
Partial List of Cationic AMPs from Arthropods and Other Invertebrates

AMP	Comments
β-Alanyl-L-tyrosine	A low MW (252), antibacterial AMP from maggots of the gray flesh fly (Salzet 2005).
Alo3	An antifungal AMP from *Phormia* spp. and *Sarcophaga* spp.
Androctonin	A scorpion AMP with two disulfide bridges. It attacks bacteria and fungi.
Attacins	Attacins A–F. First isolated in the hemolymph of immunized pupae of the silkworm moth. Attacin-related proteins have been found in *Drosophila*, *Musca domestica*, Tsetse flies, and various moths. Attacins kill *E. coli* and two other G$^-$ bacteria in the silk worm gut by inhibiting the synthesis of the outer membrane proteins by interfering with *omp* gene transcription.
Cecropins	AMPs found in the hemolymph of a silk moth, and in certain crustaceans. More effective against G$^-$ bacteria than G$^+$. Cecropins attack bacterial outer membranes.
Crustins	An 11.5 kDa AMP found in the crab, *Carcinus maenus*, as well as two species of shrimp.
Defensins	14–18 AA AMPs. Defensins are found in vertebrates and invertebrates. They are effective against bacteria, fungi, and many enveloped and non-enveloped viruses. Cells of the IS contain defensins to assist in killing phagocytized bacteria. Most defensins kill by forming pores in the target cell's membrane. Defensins have three disulfide bridges.
Diptericins (A, B, and C)	AMPs with ca. 82 AAs found in various flies. Expressed rapidly in fat body cells and thrombocytoids in response to bacterial infections or injuries.
Drosomycin	An AMP of 44 AAs found in *D. melanogaster*. Contains eight cysteines making intramolecular disulfide bridges. Synthesized in the fat body. A potent antifungal AMP, inactive against bacteria.
Gambicin	An antifungal, anti-protozoic and antibacterial AMP from the mosquito. Has more than three disulfide bridges.
Granulysin	An antibacterial AMP from sponges.
Heliomycin	An antifungal AMP from *Phormia* spp. and *Sarcophaga* spp.
Helyomycins	Lepidopteran AMPs that fight bacteria and fungi.
Hemerythrin	An oxygen-binding protein-derived AMP from annelids. It is antibacterial.
Jasplakinolide	A cyclic AMP from sponges. It attacks bacteria and fungi.
Metchnikowins	Antibacterial AMPs from the fruit fly.
Myticin A and B	Cysteine-rich AMPs from hemocytes and plasma of the marine mussel, *Mytilus galloprovincialis*. Active against G$^+$ bacteria, fungus and *E. coli*. 40 AAs.
Mytilins A and B	Cationic, cysteine-rich AMPs from *Mytilus edulis* L. Have antifungal activity.
Mytimycin	Cysteine-rich, antifungal AMP from *Mytilus edulis*. MW = 6233.5 Da.
Penaeidins	Antibacterial and antifungal AMPs found in shrimp. They have broad-spectrum fungicidal activity, but are inactive against yeasts. Active against G$^+$ bacteria. Low activity against specific G$^-$ bacteria (*Vibrio* spp.)
Perinerin	Antibacterial and antifungal AMP found in clamworm, *Perinereis aibuhitensis* Grube. 51 AAs, highly basic and hydrophobic (Pan et al. 2004)

TABLE 5.1 (continued)
Partial List of Cationic AMPs from Arthropods and Other Invertebrates

AMP	Comments
Royalisin	An AMP found in honeybees.
Tachycitin	An antifungal and antibacterial AMP from the horseshoe crab, *Limulus polyphemus*.
Tachyplesin I, II, and III	Antibacterial AMPs found in Chelicercates (Tincu and Taylor 2004).
Termicin	An antifungal AMP from *Phormia* spp. and *Sarcophaga* spp.
Thanatin	An antifungal and antibacterial AMP from hemipteran insects.
Theromyzine	An antimicrobial AMP from leech cocoons (Salzet 2005).

Note: Many more exist; see Bulet and Stöcklin (2005), Vizioli and Salzet (2002), Tincu and Taylor (2004), and Cheng-Hua et al. (2009). Also see the big table in Salzet (2005).

Chou et al. (2009) observed that different LvDscam isoforms were present in shrimp under different WSSV infection states, and hypothesized that this might be due to an immediate, nonspecific immune response, or be the result of a programmed defense strategy that tries many different candidate LvScam isoforms to find the most suitable to fight the pathogen.

Clearly, a picture is emerging that there is a hemocyte-Dscam interaction in the immune responses of certain invertebrates. This interaction can lead to highly specific Dscam expression to combat certain pathogens such as WSSV. There is also evidence for RNAi in the invertebrate immune response, and we know that miRNAs have been well established as ubiquitous genomic regulators throughout the Animal Kingdom.

Table 5.1 summarizes some of the major (non-Dscam) antimicrobial peptides found in various invertebrates.

Three major applications of invertebrate AMPs have emerged: (1) use in aqua- and mariculture to fight pathogens infecting the farmed organisms (shrimp, fish, mussels, clams, oysters, etc.); (2) to fight drug-resistant bacterial infections in humans and domestic animals (Rossi et al. 2008); and (3) for food preservation.

5.5 SUMMARY

In this chapter, we have considered how, as a general trend, increased complexity accompanies the evolution of species. We also reviewed speciation, origins of species, speciation by horizontal gene transfer, and transposons.

In order for fitness to accompany the evolution of a new invertebrate animal species, an immune system had to evolve within the species to protect it from infection by bacteria, fungi, and parasites. We have examined some of the features of invertebrate immunity, including AMPs, RAMPs, and hemocytes, and adaptive immune behavior that permits inoculation of certain invertebrate species (e.g., shrimp) against infection. Bacteriocins and antibiotics from bacteria were also described.

PROBLEMS

5.1 (A) Make a table of known bacteriocins. Give their source organism, details about the molecule, and its target organism.

(B) Can certain bacteria use their bacteriocins as AI molecules to regulate their own population densities under conditions of limited "food"?

5.2 AMPs from bacteria and invertebrates have potential use in fighting infections from antibiotic-resistant bacterial strains.

(A) How does antibiotic resistance evolve in bacteria?

(B) Comment on the proposition that extended use of AMPs will also lead to AMP-resistant bacterial strains.

5.3 Make a list of all nonstandard (unnatural) AAs that have been used to-date in the synthesis of novel, "designer" proteins using o-mRNA and o-ribosomes, or o-tRNA/synthetase methods. (See Wang et al. 2006, 2009.)

5.4 Tricolor cats (white, orange, black) are invariably female. Explain the genetic reasons for this phenomenon.

5.5 (A) Give Mendel's two laws of heredity.

(B) Give the Mendelian F1 and F2 generations' colors and genotypes when the parent generation consists of a red sweet pea flower with a dominant (**RR**) color genotype crossed with a white phenotype flower with a recessive (**ww**) color genotype.

(C) Give the Mendelian F1 and F2 generations' colors and genotypes when the parent generation consists of a red sweet pea flower with a recessive (**rr**) color genotype crossed with a white phenotype flower with a recessive (**ww**) color genotype. (*Hint*: Possible colors: red, pink, white.)

5.6 Did the MRSA bacteria that evolved from methicillin-sensitive *S. aureus* develop their antibiotic resistance from a random mutation, or by random expression of gene(s) they already carried, but were turned off?

5.7 Compare the estimated error rate in the natural duplication of the mammalian genome (by mitosis) with the error rate in duplicating bacterial genomes (e.g., that of *E. coli*). What is the reason for this discrepancy? What is its effect on the evolution of bacterial and mammalian species?

5.8 How does genome size correlate with a species' physiological complexity?

5.9 Why does the *Mimivirus*, a specific parasite of *Acanthamoeba polyphaga*, need such a large genome (1.2 Mb with ca. 1262 genes)? What virus has the smallest genome? What organism does it infect?

5.10 AMPs are found in the human urinary tract. What AMPs (if any) are found in human saliva and tears?

5.11 Triclosan is a potent wide-spectrum antibacterial and antifungal agent. It is a polychloro phenoxy phenol. It has ubiquitous use in soaps (0.1%–1.0%), deodorants, toothpastes (0.3% in Colgate), shaving creams, mouthwashes, and cleaning supplies. Discuss the unintended consequences of this widespread use: Consider its breakdown products and their effects on the environment and human health, as well as possible development of bacterial and/or fungal resistance to triclosan. Begin your research by using Wikipedia.

6 Complex Adaptive and Innate Human Immune Systems

6.1 INTRODUCTION

An important feature of nearly all complex adaptive systems (CASs) is that their general behavior patterns are not determined by centralized decision makers ("deciders"), but rather are determined by the net results of interactions between a number of independent entities (modules or variables). It is these network interactions that contribute to a CAS's complexity. Each individual entity (and class of entity) acts on the CAS with a built-in, basic set of behavioral rules. (One of the rules must be that the elements of the CAS act together.) We see this "whole is greater than the sum of the parts" property in the colonial behavior of ants, bees, and termites; migrating birds; and schooling fish. The coordinated behavior in the complex human immune system (hIS) involves a great number of different classes of hIS cells, and their biochemical rules are essentially their programmed responses to received external chemical signals, and those secreted by other hIS cells forming a hormonal/cellular "network." Even the innate IS involving the reactions of complement proteins and NK cells follows "preprogrammed" biochemical rules.

The "rules" may be viewed as a set of subsystem I/O relations, that is, a CAS is in fact made up from many subsystems. It is tempting to think that if we learn "the rules" governing subsystem behavior, we can model certain CASs with confidence, and thus be able to predict their group behavior to novel inputs. We can also see the effects of "tinkering" with innate rules on overall CAS model behavior.

In this chapter, we describe how the hIS has both a general, nonadaptive (innate) means for fighting invading pathogens (bacteria, viruses, fungi, parasites), and specific, adaptive mechanisms for combating reinfections by pathogens fought in the past. As we have seen, some sort of innate immune system is found in all multicellular organisms; however, adaptive immunity is found in vertebrates, with the exception of agnathans (jawless fish). In the previous chapter, we have seen that certain invertebrates show evidence of adaptive immunity that has a different organization, but similar function, to that of the hIS.

The hIS, because it is composed of a large number of mobile and fixed cell types that communicate by many secreted regulatory cytokines, is in the writer's opinion, second in complexity only to the hCNS (as a physiological CAS).

The hIS exhibits three basic kinds of innate, nonspecific behavior in fighting pathogens: One is mediated by the molecular complement system (CS), which consists of some 30 interacting, soluble proteins, some of which can combine, attack, and lyse invading bacteria and parasites. A second innate system is cell-mediated in which immune leukocytes known as natural killer (NK) cells, macrophages (Mϕs), and dendritic cells (DCs) directly attack pathogens, or somatic cells bearing surface evidence of internal pathogen infection. A third innate system depends on antimicrobial proteins (AMPs). AMPs are covered in Sections 6.4.6 and 5.4.

The adaptive component of hIS response depends on antigen presentation by Mϕs, DCss, B-cells, and somatic cells to naive helper T-cells (Th). By the clonal selection theory (CST), ingested pathogen coat fragments are externalized and "presented" to the naive Th cells, which then release cytokines that lead to clonal expansion of B-cells that produce specific antibodies. The CAIS uses a kind of Monte Carlo strategy to adapt to novel pathogens. It makes hundreds of thousands of different antibodies (Abs) on hundreds of thousands of different types of B-cells. When one Ab's paratope (high-affinity region) matches the molecular epitope region on an antigen protein fragment, it is this B-cell that is ultimately selected for clonal expansion to make the many specific Abs required to fight the pathogen. Through the mechanism of antigen presentation, along with the secretion of certain cytokines, free antibodies specific to a certain pathogen are eventually produced, leading to the pathogen's destruction. In addition, the CAIS produces so-called memory B- and T-cells that store the molecular information specific to making specific Abs for the pathogen in future infections. These memory cells can be reactivated by cytokines released in a fresh pathogen infection, giving the CAIS long-lived, specific immunity that expedites fighting reinfections. Memory cells bypass the inefficiency of Monte Carlo process of Ab selection. It is this adaptive synthesis of memory cells that makes immunization to exogenous allergens possible. Below, we summarize the complex behavior of the CAIS.

6.2   OVERVIEW OF THE COMPLEX ADAPTIVE hIS

As we have stated, the mammalian immune system is a thoroughly complex system, composed of many types of cells and hundreds of kinds of protein signaling molecules. Its function is to seek out and destroy invading bacteria, fungi, molds, viruses, protozoa, spirochetes, parasites, etc., that is, it protects its host animal from foreign invaders. Immune surveillance also extends under certain conditions to the detection of cancer cells and their destruction. Immune system Mϕs, neutrophils (NPs), and DCs also act to clean up necrotic body tissues damaged by injury, disease, or parasites, and also do antigen presentation.

As you will see, the hIS uses a number of mechanisms in its functioning. These include humoral (blood and lymph-borne molecular) factors, including protein antibodies (Ab), lectins, and the complement system, as well as direct cellular attack, including phagocytosis and lysis by Mϕ, NK cells, and cytotoxic

T-cells (CTL). These hIS cells participate in the direct cellular attack of hostile invaders (the "L" in CTL is because they are a form of lymphocyte).

Unfortunately, the CAIS also can respond to substances that it should not react to. These responses include allergy and autoimmunity. Certain foods and drugs taken in through the oral route can trigger a variety of adverse immune responses including nausea, headaches, hives, rashes, asthma, and anaphylaxis. Allergens such as dust, pollen, and mold spores, can also be inhaled, giving rise to rhinitis and asthma. Other substances, such as the oil from poison ivy on the skin cause well-known misery. For reasons poorly understood, the immune system can develop sensitivity to certain normal, self-, cellular proteins in the body, creating autoimmune diseases. Three well-known examples of autoimmune diseases are Type 1 diabetes, in which the immune system attacks the beta cells in the pancreas, Crohn's disease, in which the immune system attacks cells lining the bowels, creating inflammation, and Coeliac disease, where an allergy to wheat gluten also triggers an inflammatory attack on the intestinal lining. Many other kinds of autoimmune disease exist, for example, myasthenia gravis, lupus erythematosis, acute glomerulonephritis, idiopathic thrombocytopenic purpura, and autoimmune hemolytic anemia.

As we opined above, the hIS is, compared to the CNS, the second-most complex physiological system in the adult human body. Its complexity arises mostly from its many pleiotropic biochemical pathways. In order to understand the vast network of complex, causal interactions between its various components, we must understand its biochemical C^3 structure. These chemical interactions are, in general, of a feedback as well as a feed-forward nature. The adaptive immune system's components are normally self-regulating, that is, the system is homeostatic. The components include specialized immune cells, the immunoregulatory protein autacoids they secrete, and the protein antibodies manufactured. The immune system responds to biochemical signals (autacoids) from immune system complement molecules at the sites of infection, to cytokines from activated hIS cells, and to hormones from the CNS and from other tissues. Certain cellular components of the immune system are mobile and can move throughout the body's circulatory and lymphatic systems. Other immune system cells end up fixed in certain tissues and organs.

In the following sections, we will summarize the cellular components of the immune system, the major chemical signals it uses (immunocytokines, or immunoregulatory autacoids), and how these components normally work and interact. It will be seen that immunocytokines including lymphokines, interleukins, interferons, prostaglandins, etc., form a complex, hormonal, regulatory network. Many immunocytokines have multiple effects on multiple target cells (a property called pleiotropy), adding to the difficulty of formulating meaningful mathematical models of the immune system. The problems in describing the role of immunocytokines are especially enigmatic because, in general, we have an incomplete understanding of the biochemical control mechanisms governing their synthesis, and the molecular mechanisms whereby they exert their effects after binding to appropriate (or in some cases, inappropriate) membrane receptor molecules.

Although dauntingly complex, certain aspects of immune system function can be effectively mathematically modeled (Northrop 2000).

6.3 SUMMARY OF THE CELLS OF THE hIS

6.3.1 INTRODUCTION

The hIS has six major classes of leukocytes (white blood cells), which carry out its mission of protecting the body from infection and exogenous foreign substances (injected, inhaled, or eaten). These are Mφs, DCs, T-lymphocytes, B-lymphocytes, neutrophils, and NK cells. Each of these five classes of cell has several subclasses, which are based on the level of cell maturity or activation, proteins found on their cell membranes, the cells' functions, and their location in the body. Myeloblasts, besides giving rise to neutrophils, Mφs, and DCs, also differentiate into basophils and eosinophils, so named for their in vitro staining properties.

Other hIS cells such as mast cells and granulocytes also contribute to the inflammatory immune response, and megakaryocytes in the bone marrow make platelets. Platelets are 2–4 μm in diameter. They are not true cells because, similar to red blood cells, they lack nuclei. However, they have membranes with receptor proteins and contain mitochondria and RNA. Platelets also contain enzyme systems that allow them to synthesize prostaglandins, a class of nonprotein, eicosanoid immunocytokine. Platelets are necessary for blood clotting, and participate in the inflammatory immune response.

Certain immune system cells can be described as amplifiers, responding to foreign proteins, etc., by activating cells of like kind (autocatalytic stimulation) and other immune system components. Other cells can be described as *effectors*: they carry out offensive actions. Still other cells are *modulators*: they regulate the reactivity of the sensor and effector cells. For example, suppressor T-cells (Ts) down-regulate expression of Abs. The amplifiers, which include infected somatic cells, Mφs, DCs, neutrophils, and B-cells, perform antigen presentation that has the function of causing the proliferation of helper T-cells, CTL and NK cells. Certain messenger immunocytokine proteins are secreted by both the antigen-presenting cells and the receiver cells, which cause the receiver cells to reproduce, or clone themselves. Each daughter (clone) B-cell and CTL retains the specificity to bind with the presented antigenic epitope. The presented antigen is generally a piece of a coat protein of an invading bacterium or virus. It is by this amplification process that a specific immune response to an invading organism is strengthened.

The normal, total leukocyte density in human adult blood is 5,000–10,000 cells per μL. About 5.3% of these are monocytes, 30% are lymphocytes, 0.4% are basophils, 2.3% are eosinophils, and 62% are neutrophils. (Note that basophils, eosinophils, and neutrophils were named after the marker dyes used to characterize them in blood samples, not after the modern designations based on coat proteins and function.)

In general, when there is an acute viral infection (e.g., influenza), the total leukocyte count is decreased. On the other hand, when a person suffers from an acute bacterial infection, the total white cell count is increased. Other specific diseases produce changes in the ratios of specifically stained leukocytes. For example, infection by the trichinosis parasite causes a signatory increase in eosinophils, their ratio rising from about 2.3% of the total leukocyte count to about 40%–60%. Other WBC ratios stay the same, and the total WBC count increases about 30% (Collins 1968).

In the following sections, we will examine the immune system's cells in greater detail. We will first describe the specialized cells of the immune system starting with Mϕs, followed by the other major antigen-presenting cells of the hIS, the ubiquitous DCs. The various types of T-lymphocytes (T-cells; so-called because of their embryological origin in the thymus gland) are considered next. Common lymphoid progenitor stem cells give rise to NK cells, as well as T- and B-cells, and lymphoid DCs. NKs have basically the same role as CTLs, but lack the antigenic specificity possessed by CTL and B-lymphocytes. B-cells (so-called because of their embryological origin discovered in the Bursa of Fabricus of birds), and their role in manufacturing antibodies are then described. Last, but highly important in the functioning of the cellular arm of the hIS, we consider the mast cells and platelets.

Many of the immune system cells have aliases or alternate nomenclature. Early descriptions of immune system cells were based on how they were identified by various stains when prepared for viewing using light microscopy. More recent nomenclature relies on cell function, and the specific, identifying proteins found on their surfaces. We will try to identify immune system acronyms, as they arise, to minimize confusion.

6.3.2 Macrophages

Mϕs are found throughout the body. Immature Mϕs are called *monocytes*. They generally circulate freely in the blood and lymphatic system. As monocytes mature, they become more sessile, that is, they assume fixed locations in tissues and organs. Mϕs can migrate from the circulatory system and become resident in various tissues, a process called trafficking: in the liver, Mϕ are called Kupfer cells; in the brain they are microglia; in the kidneys they are mesangial cells; in connective tissue and skin they are histiocytes; in the lungs they are called alveolar Mϕs; in the lymph nodes and spleen there are fixed and free Mϕs.

The immunological functions of Mϕs are broad. They are involved in all stages of the immune response. First, they provide a rapid, "front-line," nonspecific, cellular, innate immune defense against invading bacteria, fungi, and parasites. Activated Mϕs can engulf and internalize such invaders. Internal enzymes break down the membranes and proteins of such phagocytosed bacteria into smaller molecular subunits, which can then be used to activate helper T-cells in the complex process known as antigen presentation.

Mϕs are central effector and regulatory cells of the inflammatory response. They can secrete more than 100 different effector autacoid substances. Some of these substances, such as hydrogen peroxide, lysozyme, neutral proteases, and nitric oxide, kill or damage the hostile target cells as well as adjacent normal cells (collateral damage), and cause inflammation. Other Mϕ-secreted substances act to induce inflammation, and then to repair tissues destroyed by infection and inflammation. In summary, like all immune system cells, Mϕs are complex biochemical machines. All of their products and their actions are the subject of internal and external regulatory control by the molecular messengers of the immune system network.

Mϕs are continually produced from bone marrow progenitor stem cells (monoblasts). Their rate of production is modulated by certain immunocytokines, notably monocyte colony-stimulating factor (MCSF) and granulocyte-monocyte colony-stimulating factor (GM-CSF). Those Mϕs that are induced by GCSF are larger, and have a higher phagocytic capacity than Mϕs induced by GM-CSF, which are more cytotoxic against certain tumors, express more major histocompatibility complex (MHC) class II antigen molecules on their surfaces, more efficiently kill the bacteria *Listeria monocytogenes*, and secrete more (prostaglandin E_2) PGE_2 (Northrop 2000). The very different structures and signal transduction mechanisms of the receptors for MCSF and GM-CSF appear to be evidence for different differentiation pathways and thus two, different, Mϕ populations.

Immunological studies of the coat proteins of Mϕs found in different immunological scenarios (dermatitis, gingivitis, osteoarthritis, tissue graft rejection, tumors, etc.) provide further evidence for functional subtypes of Mϕs. Even so, current practice is to classify all Mϕs by their level of activation.

For modeling purposes, we have arbitrarily divided Mϕ into three groups: inactive, primed, and fully activated. In the inactive condition, Mϕ can either be circulating or fixed in tissues such as lymph nodes. Inactive Mϕs have low oxygen consumption, little or no monokine (immunocytokine) secretion, and a low level of MHC class II gene expression. (MHC-II is a Mϕ membrane protein necessary for antigen presentation to helper T-cells [Th].) Mature, inactive Mϕs do have phagocytic activity, can respond to chemotaxic signals, and can proliferate in respond to cytokine signals. Inactivated Mϕs are primed by gamma interferon (IFN-γ) secreted from stimulated Ths.

Primed Mϕs have increased oxygen consumption (increased metabolism) and they exhibit enhanced MHC-II expression. Other immunocytokines such as IFN-α, IFN-β, IL3 (interleukin-3), MCSF, GMCSF, and tumor necrosis factor α (TNF-α) can also prime Mϕ for selected functions. Primed Mϕs can respond to secondary signals to become fully activated. In this stage, they cannot proliferate, and they have high O_2 consumption. They do maximal secretion of substances for cell killing and inflammation. On the other hand, activated Mϕs show decreased MHC class II protein production, as well as reduced antigen presentation.

Note that there is no sharp, metamorphic distinction between primed- and active-Mϕ. Among the members of a relatively isolated population of Mϕs, such as in the spleen, at any moment one can find a spectrum of Mϕ development levels and capabilities. Such diversity adds zest (and confusion) to the challenge of modeling Mϕ actions in the CAIS.

In summary, Mϕs are seen to have a key role in activating the immune response. They phagocytose and destroy invading bacteria, viruses, parasites, fungi, protozoa, etc. After internal proteolysis, the Mφs present fragments of the phagocytosed invader's protein coat on their surfaces along with their MHC-II molecules in order to activate NK and helper T-cells. NK and helper T-cells respond to a specific protein fragment or epitope bound to the Mφ's MHC coat protein, and they release certain immunocytokines that cause the specific responding cells to clonally reproduce themselves. Note that other immune cells also do antigen presentation (e.g., DCs), but the Mφs are seen to have a key role in activating helper T-cells. Mφs also secrete substances toxic to invading organisms.

6.3.3 Dendritic Cells

DCs were first described about 35 years ago in the spleen, then found in all lymphoid tissues, and finally located in most other tissues and organs, in particular, the skin, all mucosal tissues and the lining of the GI tract. DCs in lymphoid tissues arise from precursor cells that also produce monocytes and plasmacytoid DCs (pDCs) (Liu et al. 2009). According to Liu et al., DC development progresses from the Mφ and DC precursor cells to common DC precursors that ultimately produce pDCs and "classical" spleen DCs (cDCs), but not monocytes. The so-called committed precursors of cDCs (pre-cDCs) enter lymph nodes through high endothelial venules and later disperse and distribute throughout a DC network. Further expansion of cDCs involves cell division, controlled by regulatory T-cells and the fms-like tyrosine kinase receptor-3 (Flt3).

Note that the nomenclature of DCs appears variable; some workers have categorized them by their tissue locations, as well as by their signature cell-surface proteins. DCs are mobile cells that complicate the picture; however, their surface proteins (receptors, effectors) are the ultimate determinants of their functions. Many workers are studying DCs because of their role in AP. Their manipulation may lead to therapies for allergies, autoimmune diseases, and cancer.

DCs are called so because of their external morphology; "mature" DCs are covered with many tentacle-like surface projections that resemble the dendrites of neurons. The normal total DC population in the hIS is much lower than Mφs/monocytes; however, their importance lies in their potent antigen-presentation (AP) properties, and their ability to migrate (traffick). DCs are dedicated antigen-presenting cells that play an important role in mediating the adaptive (cellular) immune response (Granucci et al. 2008). DCs can activate both CD4+ (MHC-1 is used) and CD8+ (CTL) (MHC-2 is used) T-cells by AP.

In many ways DCs behave like Mφs, ingesting antigens by using receptor-mediated endocytosis, phagocytosis, and pinocytosis. All kinds of cellular and molecular debris are ingested (both self and pathogen), processed internally, then presented to T- and B-cells as peptide fragments attached to their MHC molecules. If the ingested antigens are foreign (from bacteria, viruses, fungi, parasites, cancer cells), the DC becomes "activated" and expresses co-stimulatory molecules such as B7, which binds to T-cell CD28 to effect AP.

All DCs have cell-surface, molecular pattern recognition receptors called Toll-Like Receptors (TLRs). The various TLRs have affinity to pathogen-associated molecular patterns (PAMPs). PAMPs are found in the flagellin proteins from bacteria; peptidoglycan from Gram-positive bacteria; lipopolysaccharide (LPS) aka endotoxin from Gram-negative bacteria; dsRNA from viruses, plants, and animals; also unmethylated DNA (Kimball 2009a).

There are at least two subclasses of DCs: (1) myeloid (mDCs), which, in addition to AP, secrete IL12 and IL23; and (2) plasmacytoid (pDCs), which can produce large amounts of type I IFNs. pDCs can also activate NK cells, which produce γ-IFN (Villadangos and Young 2008). pDCs primarily initiate immune responses to ingested nucleic acid fragments (Liu et al. 2009). There is some evidence that the AP by certain DCs can be suppressive in nature; these secrete IL10, which leads to the formation of regulatory T-cells (T_{reg}) that dampen the immune response. No one has yet identified (5/09) a line of specific "suppressor-DCs" that induce tolerance to self or other antigens. Tolerance-induction may be some property of immature DCs (Reis e Sousa 2006).

The origin of DCs is from hematopoietic stem cells located in bone marrow. Two major lineages have been identified (Wieder 2003): (1) Myeloid lineage DCs originate from myeloid-committed CD34 + progenitor cells. Myeloid-committed CD34(+) and CD14(–) DCs that can mature to Langerhans DCs (LaDCs) under the influence of TGFβ. LaDCs are largely found in mucosal tissue and the epidermis. Mature LaDCs can activate naive T-cells but not B-cells. (2) Lymphoid DCs also originate from CD34(+) progenitor cells in the bone marrow, and are stimulated to form by IL3. Lymphoid DCs can secrete α-IFN. Curiously, monocytes (immature MΦs) can be driven to become interstitial DCs (iDCs) by the cytokines GM-CSF and TNFα ± IL4. iDCs can induce differentiation of naive B-cells to become Ab-secreting plasma cells. iDCs may migrate to lymphoid follicles and become follicular DCs (fDCs). The interested reader is encouraged to study Akiko Iwasaki's excellent review article on mucosal DCs (muDCs) published in 2007. He makes a valiant attempt to organize the complicated surface molecular attributes of all the types of muDCs (the MHCs expressed, ILs, and other cytokines secreted), as well as their interactions with various classes of T- and B-cells, and how muDCs behave when faced with benign commensal bacteria in the gut. Inflammatory bowel diseases such as Crohn's, ulcerative colitis, or coeliac disease certainly involve muCDs, and a better understanding of their regulatory networks will eventually lead to new, cell-based therapies.

DCs also have shown a dark side for genetically susceptible individuals. They can participate in a self-renewing (autocatalytic or positive feedback) inflammatory environment that leads to autoimmune symptoms. Several scenarios of DC-mediated autoimmunity have been described by Granucci et al. (2008).

It is clear that DCs have several roles in common with MΦs, the most common of which is phagocytosis and AP. Future, dynamic hIS models should certainly include their actions. Genetically modified (GM) DCs, because of their high efficiency at AP, now appear to be candidates for cancer immunotherapy (Kirk and Mulé 2000, Wojas et al. 2003, Zhou 2005).

6.3.4 T-LYMPHOCYTES

T-cells are leukocytes originating in the bone marrow from pluripotent hemo-poietic stem cells. Their rate of production is under control by cytokine hor-mones from the immune system. They are called T-cells (TL) because they are "pre-processed" in the thymus gland (hence T-). Also, they are mostly found in lymphoid tissue, including lymph nodes, the spleen, and submucosal areas of the gastrointestinal tract. The "processing" of immature T-cells, or thymocytes, in the thymus includes selective deletion of any thymocytes that have a T-cell recep-tor (TCR) coat protein with an affinity for normal, self-antigens on body cells. Such adaptive behavior (the selective deletion theory) is thought to prevent auto-immunity if portions of normal cell proteins are accidentally presented to TL by Mφs, etc. Unfortunately, this process is not 100% effective, and many autoim-mune diseases exist (see chapter 7 in Northrop and Connor 2008).

T-lymphocytes are responsible for cell-mediated immunity, which is explained below. Three subsets of T-lymphocytes are characterized by their functions in the immune system, and further described by the proteins found on their cell surfaces. With the exception of cytotoxic T-cells, T-lymphocytes do not directly attack invading organisms. Instead, they serve as activators (amplifiers) or sup-pressors of the immune response through the protein immunocytokines that they secrete in response to antigen (Ag) presentation. Helper T-lymphocytes are gen-erally given the acronym Th, suppressor T-lymphocytes are Ts, and cytotoxic T-lymphocytes are CTL.

All T-lymphocytes have unique, external T-cell receptor (TCR) proteins that may bind to a presented antigenic epitope if affinity is high. By genomically shuf-fling protein subunits, the human body can make ca. 25 million different TCR paratopes, insuring a high probability of binding to an antigenic epitope (Kimball 2009h).

In addition to the many varieties of TCR proteins, there is a linked, five-molecule (pentameric), cell-surface, protein complex, CD3, associated with each TCR. The CD3 complex can sense when the TCR has bound to an epitope, and then it initiates a complex sequence of intracellular events collectively called T-cell activation. In addition, Th lymphocytes have the CD4 protein complex next to their TCR proteins. CD4 is a cell-surface co-receptor for MHC Class II molecules. (MHC II molecules are the type II, major histocompatibility protein molecules that exist on the surfaces of all Mφs and antigen-presenting B-cells for cell identification.) CTLs carry the CD8 protein complex next to their TCRs. Ts lymphocytes can also carry CD8 protein. CD8 is a coreceptor for MHC Class I molecules found on all body cells that do not bear MHC II molecules.

Helper T-cell activation is a complex biochemical process whereby biochemi-cal synthetic machinery in the Th cell reacts to structural changes in the intra-cellular portion of the CD3 molecular complex brought about by successful Ag presentation by an Mφ. The Mφ secretes the cytokine Interleukin-1 (IL1), which binds to receptors on the activated Th cell's (ATh) surface. Under the combined influence of IL1 and activated CD3, two subsets of ATh cells secrete certain immunocytokines. The ATh1 cells secrete the cytokines IL 2 and 3, and gamma interferon (γIFN); ATh2 cells secrete IL3, 4, 5, and 6, tumor necrosis factor α (TNFα), and GMCSF. These immunocytokines have profound effects on other cellular components of the immune system, which is described below.

Antigen presentation (see Section 6.5) by infected somatic cells having MHC Class I + Ag on their surfaces to a cytotoxic T cell (CTL) is illustrated in Figure 6.1. When the CTL's TCR has affinity to the presented Ag, and the CD8 and CD3 molecules react, immunocytokines are released by the CTL and

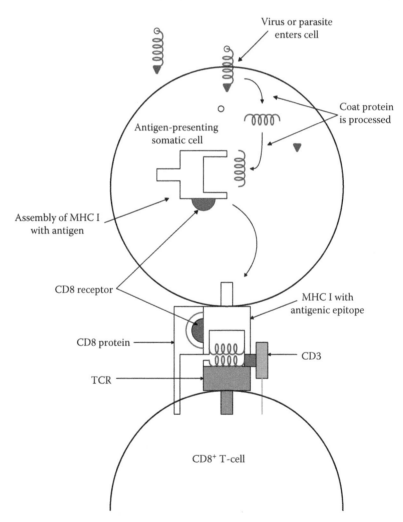

FIGURE 6.1 A schematic of antigen presentation of infecting viral coat fragments by an MHC Type I somatic cell to a CD8+ T-cell (cytotoxic T-leukocyte or CTL). (From Northrop, R.B. and Connor, A.N., *Introduction to Molecular Biology, Genomics and Proteonomics for Biomedical Engineers*, CRC Press, Boca Raton, FL, 2009.)

it is stimulated to undergo clonal proliferation, that is, it reproduces itself with the same complex TCR that has affinity for the antigenic epitope that activated it through presentation. These CTL clones can now "recognize" and bind to other somatic cells infected by the same virus or parasite and carrying the presented Ag on their surfaces. (The invader generally leaves some of its coat proteins on the cell's surface, like a criminal leaving fingerprints at the scene of a crime.) Some CTLs that bind to infected somatic cells release a protein called *perforin*, which literally bores non-repairable holes in the target cell's membrane. Ions and water pass through the holes, causing the target cell to rupture from osmotic lysis. Another means of cell killing is thought to be by the CTL inducing apoptosis in the target cell. As we have already seen, apoptosis is an internally or externally triggered, self-destruction mechanism that causes the cell and its contents to literally disintegrate (see Section 2.5.2). By killing the infected target cell, the CTL or NK prevents any internal viruses or parasites from proliferating. Once the infected cell has ruptured, other components of the immune system (MΦs, DCs, B-cells, antibodies) attack the now externalized virions or parasites. One might wonder why there appears to be two separate methods for CTLs to kill infected cells. We speculate that the evolution of redundancy insures success, and that apoptosis, operating from within the cell, causes viral nucleic acids to disintegrate, inactivating them.

6.3.5   Suppressor T-Cells

Ts, aka regulatory T-cells (T_{reg}), are yet another class of T-lymphocyte. At one point in the 1980s, the existence of Ts cells was debated by immunologists; it is now known that their populations and functions indeed do exist, and what cytokines are involved in their actions. Their exact mechanism of regulation is just beginning to be understood on a molecular level. It is not unreasonable to expect that the immune system, a complex network of cells and their signaling substances (immunocytokines, including interleukins, prostaglandins, interferons, and tumor necrosis factors), has developed mechanisms to halt the excess proliferation of CTLs, NKs, plasma B-cells, and antibodies, once the invading pathogen has been vanquished. This suppression of immune system actions is necessary to conserve immune system resources, and prevent body damage by runaway inflammatory reactions.

Ts cells are characterized by certain cell-surface proteins: CD4 and CD25 (a receptor for the IL2 α-chain), also by the forkhead transcription factor, FOXP3. Expression of cell-surface FOXP3 protein is required for regulatory T-cell development and appears to control a genetic program directing the Ts cell's destiny. Most FOXP3-expressing Ts cells are found in the larger population of major histocompatibility complex (MHC-II), CD4+ Th cells, and high levels of surface CD25 receptor protein. There also appears to be a minor population of MHC class I restricted CD8+ FOXP3-expressing Ts cells. Research on the characterization of Ts cell types is in progress. Additional Ts populations include Tr1, CD8+CD28−, and Qa-1-restricted T-cells (Sullivan et al. 2002).

IL2 has been demonstrated to be required for Ts function in vitro. Also, secondary stimulation of Tss with IL2 causes them to express IL10, a generally inhibitory, down-regulatory cytokine (see table II, de la Rosa et al. 2004). In addition, a complex, six-step, sequential pathway has been described by Chess and Jiang (2004) for the activation of CD8+ T-cells and their conversion into Ts cells. The MHC class Ib molecule (also known as Qa-1 in mice and HLA-E in humans), induces the CD8+ T-cells to become suppressors. Chess and Jiang speculated that exogenous HLA-E may one day prove effective for treating autoimmune diseases.

6.3.6   B-Cells and Antibodies

As mentioned above, B-lymphocytes are so-called because of their embryological origin in the organ called the Bursa of Fabricus, found in birds (not humans). In humans, B-cells are produced from stem cells in the bone marrow and mature in lymphoid tissue of the fetal liver. They are then released into the circulatory system where they are distributed more or less randomly throughout the body. In humans, there are normally about a total of 10^{12} B-cells. This translates into a plasma density of about 10^7 B-cells per milliliter. B-cells are morphologically identical to T-cells so they must be identified by their coat proteins. Normally, about 10%–20% of circulating blood lymphocytes are B-cells.

B-cells are key effector cells in the adaptive immune system. They have the capability of fighting infection by producing huge quantities of specific, freely circulating antibodies (Abs) with affinities to particular antigens (Ags). The Abs bind to the Ags, inactivating them, and marking them for destruction by NK cells, MΦs, and/or the complement system.

Each cell in the population of unstimulated, mature B-cells has a unique, surface-bound IgM protein antibody having great affinity to bind with a specific antigenic epitope. By the clonal selection theory, nature produces millions of different B-cells, each with unique IgM Ab molecules and having a very different affinity to some antigenic epitope. When a specific B-cell's IgM Ab (paratope) binds with a soluble antigen, such as diphtheria toxoid, the bound Ag is engulfed by the B-cell by the process of receptor-mediated endocytosis. Inside the B-cell, the Ag is now digested into fragments that are picked up by MHC II molecules, and then displayed on the B-cell's surface. A helper T-cell with a specific, complementary TCR protein binds to the antigen-presenting B-cell, as shown schematically in Figure 6.2. As in the case of MΦs, the proteins CD4 and CD3 are involved with activating the Th, which secretes the interleukin proteins, IL4, -5, and -6. The local release of these interleukins causes the B-cell presenting the antigen to reproduce clonally, copying the specific IgM antibodies (paratopes) that bound to the antigen's epitope. The activated, mature B-cell clones (also known as plasma cells) then release soluble antibodies, which can bind with free or cell-surface-bound Ag. B-cell growth is also activated by IL1 from MΦs, and by IL2. To make this process more complicated, know that there are five known different isotypes of antibodies.

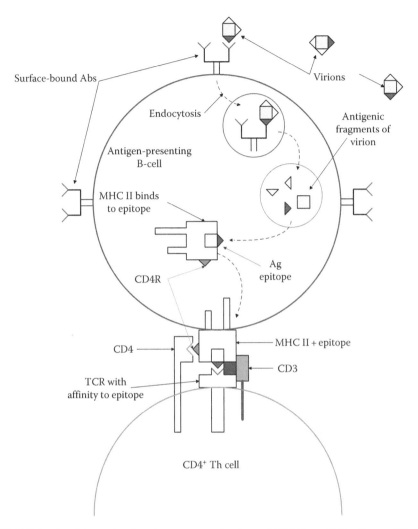

FIGURE 6.2 Schematic of antigen presentation by an MHC Type II B-cell to a CD4+ helper T-cell (Th). In order for the Th to be activated, the TH's TCR must have strong affinity to the presented antigen, and the CD4 and CD3 molecules must bind to specific sites on the MHC complex (a type of molecular "AND" logic). (From Northrop, R.B. and Connor, A.N., *Introduction to Molecular Biology, Genomics and Proteonomics for Biomedical Engineers*, CRC Press, Boca Raton, FL, 2009.)

The process of antigen presentation to Th by B-cells also leads to the production of special, clonally produced, circulating, inactive B-cells having long lives and specificity to the particular Ag. These cells are called memory B-cells (MB-cells), and they evidently provide a rapid and strong humoral response if the pathogen with the Ag is reintroduced.

The antibodies (Abs) produced by activated, plasma B-cells have a unique protein molecular structure: they are "**Y**" shaped, with the Ag-binding domain (paratope) lying between the arms of the **Y**. Molecular diversity in encoding the binding domain structure gives the possibility of well over 10^6 different Ab paratopes in the hIS at a given time (Kimball 2009h). The stem of the **Y** is made from the paired ends of two "heavy protein chains" of over 400 amino acids (AAs) in length. The arms of the **Y** are also paired; the ends of the heavy chains each have a light chain of over 200 AAs. (See Figure 6.3 for an IgG antibody schematic.) As mentioned above, there are five classes of circulating antibodies: IgA, IgD, IgE, IgG, and IgM. (Ig stands for immunoglobulin.) The heavy chains making up these Abs are called alpha, delta, epsilon, gamma, and mu,

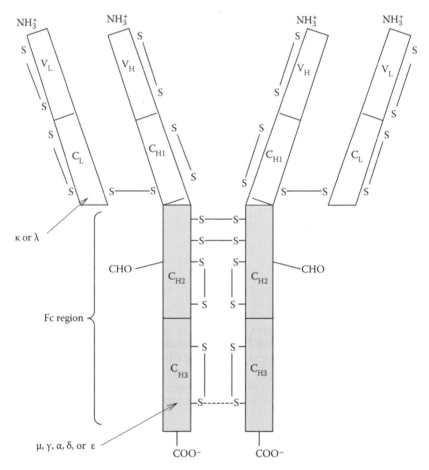

FIGURE 6.3 Schematic of a free (soluble) antibody (Ab) molecule. See text for description. (From Northrop, R.B. and Connor, A.N., *Introduction to Molecular Biology, Genomics and Proteonomics for Biomedical Engineers*, CRC Press, Boca Raton, FL, 2009.)

respectively. The light chains are called kappa or lambda. Genetically random-ized, variable regions of the light and heavy chains form the paratope sites, one in each arm of the "Y." Some Abs, however, are formed from a fusion of as many as 10 light and 10 heavy chains, and hence have 10 binding sites. The protein structure of the stem region of the Y is relatively constant, and contains a constant Fc-binding site through which NK cells, Mφs, mast cells, etc., can recognize and bind to Abs. If the Ab has bound to a cell-surface Ag, NK cells can bind to the Fc site and then secrete the protein perforin that kills the targeted cell. Mφs and DCs clean the mess up.

Abs can act in three different ways to fight invading pathogens: (1) They can directly bind to the invading pathogen, inactivating it by changing its gross structure. (2) They can activate the complement system that destroys the invader. Direct binding of Ab to Ag can lead to large clusters of Abs bound to many Ags; such a cluster is said to be an agglutination. Agglutinations can precipitate, becoming immobile, and prey for CTLs, NKs, neutrophils, and Mφs. (3) The Fc regions of Abs bound to Ags on cell surfaces form attachment sites for NK cells, which kill the cells involved, hopefully before the pathogen can multiply inside the cell.

6.3.7 MEMORY B- AND T-CELLS

The vertebrate adaptive IS has the unique ability to "remember" prior immune challenges by storing the genetic information coding the particular, success-ful Abs that fought the pathogen in unique "memory" B- and T-cells. When an activated Th cells presents its antigenic epitope to the matching B-cell, it signals the B-cell to undergo clonal expansion, producing many copies of itself and releasing many identical Abs having a matching paratope region for the Ag. A fraction of the specific B-cells are caused to cease development and not secrete Abs. They are also given long lives (protection from apoptosis). Upon reintroduction of the original pathogen, the activated Ths have a higher prob-ability of finding a B-cell match in the MB population, hence there is a swifter, larger response from the adaptive IS. MB cells protect us from repeated viral infections.

The more rapid response to repeated infection is expedited by the presence of memory T-cells (MTs). The undifferentiated or naive T-cell population encoun-ters an Ag and becomes activated and divides into many clones or daughter cells. Some of these clones differentiate into effector, helper T-cells (Th) and produce cytokines that activate the IS, or they become cytotoxic T-cells (CTL), and some form long-lived, memory T-cells (MTL). The MTLs have three distinct subtypes: (1) central memory MTLs (T_{CM}) that are thought to represent memory stem cells, (2) MTLs that strongly express genes for molecules essential to the cytotoxic function of CD8$^+$CTLs, and (3) effector memory MTLs (T_{EM}). T_{CM} cells express L-selectin and the chemokine receptor CCR7, and secrete IL-2 but not IFNγ or IL-4. T_{EM} cells do not express L-selectin and CCR7, but do secrete cytokines like IFNγ and IL-4.

6.3.8 TRAFFICKING OF hIS CELLS

All leukocytes respond to chemokines in a guided, wandering, migratory (trafficking) path in the circulatory system, lymphatic system, and in certain extra-vascular spaces. In their journeys, they respond to local concentrations of cytokines by altering the composition, expression, and/or functional activity of their trafficking molecules. In their review paper on immune cell migration in inflammation, Luster et al. (2005) stated, "After being activated, different effector leukocyte subsets must localize together in affected tissues to communicate through short-range cytokines and/or direct cell-cell contact." Therapy for autoimmune diseases and tissue transplant rejection can involve using a drug that interferes with the gathering of activated effector cells (e.g., DCs, neutrophils, T-, and B-cells). This antimigration therapy blocks essential traffic signal(s), causing the effector cells to not congregate in a critical tissue volume in high concentrations. Antimigration therapy has promise in treating tissue graft rejection, certain autoimmune diseases, asthma, psoriasis, Crohn's disease, ulcerative colitis, rheumatoid arthritis, and other inflammation scenarios (Mackay 2008).

Certain trafficking molecules in blood vessels cause leukocytes to tether to vascular endothelial cells by molecular adhesion bonds, and then roll along on their surface, dragged by blood flow, making and breaking their mechanically weak, cell/cell-surface bonds. These tether molecules include certain selectins and integrins. When the immune cell finally binds firmly, it moves by diapedesis across the blood vessel wall by a transient disassembly of endothelial cell junctions and literally squeezing through into the extravascular (interstitial) space, where it acts according to its program to received inflammatory signals. Luster et al. (2005) gave a large table of trafficking molecules involved in inflammatory disease processes, underscoring the incredible complexity of the hIS, even when it comes to IS cell migration.

T-cell trafficking is believed to be causative in allergic asthma. Medoff et al. (2008) described how airway DCs take up an inhaled allergen and then migrate to lymph nodes where they present the antigenic epitope to naive T-cells. The primed T-cells then reenter the lung where they provide surveillance for the allergenic Ag. They also re-enter lymph nodes as memory T-cells. Allergic exacerbation occurs when the allergen is re-introduced to the airway epithelia. Again, resident DCs, Mɸs, and mast cells uptake allergen and present the antigenic epitope to T-cells and MT-cells. The inflammatory cytokines released recruit and stimulate eosinophils and cause the proliferation of Th1 and Th2-cells. The cytokines released by these over-amplified responses cause mucus hypersecretion and smooth muscle cell contraction, restricting airways, hence asthma. Medoff et al. commented that specific T-cell chemokine receptors have been associated with T-cell homing on tissue-specific sites (e.g., skin, the small intestine, lymph nodes). They added that no T-cell chemokine receptor has yet been associated with airway or lung tissue.

Immune cell trafficking is perhaps best modeled by the use of agent-based modeling techniques (see Glossary and Section 11.2.8).

6.4   MAMMALIAN INNATE IMMUNE SYSTEMS

6.4.1   INTRODUCTION

All living creatures (bacteria, plants, animals) have evolved innate immunity to protect them from bacterial, viral, protozoan, fungal, parasitic, etc., infections. In animals such as vertebrates, molluscs, and arthropods, this innate immunity is sufficiently organized so we may talk of their complex, innate immune systems (CIISs). In Section 5.4, we examined some features of invertebrate CIISs in detail. Below we will concentrate on describing the human CIIS. Because of pleiotropy in certain immunocytokines, we find that the adaptive hIS is coupled to the hCIIS, and vice versa.

The hCIIS provides an immediate, pre-programmed defense against infections. However, it evidently does not confer long-lasting or protective immunity to the host. Below we examine some cells of the hIS that operate within the hCIIS, the complement system and the role of AMPs in innate immunity.

6.4.2   NATURAL KILLER CELLS

NK cells represent a small fraction of the total leukocytes in the body. NK cells are unique because they can attack and kill cells in the body that lack the normal, protective, MHC I cell-surface identification proteins, such as some cancer cells, parasites, or certain cells infected by a virus. Thus, they can form a fast, direct line of defense against cancer cells that may bear altered MHC I proteins, in theory bypassing the need for antigen presentation by the hCAIS. In a typical scenario, an NK cell uses an activating receptor on its surface to bind to a critical, cell-surface glycoprotein on a target cell. If another receptor protein on the NK cell's surface binds with a normal MHC I protein on the cell surface, the NK cell is inhibited from secreting perforin or other cell-killing autacoids. NK cells may have up to 11 variations of the MHC I binding protein, insuring that all normal variations in MHC I will be bound.

NK cell killing is also activated by the adaptive IS. This mode of killing involves antibody-dependent cellular cytotoxicity (ADCC) in which a NK cell binds to a site in the Fc region of an IgG antibody bound to an Ag on an infected cell's surface. When the NK binds to the Fc region, it makes contact with the cell's surface, and then secretes the perforin protein. It is not known whether the MHC I-induced inhibition of perforin secretion is operative in ADCC by NK cells. Presumably, both Fc binding and cell-surface contact are required for the release of perforin. Otherwise, NKs would bind to free Abs of any kind and secrete perforin at random, creating havoc to normal, neighboring cells.

NK cells are stimulated by growth hormone (GH), luteinizing hormone (LH), the interleukins IL2, IL12, IL15, and interferons α, β, and γ. Stimulated NK cells secrete γIFN and TNFα. IL10 inhibits the production of IL2-induced γIFN by NK cells.

6.4.3 NEUTROPHILS

Neutrophils, commonly called polymorphonuclear (PMN) leukocytes, are a type of granulocyte. They are called neutrophils because of the way they stain with hematoxylin and eosin (H&E) blood stain, that is, pink, while basophil WBCs stain dark blue, and eosinophils stain bright red. Neutrophils are 10–13 μm in diameter. With eosinophils and basophils, neutrophils form the class of polymorphonuclear cells, named for characteristic multilobed shape of their nuclei. Of all leukocytes, 50%–70% are neutrophils, or $2.5 - 7.5 \times 10^6$ neutrophils/mL. They have a relatively short lifespan, ca. 12 h; however, once activated, they can survive ca. 1–2 days.

Neutrophils can be classified as part of the cellular arm of the human innate immune system. They attack bacteria and fungi. They are strongly chemotactic to the cytokines, IL-8, γIFN, and C5a (a complement molecule), released by mast cells, Mφs, and activated epithelial cells. Neutrophils also express and release cytokines that amplify inflammatory reactions by other hIS cells. Neutrophils are phagocytes; they ingest and kill microorganisms. Internally, they use reactive oxygen species and hydrolytic enzymes to kill and break down the ingested microorganisms in phagosomes. Neutrophils also release a group of AMPs in three types of granules into extracellular space by a process known as degranulation. These AMPs include lactoferrin, cathelicidin, myeloperoxidase, bactericidal/permeability increasing protein (BPI), defensins, the serine proteases neutrophil elastase and cathepsin G, and gelatinase. In addition, neutrophils release complex molecules of chromatin (DNA) combined with serine proteases that form webs of fibers called NETs. The NETs trap and kill microbes extracellularly.

In humans infected with HIV and cats infected with FIV, there is neutropenia, and those remaining neutrophils have reduced chemotaxic responses to endogenous IL-8 and bacterial chemoattractants (Phillipson et al. 2006). HIV and FIV cause defective neutrophil development in the bone marrow with a resultant reduction in their granularity and chemotactic receptor expression. Administration of exogenous GM-CSF restores neutrophil granularity and function, implying a restoration of normal neutrophil development. Unlike Th and Mφs, neutrophils do not appear to be directly infected by HIV or FIV. Most probably, the immune viruses infect bone marrow stem cells and stroma cells, and there may be direct suppression of hemopoiesis by viral proteins (Phillipson et al. 2006). One of the factors leading to opportunistic infections in AIDS is the neutropenia and reduced neutrophil activity.

Recent evidence (Sandilands et al. 2005, Culshaw et al. 2008) has emerged that human and murine neutrophils can express MHC Class II and the costimulatory molecules, CD80 and CD86. These neutrophils can present MHC Class II-restricted peptides to naive Th and induce their proliferation.

6.4.4 MAST CELLS AND PLATELETS

Mast cells and platelets are important accessory immune effectors outside of the classical adaptive immune system structure (which includes B-cells, antibodies, T-cells, Mφs, DCs, and the antigen presentation process). All these effectors contribute to the inflammatory response to infection, and when viewed along with the rest of the immune system, underscore the functional redundancy of the immune system.

Mast cells are very important cells in allergic inflammatory reactions, and in particular, asthma. Mast cells are formed in tissues from undifferentiated precursor cells manufactured in the bone marrow and released into the blood. There are two types of mast cells: one is found in connective tissues, the other in mucosal sites.

Connective tissue mast cells are found mostly in the skin, while mucosal mast cells reside in the gut and lungs. Both kinds of mast cells are sessile. They collect around blood vessels, nerves, and lymphatic vessels. They are concentrated around potential points of entry of pathogens and foreign micro-particles (e.g., pollen, mold spores) into the body.

In humans, mucosal mast cells comprise a total of 1% of the cells located in the lungs and accessory tissues. About half of the mucosal mast cells are found in the intra-alveolar septa, with the other half located in the mucosa of the trachea, bronchi, and bronchioles. Mast cells contain cytoplasmic secretory granules of diverse morphology. Mast cells can release such inflammatory substances as histamine, proteoglycans (heparin and chondroitin sulfates), leukotriene-C_4, proteolytic enzymes (tryptase and chymase), serotonin (aka 5-hydroxytryptamine), IL4, IL5, IL6, TNFα, and prostaglandin-D_2. All of these substances contribute to the discomfort of an acute allergic reaction. Pain, edema (tissue swelling), inflammation, etc., are the result of mast cell stimulation. Activation of mast cells in the lungs can result in acute asthma, with bronchospasm (airway size reduction due to smooth muscle contraction), alveolar tissue swelling (vital capacity reduction), and thickened mucous that clogs small airways.

Mast cells are stimulated to release their granules by the binding of their high-affinity Fc receptors (FcRs) to the Fc region of IgE antibodies that, in turn, have high affinities to allergen molecules such as pollens or danders. There can be as many as 5×10^5 FcRs per mast cell (Guyton 1991). The actual trigger that initiates the release of mast cell products is the cross-linking of the bound IgE–Ag complexes attached to the mast cell's Fc receptors. Such a cross-linking can occur if there are high enough concentrations of antigen and free IgE Abs, and the Ag has multiple binding sites on it for the Abs. When mast cells secrete IL4, IL5, IL6, and TNFα, other immune system cells such as eosinophils and T-cells are recruited into the inflammatory scenario, which is biochemically complex.

Platelet cells have two major functions in the body: One is to form clots to stop bleeding from wounds. Clot formation is a very complicated biochemical process that we will not detail here. The other function is to participate in inflammatory reactions of the immune system. There are normally from 1.5×10^5 to 4×10^5 platelets per cubic mm of blood. Platelets are made in the bone marrow by megakaryocytes, and then enter the blood. Like mature red blood cells, they are cells without nuclei, ranging from 2 to 4 µm in diameter. Their resting shape is also discoidal, however, when activated by appropriate stimuli, they become spheroidal and develop many protruding "tentacle-like" protrusions up to 5 µm in length. They have typical phospholipid bilayer plasma membranes with many

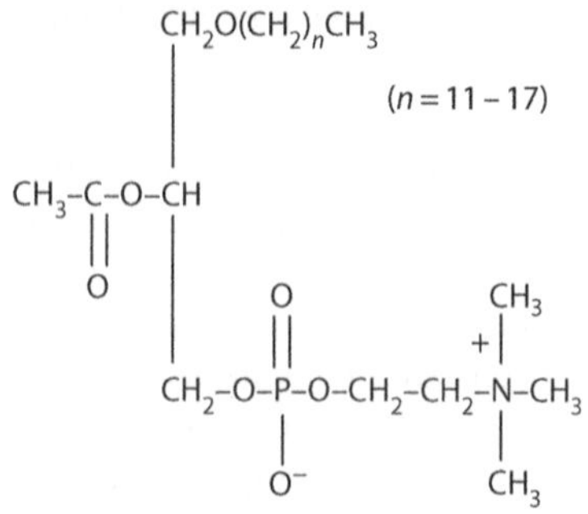

$$CH_2O(CH_2)_nCH_3$$

$$(n = 11 - 17)$$

$$CH_3-C-O-CH$$

$$O$$

$$CH_2-O-P-O-CH_2-CH_2-N-CH_3^+$$

$$O^-$$

$$CH_3$$

$$CH_3$$

FIGURE 6.4 The PAF autacoid molecule. (From Northrop, R.B. and Connor, A.N., *Introduction to Molecular Biology, Genomics and Proteonomics for Biomedical Engineers*, CRC Press, Boca Raton, FL, 2009.)

embedded receptor molecules. Their cytoplasm contains a "skeleton" of microtubules, largely formed from actin and myosin, proteins normally associated with muscles. Also within platelets are mitochondria, several types of secretory storage granules, and an endoplasmic reticulum (ER). The half-life of platelets is from 8 to 12 days. Most are destroyed by Mϕs residing in the spleen (Guyton 1991).

Platelet activating factor (PAF) is synthesized by platelets, mast cells, Mϕs, eosinophils, certain renal cells, and vascular endothelial cells. The 2D PAF molecule is shown in Figure 6.4. PAF is a pharmacologically active autacoid with many diverse functions (Hardman and Limbird 1996). It causes vasodilation and it is over 1000 times more effective than histamine or bradykinin in promoting edema. PAF promotes platelet aggregation in vitro and in vivo, and is a chemotactic factor for eosinophils, neutrophils, and monocytes, causing these leukocytes to aggregate at the source of its release. PAF also causes the contraction of smooth muscles in the GI tract, the small airways, and the uterus. PAF certainly contributes to the edema that occurs when the immune system fights a bacterial infection introduced through the skin.

6.4.5 HUMAN COMPLEMENT SYSTEM

Complement is a remarkable, concatenated, noncellular, protein-based biochemical system employed by the innate immune system to fight infections. Some version of the complement system (CS) is found in all jawed vertebrate animals (gnathostomes) (Nair et al. 2005). One goal of the CS is to create the molecular attack complex (MAC) from the individual complement proteins, C5b, C6, C7, C8, and C9, aka C5b6789. The MAC formed on the surface of an infected cell literally bores a hole in its cell membrane, lysing it, and causing it to die before any endoparasites can multiply and eclose from it.

The CS consists of more than 30 large glycoproteins, most of which are soluble in blood plasma; some are bound to the surfaces of certain types of cells. (Turn back to Figure 4.1 for a schematic of the three principal complement pathways.)

When the CS is activated, many of its component molecules break down, recombine with other complement molecules to form proteolytic enzymes, which further break down other complement molecules, in complicated chain reactions. Most of the complement glycoproteins are synthesized by liver cells; however, Mφs, fibroblasts, and epithelial cells may also produce complement components (Hardman and Limbird 1996). The organization of the complement system appears extravagantly baroque; it works, however. Its complicated operation bespeaks for the "trial-and-error" processes that must operate in molecular biological evolution.

The CS is necessarily highly regulated by both soluble and cell-surface inhibitory molecules. Its functions are strongly linked to the responses of the CAIS. For example, the somatic cell-surface-bound complement regulatory proteins, CD55 and CD59, protect mammalian cells from uncontrolled lysis by the MAC complex and the perforin protein. CD55 inhibits formation of the C3 convertase enzymes, and CD59 prevents the terminal polymerization of the membrane attack complex (Ruiz-Argüelles and Llorente 2006, Śladowski et al. 2006). The circulating proteins, clusterin (aka apolipoprotein J) and vitronectin (aka S-protein) bind to the MACs and inactivate them, preventing cell apoptosis or necrosis (ultimately, lysis) (Mabbott 2004, Chauhan and Moore 2006). Increased clusterin immunoreactivity appears to correlate with the level of PrP^{Sc} deposition in lymphoid follicles in TSE diseases (Sasaki et al. 2006).

The components of the hCS have four major roles in the immune system: (1) Certain CS molecules (C3b and C4b) are opsonins, that is, they opsonize, or coat foreign objects bound to Abs. Once opsonized, the complex of foreign object, Abs, and opsonins can be phagocytocized by hIS cells having CR3b and CR4b opsonin receptors, then be lysed and destroyed. Portions of the foreign object can then be presented on an MHC molecule in "formal" antigen presentation to Th cells. (2) Certain other CS molecules (C5a, C4a, and C3a) promote inflammation, increase vascular permeability, cause edema, and recruit phagocytic cells (neutrophils and Mφs) and activate platelets, which secrete platelet activating factor (PAF), histamine, etc. (3) The CS can lyse cells, similar to NK or CTL cells. In this convoluted scenario, C5b binds to the target cell surface, thence to C6, C7, and C8. C7 and C8 undergo conformational changes that expose hydrophobic domains, which penetrate the lipid bilayer of the cell's membrane. The C5b678 complex catalyzes the polymerization of the final, C9 component, which like perforin, makes a permanent, 10 nm diameter hole in the membrane, causing cell lysis from osmotic shock. The entire C5b6789 = C_{5b-9} molecular assembly is known as the membrane attack complex (MAC). (4) The CS participates in immune complex clearance, that is, the removal of groups of cross-linked Abs bound to Ags. This removal is accomplished by binding C3b and C4b covalently to the immune complex. CR1 receptors on erythrocytes (red blood cells) bind to the C3bC4bAbAg complex. Erythrocytes carry these complexes to the spleen and liver where they are destroyed by phagocytic cells. Thus, even red blood cells can be enlisted in the CIIS function.

There are three biochemical pathways by which CS activation is initiated or triggered: (1) the classical pathway, which begins with AgAb complexes: (2) the alternative pathway; and (3) the lectin (aka lytic) pathway, in which activation begins at the pathogen surface. In the lectin pathway, serum lectins bind to the sugar, mannan on the pathogens' surface glycoproteins, triggering complement reactions. (Glycoproteins in the classical pathway are prefixed with "C-.") All three pathways converge on an enzyme called C3 convertase. (C3 convertase in vertebrates is the C2b protein.) It is likely that the lectin pathway is evolutionarily older than the classical and alternative pathways because components of the lectin pathway can bind directly to microbial surfaces without the requirement for discrete recognition molecules such as Abs (Nair et al. 2005).

One of the reasons the CS is difficult to understand is that many of its proteins are cleaved (divided) or joined together to become active component molecules. In the case of C1, it is formed from one unit of C1q, plus two units of C1r and two units of C1s. C1q has six binding sites for the antibody IgFc. The binding affinity is low, so that at least two bound sites are required on C1q for C1r activation. This means that two adjacent IgG molecules or one distorted IgM is required. Thus random, unbound, soluble IgG or IgM will not activate complement; they have to be clumped around an antigen. When activated, C1r cleaves C1s to form the C1s protease, which in turn cleaves $C4 \rightarrow C4a + C4b$, and $C2 \rightarrow C2a + C2b$. C4b combines with C2a to form the C3 convertase, C4bC2a, which splits $C3 \rightarrow C3a + C3b$. Now, at the risk of really straining the reader's patience, we note that the complex C4aC2bC3b is formed and acts as C5 convertase, producing C5a and C5b. The molecular complex, C5b6789, is the membrane attack complex (MAC) that causes cell lysis. The proteins C3a, C4a, and C5a are mediators of inflammation and Mϕ recruitment. Details of the steps in the classical, lectin, and alternate pathways and their regulation are too complicated to describe here in detail; the interested reader should consult an online text such as Kimball (2007).

Complement is clearly important in the humoral response to infection, insofar that it works with antibodies to lyse cells, and promotes inflammation. The fact that dormant complement molecules are always present in the blood means that the complement response needs little time to generate a local defense against an introduced pathogen. Protein fragments from lysed pathogens can be phagocytosed, and presented as antigens to Th cells by B-cells, Mϕs, and DCs, thus recruiting the adaptive responses of the immune system.

Because the activated complement proteins can cause extensive collateral damage to "good" cells, the inflammatory action of the CS is tightly regulated in order to keep it localized. Regulatory actions are known to act at three points: (1) A C1 inhibitor protein (C1INH) binds to free C1 and inhibits its spontaneous activation. C1INH is released from C1 upon activation by immune complexes (Ag*Ab). It also limits the activation of C4 by inhibiting the C1r and C1s proteases. (2) Decay accelerating factor (DAF) found on cell surfaces, C4 binding protein (C4bp), and Factor H all act to speed the breakdown of C3 convertases, breaking the chain of reactions leading to MAC. (3) Two proteins found on cell surfaces, CD59 and homologous restriction factor (HRF), inhibit the binding of C9 to C5b678 to form the MAC. The factors controlling these inhibitors are not known.

Because of its complexity and unknown rate constants, the human complement system (hCS) has hardly ever been included in dynamic mathematical models of immune system function (Gruber 2005). One paper by Chen et al. (2006) described an iterative immune algorithm based on the "classical" complement activation pathway (ODEs were not used). Their paper was written from the computer science viewpoint of genetic algorithms (standard genetic algorithm and clonal selection algorithm), rather from the viewpoint of molecular biology/physiology. They did not use mass action kinetics; rather reactions were modeled by positive and reverse "bind operators" and a "cleave operator." However, the Chen et al. model included all known molecular components of the classical complement pathway. Another more comprehensive model of the hCS was devised by Bellur (2004), described in his MS thesis. Bellur used the C++ language. His model included chemotactic attraction of phagocytes to an infection site, promotion of opsonization, generation of inflammation, and removal of immune complexes from the circulation. Disease scenarios involving certain bacteria were modeled. Object-oriented programming was used to develop a random simulation. Nonlinear ODEs and mass-action kinetics were not used.

6.4.6 HUMAN AMPS IN INNATE IMMUNITY: THE URINARY TRACT

Not all combat against invading pathogens is carried out by the relatively slow hAIS, and the more rapid response of the molecular hCS. Humans, too, use AMPs to try to maintain sterility in locations such as the urinary tract (Zaslof 2007). Cathelicidin is a linear peptide that folds into an α-helix on contact with a membrane; it is expressed on all epithelial cell surfaces, and by circulating leukocytes (neutrophils, monocytes, NK cells, and γ,δ T-cells). Cathelicidin attracts neutrophils and monocytes, and interacts with fMLP receptors on these cells, evidently stimulating additional AMP expression. Both cathelicidin and defensins are expressed along the human urinary tract.

Human α-defensins contain 18–45 AAs, and are effective against bacteria, fungi, and many viruses. They kill bacteria by making holes in their membranes. In humans, neutrophils secrete α-defensins HNP1–4, and Paneth cells in the small intestine secrete defensins HD5 and HD6 into the lumen, protecting intestinal stem cells and influencing the composition of intestinal commensal bacteria. About 40 β-defensins are widely expressed throughout human epithelia. Both the α- and β-defensins attract immature DCs, a component of the adaptive IS; β-defensins bind to the CCR6 receptor. Human β-defensin-1 (HBD1) is constitutively produced by cells in the kidneys' loops of Henle, the distal tubules, and the collecting ducts. Thus, it protects the kidneys from ascending infectious bacteria (Zasloff 2007). This and other AMPs are present in an AMP-rich biofilm on the surface of the tubular epithelia, and thus can kill or disable pathogens that try to colonize on the epithelia, protecting the urinary tract. Other AMPs in the urinary tract include the constitutively secreted Tamm–Horsfall glycoprotein, and lactoferrin, which damages microbial membranes.

If pathogens overwhelm the constitutive defenses by HBD1 and cathelicidin, the growing infection stimulates the inducible HBD2 defensin, which attracts local leukocytes. In a positive feedback process, attracted Mφs can secrete cytokines such as IL-1 and TNF, which in turn further stimulate the expression of inducible AMPs, such as lipocalin, by epithelial cells. Further inflammation results in the secretion of IL-8, which leads to the recruitment of more neutrophils to fight the infection. The AMP, lipocalin, provides antimicrobial defense by sequestering bacterial siderophores. If the AMP system fails, severe bacterial infection occurs in which the complex adaptive IS and complement system take over, and in which there is generally tissue destruction.

Of course the urinary tract is not the only place the human innate immune system employs AMPs. Salzet (2005) reported that during coronary bypass surgery, a new family of serine-rich, bi-phosphorylated AMPs was isolated from patient plasma. These new AMPs were fibrinopeptide A, apolipoprotein CIII, peptide B, and enkelytin.

6.5 ANTIGEN PRESENTATION

Antigen presentation (AP) is a complex biochemical process whereby the complex adaptive immune system (CAIS) activates itself to fight specific invading microorganisms, including viruses, bacteria, parasites, molds, etc., and then stores the information to clonally reproduce specific Abs. As you will see, there are several AP scenarios.

Antigens (Ags) are by definition molecules or, more precisely, parts of molecules that elicit an immune response involving antibodies (Abs) in the body. Ags are generally proteins, glycoproteins, or polysaccharides. Nonprotein molecules called haptens can also cause an immune response. This response is generally thought to be due to the hapten reacting chemically with certain protein tissue molecules; the product of the reaction is the actual antigen. Antigens can be of exogenous or endogenous origin. Exogenous antigens can be inhaled proteins from animal dander, dust mites, pollen, or mold spores. Exogenous antigens can also be eaten; nuts and shellfish are common foods to which people are allergic. Another route for internalization of antigens is through the skin and mucous membranes. These can include injected antigens such as used in allergy desensitization, pollen injected by a thorn stick, bacteria entering through a cut or abrasion, parasites that are injected by insects or that burrow into the skin, and poison ivy oil. Proteins from the cell walls of internalized bacteria, viruses, spirochetes, parasitic worms, etc., can activate immune responses. Lectin proteins are involved in these responses.

Endogenous antigens include those which originate within the cells of the body. In the case of autoimmunity, a normal, somatic cell protein is misidentified by some component of the immune system as a foreign, invading protein. In other cases, cancer cells may have mutant proteins on their surfaces that the immune system can recognize and attack.

Antigen presentation is a crucial step in the amplification of specific immune responses by the CAIS. All antigen presentation involves the binding of an antigen molecule to a large protein molecule present on the antigen-presenting cell's surface. This large molecule is called the major histocompatibility complex (MHC) molecule, or equivalently, the human leukocyte antigen (HLA) molecule. Part of the MHC molecule projects through the antigen-presenting cell's plasma membrane into its interior. There are two major types of MHC molecule: Type I and Type II. Nearly all nucleated cells in the body carry MHC I proteins, and can present antigens derived from internal foreign proteins to Th cells. The Type I MHC molecule has three major components: (1) A transmembrane protein that is exposed at the cell surface. The outermost portion is made from two alpha helices that form a groove between them. (2) A short peptide molecule is attached in the groove between the alpha helices. (3) A β_2-microglobulin molecule is also attached to the α helices. To make matters more complex, humans have three subtypes of MHC I molecules, designated as HLA-A, HLA-B, and HLA-C. The genes coding the structures of these molecules are inherited, so if a person is heterozygous for HLA-A, HLA-B, and HLA-C, then we will observe six different MHC I proteins in such a person. MHC I proteins are found on all endogenous, nucleated cell surfaces in the body, other than Mϕs, neutrophils, and certain B-lymphocytes.

The MHC I processing pathway begins inside the MHC I cell, where peptides from a virus' coating or a parasite that has entered the cell are broken up by proteolytic enzymes from proteasomes in the cell's cytosol. The peptide fragments, between 8 and 18 amino acids, are then transported into the lumen of the rough endoplasmic reticulum (RER). Special transporter molecules move the peptide fragments inside the RER lumen where ribosomes form the MHC I molecules around them. The assembled MHC I–peptide complex is then transferred to the cell's plasma membrane by Golgi apparatus. The MHC I–peptide complex is externalized through the plasma membrane; however, one transmembrane protein "root" projects back through the plasma membrane into the cell's interior. The three subunits in the externalized, active MHC I molecule are (1) the "presented" antigenic peptide, (2) the β_2 microglobulin, and (3) the transmembrane polypeptide.

Ag presentation from cells carrying active MHC I molecules is made to T-leukocytes that carry the CD8 surface glycoprotein, as well as receptor proteins for the MHC I complex. CD8+ leukocytes are generally cytotoxic T-lymphocytes (CTL), although some regulatory T-cells also carry CD8. The T-cell receptor (TCR) protein for MHC I molecules, and adjacent CD8 protein molecule protrude through the surface of the CTL. Antigen presentation is only possible if the TCR protein has an affinity for the presented, antigenic peptide nestled in the MHC I's cleft, and the CD8 protein on the CTL binds to the MHC I side site. (See Figure 6.1 for a schematic illustration of this process.) Once the CD8 and TCR proteins have bound to the antigen-presenting cell's MHC I–peptide epitope complex, an internal, biochemical "message" is sent to the CTL, activating it to destroy the antigen-presenting somatic cell. Destruction is necessary to prevent eclosion of reproduced virions or parasites from within the cell. It is destroyed for the "greater good" of the organism. The activated CTL either perforates the cell membrane of the infected cell with a special protein it secretes, or it sends chemical messengers to the infected cell that cause it to self-destruct by apoptosis.

The affinity match between the presented antigen and a TCR protein is coincidental, that is, many millions of possible TCR configurations are randomly generated on CD8⁺ CTLs. The CTL receiving the message is also stimulated to divide, exactly reproducing a clone of CTLs with the TCR protein specific for the presented antigen. In this way, the immune system greatly amplifies its ability to fight specific, internal viral or parasitic invaders.

A second kind of antigen presentation is done by cells that carry the Type II MHC protein. The MHC II protein is present on the surface of mature Mφs and other phagocytic cells such as neutrophils, certain B-lymphocytes (B-cells), and DCs. These cells can engulf, phagocytose, or endocytose an entire bacterium or virus, as well as fragments of cell debris. Once internalized, the foreign material is broken down enzymatically to smaller peptide fragments that are taken up by MHC II molecules, which in turn migrate to the cell surface with the presented peptide fragment antigen (Ag), or epitope. When the antigen-presenting Mφ, neutrophil, or DC encounters a helper T-cell with a TCR with affinity to the presented Ag, and the Th's CD4 protein binds to a specific site on the MHC II molecule, signals are passed to the Th to activate it to secrete various immunocytokines that in turn activate the adaptive (cellular) arm of the immune system. In particular, activated helper T-cells (AThs) secrete IL2, 3, 4, 5, 6, 7, 10, and 13, also TNFβ, γIFN, and GM-CSF. Antigen presentation by a Mφ to a Th cell is illustrated schematically in Figure 6.5.

The antigen-presenting B-cells have B-cell receptor (BCR) proteins bound on their outer surfaces. The BCRs are surface-bound, IgG antibodies; they, too, have enormous variability in their affinities for antigenic epitopes. When a certain BCR has a strong affinity to an epitope, it binds to that epitope, and then is internalized where the Ag protein is broken down, and pieces of it are bound to MHC II proteins, which are then externalized on the B-cell. If a Th has a T-cell receptor (TCR) protein with affinity to the peptide epitope nestled in the B-cell's MHC II protein, it binds to the antigen-presenting B-cell. A CD4 molecule on the Th must also bind to the side of the MHC II molecule to activate the Th, which then secretes cytokines that cause the antigen-presenting B-cell to reproduce clonally with its BCRs that are specific for the Ag in question. This process is a molecular "AND" logic; it is shown schematically in Figure 6.2. The clone of plasma B-cells grows identical BCRs with the affinity to the presented epitope. These BCRs are ultimately released as free Abs. Since the original BCR bound to this epitope, the Abs released are also specific for it, and contribute to an amplified, humoral, immune defense.

6.6   AUTACOIDS: IMMUNOCYTOKINES, PROTEINS, AND GLYCOPROTEINS SECRETED BY hIS CELLS

As we will describe, under certain conditions, cells from the immune system as well as other cells secrete cytokines that can affect immune system cells. Collectively, we call these intercellular messengers *autacoids*, a term coined by Hardman and Limbird (1996). The word autacoid is from the Greek *autos* = "self,"

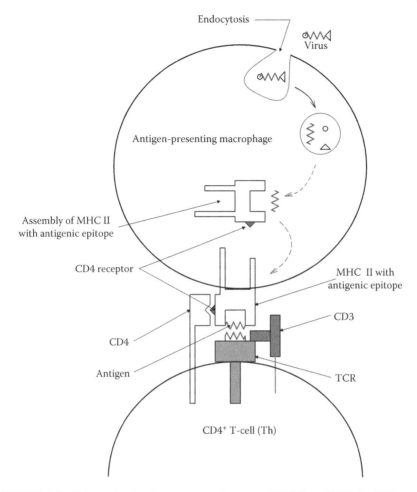

FIGURE 6.5 Schematic of antigen presentation by an MHC Type II Mφ (or DCs) to a CD4+ Th cell. Note that full Th activation requires not only that the TCR on the Th have affinity to antigenic epitope, but also that the Th's CD4 and CD3 bind to sites on MHC II. (From Northrop, R.B. and Connor, A.N., *Introduction to Molecular Biology, Genomics and Proteonomics for Biomedical Engineers*, CRC Press, Boca Raton, FL, 2009.)

and *akos* = "medicinal agent or remedy." There are currently about 54 immuno-cytokines (aka, cytokines) known. The cytokines include the interleukins (about 44 are named at this writing, and more are being discovered every year). The immune system also uses interferons (3), certain prostaglandins, tumor necrosis factors (2), and various cell growth stimulating factors. In addition, there are certain cluster of differentiation (CD-x) proteins fixed on cell surfaces, and many cell-surface receptor proteins. Any one of these many and diverse humoral substances can collide with an immune system cell where it can bind to its specific surface receptor protein. Once bound, a molecular transduction mechanism causes the cell to internally synthesize the same or other immunocytokines, more of the same or other membrane receptors for immunocytokines, proteins used in antigen presentation (TCRs, MHC, CD3, CD4, CD8, etc.), antibodies (if the cell is a B-lymphocyte), complement components, perforin, etc., or the binding can lead to the inhibition or switching off of an ongoing biochemical synthesis by the target cell.

Individual immunocytokines can posses the property of pleiotropy, in which a cytokine can have different effects on different target cells having receptors for it. Also, several different cytokines can have the same biological function on the same target cell (this is called degeneracy).

Presumably suppressor T-cells (Ts) synthesize immunocytokines (e.g., IL-10) that turn-off or down-regulate the cellular synthetic machinery in the immune system that was activated in the initial stages of the immune response. It will be necessary to understand the kinetics of their production and their half-lives in vivo in order to incorporate immunocytokines into any quantitative, dynamic model of the immune network. Sadly, much of this detailed information is nonexistent, or is based on in vitro studies.

The interleukins (ILs) are high molecular weight glycoprotein autacoids secreted by leukocytes for purposes of signaling other leukocytes and somatic cells. The ILs have diverse functions such as cell attraction (chemotaxis), inducing target cells to secrete other ILs and other immune system autacoids and to manufacture more receptors for ILs. They also figure in the maturation of hIS cells such as MΦs and T-cells. Most ILs are stimulatory in function, there are a few, however, that inhibit or down-regulate certain immune system functions.

There is some disagreement in the literature on the exact number of interleukins. Several tables list 35 (e.g., Wiki 2010), others list more, counting subtypes. In our opinion, about 44 interleukins have been defined at this writing (March 22, 2010) (counted in the 2006 Genenames IL table, including subtypes). The Genenames 2006 table also gives IL receptors (ILRs), and the human chromosomes on which they both are coded. However, the ongoing research on the immune system reveals the identity of new cell-surface proteins, receptors, and immunocytokines at a rate such that our knowledge base for the immune system is never in the steady state; new discoveries arise at an amazing rate. In addition, IL naming has in some cases lacked coordination, resulting in some confusion between initial IL names and those finally approved by The International Union of Immunological Societies' Interleukin Nomenclature Committee. For example, there is now no official IL14 and IL30. IL30 in Table 6.1 is now officially IL27. IL14 no longer appears in the 2006 Genenames table. The COPE table of interleukins (Ibelgaufts 2008) lists 49 ILs, including the proscribed IL14 and IL30 designations. The COPE table has an update of September 2008. It is probably safe to say that including subtypes, there are currently between 44 and 49 ILs.

It is clear that our knowledge about the interleukins, their origins, and functions is continually changing. Table 5.1, updated from Northrop and Connor (2008), summarizes the known (and partially known) ILs. It gives their sources, their target cells, and their effects. To effectively signal, or carry information to

TABLE 6.1

Table of Human Interleukin Cytokines: 46 ILs, Including Subtypes, Are Listed

Interleukin	Sources	Targets	Effects
IL1α, IL1β also, IL1F5–10	Activated Mφs: IL1α is bound to Mφ surface, IL1β. Both bind to same receptor. Also endothelial cells, B-cells, and fibroblasts	TLs, B, Mφs, neutrophils, bone marrow cells, fat cells, bone osteoclasts, brain cells, adrenal cells, vascular endothelial cells, smooth muscle cells	B-cell proliferation, TL production of other cytokines, fever, cachexia
IL2: aka T-cell growth factor (TGF)	Activated CD4⁺ T-cells, CTL and large granular lymphocytes (LGL)	IL2 receptors on Th, CTL, Mφ, B-cells	Proliferation of CD4⁺Th cells, also CD8⁺ CTL. Stimulated production of IFNγ by TCs. Stimulated production of NK and LAK cells, also Mφ to secrete IL1 and TNFα. Edema
IL3	Mast cells, activated CD4⁺Th	IL3 receptors on mast, B-cell and some mature granulocytes	Hematopoietic growth factor, aka multicolony-stimulating factor and mast cell GF. Stimulates B-cell differentiation; inhibits LAK cell activity
IL4	15–19 kDa glycosylated protein dimer. CD4⁺ Th, CD8⁺ memory T cells, mast cells and basophils	T-cells, B-cells, Mφ, fibroblasts and endothelial cells	Induces CD4⁺ Th to differentiate into Th2 cells, suppresses development of Th1 cells. Acts as growth factor for B-, T-, and mast cells. Stimulates MHC2 expression on B-cells, promotes plasma B cells to make IgE1 and IgE Abs
IL5	45 kDa protein dimer. CD4⁺Th2, NK and mast cells	Eosinophils, B-cells, mast cells	Activation and differentiation of eosinophils. Proliferation of immature B-cells. Stimulates Ig class switching to IgA. Stimulates mast cells
IL6: aka IFNβ2 and other aliases	212 AA glycoprotein, 26 kDa, made by: Mφ, mast cells, T-cells, fibroblasts, neutrophils	Stimulated B-cells, Mφ, hepatocytes, CD4⁺ and CD8⁺ Th cells, fibroblasts	Stimulates acute-phase protein synthesis by liver. B-cell GF; induces B maturation into Ab-secreting plasma cells. T-cell activation and differentiation and stimulates their production of IL2 and IL2Rs. Inhibits production of TNF
IL7: aka T-cell GF	Glycoprotein, 25 kDa. Bone marrow cells, thymus stromal cells	B- and T-cells	Stimulates development of pre-B and T cells and early thymocytes

(continued)

TABLE 6.1 (continued)
Table of Human Interleukin Cytokines: 46 ILs, Including Subtypes, Are Listed

Interleukin	Sources	Targets	Effects
IL8	Activated T-cells, Mφ, and lymphocytes	Mφ, neutrophils, and basophils	Chemotactic substance, stimulates production of inflammatory leukotrienes
IL9	CD4$^+$Th2 cells, some Bs	CD4$^+$Th cells, CD8$^+$ T-cells, B-cells, bone marrow precursor cells	Inhibits lymphokine production by CD4$^+$Th cells producing IFNγ. Stimulates growth of CB8$^+$T cells, promotes production of immunoglobulins by B-cells and proliferation of mast cells
IL10: aka cytokine synthesis inhibitory factor	18 kDa glycoprotein made by CD4$^+$Th, activated CD8$^+$Th and activated B-cells, Ts cells	T-cells, NK cells, Mφs, CTLs	Inhibits IFNγ production by activated T-cells. Inhibits IL2-induced IFNγ production by NK cells. Inhibits IL4- and IFNγ-induced MHCII expression on Mφs. Reduces CTL proliferation
IL11	26 kDa protein produced by bone marrow stromal cells, Mφs, and some fibroblasts	Stimulated. B-cells, Mφ, hepatocytes, CD4$^+$ and CD8$^+$ Th cells, fibroblasts	Mimics activity of IL6. Stimulates Th-dependent, B-cell immunoglobulin secretion, increases platelet production and induces IL6 expression by CD4$^+$T-cells
IL12: NK stimulatory factor (NKSF) (IL12A and B)	Disulfide-linked dimer; 40 and 35 kDa Ea. From activated B-cells, and antigen-presenting cells (Mφ and B)	Th1, CTL, NK, T- and B-cells	Activates Th1 cells. Increases rate of production of CTL, NK, and lymphokine-activated killer (LAK) cells, increases NK cytotoxicity, induction of IFNγ production by NK and T-cells. Inhibition of IgE synthesis by B-cells
IL13 P600	12–17 kDa protein made by activated Th cells	Mφ, B-cells	Inhibits Mφ activation and release of IL1β, TNFα, IL6, and IL8. Enhances Mφ and B-cell differentiation and proliferation, increases CD23 expression, induces IgG$_4$ and IgE class switching
No IL14			
IL15	114 AA protein of 15 kDa. From activated Mφ, epithelial cells, and fibroblasts	NK, B- and T-cells	Stimulates NK, B- and T-cells. Also stimulates CTK and LAK cell activity and Ig production. Also an attractant for T-cells. (Actions similar to IL2.)

IL16 lymphotactin LCF	Four linked, homotetrameric chains. 56 kDa. Made in CD8+T-cells and eosinophils. Released in resp. to histamine and serotonin	All cells carrying CD4 protein	Attracts CD4+T, eosinophils, and some Mφ. CD4 is receptor for IL16. Increases MHCII expressed on monocytes. Inhibits HIV replication in monkey and human CD4+T cells
IL17A-D,F:	17.5 kDa protein made by activated CD4+T-cells	Certain immune system cells	Induces the release of IL6, IL8, G-CSF and PGE_2 (inflammatory cytokines)
IL18: IGIF	Mφ	Th1-cells	Induces IFNγ production
IL19: aka MDA1	IL10 homolog. Distinct population of keratinocytes. Monocytes	Binds with IL20 receptors on CD4+T-cells	Up-regulates IL4 expression in CD4+T-cells. Induces IL10 that down-regulates IL19
IL20: aka IL10D	20.1 kDa, 176 AAs. IL10 homolog. Distinct population of keratinocytes. Monocytes	Distinct population of keratinocytes	Induces hyperproliferation of keratinocytes that express IFNα
IL21	18.6 kDa. T-cells	CTL, NK and B-cells	Stimulates CTL and NK cells. Suppresses growth of metastatic melanoma. Augments therapeutic antibody-mediated tumor lysis
IL22: IL-TIF	20 kDa. IL10 homolog. Activated Th1 cells, leukocytes, hematopoietic stem cells	Pancreatic acinar cells. Intestinal epithelial cells	Up-regulates MHC1 Ag expression. No IL22R on activated B-cells
IL23A: aka IL23, P19	20.7 kDa. Heterodimer, IL10 homolog	Expressed mainly by dermal cells	Promotes proliferation of naïve and memory T-cells and stimulates their IFNγ production
IL24: aka MDA7	23.8 kDa. IL10 homolog. Lymphocytes, monocytes, Mφ, leukocytes, melanocytes, thymus, spleen, skin, mammary gland	Tumor cells	Selective tumor apoptosis. Induces dose-dependent death in melanoma cells
IL25	Not assigned		
IL26: aka AK155	19.8 kDa. Heterodimer. IL10 homolog. T-lymphoblasts. Activated NK, and memory CD4+ T-cells		

(continued)

TABLE 6.1 (continued)
Table of Human Interleukin Cytokines: 46 ILs, Including Subtypes, Are Listed

Interleukin	Sources	Targets	Effects
IL27	Not assigned		
IL28A: IFNλ2	22.3 kDa. Monocytes	Intestinal epithelial cells	Increased IL8 production, increased expression of antiviral proteins: myxovirus resistance A and 2′,5′-oligoadenylate synthetase. No influence on Fas-induced apoptosis. Decreases cell proliferation
IL28A: IFNλ2 IL28B: IFNλ3	21.7 kDa. Monocytes	Intestinal epithelial cells	Increased IL8 production, increased expression of antiviral proteins: myxovirus resistance A and 2′,5′-oligoadenylate synthetase. No influence on Fas-induced apoptosis. Decreases cell proliferation
IL29: IFNλ1	21.9 kDa. Monocytes	Intestinal epithelial cells	Increased IL8 production, increased expression of antiviral proteins: myxovirus resistance A and 2′,5′-oligoadenylate synthetase. No influence on Fas-induced apoptosis. Decreases cell proliferation
No IL30			
IL31			Role in skin inflammation?
IL32			Induces monocytes and Mφ to secrete TNFα, IL8, and CXCL2
IL34			
IL35	Regulatory T-cells		Suppression of Th cell activation

immune system cells, all interleukins (and in general, all cell signal substances including hormones and neurotransmitters) must be eventually broken down to inactive forms and their amino acids recycled metabolically. In vivo IL half-lives are generally unknown, but necessarily must be on the order of hours to days in order for the stable dynamic functioning of the immune system. A knowledge of IL half-lives is necessary to effectively formulate a valid, dynamic, mathematical model of the immune system.

Interferons are another important class of immune system circulating protein. Three types of interferon molecules are currently known: interferon alpha (IFNα), IFNβ, and IFNγ. IFNα is thought to have 15 subtypes, which presumably act alike, so we can deal with generic IFNα. IFNα is made by virus-infected monocytes and lymphocytes, IFNβ comes from virus-infected fibroblasts, and activated T- and NK cells release IFNγ, which has three different isoforms (Held et al. 1999).

All three IFNs engage in antiviral activity; they increase NK and CTL cell activity and increase MHC I expression on cells. Specifically, IFNα and IFNβ activate NK and CTLs, stimulate B-cell differentiation into plasma cells (which release soluble antibodies), induce MHC I protein kinase, and cause other internal biochemical events that inhibit viral messenger RNA translation, preventing viral multiplication in the target cell. IFNα is the first cytokine to be used effectively in clinical trials as an exogenous immunotherapy agent; it has been approved for the treatment of several types of human cancer. It has been demonstrated that IFNα has a direct antitumor action: it down-regulates oncoproteins, induces tumor suppressor genes, has antagonistic effect against the action of growth factors, induces cell suicide (apoptosis), inhibits angiogenesis (growth of new blood vessels that nourish the tumor), as well as increases MHC I production and tumor-associated antigens (Northrop 2000).

Because they cause immune suppression, exogenous IFNβ1a and IFNβ1b have been successfully used to treat the autoimmune disease, multiple sclerosis (MS). Both interferons slow the progress of disability, reduce the rate of relapses, and reduce the severity and number of MRI-detected lesions. They do not cure MS, however.

IFNγ increases expression of MHC II molecules and receptors for the Fc region of antibodies (FcγR) on Mφs, increases the activity of neutrophils and NK cells, promotes T- and B-cell differentiation, and increases IL1 and IL2 synthesis. In addition, IFNγ increases type IgG2a antibodies while suppressing IgE, IgG_1, IgG_{2b}, and IgG_3 synthesis. IFNγ also inhibits proliferation of Th2 cells, but not Th1 cells.

One limitation to IFN therapies is that exogenously administered IFNs have the risk of loosing their effectiveness over time due to the production of host antibodies (Northrop 2000). It is ironic that a potentially effective immunotherapy may be thwarted by the host's immune system.

Two TNF proteins are known: TNFα and TNFβ. The principal effect of the TNFs is to cause inflammation. TNFα or Cachectin is a 17 kDa, soluble protein trimer. It is a 185 AA glycoprotein hormone, cleaved from a 212 AA peptide on the outer surface of Mφs. It is the product of activated Mφs, fibroblasts, mast cells, and some T- and NK cells. Peritoneal mast cells have preformed, reserve TNFα available for immediate release upon appropriate stimulation. TNFα can induce fever, either by stimulating the release of pyrogenic prostaglandins (PGs), or by causing the release of IL1, which is also pyrogenic. TNFα plays a major positive role in fighting local infections. It induces the production of acute-phase proteins, mobilizes neutrophils, activates T- and B-cells, increases the release of antibodies and compliment, increases the adhesion of platelets to blood vessel walls, and increases the extravasation of lymphocytes and Mφs (diapedesis) to fight the infection in intracellular space. These actions result in the phagocytosis of the pathogens, local vessel occlusion, and the drainage of cells, debris, and fluid into the lymphatic system. These actions, carried out with other relevant cytokines, lead to the removal of the infecting pathogen, and eventual tissue repair. TNFα also can bind to TNF receptors on tumor cells and kill them. Some tumor cells shed their TNF receptors, which become soluble and bind to TNFα, inactivating it. Excess TNFα can also cause the body to create antibodies to it, yielding a similar deactivation. The exact mechanism of infected cell killing by TNFα is not known.

High levels of TNFα cause pain, systemic edema, hyperproteinemia, and neutropenia. TNFα binds to TNF receptors on virus-infected cells, causing apoptosis and cell death. Acute infections can lead to the overproduction of TNFα, leading to fever, immune suppression, septic shock due to the loss of blood volume to extracellular space, fatigue, anorexia, and cachexia (wasting of the body tissues).

TNFβ, or lymphotoxin, is made by activated CD4+ and CD8+ T-cells. It binds to the same receptor sites as TNFα. It has similar properties to TNFα, and induces apoptosis in many types of virally infected cells, tumor cells, and damaged cells. Certain endotoxins, such as that from *Staphylococcus* bacteria, cause high production of TNFs that contribute to the onset of toxic shock syndrome. Chronic, high production of TNFs may be responsible for the cachexia observed in many chronic parasitic infections and some cancers.

The prostaglandins and other eicosanoids are part of a family of nonprotein autacoids that includes the leukotrienes, thromboxanes, and prostacyclin. Collectively, they are called eicosanoids because they are synthesized from 20-carbon fatty acids that are in turn derived from the 20-carbon arachidonic acid (see Figure 6.6). The metabolites of arachidonic acid are varied, and have diverse pharmacological effects. Subtle changes in eicosanoid structure can produce dramatic alterations in their bioactivity. Receptors for eicosanoids are highly specific for affinity and effect.

Two major biosynthetic pathways exist for the products of arachidonic acid: The prostaglandins (PGs), prostacyclin, and thromboxanes are made through the cyclooxygenase synthetic route. The second major pathway of arachidonic acid metabolism begins with lipoxygenase. This pathway leads to the biosynthesis of various leukotrienes. These pathways are shown schematically in Figure 6.7.

In discussing the immune system's use of eicosanoids as effectors and signaling substances, we will be restricted to the consideration of those autacoids

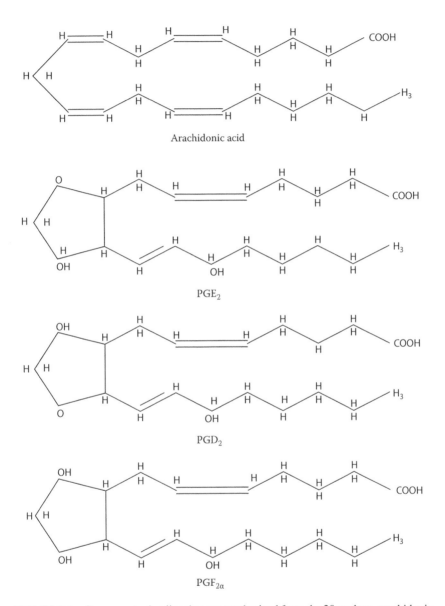

Arachidonic acid

PGE$_2$

PGD$_2$

PGF$_{2\alpha}$

FIGURE 6.6 Some prostaglandins that are synthesized from the 20-carbon, arachidonic acid molecule. These eicosanoids include PGE$_2$, PGD$_2$, and PGF$_{2\alpha}$. (From Northrop, R.B. and Connor, A.N., *Introduction to Molecular Biology, Genomics and Proteonomics for Biomedical Engineers*, CRC Press, Boca Raton, FL, 2009.)

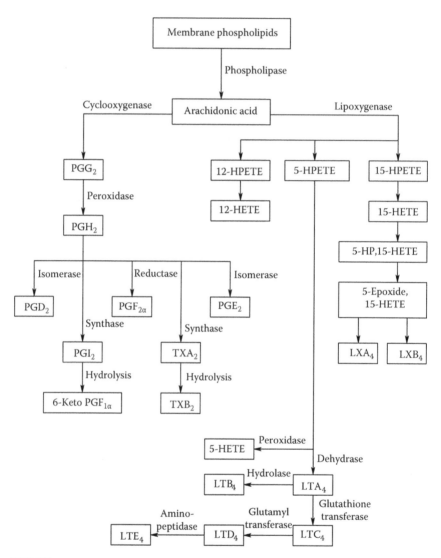

FIGURE 6.7 The cyclooxygenase and lipooxygenase pathways whereby arachidonic acid is converted to various prostaglandins, leukotrienes, and other 20-carbon autacoids. Important molecules: PG_ = prostaglandins, TXA_2 = thromboxane A_2, x-HPETE = x-hydroperoxy-eicosatetraenoic acid, x-PETE = x-monohydroxyeicosatetraenoic acid, LTX_4 = leukotriene X_4, and LXA_4 = tetraene trihydroxy lipoxin A_4. (From Northrop, R.B. and Connor, A.N., *Introduction to Molecular Biology, Genomics and Proteonomics for Biomedical Engineers*, CRC Press, Boca Raton, FL, 2009.)

considered by the authors as having a direct effect in the functioning of the immune system. First, it is relevant to remark that the eicosanoid immunocytokines are short-lived. They are chemically unstable and are rapidly broken down in the body (mostly in the pulmonary circulation) to inactive forms. Their half-lives are thus on the order of minutes, and they must be continually synthesized to have effective concentrations.

PGE_2 is formed by Mφs and granulocytes; PGD_2 comes from mast cells. Both of these PGs are formed through the cyclooxygenase pathway, followed by an arcane series of biochemical reactions. PGE_2 causes the relaxation of smooth muscle, both in blood vessels and in the lungs; thus, it acts as a bronchodilator when given as an exogenous aerosol. PGE_2 and PGI_2 enhance edema formation, the infiltration of leukocytes into the site of infection/inflammation and they potentiate the production of pain by bradykinin (Hardman and Limbird 1996).

Most importantly, PGE_2 also acts as an immune response suppressor substance. Both PGE_2 and PGD_2 decrease the rate of release of proteases from stimulated granulocytes, thus limiting the inflammatory reaction. In addition, PGE_2 inhibits histamine release from basophils and generally inhibits the release of activating immunocytokines from sensitized T-cells and B-cells. It also inhibits CTLs. PGE_2 simulates the release of the hormone ACTH from the anterior pituitary gland. ACTH in turn stimulates the formation of certain adrenocortical hormones, notably cortisol. Cortisol is a potent anti-inflammatory hormone that has the following effects on the inflammatory immune response (Guyton 1991): (1) It lowers fever, mostly by suppressing the release of IL1 by Mφs. (2) It suppresses the reproduction of T- and B-cells, reducing the number of circulating Abs. (3) It reduces the chemotaxis of Mφs, NKs, and CTLs into the inflamed area, thus reducing cell death and the release of inflammation-producing autacoids. (4) It decreases capillary permeability, hence edema. (5) It stabilizes and strengthens lysosomal membranes so that when cells are ruptured, there is reduced spread of proteolytic enzymes causing reduced collateral damage to adjoining cells.

It is well known that aspirin (acetylsalicylate) acts as an antipyretic (fever reducer) and anti-inflammatory substance. Aspirin inhibits the function of the cyclooxygenase enzymes responsible for the conversion of arachidonic acid to the various prostaglandins. In particular, certain cytokines (IL1β, IL6, IFNα, IFNβ, and TNFα) induce PGE_2 production by the circumventricular organs near the hypothalamic area in the brain. PGE_2 triggers the hypothalamus to cause the body temperature to rise by readjusting the body's heat balance. Aspirin and other nonsteroidal anti-inflammatory drugs (NSAIDs) block the production of PGE_2, hence lower fever.

We know that chronically elevated cortisol levels (as in the case of prolonged periods of stress) can lead to blood sugar imbalances (Rizza et al. 1982), decreased bone density (Merck 2007), high blood pressure (Whitworth et al. 1995), and the accumulation of abdominal fat (Epel et al. 2000). Because NSAIDs block the production of PGE_2, and PGE_2 is a precursor of cortisol, could NSAIDs be a potential therapy for persons suffering from chronic stress? Beware of the LUC.

The final immunoactive peptide we shall mention is vasoactive intestinal peptide (VIP), a pleiotropic hormone found throughout the body (see Glossary). VIP has general dampening effects on the hIS, causing the suppression of inflammation and expediting the formation of memory T- and B-cells. VIP is released by neuro-secretory cells of the autonomic nervous system; it has a relatively short half-life, ca. 2 min. It stimulates the release of IL-10 by certain immune cells (see the review by Delgado et al. 2004).

For a brief summary of all cytokine functions, see the review by Rameshwar and Bardaguez (2009).

6.7 DISCUSSION: HOW THE hIS IS ADAPTIVE

In Section 6.3, we summarized the immune system's many cells, and the more important cytokine and autacoid proteins having strong signaling and/or effector actions in the immune network. Clearly, most of these immunocytokines are stimulators or activators that increase the degree of immune response at a number of levels. Very few immunocytokines have definite, known inhibitory actions in the immune network; these include IL10, PGE_2, and VIP. This apparently disparity between the number of activators and suppressors challenges our understanding of how the overall immune response works. However, all immunocytokine molecules have finite lifetimes in vivo; they are broken down by enzymes so that their effects as immune system activators naturally decay in time. Thus, if the stimulus (e.g., a bacterial endotoxin) for immunocytokine production disappears because of successful combat by the immune system, immunocytokine production decreases, and their titer is reduced enzymatically, leaving the immune network in a resting state. hIS cells also have finite lifetimes, eventually undergoing apoptosis.

Another possible natural scenario to explain why there are so few suppressor immunocytokines may lie in the concept of anti-idiotypic antibodies. Vigorous production of identical Abs for a particular antigen, **X**, leads to the demise of the pathogen associated with it. An antigen-presenting, phagocytic, B-cell, by chance, has a surface antibody with an affinity to the Abx's **X** binding site, that is, the particular, unique, IgG on the B-cell's surface has a site that looks like **X** to the Abx. Thus the bound Abx is phagocytosed by the B-cell, broken down enzymatically, and then the protein **X** binding site is combined with MHC II and externalized to the B-cell's surface where it is presented to a helper T-cell (Th). The Th is stimulated by the process of antigen presentation to release immunocytokines that stimulate clonal expansion of the antigen-presenting B-cell with its particular IgG with affinity to the Abxs. Eventually, the B-cell daughters change to plasma cells and release soluble Abx with affinity to Abx. These Abxs combine with the binding sites on the Abxs, inactivating them, thus damping the humoral immune response and inflammation, etc. One can speculate that if there are enough anti-idiotypic Abxs in circulation, anti-anti-idiotypic antibodies might be made to suppress the Abxs. Complexity is in the eye of the beholder.

6.8   CATALYTIC ANTIBODIES AS THERAPEUTIC AGENTS

Catalytic antibodies (CAbs), also known as abzymes, are an ingenious way of using the human complex adaptive immune system (hCAIS) to bioengineer Abs that can catalyze specific chemical reactions that have therapeutic value. The biosynthesis of CAbs is a true molecular/biological engineering application of the human (or mouse) CAIS to create designer Abs (Lerner et al. 1991). These CAbs can be used to fight cancer, infections by pathogens, and even counteract weight gain, drug addiction, genetic diseases, and autoimmune diseases.

Enzymes and CAbs are catalysts, that is, they are chemicals that reduce the free energy of the transition state of a biochemical reaction, causing the reaction product to form more rapidly. According to Ali et al. (2009), "They obtain much of their catalytic efficiency from tight binding of the transition state for the reaction. This binding energy stabilizes the transition state of the substrate, which reduces the activation energy for the chemical modification of bound substrate."

Often, CAbs are used in conjunction with prodrugs. It has been estimated that only one out of 10^7–10^9 different endogenous human Abs will form a bond with a foreign antigen (Wentworth and Janda 2001). It is possible to inoculate with a specific, bioengineered, antigenic hapten molecule that will cause the CAIS to generate a population of CAbs, which then can be harvested and used to catalyze specific biochemical reactions. The hapten molecule is generally chosen to be the transition state analog (TSA) of a biochemical reaction that we wish to accelerate or catalyze, coupled to a carrier molecule. The reaction's transition state is the intermediate molecular configuration or state having the highest chemical energy. In an irreversible reaction, colliding reactant molecules in the transition state will always go on to form product(s). (The TSA mimics the transition state [TS] in both shape and electronic configuration, but has a much higher chemical stability than that of the true TS.) Both natural enzymes and CAbs act to stabilize a reaction's TS.

CAbs were first prepared by Slobin (1966) to catalyze the specific hydrolysis of ortho nitrophenol esters. In their review paper, Ali et al. (2009) commented, "Catalytic antibodies offer new possibilities for their potential therapeutic applications because of a high degree of reaction specificity, greater affinity towards transition state analog and their latent ability to block unwanted protein-protein interactions."

In one example of a CAb application cited by Ali et al., CAbs were developed with two active sites, one site had an affinity to a unique surface protein on certain cancer cells, the other catalyzed the cleavage of an anti-cancer prodrug into its active form at the cancer cell surface. This targeted chemotherapy requires that the CAbs are given first and are allowed to bind with the target Ca cells, some unbound CAbs are allowed to circulate. Next, the anti-cancer prodrug is given; it is activated by the CAbs bound to the Ca cell surfaces. Thus, there is a high local concentration of active anti-cancer drug, while the low concentration of free CAbs elsewhere produces a generally low concentration of the active anti-cancer drug, sparing the patient the many side effects of chemotherapy. The unconverted, inactive prodrug is generally not toxic.

The same strategy was used experimentally (Paul et al. 2003) to bind engineered CAbs to the gp120 surface protein on HIV virions (see Figure 8.19). The enzyme carried on the CAbs catalytically cleaved the gp120, distorting the 3D structure of the HIV envelope protein, thus interfering with HIV binding to CD4+ Th cells and preventing the HIV virions from infecting them.

There are three main methods of CAbs production: polyclonal, hybridoma, and phage-display (Wentworth and Janda 2001). In the polyclonal method, the subject is hyper-immunized with the transition-state analog (TSA) hapten, causing production of the desired CAb population, mixed in the blood with all the other Abs of the hIS. No attempt is made to purify the CAbs so generated. This process is fast and cost-efficient.

The method of choice for specific CAb production is the hybridoma method. This process begins with hyperimmunization with the TSA-hapten; then the specific CAb-producing B-cells are isolated from the spleen. Next, these B-cells are immortalized by fusion with a mouse cancer cell line, in vitro. The hybrid cells thus formed secrete the desired monoclonal CAbs programmed by the B-cells. Because this process is done in vitro, the CAbs formed are homogeneous, easily purified, and harvested. A flowchart outlining the hybridoma method of monoclonal CAb production is shown in Figure 6.8.

The phage-display method is more complicated, and will not be described here. Wentworth and Janda (2001) give a flowchart comparing the steps in the three methods of producing specific CAbs. The reader interested in CAb production should see the review papers by Wentworth and Janda, and Tanaka (2002).

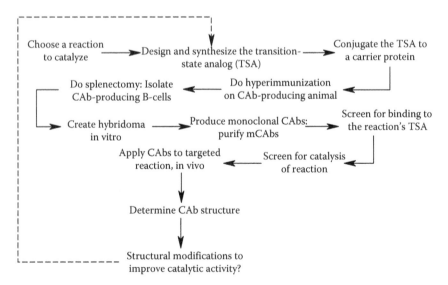

FIGURE 6.8 A flowchart outlining the hybridoma method of monoclonal CAb production. See text for description.

The amazing fact that CAbs can occur naturally was demonstrated in the blood of asthmatic patients (Paul et al. (1995, 2003), cited by Nevinsky et al. 2000); natural CAbs were found in asthmatic patients who had specific proteolytic activity for VIP (vasoactive intestinal peptide). Paul et al. (1995) found that immunizing mice with anti-VIP CAbs resulted in the induction of mouse asthma. It is known that a deficit of VIP in tissues of the respiratory system plays a major role in the pathophysiology of bronchial asthma. Nevinsky et al. suggested that the dysfunction of the respiratory tract in asthma may stem from the protease activity of these VIP-specific CAbs.

Nevinsky et al. (2000) also cited evidence that endogenous RNA-hydrolyzing CAbs are found in patients with certain autoimmune diseases. Another source of CAbs is human breast milk. These CAbs may protect neonates during their first months against viral and bacterial infections with their DNAse and RNAse activities. In this respect, they act like AMPs (see Sections 5.4.5 and 6.4.6).

In a study by Lacroix-Desmazes et al. (2005), plasma was withdrawn from patients with sepsis or septic shock, and IgG Abs were isolated. They found that the presence in plasma of IgGs having natural serine protease-like hydrolytic activity was strongly correlated with survival from sepsis. These authors commented that in some patients, these endogenous CAbs may participate in the control of disseminated microvascular thromboses, perhaps playing a role in the recovery from a disease. They found that IgG from three of the surviving patients hydrolyzed clotting factors 8 and 9. The big questions are through what mechanism(s) did the natural IgG, CAbs, specific for serine protease-like hydrolytic activity, arise? What cytokines keep the B-cell clones for them activated?

We are challenged to come up with candidate biochemical/genetic mechanisms that are responsible for the evolution of natural CAbs. Presumably, the mechanisms of antigen presentation are required for the biosynthesis of clones of both natural and designer CAbs (see Section 6.5). Has molecular evolution allowed the gradual design of natural CAbs? What mechanisms down-regulate specific CAbs? What cytokines are involved, and are there specific anti-CAb Abs? Certainly the existence of natural CAbs in the hIS underscores the marvelous complexity of this nonlinear complex system.

While exogenous CAbs have unquestioned therapeutic value in a wide range of human conditions and diseases, it was recognized in 1990 by Blackburn et al. that they can also form the basis of potentiometric molecular recognition sensors. These authors developed a prototype sensor using modified pH electrodes and a CAb specific for the hydrolysis of the analyte, phenyl acetate. The catalyzed hydrolysis produced H^+ ions that were sensed by the pH system. The pH EMF could be related linearly to the log of the phenyl acetate concentration [PA] over a range of $20–500\,\mu M$, with a detection limit of $5\,\mu M$ [PA]. Alternative applications of CAbs in other biosensor applications were described in their paper.

6.9 SUMMARY

The hIS is seen to be a very complex biochemical/physiological system, certainly orders of magnitude more complex than the complex IS of the fruit fly,

Drosophila. Instead of just having hemocytes for immune functions, the hIS has many specialized cell types: Mφs, DCs, helper T-lymphocytes, suppressor T-cells, cytotoxic T-cells, B-cells and antibodies, memory B- and T-cells, NK cells, neutrophils, mast cells, and platelets. In addition, many of these immune cell types have levels of activation to cytokine messages. The many cytokines in the hIS, and their sources and targets are described in Section 6.6.

The human complement system, nominally called the innate arm of the hIS, was seen to be actually coupled to the adaptive arm of the hIS. Finally, we describe some AMPs unique to the hIS.

In summary, the complexity of the human immune network is imposed by the large number of participating cells (and their graded states of maturity or activation), the many immunocytokines and their often pleiotropic actions, and the variable causal interconnections between the IS's components (the "rules" change as cells mature). Complexity is further produced by the antimicrobial peptide (AMP) branch of the hIIS, and its regulation. Cytokines provide the signals that lead to immune system activation or suppression. They may also induce AMP expression. The fact that many cytokines exhibit pleiotropy makes our understanding of the mechanisms of immunoregulation more difficult. In addition to the interleukins, certain immune system cells and other cells secrete other regulatory molecules such as interferons and tumor necrosis factors—all proteins—and the eicosanoid prostaglandins (not proteins).

As you have just seen from the text above, the complexity of the immune system is daunting. Its functional defects (e.g., allergies, anaphylaxis, and autoimmune diseases) are generally treated by "sledge-hammer pharmacology"—drugs that massively suppress the overall immune responses and inflammation. Only recently have drugs such as Singulair™ appeared, which selectively block leukotriene receptors to lower the inflammatory response to allergens. As our understanding of this complex system increases, there will be many more targeted immunotherapies (including for cancer and autoimmune diseases) developed in the near future, such as drugs to block hIS cell trafficking.

7 Complexity in Quasispecies and MicroRNAs

7.1 QUASISPECIES: VIRAL AND BACTERIAL EVOLUTIONARY MODELS

7.1.1 INTRODUCTION

Quasispecies (QS) are well-mixed "clouds" of viral or bacterial genotypes that appear in a population at a steady-state, mutation-selection balance (Bull et al. 2005). Viral quasispecies (VQS) have been considered to be complex adaptive systems (CASs) (Ruiz-Jarabo et al. 2000). RNA viruses, because of their high mutation rates, are used both experimentally (in vivo), and in stochastic mathematical models as a prototypical QS to study the dynamics of evolution in binary systems, mutation-selection balance, and also the mechanisms of catastrophic population collapse.

Manrubia et al. (2005) stated, "The high error rate inherent to RNA virus replication is probably due to the absence of proofreading activities of RNA replicases and retrotranscriptases. Although most of the mutations produced upon replication have a negative effect on fitness, advantageous mutants [by definition] reproduce faster than deleterious, such that a mutant-selection equilibrium arises." It also appears that viral mutations can enable a VQS to jump host species, landing in a new host with a naive CAIS, allowing it to surge ahead in the fitness race. The concept of a mutation-selection balance is a fundamental principle of population genetics. In an ecosystem, there is a steady-state equilibrium distribution between fit variants and less fit variants, both created by natural selection.

In VQS modeling, the steady-state, final distribution of genotypes in populations is generally of interest, as well as how the populations approach this equilibrium, given an initial distribution of genotypes, mortality parameters, and environmental conditions. Mutations introduce new genotypes with various fitnesses, while natural selection causes the more fit individuals to increase in frequency at the expense of less fit variants. Bull et al. (2005) remarked that "Ultimately, a population reaches a genotype distribution at which these two forces exactly cancel each other, leaving the genotype distribution unchanged. This stable assemblage of genotypes has been called both a mutation-selection balance and a quasispecies."

Clearly, VQS system models may approach equilibrium asymptotically, and so for practical reasons a large, finite number of generations, F_N, is used. Because stochastic parameter selection is used in VQS models, the steady states are in fact noisy—they vary around their means as $N \to \infty$. By running many identical simulations, the noise can be reduced by ensemble averaging.

Bacterial evolution has also been studied experimentally and also modeled from a QS viewpoint. Elena and Lenski (1997) measured the relative fitnesses of mutant strains of *Escherichia coli* to the wild type. They generated 255 mutant strains of *E. coli* that differed from the wild type by one, two, or three mutations. Tagkopoulos et al. (2008) used in vivo studies as well as computer modeling to simulate the interaction of *E. coli* QS with their environment, and their evolutionary responses. They were able to show that *E. coli* transcriptional responses showed an "associative learning" paradigm leading to "predictive behavior" relative to temperature and oxygen changes. When *E. coli* travel from a high pO_2, low T, outside environment into a mammalian gut, $pO_2\downarrow$ and $T\uparrow$. Tagkopoulos et al. found that many of the *E. coli* genes down-regulated by $T\uparrow$ were the same as those down-regulated by $pO_2\downarrow$, suggesting that the bacteria use the temperature step as a signal predicting low-pO_2 conditions to come. When the bacteria were placed in an environment where the relationship between T and pO_2 was reversed, they observed a strong selection pressure for individuals that managed to uncouple the linkage between the two responses. This so-called learning exhibited by bacteria is not the same as that which takes place in metazoans with a CNS, but rather involves changes in the bacterial genetic network that takes place over generations rather than the course of a single bacterial lifetime. It may involve adaptive flexibility in operon regulatory networks.

7.1.2 VQS MODELS

Manrubia et al. (2005) described the structure of their VQS computer model. Every simulation started with a single "virion" with fitness value W ($1 \leq W \leq \infty$). At each generation (replication round, F_k), the virion produces n descendants which go on to replicate in the F_k generation ($k = 1, 2, 3, \ldots, r$). The process is repeated for r generations (F_r), then stops. The number n of descendants of each individual virion is determined by a Poisson distribution with mean W. Thus the replication rate is proportional to fitness. The average number of mutations per replication is m, and the actual number of mutations, k, also follows a Poisson distribution with mean, m. Each individual mutation can have a neutral, positive, or adverse effect on fitness with probabilities $(1 - p - q)$, q, and p, respectively. These authors assumed that there were no epistatic interactions among the mutations, and that the total change in fitness, ΔW, is given by the sum of the individual effects of the mutations occurring in the same sequence. Thus,

$$\Delta W = \sum_{i=1}^{k} \delta W_i \qquad (7.1)$$

where the individual fitnesses δW_i are summed over the actual number of mutations, **k**. The absolute value of each δW_i is obtained from an exponential probability distribution with unity average, sic:

$$Pr(|\delta \mathbf{W}|) = \exp(-\delta \mathbf{W}) \tag{7.2}$$

Manrubia et al. went on to describe the sign of the change to be negative with probability **p** and positive with probability **q**. Virions with fitness <1 are deleted from the system, that is, their genome is made extinct. $\mathbf{N_m}$ is the maximum number of virions per lytic plaque. In their paper, these authors studied the role of **r** and **m** on their models' behavior.

VQS models have also explored the interaction of a general VQS system with an adaptive immune system (Kamp et al. 2003), and the fate of HIV in the hIS where there is vast spreading of viral genomes, well over what happens with an ordinary viral infection (Kamp and Bornholdt 2002). Wilke (2003) at the Digital Life Laboratory at Caltech, Pasadena, also described a complicated, detailed, stochastic model of VQS evolution.

Eigen (1993) found that there were VQS modeling scenarios in which a minute increase in the mutation rate could trigger a large change in the genotype composition in the population. This tipping point is characteristic of CNLSs, and was called an error catastrophe.

Gómez et al. (1999) examined large numbers of cDNA clones of the hepatitis C virus (HCV) grown from single isolates and found "unquestionable proof that the viral genome cannot be defined by a single sequence, but rather by a population of variant sequences closely related to one another. This way of organizing the genetic information is referred to as quasispecies." Figure 7.1 illustrates schematically the growth of a "typical" VQS from an original (F0 virion), defined cDNA sequence. In the F1 generation, one mutation is seen (black dot); in F2, we find the

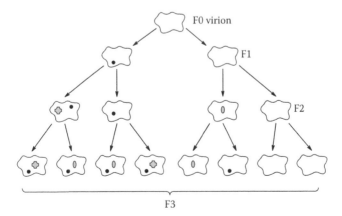

FIGURE 7.1   Schematic illustrating the growth of a "typical" VQS from an original (F0) virion to the F3 generation.

original F0 RNA, plus the F1 (dot) mutation, plus a mutated F1 (gray cross + dot), plus a new mutation of the original cDNA (gray oval). The F3 generation has five different genomes, as shown in the figure. Note that two viral cDNAs are the original (F0) cDNA.

Gómez et al. described the *complexity of quasispecies* generally as the amount of nonrepetitive genetic information stored in a genome. There is low redundancy in the cDNA (or cRNA) of viruses. Complexity in VQS also is related to their mutation frequency and polymorphism among the variants.

In a more recent in vivo study, Frost et al. (2005) examined the mechanisms whereby HIV1 can rapidly escape from neutralizing Abs in the hIS. They compared the pattern of evolution of the HIV1 *env* gene between individuals with recent HIV infection whose VQS had either a low or a high rate of escape from hIS Ab responses. They found that the HIV1 *env* gene coded many small AA substitutions over the entire viral protein envelope. Frost et al. stated that neutralizing Ab responses exert a "soft" selection pressure by affecting the relative fitness of different strains present in the VQS. They also speculated that cellular immune responses (NK, CTL) may exert "hard" selection pressure in which the absolute fitness of the HIV1 population is affected.

Frost et al. concluded, "Identification of individual amino acid changes contributing to escape is complicated by diversity between individuals in terms of both the genetic sequence of the infecting virus and the composition of neutralizing antibodies." Not only do we have an evolving VQS, but also the complex responses of the adaptive hIS. They are of the opinion that identifying which viral mutations contribute to escape, which act as compensatory mutations, and which emerge because they are linked to advantageous mutations is an important area for further research. In silico HIV/hIS war scenarios should be run to explore unexpected developments in the effectiveness of the hIS.

7.1.3 BQS Models

As we have seen above, one of the more interesting aspects of QS modeling is examining the interaction of two, competing QSs in a closed environment, and how the QSs respond to changes in their environmental parameters. Weitz et al. (2005) examined a model in which a BQS and a λ bacteriophage (VQS) interact in a (model) chemostat with a fixed washout rate and inflow resource concentration. (λ bacteriophage infects and kills *E. coli* through the bacteria's LamB receptor. A mutation in the *lamB* gene may prevent the bacteria from phage infection.)

Another bacterial QS scenario that has been studied is the evolution in vivo of the Gram-negative stomach bacteria, *Helicobacter pylori* (Suerbaum and Josenhans 2007). *H. pylori* colonizes the stomachs of more than one-half of the world's population; it is one of the world's most successful microorganisms. *H. pylori* is responsible for stomach ulcers, lymphoma, and gastric adenocarcinoma.

H. pylori has an elevated mutation rate, but unlike viruses, an additional method *H. pylori* uses to achieve genetic variability to insure its survival is inter-strain recombination (the closest thing to sex for bacterial). Suerbaum and Josenhans are of the opinion that a decline in prevalence of *H. pylori* in Western countries could make a recombination-based method of genomic adaptation ineffective, giving rise to weak infections (i.e., those easily overcome by the host's IS), and ultimately causing Eigen's QS error catastrophe, where *H. pylori* infections disappear from the host (human) population. Perhaps computer modeling the *H. Pylori* BQS versus the adaptive IS, and/or certain antibiotics could shed light on this scenario.

Related to the stochastic models of BQS is the synthetic, "predator–prey" ecosystem model of Balagaddé et al. (2008). They used two *E. coli* populations that communicated bidirectionally through secreted AI molecules and quorum sensing. Here, relative fitness was determined by the dynamics of the genes regulated by quorum sensing, not by stochastic mutations. Both experimental and modeling studies were done. Damped oscillations were shown in the two bacterial populations before they reached steady states.

7.2 ROLE OF MICRORNAs IN THE REGULATION OF GENE EXPRESSION AND DISEASE

7.2.1 INTRODUCTION

In this section, we consider the role of miRNAs in the complex processes of post-transcriptional gene product regulation. Defects in the regulation of miRNAs are seen to be associated with certain diseases.

miRNAs are single-stranded RNA molecules that are ca. 21–24 Nts long. Most miRNAs are generated from primary transcripts of DNA genes produced by RNA polymerase II, the same RNAP that transcribes protein-coding genes. However, some miRNAs in repetitive regions of the genome are transcribed by RNAP III. Some miRNA primary transcripts encode only a single, mature miRNA, while other gene loci contain clusters of miRNAs that appear to be produced from a single primary transcript. About 50% of all human miRNA genes are found within the introns of protein-coding genes, while others are located in the exons of untranslated genes or in the "junk DNA" (introns) (Boyd 2008). There are over 328 miRNAs in the human genome, and 199 in the plant, *Arabidopsis thaliana* (Chen and Rajewsky 2007). In fact, there may be over 10^3 miRNAs in the human genome (Berezikov et al. 2006). We are just beginning to understand their complex role in the regulation of gene products.

Once transcribed, primary miRNAs are preprocessed by a nuclear protein complex called a "microprocessor," which contains the Drosha type III RNase, the double-stranded RNA-binding protein DGCR8, plus some other proteins. The microprocessor cleaves the hairpin-loop pre-miRNAs away from the rest of the primary transcript, permitting the pre-miRNAs to pass through the nuclear membrane into the cytoplasm. There, the pre-miRNAs are processed by another protein complex containing the Dicer type III RNase that cleaves the miRNAs to their mature form.

The mature miRNAs become associated with a final protein complex called the miRNA-induced silencing complex (miRISC), which typically contains Dicer, Argonaute family proteins, and other proteins. These complexes do the silencing (turning off gene expression) of gene transcripts (mRNAs) specified by the miRNAs.

The steps leading to the production of miRNAs are evidently actively regulated; but what regulates the regulators? The regulation of gene expression is a very complex, degenerate process; there are multiple, interacting regulatory pathways. For a glimpse into this complexity, read the review papers by Boyd (2008), Wang et al. (2007), Chen and Rajewsky (2007), and Bushati and Cohen (2007).

The regulation of gene expression, especially in development, is a very complex process; both the timing and quantity of gene products must be controlled. An emerging body of information on this process has implicated miRNAs, as well as various transcription factor (TF) switches. Since the initial discovery of two miRNAs in the genome of the worm, *Caenorhabditis elegans* (Lee et al. 1993), hundreds of miRNAs have been discovered in almost all metazoan genomes, plant and animal (He and Hannon 2004, Esquela-Kersher and Slack 2006).

For an example of how miRNAs are highly conserved, consider miR-1, a miRNA critical for normal muscle development. miR-1 is nearly identical in sequence in *Drosophila*, *C. elegans*, and vertebrates, and exhibits skeletal muscle and heart-specific expression in these species. Chang and Mendell (2007) stated, "Over-expression or inhibition of miR-1 promotes or inhibits mammalian in vitro muscle cell differentiation, respectively."

7.2.2 MECHANISMS OF miRNA REGULATION OF GENE EXPRESSION

miRNAs negatively regulate their targets in at least two ways: (1) miRNAs bind with near perfect complementarity to target protein-coding mRNA molecules and induce their cleavage (by the miRISC complex). This direct cleavage of mRNA targets is commonly found in plants as well as mammals. (2) The second mode of negative regulation in animals is by binding to imperfect complementary sites within the 3′ untranslated regions (UTRs) of their mRNA targets. Thus, they repress target gene expression posttranscriptionally; a miRISC complex is used (Esquela-Kersher and Slack 2006).

According to Boyd (2008), most miRNAs form imperfect complementary stem–loop structures and pair imperfectly with sites in the 3′ UTR of their target mRNAs, and can apparently act by decreasing target mRNA levels in a graded manner, or by directly inhibiting translation. In some cases cited by Boyd, miRNAs have been proposed to act relatively pleiotropically and to inhibit multiple targets. Thus, pleiotropy appears to be present in the miRNA system. Many protein-coding genes (PCGs) have predicted target sites for several different miRNAs in their 3′ UTRs, suggesting the possibility for a combinatorial action of multiple miRNAs to maximize inhibition of gene expression. (In such cases, gene inhibition may be graded, rather than binary.) Boyd goes on to tell us, "Preliminary *in silico* estimates are that about one-third of all protein-coding genes in the human genome may be regulated in part by microRNAs."

Some proposed mechanisms of miRNA action include the following: (1) Interference with recognition of the 5′ cap of the mRNA, preventing translation initiation. (2) Cleavage of mRNA (Wang et al. 2007). (3) miRNA*RISC interferes with mRNA translation by binding to the 3′ UTRs of the mRNA and interferes with translation initiation factors, or it destabilizes the mRNA via de-adenylation of the poly-A tail. (4) miRNA*RISC may also interrupt the continuation of translation by forming a stable complex with polyribosomes (recruitment of the ribosome protein eIF6).

We note that evidence exists for a single miRNA molecule down-regulating hundreds of identical mRNA transcripts, that is, a single miRNA can act on multiple, specific targets (Chang and Mendell 2007).

7.2.3 PHYSIOLOGICAL FUNCTIONS OF miRNAs

A key role played by miRNAs is in the temporal coordination of development. They also are effective regulators of cell proliferation and differentiation. An embryo requires coordinated cell growth for development, but clearly a liver, kidney, femur, etc., must be "just the right size." Signals from miRNAs stop tissue growth at just the right time/size. (What regulatory pathways are involved?) We are fortunate that miRNAs are highly conserved in the Animalia. Thus, a variety of animals (other than humans) can be used to investigate miRNA in cell physiology, including fruit flies, mice, chicks, etc. Bushati and Cohen (2007) reported that mice deficient in Dicer, an enzyme required to preprocess miRNAs, were miRNA deficient and their embryonic development was arrested in gastrulation, before neural axis formation. Studies have shown that embryonic stem cells were impaired in their ability to proliferate in Dicer-deficient mice. An example of the physiological role of one miRNA was given by Bushati and Cohen: an excess of pancreatic islet-specific miR-375 inhibits glucose-induced insulin secretion by beta cells. However, depletion of miR-375 increases myotrophin levels and enhances glucose-stimulated insulin release. Another in vitro study cited by Bushati and Cohen showed that miR-134 regulates dendritic spine size in hippocampal neurons; it inhibits translation of the *Limk1* gene.

Programmed cell death (PCD), or *apoptosis*, is an evolutionarily conserved process that allows animals to remove cells that are internally damaged (damaged DNA, proteins, organelles), remove senescent cells so they can be replaced with new ones, and in development, remove fetal tissues such as finger webs. There are several triggers for PCD, once it has been initiated, caspase proteins cleave internal structural and functional elements of the cell. Evidence for the role of miRNAs in PCD come from the *Drosophila* eye. The absence of the miRNA gene, *miR-14*, leads to an increase in the PCD effector, Drice, suggesting that miR14 is in fact an inhibitor of apoptosis (by turning off caspase expression) in fruit flies (Wang et al. 2007). Similarly, the *bantam* gene encodes an miRNA that when over-expressed, suppresses PCD in the fruit-fly retina.

Bushati and Cohen (2007) noted that several miRNAs have been shown to be part of regulatory feedback loops. (This should not be surprising, because all powerful genomic down-regulators themselves must be under feedback regulation.) In one example, the development of the compound eyes of *Drosophila*, miR-7 represses expression of a protein, Yan. A transient epidermal growth factor signal leads to the degradation of Yan. Decreased Yan concentration represses the expression of miR-7; lower miR-7 concentration thereby increases the expression of Yan, partially restoring its concentration. (This is actually negative feedback on the protein concentration.) These authors also describe a positive feedback (PFB) loop in a *C. elegans* miRNA. In secondary vulval precursor cells, a protein, LIN-12/Notch, directly activates transcription of miR-61. miR-61 directly represses the protein Vav-1, which in turn can repress expression of LIN-12, thereby forming a PFB loop on the LIN-12 concentration.

7.2.4 MIRNAS IN DISEASE

7.2.4.1 Viruses

It is well known that for their survival, most viruses are adept at inhibiting their host cell's apoptosis mechanisms (PCD). Wang et al. (2007) stated, "Recently it has been discovered that the Herpes Simplex Virus-1 inhibits apoptosis through a latency-associated miRNA (miR-LAT) that modulates TGF-β signaling. By using miRNAs instead of proteins in the inhibition of apoptosis, viruses are able to survive as well as evade immune detection." (Note that when a virally infected cell undergoes PCD, lots of signature viral debris is released, then "cleaned up" by antigen-presenting dendritic cells and macrophages, then presented to Th cells, etc. Hence viral debris, antigen presentation, and hIS activation will be kept low when PCD is inhibited. The viruses will have plenty of time to use the host cell's resources to multiply.) Wang et al. go on to point out that human adenoviruses can block host cell miRNA expression, thus suppressing the very antiviral miRNAs that are meant to stop adenovirus replication.

Boyd (2008) noted the fact that many viruses express short, noncoding RNA transcripts from their genomes and may use them to modify their host cell's miRNA genomic machinery. There is evidence that the γ-herpesvirus, Epstein–Barr (a DNA virus), encodes two clusters of miRNAs: one group in the introns of the *BART* gene, another in the 3' and 5' UTR of the *BHRF1* gene. Possible targets for these viral miRNAs are systems regulating cell proliferation, apoptosis, and immune signaling. Other DNA viruses such as Kaposi's sarcoma virus, cytomegalovirus, and the SV40 DNA tumor virus all express miRNAs from their genomes. Boyd stated, "The single SV40 miRNA is necessary for regulating the levels of early genes in the viral life-cycle, and contributes to virally infected cell resistance to cytotoxic T-cell lysis."

Boyd reported that oddly, RNA viruses such as yellow fever, HCV, and HIV-1 do not seem to make miRNAs. However, HCV makes use of the host's hepatocyte-expressed miR-122 to increase the replication rate of its own genome.

Some viruses appear to block activity of host miRNA pathways altogether. Inhibitors of RNA-mediated gene silencing pathways have been reported in influenza A, vaccinia, HCV, HIV-1, and Ebola viruses (Boyd 2008).

7.2.4.2 Cancer

Some miRNAs appear to act as oncogenes, and are expressed at elevated levels in certain cancers. Increased expression of the miRNA gene, *miR-21*, appears to contribute to decreased apoptosis in malignant cells found in glioblastoma multiforme, cervical cancer, breast cancer, and many other solid tumor types (Boyd 2008). Boyd stated, "Similarly, the locus containing seven microRNAs of the miR-17-92 polycistron cluster at human chromosome 13q31 is amplified in a variety of B-cell lymphomas, nasal-type NK/T cell lymphoma, and solid tumors." He goes on to comment that the targets and mechanism(s) of miR-17-92 action are not yet known. Boyd cited evidence that some miRNAs can act as tumor suppressors. Deletion or mutation of these miRNA genes facilitates oncogenesis.

Certain miRNAs are known to have tumor-suppressive activity. The loci for *miR-15a* and *miR-16-1* (on human chromosome 13q14) are deleted in over 50% of B-cell chronic lymphocytic leukemia (CLL) cases, as well as in other malignancies. Evidence suggests that one or both of these miRNAs contributes to tumorigenesis by targeting the anti-apoptotic gene, *BCL2*. In this putative scenario, loss of function of miR-15a and miR-16-1 promotes high expression of Bcl2, hence higher-than-normal survival of CLL-infected cells. Another oncogenic miRNA is miR-155; its over-expression occurs in various tumors including B-cell lymphoma, breast, thyroid, and colon cancers. The miR-17 cluster of miRNAs is greatly over-expressed in various types of cancer. The miR-17 cluster of five miRNAs is found at human gene locus 13q31-32. These microRNAs include miR-17-5p, miR-19a, miR-20a, miR-19b-1, and miR-92. Over-expression of the miR-17 cluster is an action of the c-Myc oncogenic transcription factor, a protein widely found in a large percentage of human malignancies (Chan and Mendell 2007).

Esquela-Kerscher and Slack (2006, fig. 2) illustrated the role of miRNAs in normal tissues, functioning as tumor suppressors, and miRNAs acting as oncogenes. Also, in this excellent review paper is a table of miRNAs associated with human cancers, including the gene loci, cancer association, function, and references; over 13 oncogenic human miRNAs are listed.

7.2.4.3 Other Diseases

Tourette's Syndrome (TS), a neuropsychiatric disorder, involves a miRNA dysregulation. A binding site for miR-189 lies in the 3′UTR of the mRNA transcribing the gene *SLITRK1*. This binding site is mutated in some TS patients; a **GU** RNA base pair is replaced with an **AU** pair. This SNP leads to stronger regulation by miR-189. Note that many rare mutations lead to TS.

Fragile X syndrome represents one of the most common forms of mental retardation; it is caused by the loss of function of the *Fragile X Mental Retardation 1* (*FMR1*) gene. In most cases, the causative factor is a base triplet repeat expansion in the 5′UTR of *FMR1*. This mutation leads to hypermethylation of the *FMR1* promoter and subsequent gene transcription silencing. FMRP, the product of the *FMR1* gene, has an important role as a negative regulator of local protein translation within dendrites (Jin et al. 2004). Recent data support the theory that miRNAs are intimately involved in FMRP-mediated translational suppression, perhaps by acting on certain target mRNAs. Further research should clarify this issue.

7.2.5 EVOLUTION OF GENE REGULATION BY MIRNAS

miRNAs act at the level of posttranscriptional control of gene expression in all known plant and animal genomes (Chen and Rajewsky 2007). Pretranscriptional gene regulation is effected by transcription factors (TFs) which activate (or suppress) transcription by binding to cis-regulatory sites, which are usually located upstream of protein-coding genes. Other gene regulatory mechanisms also act at the transcriptional or posttranscriptional level, including mRNA splicing, polyadenylation and localization, chromatin modifications (e.g., methylation and demethylation), and protein editing by enzymes. In humans, we speculate that the ca. 10^3 miRNAs make up ca. 4% of the gene repertoire and may regulate >30% of all protein-coding genes.

Wray et al. (2003) suggested that repressors of gene expression should evolve faster than activators, as there are many ways to down-regulate a gene but relatively few ways to activate it. Thus, it might be expected that miRNA binding sites evolve faster (in numbers and variety) than TF binding sites. However, the situation is more complicated because of pleiotropy: TF binding sites are fuzzy; the same TF can bind to many similar DNA sequences, perhaps with different binding affinities. However, many miRNA binding sites have exact specificity, either to the first 6–8 bases from the 5′ end of the miRNA*RISC complex in animals, or to the entire mature miRNA in plants. Chen and Rajewsky (2007) commented, "...under neutral evolution, one would expect that it is more difficult to destroy a functional TF binding site than to create a new one, whereas the converse would be expected for miRNA binding sites."

Chen and Rajewsky have commented that few new TF families have evolved since the evolutionary divergence between plants and animals. They pointed out that the situation is different for miRNAs; from a combination of bioinformatic and gene sequencing studies, it is evident that the process of new miRNA creation is both active and ongoing, at least on an evolutionary timescale. A major goal of genomics is to describe and understand how TFs, miRNAs, signaling pathways, and other regulators are connected to effect gene regulation, and just as importantly, what caused them to be wired that way. Seeking answers to these problems is made difficult by the fact that many regulatory pathways are pleiotropic, others are degenerate, and all are complex.

7.3 SUMMARY

In this chapter, we have examined two different complex systems, viral and bacterial QS, and the role of microRNAs in posttranscriptional gene regulation. It appears that viral and bacterial mutations can enable a QS to jump host species, landing in a new host with a naive CAIS, allowing it to surge ahead in the fitness race. The concept of a mutation-selection balance is a fundamental principle of population genetics. In an ecosystem, there is a steady-state equilibrium distribution between fit variants and less fit variants, both created by natural selection. VQS and BQS systems can easily be modeled.

microRNAs have evolved as a means of posttranscriptional control of gene products in metazoans. Ongoing research has shown that they have an important role in the etiology of diseases including certain forms of cancer and mental retardation. The elaboration of the complex system that regulates more than 10^3 microRNAs in humans promises to be a big challenge in genomics.

8 Introduction to Physiological Complexity: Examples of Models of Some Complex Physiological Systems

8.1 INTRODUCTION

It is axiomatic that medical doctors, dentists, nurses, emergency medical personnel, and biomedical engineers should be educated in human physiology. It is also imperative that workers studying physiological systems (PSs) undergo training in complex systems analysis. Needless to say, PSs are generally complex, regardless of scale.

It is reasonable to identify 12 vital physiological regulatory systems in mammals: (1) cardiovascular (circulatory), (2) CNS/PNS, (3) endocrine, (4) gastrointestinal (digestive), (5) glucoregulatory, (6) immune, (7) integumentary, (8) muscular (striated, smooth, cardiac), (9) renal/urinary, (10) reproductive, (11) respiratory, and (12) skeletal. Distinctions between these systems are often fuzzy; they all interact to some degree, some more than others, and all depend on the molecular biological regulatory systems inside their component cells. For example, the skeletal system supports the body and permits coordinated motion in conjunction with the muscular system; it also is responsible for hematopoiesis through certain stem cells located in the bone marrow.

It is clear that PSs are inherently complex on every level of scale (Burggren and Monticino 2005). Complexity exists because they are nonlinear; their models have many nodes and directed branches, branch gains can vary parametrically, every PS interacts with many other PSs, and its functioning is based on many, coupled, intracellular biochemical reactions. In addition, there may be a lack of high predictability in system output(s) (chaotic behavior), and initial conditions can affect system behavior.

We must ask some questions in considering complex PSs: Why have complex physiological regulatory mechanisms evolved? Has natural selection caused the evolution of complexity, as Bonner (1988) has asserted (in particular, regulatory complexity)? If not, what has? Can we describe and model the complex relationships between the mechanisms that produce variable physiological, morphological, and behavioral traits during development, and the adaptive significance of these traits (Reiber and Roberts 2005)?

It is easy to argue that a more complex animal has improved resources for survival (higher fitness), ranging from a better immune system, to a CNS that can remember dangers and food sources, regulate adaptive behaviors (e.g., feeding, hibernation, aestivation, dormancy, mating, and fight or flight), and process enhanced sensory information from more complex sensors. The evolution of larger physical size facilitates prey capture, or allows the animal to evade predation. In the following text, we focus our attention on human PSs.

8.2   PHYSIOLOGY DEFINED

Guyton (1991) defined *physiology* as the "...study of function in living matter, attempting to explain the physical and chemical factors that are responsible for the origin, development and progression of life." Guyton goes on to remark that each type of life (e.g., bacteria, plants, cells, animals, insects, fish, and humans) has a sub-speciality of physiology associated with it. In addition, we have systems, regulatory, and medical physiology, to cite some other major specializations of this broad discipline.

Whether examining a bacterium or a human, it is evident that function can be broken down into a series of biochemical events, which generally take place intracellularly; however, some reactions occur on cell membrane surfaces, while others take place in the extracellular volume or in circulating blood. The sum of these biochemical events is responsible for homeostasis (see Section 6.3).

Homeostatic maintenance of a physiological (output) parameter is a dynamic process; it is generally the result of a number of interacting, nonlinear, negative feedback loops, from which nature forms a robust regulator to stabilize the parameter against external and internal influences that would force it into a range that would harm the organism and decrease the probability of propagating a species. The importance of physiological regulation cannot be overestimated. Every clinically and experimentally measurable physiological parameter is under some sort of internal feedback regulation. Regulation is necessary because a living organism is continually being perturbed by physical and chemical changes (challenges) in its external environment. Regulation makes a PS robust to these disturbances.

For example, a genetically and hormonally activated internal process, such as the biosynthesis of something as basic as an eggshell in birds, creates a tremendous internal sink for calcium ions, which must be regulated to compensate for the loss. Because of its critical role in muscle contraction/relaxation, insulin secretion, etc., the vertebrate Ca^{++} regulator is fairly "stiff" (robust against external disturbances) (Northrop 2000). It has only one source (diet), several storage compartments (bones, muscles, chelates in the blood), and several of sinks (e.g., eggshell, bones, muscles, feces, urine). Blood [Ca^{++}] regulation in humans is effected by several hormones and vitamins (Guyton 1991). These complex processes are illustrated in Figures 8.1 through 8.3. Note that osteoblast cells break bones down to release Ca^{++} from the bone compartment, while osteoclast cells make new bone.

In this section, let us make the distinction clear between physiological regulators and control systems, and the external (exogenous) control of PSs. A physiological regulator acts to maintain a constant level of the regulated parameter (blood calcium, for example) in the face of fluctuating dietary input and the biosynthesis of bone or eggshell. Its action is completely involuntary, that is, we cannot consciously change the output. A regulator has a virtual, internal, constant, setpoint that is ideally the nominal parameter value. The setpoint can be explicit or implicit (Jones 1973). Most physiological, homeostatic feedback systems are regulators. Regulators are a subset of feedback control systems.

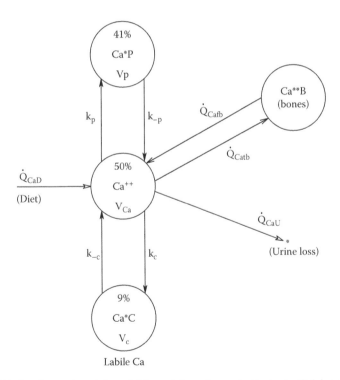

FIGURE 8.1 Compartmental model for calcium storage in the body. Calcium exists in ionic form in blood, extracellular fluid (ECF), and tissues in the ionic form (Ca^{++}), as a chelate in blood and ECF (Ca*C), and bound to protein in blood and ECF (Ca*P). (From Northrop, R.B., *Signals and Systems in Biomedical Engineering*, 2nd edn., CRC Press, Boca Raton, FL, 2010.)

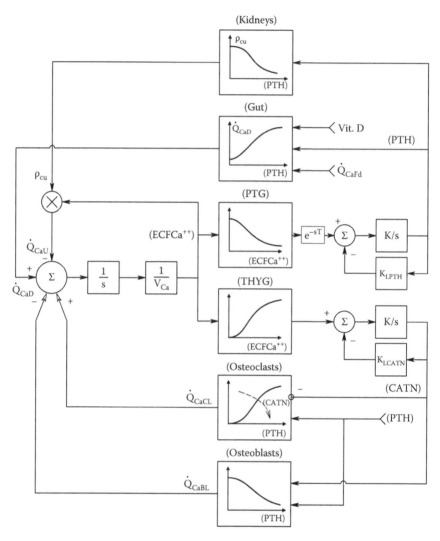

FIGURE 8.2 Block diagram describing calcium metabolism. (Extracellular Ca^{++} exchange with protein and chelate is neglected.) (From Northrop, R.B., *Signals and Systems in Biomedical Engineering*, 2nd edn., CRC Press, Boca Raton, FL, 2010.)

An example of a physiological regulator that can be overridden is the system that regulates our rate of breathing, hence blood hemoglobin oxygen saturation and pCO_2. We can override its regulatory function (i.e., control it) by conscious thought. For example, we can hold our breath when swimming under water. Another regulator that can be overridden is the ocular tracking system. (A moving visual object's image is kept centered on the eyes' foveas by the ocular tracking regulatory mechanism. However, we can override this regulator and consciously control our gaze.) In addition, biofeedback conditioning makes it possible

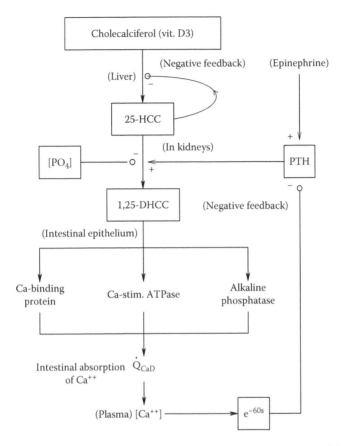

FIGURE 8.3 Compartmental model for the role of vitamin D_3 and parathyroid hormone (PTH) in regulating plasma $[Ca^{++}]$. Causal negative feedback paths are shown by circles and minus signs. (From Northrop, R.B., *Signals and Systems in Biomedical Engineering*, 2nd edn., CRC Press, Boca Raton, FL, 2010.)

to consciously modulate other regulated physiological parameters, such as heart rate and skin temperature.

The output parameters of many physiological regulators and control systems can also be modulated by the application of exogenous drugs and hormones. The purpose of such external medications is to correct for internal maladjustment of a regulator caused by disease, injury, etc., and to reestablish a normal balance of physiological parameters in the body. For example, exogenous sodium nitroprusside (SNP) can be administered by IV injection to lower (regulate) mean arterial blood pressure; a closed-loop control system is used (Northrop 2000).

Systems biology is a field of endeavor that seeks to describe quantitatively how biological systems function at the systems level (i.e., physiologically). Think of it also as quantitative physiology. Rather than break a biological system down into its component parts and examining each component separately (often an imprudent approach when dealing with complex systems), it uses a holistic approach to study how each biological subsystem is integrated into the overall network of command, communication, and control (C^3) in an organism (Alon 2007). All levels of interaction are considered: cell signaling circuits, metabolic pathways, organelles, membrane receptors, cells, PSs, and organisms. The models created in systems biology are very large and complicated. The mathematical tools of diffusion and chemical mass-action are often used in model formulation; however, there is now a trend to use agent-based modeling (ABM) of PSs (ABM is described in Section 11.2.8).

One successful experimental approach in systems biology is the use of massively parallel protein and nucleic acid analytical microarrays to quantify changes in mRNAs and proteins during cellular processes, such as the embryonic differentiation of stem cells. The recent development of high-throughput, array technologies has allowed systems biologists to simultaneously examine experimentally many more system states, and avoid the errors inherent in trying to characterize a complex, nonlinear, biological system using reductionism. With the collection of massive amounts of quantitative data from arrays, systems biologists can formulate more robust mathematical models to describe a system's behavior in response to changes in its environment and human-generated inputs. For example, what molecular inputs are required to make an adult somatic cell (e.g., an epithelial cell) revert to a stem cell having embryonic, pluripotent status, and then differentiate into a desired tissue type? The elaboration of these processes is the "holy grail" of regenerative medicine.

8.3   HOMEOSTASIS

We shall divide our consideration of homeostasis into cellular homeostasis and organismic homeostasis. Homeostasis is the biochemical regulatory juggling act that all living cells and organisms must perform in order to remain alive. Broadly posed, homeostasis is the dynamic maintenance of each of a number of biochemical and physical parameters within safe bounds so a cell or organism can carry on its normal life functions and reproduce its kind (if that is what it has been programmed to do genetically). Homeostasis involves gene expression and biochemical reactions (intracellular, on cell membranes, and among cells). The reason homeostasis is complex is that there are many, many interactions between the systems that regulate the parameters. Much of physiology details the homeostatic regulatory mechanisms in organisms and cells.

The important internal parameters regulated by cells include (but are not limited to) osmotic pressure requiring the active adjustment of certain intracellular ion concentrations (e.g., Na^+, K^+, Ca^{++}, Cl^-, and HCO_3^-), as well as H_2O. H_2O passes through various aquaporin transmembrane proteins (Agre et al. 2002), often under hormonal regulation. The transmembrane Na^+-coupled glucose water pumps (SGLT1) evidently actively pump water and glucose into *Xenopus* oocytes. (The inward flow of sodium ions down concentration and electrical potential gradients powers these water pumps [Loo et al. 2002, Zeuthen 2007]).

Also, energy substances must be imported by cells and metabolic wastes eliminated, intracellular pH must be regulated, certain ranges of ionic concentrations are required (for Na^+, K^+, Ca^{++}, Mg^{++}, Fe^{++}, Zn^{++}, Cl^-, $PO_4^=$, HCO_3^-, etc.) for normal metabolic activities and motion (of cilia, cell membranes). Cells must also regulate myriads of chemical reaction pathways (e.g., glycolysis, the Krebs cycle, fermentation, photosynthesis, protein synthesis, synthesis of regulatory RNAs): genes are expressed and repressed; proteins are made and then broken down; amino acids and lipids are synthesized and metabolized; carbohydrates, fats, proteins, and photons (in chlorophyll containing organisms) are used for energy (e.g., the synthesis of ATP, used to power cellular biochemical reactions). ATP synthesis involves highly regulated metabolic pathways such as glycolysis, the Krebs cycle, and photosynthesis.

Cellular homeostasis necessarily must be under genetic control. The protein enzymes that catalyze the many biochemical reactions are continuously being destroyed, and must be continuously remade by ribosomes using mRNA templates from their corresponding genes. Expression of genes themselves is tightly regulated by a number of feedback loops and regulatory motifs (see Section 4.5) involving transcription factor (TF) proteins and microRNAs, themselves regulated, etc., loops within loops.

Organismic homeostasis in multicellular plants and animals is no less a complex biochemical juggling act than inside individual cells; the scale is just larger. In higher animals, groups of cells are organized into specialized tissues and organs, for example, muscles, neurons (the CNS), endocrine glands, organs (skin, liver, pancreas, spleen, stomach, heart, lungs, intestines, kidneys, bones, eyes, ears, etc.), also into distributed systems (e.g., the immune system, the hematopoietic system). Organismic homeostatic balance is largely effected through the circulatory system (endocrine hormones), and also regulated through the autonomic nervous system. Important molecules (glucose, water, endocrine hormones) and ions such as Na^+, K^+, Ca^{++}, Mg^{++}, Cl^-, HCO_3^-, and H^+ are regulated in the blood, as is the transported O_2 and CO_2. Of course, blood pressure and flow themselves are regulated, both hormonally and by the autonomic nervous system; more regulatory loops within regulatory loops. Homeostatic regulation of body temperature can involve behavior, as well as physiological and biochemical actions.

8.4 STRUCTURE AND FUNCTION: SOME EXAMPLES OF COMPLEX PHYSIOLOGICAL REGULATORY SYSTEMS AND THEIR SIMPLIFIED MODELS

8.4.1 INTRODUCTION

In the sections below, we consider three examples of complex PSs about which enough is known of their structure and function to enable us to formulate reasonably good dynamic models that can predict their behaviors. They are (1) the human glucoregulatory system, (2) the human immune system (hIS) versus HIV/AIDS, and (3) the hIS versus cancer. Our models of complex systems necessarily must be approximations; however, their in silico accuracy must be validated with in vivo data.

Motivation for studying the glucoregulatory system in man has been largely driven by the need to manage the dysregulatory conditions such as hypoglycemia, type I and II diabetes, and obesity. We will examine the major features of the normal human glucoregulatory system, and show how the major regulated parameter, the plasma (or blood glucose) concentration, responds to diet and an intravenous glucose tolerance test (IVGTT).

The second complex system we have modeled is the human immune system's (hIS's) response to HIV infection, and how modeled drug therapy affects the development of AIDS. Since the early 1990s, the spread of HIV/AIDS has stimulated research in the hIS, and there now is much data available about its structure and function.

The third system we have modeled is the hIS's responses to growing cancer cells. The motivation here was to see how the hIS might be manipulated to specifically target cancer cells, instead of using generally toxic chemotherapy drugs and radiation treatments.

8.4.2 GLUCOREGULATION

8.4.2.1 Introduction

In this section, we will describe and mathematically model the normal regulation of blood glucose in humans. The roles of the pancreatic hormones insulin and glucagon (GLC) in maintaining normoglycemia will be given, and the sources and sinks of glucose in the body, including storage in the liver and muscles as the polymer glycogen (GN) will be examined. We will stress dynamics wherever possible. The condition of type I diabetes mellitus and exogenous insulin therapy will be considered and modeled.

The reason D-glucose is important in living organisms is that it is the principal source of energy for aerobic cell metabolism, anaerobic glycolysis, and alcoholic fermentation. Figure 8.4 illustrates a schematic of a D-glucose molecule; it has a molecular weight of 180.12 g/mol.

In humans, there are a number of biochemical pathways whereby other sugars, starches, fats, and proteins in the body can be converted to glucose. However, there are three main sources of glucose in the body: (1) diet; (2) gluconeogenesis, where intracellular molecules such as glycerol, lactate, alanine, pyruvate, oxaloacetate, and dihydroxyacetone are converted into complex, multistep, enzymatically regulated processes to glucose; and (3) the breakdown of (stored) liver and muscle GN to glucose. The latter two steps are stimulated by the pancreatic hormone, GLC. (Glucose from the liver is released

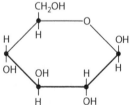

FIGURE 8.4 The D-glucose molecule. (From Northrop, R.B., *Endogenous and Exogenous Regulation and Control of Physiological Systems*, Chapman & Hall/CRC Press, Boca Raton, FL, 2000.)

into the circulation, while glucose from muscle GN is mostly used directly by muscle cells.)

Once in the circulatory system, there are several sinks for D-glucose: (1) It can be stored in the liver as the high molecular weight polymer, GN, and to a lesser degree in muscle cells as muscle GN. (2) It can be lost in the urine if the blood glucose concentration rises above a kidney transport threshold (about 1.8 g/L). (3) It diffuses into insulin-insensitive cells (such as neurons) at a rate determined by the concentration gradient of glucose across the cell membranes. (This gradient is proportional to the difference in concentration of extracellular fluid (EF) and intracellular fluid (IF) glucose.) (4) It diffuses into insulin-sensitive cells. The diffusion constant for glucose into insulin-sensitive cells increases monotonically with the concentration of insulin, that is, the diffusion constant is adjusted parametrically by insulin, which binds with receptor sites on the insulin-sensitive cells and activates the glucose transport protein in their cell membranes, increasing glucose permeability by as much as 10- to 20-fold.

The glucoregulatory system (GRS) is principally regulated by two hormones, insulin and GLC. (Note that there are a number of molecules other than the blood glucose concentration [BG] that affect the insulin secretion rate, $\dot{Q}_I$, by pancreatic β-cells.) The GRS has one external (diet) and two internal inputs (from the liver and gluconeogenesis), and four sinks. When the blood concentration of insulin [BI] is low, little glucose can enter the insulin-sensitive cells or be stored in the liver as GN. Thus the blood glucose concentration [BG] rises, raising the osmotic pressure of the blood. Water leaves the cells, and a condition of intracellular dehydration occurs. The higher osmotic pressure difference across cell membranes triggers the need to drink water, thence to urinate, major symptoms of diabetes.

If excess insulin is present, too much glucose enters the insulin-sensitive cells and the liver, and the [BG] falls. If the [BG] falls too low, the CNS cells are deprived of their energy source, and a person can experience nervous irritability, convulsions, loose consciousness, or die. Injection of excess insulin can lead to this hypoglycemic condition.

8.4.2.2 Molecules Important in Glycemic Regulation

The pentose sugar, glucose, has two isomers; the one used in nature is D-glucose, or dextrose. D- and L-glucose are optically active. D-Glucose is called thus because when linearly polarized light is passed through a length l meters of an aqueous solution of D-glucose, the polarization angle of the emerging light is rotated clockwise (seen looking toward the source) by a degree angle, $\theta = 1 C [\alpha]_\lambda^T$, where C is the concentration of the solution (e.g., in g/L), and $[\alpha]_\lambda^T$ is the specific optical rotation of glucose in degrees/(m × g/L) evaluated at a known temperature T and wavelength λ. If the L-glucose isomer is ingested, it cannot be metabolized by the body.

As we noted above, the two most important glucoregulatory hormones are insulin and GLC. These proteins are manufactured by islet cells in the pancreas; insulin by the beta cells, and GLC by the alpha cells.

The human insulin molecule (I) consists of two sulfur-linked amino acid (AA) chains with a total of 51 AAs, and it has a molecular weight of 5808. Many molecules modulate the rate of insulin release by the beta cells, the most important of which is glucose. $\dot{Q}_I$ is also up-regulated by the presence of the AAs arginine, lysine, and alanine, as well as the gastrointestinal hormones gastrin, secretin, cholecystokinin, and gastric inhibitory peptide (GIP). $\dot{Q}_I$ is also increased by the hormones: growth hormone (GH), cortisol, GLC, progesterone, estrogen, and epinephrine. Insulin release rate is down-regulated by the hormones leptin, somatostatin (from the δ-cells in the pancreas), and α2-adrenergic nervous stimulation.

GLC is a pancreatic, endocrine, protein hormone consisting of a single chain of 29 AAs, with a MW of 3485. The molecular structures of both human insulin and GLC are known exactly (Guyton 1991).

Both hormones are secreted into the hepatic portal venous system via the pancreaticoduodenal vein. Thus, liver cells are the first to receive the secreted insulin and GLC, and generally see higher concentrations of these hormones than are present in the systemic circulation. An increased secretion rate of GLC, $\dot{Q}_{GLC}$, is caused by a decreased plasma glucose concentration [PG], increased epinephrine and norepinephrine (sympathetic nervous system activity), increased plasma AAs (from a meal), increased cholecystokinin and acetylcholine.

$\dot{Q}_{GLC}$ is down-regulated by somatostatin, insulin, and directly by high [BG]. GLC secretion is also down-regulated by Zn^{++} and the inhibitory neurotransmitter, gamma amino butyric acid (GABA) (Gromada et al. 2007).

GN is a polymer made in the liver and muscle cells by enzymatically linking many uridine diphosphate glucose (UDG) molecules together. It allows glucose molecules to be efficiently stored until needed. A typical GN molecule can have a MW of 5 million or more. UDG is made from glucose 1-phosphate (G1P), which in turn is made from glucose 6-phosphate (G6P). G6P is made directly from D-glucose that diffuses into the liver cells.

Why does nature store glucose in the liver cells as a polymer? The answer lies in osmotic pressure, P_{osm}. One milli-osmole of glucose weighs 0.180 g. One milli-osmole of GN may weigh as much as 5 E3 g. This means that 28,000 glucose molecules joined as one GN molecule will raise the intracellular P_{osm} the same amount as a single, free, intracellular glucose molecule. Thus, intracellular storage of glucose as a very high molecular weight polymer does not impose any excessive rise in liver cell intracellular P_{osm}. See Figure 8.5 for an illustration of a GN (polymer) molecule and the biochemical pathways by which it is formed and broken down.

Muscle GN provides a local, fast source of energy for sudden bursts of muscle activity. If muscles work anaerobically, lactate (lactic acid) is produced by glycolysis. This diffuses out of the muscle cells into the blood; thence to the liver cells, which take it up and convert it to pyruvate; thence to G6P; and thence to free glucose (Guyton 1991). We will examine the chemical kinetics of GN storage and glucose release from liver cells in the sections below.

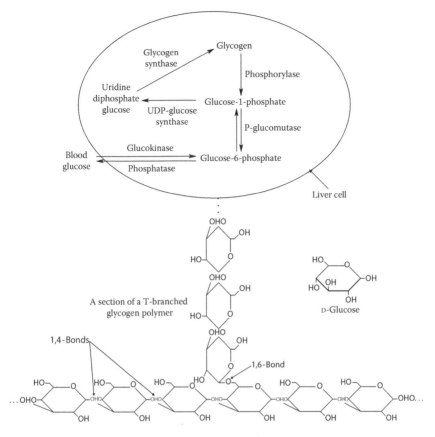

FIGURE 8.5 Top: A liver cell showing the reaction pathways and enzymes used in converting blood glucose to the polymer, GN. Bottom: Portion of a GN molecule. (From Northrop, R.B. and Connor, A.N., *Introduction to Molecular Biology, Genomics and Proteonomics for Biomedical Engineers*, CRC Press, Boca Raton, FL, 2009.)

8.4.2.3 Hormone Leptin

In 1994 a new, pleiotropic protein hormone called leptin was described that has far-reaching implications in the regulation of the amount of excess dietary calories stored as fat in fat cells (adipocytes) versus the amount of glucose stored as GN in the liver and muscles (Considine et al. 1996, Remesar et al. 1996, Montague et al. 1997, Clement and Vaisse 1998). Leptin, a 16 kDa protein, was first discovered in congenitally fat mice. Leptin has 146 AAs with a complex, helical, folded structure not unlike interleukins 2 and 4, and GH. Leptin is evidently secreted by all adipocytes (fat cells) in the body. There are known leptin receptors on certain hypothalamic neurosecretory cells and on neurons in the arcuate nucleus where the enigmatic neuropeptide Y (NPY) is synthesized. Perhaps the most significant location for leptin receptors is on the pancreatic beta cells that secrete insulin. (There may be leptin receptors on certain hepatocytes, as well.)

Apparently, one important purpose of leptin is to regulate the amount of stored body fat. Obesity is a condition that affects about one-third of the American population. It can lead to many health problems, for example, diabetes, heart disease, high blood pressure, stroke, and arteriosclerosis. Exogenous leptin has been used successfully to cause obese mice genetically lacking the ability to produce the hormone to lose weight. Thus, there is hope that it can be used in conjunction with other drugs to effect fat reduction in humans. The connections in the leptin feedback systems are described below.

Enough is now known about leptin in order to try to describe its regulatory pathways and its role in glucoregulation (Camisotto et al. 2005). The regulation of carbohydrate and lipid metabolisms involves the complex interaction between many hormones and their receptors, plus appetite and diet. Normally, carbohydrates and fats are digested and then processed in several interacting biochemical pathways. An excess of carbohydrates leads to the synthesis of fatty acids in liver cells. These fatty acids are then transformed into triglycerides, which are released into the blood in the form of lipoproteins. Under the influence of insulin, the lipoproteins are converted into fatty acids in the capillary walls of adipose tissue by the enzyme, lipoprotein lipase. These fatty acids are taken up by the fat cells and again stored as triglycerides. It is the excess of triglycerides stored in fat cells that produces obesity.

The stored triglycerides can be broken down to free fatty acids and released into the blood under the influence of certain hormones and the enzyme, hormone-sensitive triglyceride lipase (HSTL). Insulin actively suppresses the action of HSTL in fat cells. However, other hormones activate HSTL, raising fatty acids in the blood, which can be used for energy. These hormones include epinephrine (EP), norepinephrine, cortisol (a glucocorticoid), GH, and tri-iodothyronine (TI3). (All of these hormones do not necessarily act directly on HSTL.)

It is known that the primary source of leptin is white adipose tissue (WAT) cells (body fat), but it is also released in lesser degree by brown adipose tissue, the placenta, ovaries, skeletal muscle, stomach, mammary epithelial cells, bone marrow, the pituitary, and hepatocytes.

Leptin is a pleiotropic protein hormone, affecting different types of cells in different ways. For example, it suppresses insulin release by β-cells, it decreases the rate of production of the neurohormone, NPY, and it down-regulates lipogenesis. Leptin is also important to the normal functioning of the hIS; behaving like a cytokine. Faggioni et al. (2001) have shown that leptin plays a role in innate and adaptive immune responses. Leptin levels increase sharply during infection and inflammation. A leptin deficiency increases susceptibility to infections and inflammation, and is associated with dysregulation of cytokine production, and causes a defect in hematopoiesis. Elevated leptin affects T-cell phenotype, producing a higher Th1/Th2 cell ratio.

The rate of leptin release from WAT (fat) cells, $\dot{Q}_L$, is raised by insulin, NPY, tri-iodothyronine, epinephrine, cortisol, and the adipose tissue mass (Remesar et al. 1996). Cammisotto et al. (2005) reported that in WAT of rats, $\dot{Q}_L$ effectively saturated at a glucose concentration [G] of 5 mM. They found that WAT cells secreted leptin in the absence of glucose or other substrates. Five mM [G] increased this secretion, which was doubled in the presence of insulin. Similar results were obtained when glucose was replaced by 5 mM pyruvate or fructose. The amino acids L-glycine and L-alanine mimicked the effect of glucose on basal leptin secretion, but completely prevented leptin secretion increase by insulin. Strangely, $\dot{Q}_L$ was increased by insulin when glucose was replaced by the AAs: L-aspartate, L-valine, L-methionine, and L-phenylalanine, but not by L-leucine. Cammisotto et al. concluded, "1) Energy substrates are necessary to maintain basal leptin secretion constant. 2) High availability of glycolysis substrates is not sufficient to enhance leptin secretion, but is necessary for its stimulation by insulin. 3) Amino acid precursors of tricarboxylic acid cycle intermediates potently stimulate basal leptin secretion per se, with insulin having an additive effect; and 4) Substrates need to be metabolized to increase leptin secretion."

Elevated leptin concentration in the pancreas, in turn, lowers or suppresses the rate of release of insulin by pancreatic β-cells (at a given plasma glucose concentration), forming a negative feedback loop. A lower insulin secretion rate means lower portal and plasma insulin concentrations than normal. Lower insulin concentrations mean less glucose enters the liver cells and other insulin-sensitive cells. Thus, less GN is stored, and less fatty acids are made. Lower insulin concentration leads to increased secretion of the hormone GLC. Higher GLC means that more fatty acids are released from adipocytes into the blood. Elevated GLC also inhibits the storage of triglycerides in the liver.

The rate of leptin release from WAT (fat) cells, $\dot{Q}_L$, is also decreased by free fatty acids (FFA), cyclic AMP (cAMP), catecholamines, thyroid hormone (TH), melatonin (during the night), and reduced mass of WAT.

In mice, chronic obesity appears to be related to either the lack of ability to produce leptin, or the lack of leptin receptors in the hypothalamic neurons that secrete NPY. Similar conditions may occur in obese humans.

8.4.2.4   Neuropeptide Y

NPY is actually one of a family of three peptides having similar structures and pleiotropic actions in the body. The family consists of NPY, peptide YY (PYY) and pancreatic polypeptide (PP) (Sandeva et al. 2007). NPY itself is a 36 AA peptide that is highly conserved throughout evolution; it is the most abundant neuropeptide in the mammalian CNS (Pedrazzini et al. 2003). (We have described the role of NPY in appetite control in the previous section.) Neural tissue is the major source of NPY in the human body; it is widely distributed and released from specific kinds of CNS neurons (hippocampus, inner layer of olfactory bulb, striatum, septum, amygdala, and basal forebrain), and also peripheral neurons including noradrenergic, non-cholinergic, and sympathetic, also sensory neurons (Sandeva et al. 2007).

NPY is in general, a pleiotropic, inhibitory neuromodulator/neurotransmitter. According to Sandeva et al., NPY exhibits "…anxiolytic, anti-stress, anti-convulsant, and anti-nociceptive actions, in addition to its hypertensive, and potent appetite-stimulating effects and its capacity to shift circadian rhythms." These authors also noted that NPY is also a potent down-regulator of alcohol consumption. NPY is found in large vesicles in the endings of certain sympathetic nerve fibers in the body. NPY from this source evidently acts as a potent, long-acting vasoconstrictor, having synergistic action with norepinephrine (Guyton 1991). The hypertensive effect of exogenous NPY is largely due to its potentiation of norepinephrine-induced vaso-constriction, as demonstrated in isolated blood vessels from rabbits. NPY also pro-motes sleep, and NPY agonists may provide a novel treatment strategy for affective disorders (Obuchowicz et al. 2004).

NPY-related peptides bind to a large and heterogeneous family of G-protein-coupled receptors, Y1–Y5 and y6. The Y1 receptor is involved with the vascular and anti-nociceptive effects of NPY, as well as its psychological functions such as decreased anxiety and depression. Receptor Y1 also figures in the appetite stimulation by NPY. When NPY acts as a neuromodulator, it is through the Y2 receptors that are located presynaptically and inhibit the release of neu-rotransmitter. The pleiotropy of NPY (why it is enigmatic) is illustrated by its action on Y2 receptor neurons (increase blood pressure) versus its action on Y1 receptor neurons (decreases BP). Y2 receptors are also associated with angio-genesis (important in tumor proliferation), and the effects of NPY on circadian rhythms.

We noted above that elevated leptin concentration in the blood decreases the rate of NPY release from hypothalamic neurosecretory cells in the brain. Thus we see there are two known negative feedback loops regulating leptin secretion, leptin/insulin and leptin/NPY. Although the regulated variable appears to be leptin, leptin evidently regulates the average dietary calorie input through NPY. Excess calories over time, if not balanced by metabolic loss due to exercise, cer-tainly lead to excess fat, as we all know. In mice, leptin is required for female and male fertility. Puberty in humans is linked to a critical level of body fat (hence leptin concentration). Low levels of fat (as in anorexia) are associated with cessa-tion of the ovarian cycle. These many actions of NPY are examples of hormonal pleiotropy.

Ghrelin is a 28 AA peptide hormone released from the duodenum that stimu-lates the release of GH from the pituitary. Ghrelin also stimulates the release of NPY from neurons in the brain's arcuate nucleus. Leptin, however, inhibits NPY release in the hypothalamus.

In conclusion, a complex relationship exists between NPY and the hypotha-lamic/pituitary/adrenocortical axis. This network includes positive feedback between NPY and adrenal corticosteroids, and negative feedback between NPY and corticotrophin-releasing factor and for leptin release.

We have seen that the blood glucose regulatory system is far from a simple glucose-insulin-glucagon-hepatic storage sources and sinks model. Many other important hormones also affect glucoregulation, such as leptin and NPY, just to mention two. The temptation has been to consider their effects as secondary, treat their concentrations as constant, and model the core glucoregulatory module, as we have done below.

8.4.2.5 Normal Blood Glucose Regulation System

We will start describing basic glucoregulation by first discussing the protein hormone, insulin. Insulin is produced by the pancreatic beta cells primarily in response to elevated plasma (blood) glucose concentration. To underscore the complexity of glucoregulation, we point out that the rate of insulin release from the pancreas, $\dot{Q}_I$, is increased by a number of molecules in the pancreatic blood supply other than glucose. Specifically, the amino acids alanine, glycine, and arginine, the digestive hormones gastrin, secretin, GIP, and cholecystokinin. It is also raised by epinephrine, GLC (from the pancreas), GH, cortisol, progesterone, and estrogen. Not unexpectedly there are substances that reduce $\dot{Q}_I$. These include the pleiotropic hormone leptin from fat cells, somatostatin, and the neurotransmitter acetylcholine released from $\alpha2$-sympathetic nerve endings on the beta cells.

The release rate of insulin into the portal circulation can be modeled by the system shown in Figure 8.6.

A fast, rectified, single-time constant path is in parallel with a slow, two-pole, lag path. The former path describes the dynamics of immediate, stored

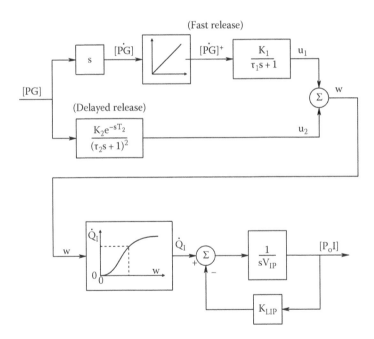

FIGURE 8.6 Block diagram illustrating the nonlinear dynamics governing insulin release from pancreatic β-cells as a function of plasma glucose concentration [PG]. [PoI] = portal insulin concentration. The upper, fast-release path obeys unidirectional rate sensitivity. Although many factors affect insulin release rate, this simple model considers only a [PG] input, and portal insulin concentration as the output [PoI]. (From Northrop, R.B., *Endogenous and Exogenous Regulation and Control of Physiological Systems*, Chapman & Hall/CRC Press, Boca Raton, FL, 2000.)

insulin release by β-cells in response to a sudden increase in [BG]. The two-time constant, slow path with the transport lag term models the metabolic activation of beta cells to synthesize new insulin for release. At a normal, resting [BG] = 90 mg/dL, the beta cells secrete $\dot{Q}_I = 10$ ng/min/kg body weight (BW) of insulin. At very high [BG] > 500 mg/dL, $\dot{Q}_I$ saturates at about 20× normal, or 200 ng/min/kg BW.

Once secreted, insulin is inactivated by several biochemical processes. It is broken down by proteolytic enzymes when it binds to insulin receptor proteins on cell surfaces. Liver and kidney cells also contain enzymes that destroy free insulin. The liver is the principal site of insulin inactivation. Liver glutathione-insulin transhydrogenase (GIT) cleaves the three S–S bonds in the insulin molecule, rendering it inactive and subject to further proteolysis (Izzo 1975).

If an IV bolus injection if insulin is given to normal subjects, the blood insulin concentration [BI], falls from its peak value at t = 0+ to the normal concentration with a decay that can be described by three exponential terms with different time constants, 1/a, 1/b, and 1/c s, that is,

$$[BI](t) = I_o + Ae^{-at} + Be^{-bt} + Ce^{-ct} \qquad (8.1)$$

This response is considered to be good evidence for a three-compartment, pharmacokinetic (PK) model for insulin; the three compartments are probably the blood (plasma), the extracellular fluid, and the liver cell volume.

The fast phase of insulin decay is seen to take about 20 min for the concentration to go from an initial 400 to 20 µU/mL (Sherwin et al. 1974).

No surprise, insulin has other hormonal effects beside its major role of facilitating the diffusion of extracellular glucose molecules into insulin-sensitive cells. Insulin inhibits gluconeogenesis; it does this by decreasing the quantities and activity of the liver enzymes necessary for gluconeogenesis. Insulin acts to decrease the rate of release of amino acids from muscle and other non-liver tissues, thus reducing the available pool of precursor molecules used in gluconeogenesis. Insulin also promotes the conversion of excess intracellular glucose into fatty acids.

These fatty acids are converted to triglycerides in low-density lipoproteins, which are transported to adipose tissues where they become fat. High portal [I] inhibits the enzyme, glucose phosphatase in liver cells, an enzyme that removes the phosphate group from glucose-6-phosphate (G6P), converting it to glucose that can diffuse from the liver back into the blood. High portal [I] also activates the liver enzyme phosphorylase. Phosphorylase makes G6P from glucose that has diffused into liver cells, thereby trapping the glucose inside the liver cells. G6P is then enzymatically converted to glucose-1-phosphate (G1P), then to uridine diphosphate glucose, then to GN. Thus elevated portal insulin concentration increases the rate at which liver GN is formed, and also inhibits its breakdown.

The hormone GLC, as we have previously discussed, is made by the alpha cells of the pancreatic islets of Langerhans. Its release rate is increased by low [BG], among several factors. Normal plasma concentration of GLC is between

100 and 200 ng/L. One of the major effects of GLC is to raise the rate at which GN is broken down, releasing glucose stored in the liver cells back into the blood and ECF. GLC also stimulates the process of gluconeogenesis, which also raises [BG] over the long term. The steps by which GLC stimulates the breakdown of GN are well-known. After combining with GLC receptor proteins on the hepatic cell membrane, the enzyme adenyl cyclase is activated. Adenyl cyclase catalyzes the formation of cyclic AMP (cAMP), which activates the following series of reactions:

protein kinase regulator protein $\rightarrow$ protein kinase $\rightarrow$ phosphorylase **b** kinase

$\rightarrow$ phosphorylase **b** $\rightarrow$ phosphorylase **a**

Phosphorylase **a** is the active enzyme that cleaves a GN polymer unit, forming G1P. G1P is in turn converted to G6P by another enzyme, and then phosphatase, activated by GLC, removes the phosphate group and the resulting D-glucose diffuses out of the liver cell into the circulation. Note that the outward flow of glucose occurs during periods of fasting as the hepatic glucose regulator attempts to maintain normoglycemia.

Gluconeogenesis is a complex metabolic process activated by elevated [GLC]. Several metabolic pathways are involved, and many enzymes are necessary for the multistep reactions. The details of these reactions are beyond the scope of this text. However, high [GLC] causes free amino acids in the blood to be taken up by liver cells where they are converted to glucose. High [GLC] also activates the enzyme that converts pyruvate to phosphoenolpyruvate; this conversion is a key step in gluconeogenesis. Note that the gluconeogenesis processes are catabolic. Under extreme conditions of starvation, the body, to attempt to maintain normoglycemia, first converts fats, then proteins to glucose through various gluconeogenesis pathways. In extreme cases of starvation, what is left is "skin and bones"; we have all seen pictures of victims of prison camps, etc., that underscore this grim, but life preserving process.

Figure 8.7 illustrates the basic relationship between [PG] and $\dot{Q}_{GLC}/\dot{Q}_{GLC90}$, the normalized rate of release of GLC from the pancreatic alpha cells. We have assumed simple, first-order, loss kinetics for the breakdown of [GLC] in the absence of detailed pharmacokinetic data. As in the case of insulin, there are several factors that modulate $\dot{Q}_{GLC}$ beside [PG]. High concentrations of amino acids (especially arginine and alanine) in the blood after a high protein meal stimulate the release of GLC. GLC promotes the rapid conversion of the amino acids to glucose, raising the [PG].

The delta cells of the pancreatic islets of Langerhans secrete the 14 amino-acid peptide hormone, somatostatin. Somatostatin is secreted in response to almost all factors related to eating. It has a very short half-life of about 3 min. Somatostatin acts to suppress the secretion rates of both insulin and GLC. It also has a general slowing effect on all aspects of the digestion process. Guyton (1991) speculated that somatostatin's role is to extend the period of time over which food nutrients are digested and assimilated into the blood. By suppressing the release of GLC

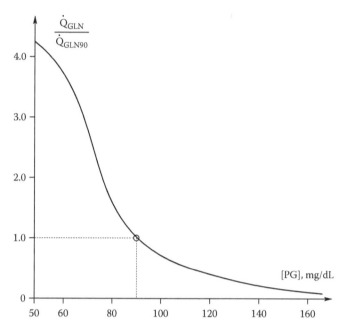

FIGURE 8.7 Normalized rate of GLC secretion as a function of steady-state plasma glucose concentration. The half-life of free GLC is ca. 10 min. See text for discussion. (From Northrop, R.B., *Endogenous and Exogenous Regulation and Control of Physiological Systems*, Chapman & Hall/CRC Press, Boca Raton, FL, 2000.)

and insulin, somatostatin slows utilization of the absorbed nutrients by the tissues, making them available over a longer period.

8.4.2.6 Simple, Compartmental Model for Normal Blood Glucose Regulation

As we have seen, the human glucoregulatory system consists of sources and sinks for glucose, some of which are primarily regulated by the hormones insulin and GLC, which are made in the pancreatic beta and alpha cells, respectively, and are secreted into the hepatic portal blood supply.

First, we examine two of the sinks for plasma glucose. We will model the dynamics of glucose entry into insulin-sensitive and non-insulin-sensitive cells. These sinks are called glucose utilization, and are modeled by the systems shown in Figure 8.8A and B.

In Figure 8.8A, the diffusion parameter, $K_{DISC}(I)$, is a function of insulin concentration, that is, it is parametrically regulated. Extracellular insulin binds with receptor proteins on insulin-sensitive cell surfaces, and within seconds, causes the rate that glucose enters insulin-sensitive cells, $\dot{Q}_{GISC}$, to increase. Insulin effectively increases the diffusion constant, $K_{DISC}(I)$. Insulin also causes increased cellular permeability to many amino acids, K^+, Mg^{++}, and PO_4^{3-}, as well as causing slower changes in intracellular enzymes.

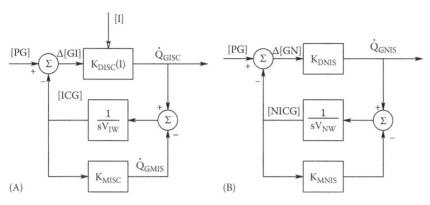

FIGURE 8.8 (A) Block diagram describing the dynamics of glucose utilization dynamics by insulin-sensitive cells with first-order linear dynamics. The plasma insulin concentration [PI] parametrically increases K_{DISC}. See text for description. (B) Block diagram modeling glucose utilization by non-insulin-sensitive cells with first-order linear dynamics. K_{DISC} is constant. (From Northrop, R.B., *Endogenous and Exogenous Regulation and Control of Physiological Systems*, Chapman & Hall/CRC Press, Boca Raton, FL, 2000.)

To describe glucose utilization by insulin-sensitive cells, $\dot{Q}_{GISC}$ is subtracted from the central extracellular glucose rate summer. To derive the dynamics of glucose utilization by insulin-sensitive cells, we subtract the intracellular rate of glucose consumption by metabolism from $\dot{Q}_{GISC}$, then divide the difference by the intracellular water volume, V_{IW}, then integrate to obtain the intracellular glucose concentration [GIC]. We assume simply that the rate of intracellular metabolism is equal to K_{MIC} [GIC] mg/min. The metabolic rate constant, K_{MIC} is a function of temperature, thyroid hormone concentration, epinephrine concentration, and the mechanical work load if the cells are muscle; for simplicity, we will treat it as a constant. Reduction of the block diagram yields a pseudo-linear transfer function relating $\dot{Q}_{GISC}$ to [PG]. (*Note*: $K_{DISC}(I) = H_i K_{DISC}$.)

$$\frac{\dot{Q}_{GISC}}{[PG]}(s) = \frac{H_i K_{DISC}(s\tau_c + 1)}{(1 + H_i K_{DISC}/K_{MIC})[s\tau_c/(1 + H_i K_{DISC}/K_{MIC}) + 1]} \qquad (8.2)$$

where

$\tau_c = V_{IW}/K_{MIC}$ min

H_i is a first-order, saturating Hill function

Note that insulin acts parametrically to increase K_{DISC}, and a number of factors can influence K_{MIC}, hence intracellular metabolism. H_i is given by

$$H_i = c_o + \frac{c_m K_i[PI]}{1 + K_i[PI]} \quad \text{Hill function for insulin}$$

activation of K_Ds and enzymes. [PI] is plasma insulin conc. (8.3)

Note that H_i ranges smoothly between c_o and $(c_o + c_m)$ as [PI] increases.

For the case of non-insulin-sensitive cells (including CNS neurons, red blood cells, intestinal epithelial cells, and kidney tubule epithelial cells), illustrated in Figure 8.8B. The architecture of the block diagram is the same as for insulin-sensitive cells, except that the diffusion constant, K_{DNIS}, is a constant, that is, it is not a function of [PI]. $\dot{Q}_{GNIS}$ is the rate at which extracellular glucose enters the non-insulin-sensitive cells; it also must be subtracted from the central extracellular glucose rate summer. After reduction, we find the transfer function:

$$\frac{\dot{Q}_{GNIS}}{[PG]}(s) = \frac{K_{DNIS}(s\tau_{cn}+1)}{(1+K_{DNIS}/K_{MNIS})[s\tau_{cn}/(1+K_{DNIS}/K_{MNIS})+1]} \qquad (8.4)$$

where $\tau_c = V_{NW}/K_{MNIS}$ min. The parameters K_{DNIS} and K_{MNIS} are considered constants in this case as the CNS's metabolism is relatively constant, and is independent of insulin and GLC.

Figure 8.9 shows a basic systems block diagram summarizing normal blood glucose regulation. The system producing the hormone GLC is modeled by simple, first-order loss kinetics. A static, nonlinear function provides the GLC rate input to the GLC loss dynamics ODE.

The normal pancreatic insulin system is modeled by a single time-constant, fast-release component that is only responsive to positive rates of increase of glucose concentration. A second, slow, insulin component has a damped, second-order response to glucose. The sum of the fast and slow insulin release rate systems is further conditioned by a first-order lag system.

A single, central glucose compartment, pooling the plasma volume of the circulatory system and the extracellular fluid volume is used for simplicity. The inputs or sources to the glucose rate summer are from diet, from the liver, and from gluconeogenesis. The sinks of glucose include the liver, the kidneys, and cellular utilization (uptake by the insulin-sensitive cells, and uptake by the non-insulin-sensitive cells).

The difference between the outward flow rate of glucose from the liver ($\dot{Q}_{HGR}$) and the inward flow rate from the plasma ($\dot{Q}_{HGS}$) is called the net hepatic glucose balance (NHGB). NHGB can be positive, zero, or negative.

Negative NHGB occurs when the liver is actively storing glucose as GN. (The convention of NHGB > 0 for $\dot{Q}_{HGR} > \dot{Q}_{HGS}$ models by Cobelli and Mari (1983)). Note that liver cells can store GN up to 5%–8% of their weight. Liver GN has an average molecular weight of 5 million. Muscle cells can also store GN; up to a maximum of 1%–3% of their weight (Guyton 1991). However, glucose from muscle GN is generally directly metabolized inside the muscle cells.

The dietary glucose input rate, $\dot{Q}_{GD}$, for a complex meal containing both carbohydrates, sugars, and proteins is generally bimodal. Sugars and carbohydrates are digested first and converted to glucose, followed by the amino acids released by the digestion of proteins. Some amino acids are converted to glucose.

The loss rate of glucose in the urine can be approximated by a simple, piecewise-linear relation. If we assume a constant glomerular filtration rate of

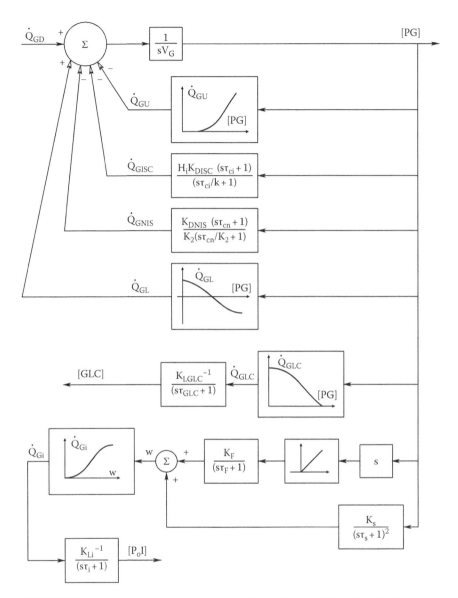

FIGURE 8.9 A simple glucoregulatory system block diagram. There is a single, central glucose compartment with glucose loss rates in urine, into insulin-sensitive cells, into non-insulin-sensitive cells, and a hepatic glucose flux (storage or release) that depends not only on [PG] but also on the hormones insulin, GLC, leptin, etc. The production of plasma GLC and portal insulin are also modeled. Glucose input rate from diet is shown. See text for description. (From Northrop, R.B., *Endogenous and Exogenous Regulation and Control of Physiological Systems*, Chapman & Hall/CRC Press, Boca Raton, FL, 2000.)

125 mL/min, the kidneys reabsorb all blood glucose molecules in the filtrate up to a [PG] of about 160 mg/dL. Above 160 mg/dL, the kidneys loose more and more glucose in the filtrate as [PG] increases. For [PG] > 320 mg/dL, the loss rate of glucose in the urine, $\dot{Q}_{GU}$, increases linearly with the tubular load (TL) of glucose in the renal artery. The tubular load of glucose in mg/min is simply

$$\text{TL (mg/min)} = [PG](\text{mg/dL}) \times 1.25(\text{dL/min}) \tag{8.5}$$

The loss rate of glucose through the kidneys, $\dot{Q}_{GU}$, can thus be approximated by the linear equation

$$\dot{Q}_{GU} = -320 + 1.25[PG] \text{ mg/min}, \quad \text{for } [PG] \geq 325 \text{ mg/dL, and}$$

$$= 0 \quad \text{for } [PG] < 325 \text{ mg/dL} \tag{8.6}$$

This relation is shown in Figure 8.10.

The dynamics of NHGB can be described using chemical mass action kinetics to describe the liver reactions that form and break down GN. Five, nonlinear ODEs are written whose rate constants are functions of the concentrations of insulin and GLC. The ODEs are nonlinear because all states are nonnegative. Positive NHGB is defined here as the rate of mass diffusion of glucose from the liver cells compartment to the plasma compartment, that is, positive NHGB makes [PG] and [BG] rise.

$$\text{NHGB} \equiv K_{DL}(I)\{[LG] - [PG]\} \text{ mg/min} \tag{8.7}$$

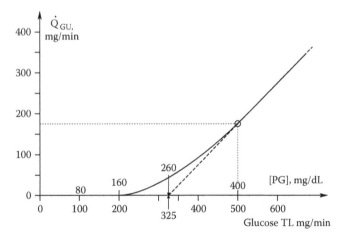

FIGURE 8.10 Average rate of glucose loss in the urine as a function of tubular load (TL) as well as [PG], assuming a constant glomerular filtration rate. (From Northrop, R.B., *Endogenous and Exogenous Regulation and Control of Physiological Systems*, Chapman & Hall/CRC Press, Boca Raton, FL, 2000.)

The five ODEs we use to describe the dynamics of hepatic GN storage and release are

$$[\dot{LG}] = K_{DL}(PI)\{[PG]-[LG]\} + K_{PH}(GLN)[G6P] - K_{GK}(I)[LG] \, g/(L \, s) \quad (8.8A)$$

$$[\dot{G6P}] = K_{GK}(PI)[LG] - K_{GLS}[G6P] - K_{PH}(GLN)[G6P] - K_{61}[G6P] + K_{16}[G1P] \quad (8.8B)$$

$$[\dot{G1P}] = K_{PHL}(GLN)[GN] - K_{16}[G1P] + K_{61}[G6P] - K_{1U}(I)[G1P] \quad (8.8C)$$

$$[\dot{UDG}] = K_{1U}(I)[G1P] - K_{UG}[UDG] \quad (8.8D)$$

$$[\dot{GN}] = K_{UG}[UDG] - K_{PHL}(GLN)[GN] \quad (8.8E)$$

These reactions are illustrated in Figure 8.11. Note that the five states are non-negative, and there is a saturation limit on [GN], the GN concentration, at 8% of the liver mass. Note that the rate constants for certain enzymes are generally increasing functions of the concentrations of the respective hormones that modulate them. The precise nature of the parametric modulation of the rate constants is not known, but we expect them to be sigmoid in nature. For example, the rate constant for glucokinase, K_{GK}, is increased by insulin. So we have modeled K_{GK} with an nth-order Hill function:

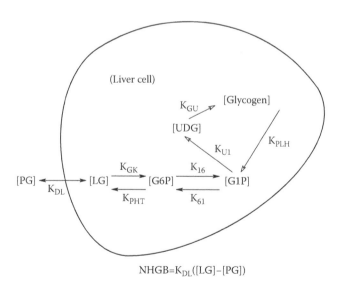

NHGB=$K_{DL}([LG]-[PG])$

FIGURE 8.11 Schematic of the five reactions involved in NHGB. See the five mass-action ODEs, Equations 8.8A through E in the text. (From Northrop, R.B., *Endogenous and Exogenous Regulation and Control of Physiological Systems*, Chapman & Hall/CRC Press, Boca Raton, FL, 2000.)

$$\mathbf{K_{GK}}(\mathbf{I}) = \left\{ K_{GKmin} + \frac{(K_{GKmax} - K_{GKmin})[PI]^n}{\Phi^n + [PI]^n} \right\} \qquad (8.9)$$

The exact values for the threshold, Φ, the exponent $\mathbf{n}$, and the minimum and maximum $\mathbf{K_{GK}}$ must be determined experimentally. (Note that a number of other algebraic forms giving a sigmoid modulation of $\mathbf{K_{GK}}$ with [PI] are available; these include hyperbolic tangent functions and exponential functions.) While the five ODEs above are physiologically accurate, they are complicated and involve rate constants, the values of which are unknown. Thus we sought to represent NHGB by a simpler, heuristic, description. Figure 8.12 illustrates a simplified, second-order (2-state) nonlinear system that can be used to describe NHGB. The simplified ODEs and relations describing NHGB are

$$\mathbf{H_i} = c_0 + \frac{c_m K_i[PI]}{1 + K_i[PI]} \quad \text{Hill function for insulin activation}$$

$$\text{of } K_D s \text{ and enzymes. [PI] is plasma insulin conc.} \qquad (8.10A)$$

$$\mathbf{H_{glc}} = b_0 + \frac{b_m K_{G14}[GLC]}{1 + K_{G14}[GLC]} \quad \text{Hill function for glucagon activation of enzymes}$$

$$(8.10B)$$

$$[\dot{LG}] = \mathbf{H_i} K_{DL}\{[PG] - [LG]\} + \mathbf{H_{glc}}[rGN]$$
$$- f_i K_S[LG] \quad \text{g/ min Free liver glucose} \qquad (8.10C)$$

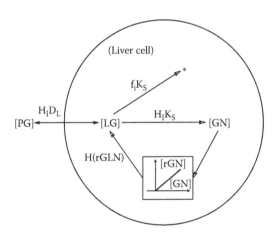

FIGURE 8.12 A simplified, nonlinear, second-order model for NHGB. See text for description. (From Northrop, R.B., *Endogenous and Exogenous Regulation and Control of Physiological Systems*, Chapman & Hall/CRC Press, Boca Raton, FL, 2000.)

$$[LG] = \frac{LG}{V_l} \quad \text{g/L Conc. of free liver glucose} \qquad (8.10D)$$

$$[\dot{GN}] = [LG]H_iK_S - r[GN]H_{gln} \quad \text{g/min Rate of glycogen formation} \qquad (8.10E)$$

$$rGN = \text{IF GN} < 0 \text{ THEN 0 ELSE GN} \quad \text{g. Glycogen is nonneg.} \qquad (8.10F)$$

$$NHGB = H_iK_{DL}\{[LG] - [PG]\} \quad [GN] = \frac{d[GN]}{dt}, \text{ etc.} \qquad (8.10G)$$

As noted previously, the normal pancreas releases insulin in response to elevated plasma glucose concentration in two steps: Figure 8.6 illustrates a block diagram describing the fast and sustained release dynamics for insulin. There is a rapid, transient release of insulin stored in beta cells in response to a positive rate of increase in [PG], and there is a slower, sustained release involving intracellular synthesis of insulin in response to elevated [PG]. The final feedback system describes the simple, first-order destruction of insulin. Insulin distribution dynamics are ignored.

GLC release can be modeled more simply by a first-order loss process. GLC release increases with low plasma glucose concentration [PG]. The release rate of GLC is modeled simply by

$$[\dot{GLC}] = \frac{K_{G11}}{(1 + K_{G12}H_i[PG])} - h[GLC] \quad \mu\text{g/min} \qquad (8.11A)$$

$$[GLC] = \frac{GLC}{V_{GLN}} \quad \mu\text{g/L concentration} \qquad (8.11B)$$

It was important to validate the normal glucoregulatory system model described above by simulation and comparison with physiological data. The Simnon™ program, **Glucose5.t**, given in Appendix A.3, uses only 12 states, and allows one to investigate the "normal" glucoregulatory responses to an IVGTT. Many of the constants were chosen by trial and error to validate the system's responses to known physiological conditions. In this model, we have added three compartments (states) describing insulin distribution, after the model of Cobelli and Mari (1985). Normal pancreatic insulin is secreted into the hepatic portal venous system, from which it directly affects liver cells. Insulin then diffuses into the circulatory system, thence into interstitial volume, where it can bind with the receptors on insulin-sensitive cells.

Figure 8.13 shows the model's response to a (simulated) IVGTT. The traces are #1 = [PG], 2 = [BI], 3 = [GLC], 4 = [NHGB], 5 = zero, 6 = 0.8 g/L [PG]. [PG] is to scale, other quantities are not to scale in order to fit them on the PG vertical

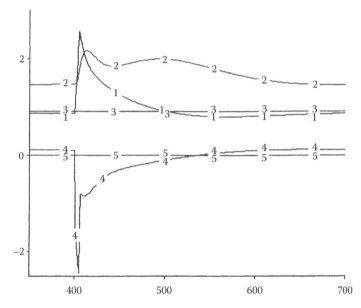

FIGURE 8.13 Response of the 12-state, nonlinear Simnon model of normal glucoregulation, `glucose5.t` (in Appendix A.3) to a bolus IVGTT given at t=400. Vertical axis: g/L for [PG]. Horizontal axis: time in minutes. Trace 1=[PG], trace 2=blood insulin concentration [bI], trace 3=0.9 (constant), trace 4=NHGB, trace 5=0. (From Northrop, R.B., *Endogenous and Exogenous Regulation and Control of Physiological Systems*, Chapman & Hall/CRC Press, Boca Raton, FL, 2000.)

axis. A positive [NHGB] signifies glucose release by the liver in response to GLC. Negative NHGB means the liver is taking up glucose and storing it as GN. In the model in Appendix A.4 we have simplified the NHGB equations to two, nonlinear ODEs, rather than the five described above. We have also made the rate of breakdown of GN independent of the GN concentration [GN], assuming an excess of GN stored.

Figure 8.14 illustrates the modeled distribution of endogenous (pancreatic) insulin between the plasma, portal, and interstitial fluid compartments in response to the IVGTT. Also plotted as trace 5 is the rate of glucose loss through the urine, $\dot{Q}_{GU}$. In Figure 8.15, we see the effect of an exogenous, IV bolus dose of insulin on [PG], [BI], [GLC], and [NHGB]. Note that the insulin causes the [NHGB] to go transiently negative under fasting conditions. The rate of glucose loss into the liver and insulin-sensitive cells causes the [PG] to drop transiently, and paradoxically, [GLC] also drops in response to the insulin bolus.

Although the Simnon model above for normal glucoregulation is parsimonious, it gives realistic results. The model includes a switch that turns off hepatic glucose release during fasting when liver GN is exhausted. This is shown in Figure 8.16. Note that at $t \cong 360\,\mathrm{min}$, [GN]→0, and consequently NHGB→0.

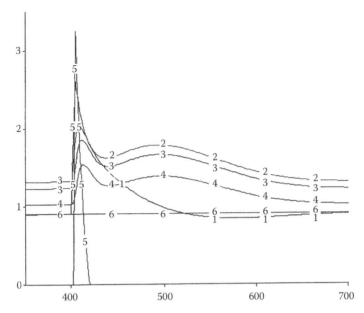

FIGURE 8.14 Additional results of the IVGTT simulation of Figure 8.13. Same axes. Trace 1 = [PG], trace 2 = portal insulin concentration [PoI], trace 3 = [bI], trace 4 = interstitial fluid insulin concentration [ifI], trace 5 = rate of glucose loss in urine, trace 6 = 0.9 g/L (constant). (From Northrop, R.B., *Endogenous and Exogenous Regulation and Control of Physiological Systems*, Chapman & Hall/CRC Press, Boca Raton, FL, 2000.)

[PG] drops rapidly, and [GLC] rises in response. [PG] plateaus at about 0.6 g/L because of a fixed gluconeogenesis rate of 0.5 g/min. (It would be more realistic to have gluconeogenesis as a lagged, increasing function of [GLC], but we used a fixed value in the model for simplicity.)

To simulate Type I diabetes under exogenous feedback regulation, we disabled insulin secretion in our model and used a controller to inject or infuse exogenous insulin into the model patient. Three controller subroutines are embedded in the Simnon program, **Glucose6.t**, in Appendix A.4; a proportional + integral (PI), a proportional + integral + derivative (PID), and a simple, integral pulse frequency modulation (IPFM) controller (Northrop 2000). Figure 8.17 illustrates the simulated, type I diabetic's response to an IVGTT (bolus injection) using the PI controller, as well as an IV insulin bolus injection.

8.4.3 HIS VS. HIV

8.4.3.1 Introduction

As we have seen in Chapter 6, the human immune system has considerable complexity, perhaps second only to the hCNS. Just to try to verbally describe its structure and function is daunting. The complexity of the hIS arises not only from the large number of cell types contributing to immune responses (e.g.,

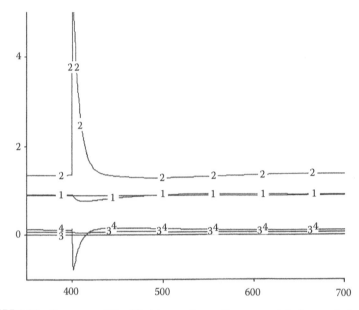

FIGURE 8.15 Response of the 12-state, nonlinear Simnon model of normal glucoregulation, `glucose5.t` (in Appendix A.3) to an IV bolus of insulin given at t=400. Same axes. Trace 1=[PG], trace 2=[bI], trace 3=GLC concentration [glc], trace 4=NHGB. Note that the exogenous insulin causes the liver to take up glucose. (From Northrop, R.B., *Endogenous and Exogenous Regulation and Control of Physiological Systems*, Chapman & Hall/CRC Press, Boca Raton, FL, 2000.)

macrophages [Mφs], natural killer cells, cytotoxic T-cells, dendritic cells, helper T-cells, B-cells, memory B-cells, other memory cells, mast cells, neutrophils, granulocytes, and basophils), but also from the bewildering number of immuno-cytokines (immune system autacoids), including the molecules of the complement system, molecules secreted from platelets and mast cells, and the immunoglobu-lins (antibodies and anti-idiotypic antibodies).

Complexity is also engendered from the pleiotropy of the immunocytokines, as well as their number. The effects they have on their target cells vary with the level of maturity and activation of the cells, and the prior reception of other cytokine signals by these cells. The hIS is also affected by the hormone leptin, secreted principally by WAT (fat cells). Low leptin concentrations (such as in star-vation) have been shown to cause dysregulation of cytokine production, leading to susceptibility to infection and disease (Faggioni et al. 2001).

In order to accurately model the dynamics of immune system cell develop-ment, growth, control, action, and death, we need to know the rates that immune system cells are normally produced in the bone marrow, the rates that they can undergo clonal expansion under the influence of various immunocytokines, and the normal half-lives of the cells. We also need to know the dynamics governing the controlled, intracellular manufacture of immunocytokines, the dynamics of

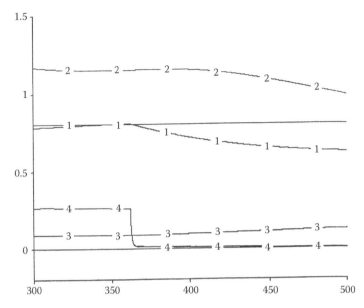

FIGURE 8.16 Simulated results using the model, `glucose5.t`, showing liver GN exhaustion at t = 360 min. under fasting conditions. Note that the [PG] (trace 1) begins to decay. [bI] (trace 2) also decays, while [glc] (trace 3) begins to increase. NHGB (trace 4) drops abruptly to zero as no further liver GN can be converted to glucose and released. (From Northrop, R.B., *Endogenous and Exogenous Regulation and Control of Physiological Systems*, Chapman & Hall/CRC Press, Boca Raton, FL, 2000.)

their release, and their normal half lives. Immunocytokines combine with receptors on cell surfaces at rates that are generally assumed to be governed by the laws of mass action. However, these rates must include saturation functions to account for the finite number of cell-surface receptors. As we have already shown, modelers generally use the hyperbolic Hill function format, or hyperbolic tangent functions (tanh(ax)) to emulate rate saturation.

Cytokine molecules binding to receptors trigger the internal biochemical machinery of the cell to either manufacture or stop manufacturing biochemical products, including transcription factors (TFs). How this signal transduction works is currently being studied. Every cell has a finite number of receptors for certain cytokines, hence a cytokine message's effect can saturate, that is, a cell can make a product at a maximum rate, given 100% saturation of its receptors by an excess of cytokine. These saturation effects must also be incorporated in an accurate model. A good, validated, mathematical model should be able to predict how well a proposed drug or immunotherapy will work in killing the pathogen or restoring normal immune system homeostasis, before resources are committed to animal trials.

In formulating a simulation of an immune system event, we must make certain assumptions about where in the body it is assumed to take place. It is unsuitable

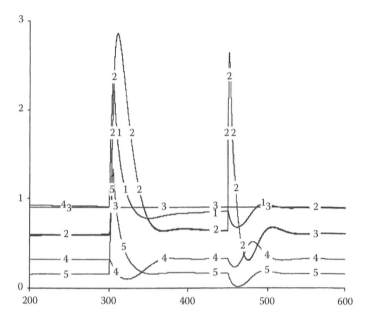

FIGURE 8.17 Simulated behavior of the glucoregulatory plant model of Figure 8.9 using an external, continuous PI controller to drive an IV insulin infusion pump. (the model has no beta cells). (The 14-state Simnon model is `glucose6.t` in Appendix A.4.) Vertical axis: g/L of [PG], only. Horizontal axis: minutes. Trace 1 = [PG], trace 2 = [bI], trace 3 = regulator set point (0.9 g/L), trace 4 = [GLC], trace 5 = ru (rectified controller output). An IVGTT of 30 g/min. was given for 5 min at t = 300. An IV bolus injection of insulin was given at t = 450, producing an interesting transient in the GLC concentration [GLC] and causing the [PG] to dip, and then return to the set point. Euler integration was used with Δt = 0.0005. Controller gains: K2 = 0.01, K3 = 0.75. (From Northrop, R.B., *Endogenous and Exogenous Regulation and Control of Physiological Systems*, Chapman & Hall/CRC Press, Boca Raton, FL, 2000.)

to model the body as a whole because of the differences between various organs and tissues, and the fact that most infections are localized in a particular organ or tissues, for example, the lymphatic system. Also, we have seen that certain types of immune system cells (e.g., certain MΦs, mast cells) are sessile, that is, they are fixed in some tissue or organ and therefore are not "well mixed," a requirement for mass-action kinetics assumptions. They would participate in an immune reaction only if the pathogen being fought were nearby in a relatively constant concentration.

For example, the site, or modeling scenario, can be in a "unit volume" surrounding a cancer beginning to grow in the lungs, or it can be in a unit volume of blood, in a lymph node, in a small volume of tissue surrounding an injury, or in the pleural cavity. In a model, the immune system cell states can be considered to be an absolute number in the defined volume, or a density (number per cubic mm). Cytokines can be represented as a molecular population density (pd) (molecules/mm³), or by weight, for example, pg/mm³, or by mol/L, or by "units"/mm^3 (an operational definition). Once secreted, cytokines diffuse away from their source, and then either bind to their receptors or not. Cytokines are broken down enzymatically, contributing to their loss rate in the scenario volume.

An immune system model generally consists of a large set of coupled, non-linear ODEs based on mass-action kinetics based on the probability of interaction between receptors on cells and local, soluble immunocytokines. The model should include the effects of certain cytokines on cell growth, maturation, clonal expansion (if it occurs), and cell function.

There is a natural tendency in modeling a complex system to want to simplify it, yet preserve its essential behavior. (We did this above in considering the glucoregulatory system.) Such simplification can follow the best of intentions. For example, suppose that a relation between the advent of a certain cytokine and the induced manufacture of a second cytokine by an IS cell has been observed qualitatively in vivo or in vitro. An ODE can be formulated to describe this process. One might argue that if we do not know the numbers for the rate constants, decay constants, etc., we should not include the putative relation in the model. However, if the product cytokine is known to be critical in other aspects of the model, the process is considered important to the overall model. Even though numbers for rate constants may not have been measured on living systems for the process, reasonable estimates of rate constants and other parameters can be made for the speculative ODE for purposes of validation. If the ODE and the trial numbers do not contribute to the accuracy of model validation, then the numbers can be revised and the model run again. If no improvement is seen over a reasonable range of parameter values, then the ODE probably is not relevant to the phenomenon being simulated, and should be deleted from the model. Models are meant to be tinkered with.

Model simplifications may be made for a less pure motive, that is, we do not want the complex system to confuse or bewilder our intended audience, or ourselves. In any case, such reductionism may rob the model of realistic, functional detail. High-order, coupled, nonlinear systems often exhibit unexpected or counter-intuitive (chaotic) behavior, which will not be seen in a reduced model, but which can be verified in vivo. Thus, it should be a paramount goal of anyone seeking to model immune function to include as much validated detail as possible. We urge the modeler to eschew reductionism. Models are inexpensive compared to animal studies; the computer does the work. They are an ideal platform on which to test hypotheses on immune system interactions. Remember, however, a model is not the end, but a means to obtain a greater understanding of a complex system. A model must be validated in order to be an effective predictor.

8.4.3.2 Models of HIV Infection and AIDS

The realization in 1983 that the human acquired immune deficiency syndrome (AIDS) was the result of a retrovirus that attacked specific cells in the human immune system has led to intense research on the molecular biology and physiology of the immune system, and on the different varieties of the human immune deficiency virus (HIV). The HIVs are retroviruses in which the genetic information for their reproduction is coded in a single-strand RNA base sequence, rather than in DNA. HIV, like certain other retroviruses, infects cells in the human immune system. HIV is unique, however, in that it causes a gradual, and generally fatal loss of effectiveness of the immune system's ability to fight infections. AIDS, the acquired immune deficiency syndrome, is characterized by a terminal illness from an "opportunistic infection," such as *Pneumocystis carinii* pneumonia, Kaposi's sarcoma, aspergilliosis fungal infection (usually of the lungs or brain), tuberculosis, and syphilis.

It should be pointed out that other retroviruses attack immune system cells. For example, the feline leukemia virus (FeLV) and the human T-lymphotropic virus 1 (HTLV-1) both infect an immune system's T-cells causing leukemia and immune suppression. The Visna virus of sheep attacks their monocytes/Mϕs, causing progressively debilitated immune function and CNS problems. The Visna virus and human HIV are structurally similar, yet Visna does not attack sheep's T-cells (Pauza 1988).

Significant in human HIV infection is the fact that HIV attacks, and is found within, both CD4+ helper T-cells and monocyte/Mϕs and related cells in the CNS. In about 95% of the human population, untreated HIV infection is characterized by a slow, progressive decline in the number of CD4+ T-cells (helper T-cells or Th) over a period of 5 years or so. Eventually, the ability of the immune system to respond successfully to opportunistic infections is sufficiently impaired, so that the patient is killed by such an infection. Normal CD4+ Th density in the blood is about 800 cells/mm^3. When the Th density has fallen to about 200 per mm^3, the patient may be considered to have AIDS. The cause of AIDS is certainly not as simple as a decreased count of Th cells, however. As you will see, there are many other factors that contribute to the loss of immune function due to HIV infection, which is a complicated process operating on a complex system.

Significantly, from 5% to 10% of the human population infected with HIV maintain normal CD4+ Th cell counts and do not develop AIDS for over 10 years after infection, or the progression of the disease is very slow. The reason for this resistance has been recently discovered, and hopefully, may lead to effective immunization and treatment for HIV infection.

Figure 8.18 illustrates a schematic, cross-sectional structure of a "typical" HIV virion. It is a very efficient, compact, self-replicating, "molecular nanomachine." The outside of an HIV virion is a 20-sided volume (icosohedron) about 100 nm in diameter. It has an outer protein membrane, and a nucleocapsid, or core containing two, identical, single strands of RNA (ssRNA), each of about 9000 base pairs that code its structural proteins and reproductive enzymes. Also inside of the nucleocapsid are the HIV's "start-up" reproductive enzymes including reverse transcriptase (RT=p66), protease (PR=p9), and integrase (IN=p32). Once inside the host cell, the reverse transcriptase makes a DNA copy of the virion's RNA using the host cell's nucleotides. This viral DNA (vDNA) copy is inserted in the host cell's DNA by the integrase. When activated, the host cell makes HIV RNA from the new, viral DNA (vDNA), using the host cell's RNA Polymerase II (RNAP2). The host cell is also tricked to make HIV proteins from the new, vDNA

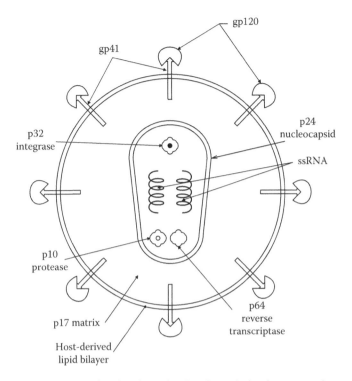

FIGURE 8.18 Cross-sectional schematic drawing of the important features of a "typical" HIV virion. The outer capsule is configured in an icosahedron ca. 100 nm in diameter, shown here as a circle. (From Northrop, R.B., *Endogenous and Exogenous Regulation and Control of Physiological Systems*, Chapman & Hall/CRC Press, Boca Raton, FL, 2000.)

in its genome. New HIV virions are assembled within the host cell and then released by the process of eclosion. Portions of the host cell's plasma membrane, including some host cell surface proteins such as MHC I become the eclosed virions' outer membranes.

On the outer surface of the HIV membrane is a dense array of about 72–80 glycoprotein 120 (gp120) molecules with which the virion makes contact with its host cells. Part of gp120 is known to have a strong affinity for a binding site on the CD4 glycoprotein found on the surface of helper T-cells (Th) and MΦs. (Recall that CD4 is a critical cofactor in antigen presentation (AP) by certain B-cells and MΦs. It combines with a site on MHC II + Ag to trigger Th and B-cell activation.) The gp120 "cap" is attached to the virion by a "stem" of glycoprotein 41 (gp41) that penetrates the virion's coat (the entire "lollypop" assembly is called gp160). The cytosol of the virion is filled with a matrix protein, p17.

The mechanisms of introduction of HIV into the body are well-documented and will not be discussed here. However, we will examine the mechanisms by which HIV virions enter CD4+ Th cells and CD4+ MΦs, and infect them. The first step in HIV replication is the binding of the virion to the cell it will infect. Until recently, it was believed that HIV binding to its host cell was due to the affinity of its gp120 to CD4 surface proteins, found on Mφs and CD4+ Th cells. In 1996, it was reported that β-chemokines (chemokines are attractants of immune system cells secreted by other immune system cells) are major suppressors of HIV infection (Harden and D'Souza 1996). Subsequent studies showed that the excess β-chemokines combined with and thus saturated and blocked their receptors on Th and Mφs. Finally, it was discovered that HIV binding and entry required not only the CD4 surface protein, but also one of several chemokine receptors adjacent to the CD4 molecule. Apparently gp120, once bound to CD4, sends out a *v3 loop*, which has affinity for an adjacent chemokine receptor. This second binding triggers the inclosion of the virion. The virion's coat with its attached gp120 molecules and some free gp120, remain outside the infected cell while the core with its RNA and enzymes enters through a hole in the cell's plasma membrane, which then reseals.

Once inside the cell, the virion's core capsule uncoats, and the RNA and enzymes are released into the infected cell's cytosol. The RNA is copied into vDNA by the viral reverse transcriptase. The vDNA is then transported into the host cell's nucleus, where the viral enzyme, integrase, inserts the vDNA into the host's DNA genome, where it becomes an addendum to the host cell's genes. The vDNA remains dormant until some exogenous factor activates its reading and protein synthesis, which leads to the reassembly of viral enzymes, vRNA, a core coat, and the eventual eclosion of many new virions. Mass eclosion of virions kills the infected cell. HIV eclosion is not the only cause of death of HIV-infected cells, however.

Let us go back and examine the role and importance of the chemokine receptor cofactors for HIV binding and entry. Mφs carry several chemokine receptors, some of which have affinity for part of the HIV's gp120 molecule. The Mφ CCR5 chemokine receptor normally binds with RANTES (which stands for regulated on activation, normal T-cell expressed and secreted, the ultimate acronym). Mφ CCR5 also binds with an HIV1 gp120 site. Helper T-cells carry a chemokine receptor called CXCR4, which normally binds to stromal cell-derived factor 1 (SDF-1). The CXCR4 receptor also has an affinity to HIV1's gp120. An HIV virion is called M-tropic or T-tropic depending on whether the *v3* loop of gp120 has a major affinity for Mφ CCR5 or T-cell CXCR4, respectively. What controls this tropism switch is not known.

Back in the late 1980s, research was focused on "confusing" the HIV by the injection of soluble CD4 molecules, (sCD4) or CD4 molecules bound to erythrocyte membranes or other molecules. In theory, the excess sCD4 would bind to the free virion's gp120 molecules, preventing HIV from binding to normal CD4 on Mφs and T-cells. This strategy was largely ineffective in vivo because the sCD4 combined with its normal binding site on Mφ and B-cell MHC II, and presumably blocked normal AP by Mφs and B-cells, severely weakening the immune response to all pathogens. This is a perfect example of the LUC. Free, soluble gp120 in the blood could combine with the sCD4, giving a null result. HIV virions already inside Mφs and CD4+ Th in the process of assembly could not be affected by sCD4. In hindsight, sCD4 was a bad idea (the LUC again).

The discovery of the chemokine cofactors required for HIV binding and entry opened up a whole new avenue of research for fighting HIV infection. The chemokine receptors are evidently not as vital for the operation of the immune system, as is CD4, but are vital to HIV. Thus, a new therapy based on selectively blocking the *v3* loop from combining with its M- or T-chemokine receptors appeared possible.

We note that there are a large number of chemokines, and not all their receptors have affinity to the *v3* loop of gp120. Chemokines generally attract immune system cells to sites of inflammation. There are about 16 known chemokines that attract one or more kinds of leukocytes. Some chemokines such as lymphotactin, attract only T-cells and NKs (Harden and D'Souza 1996). Other chemokines like RANTES attract Mφs, eosinophils, basophils, T-cells, and NKs, but not neutrophils. IL8 attracts neutrophils and T-cells, while complement C5a attracts neutrophils, Mφs, and eosinophils, and no others. Such complexity appears to be the rule as we try to gain an understanding of the complex human immune system's functions.

There are a number of theories why HIV infection is so hard to control, let alone cure. One involves the large number of soluble gp120 (sgp120) molecules in the blood of an infected person. As we have seen, gp120 has affinity for CD4 protein. If free, sgp120 molecules bind with an uninfected Th cell's CD4 molecules, AP can be blocked (Mann et al. 1990). Another scenario involves antibodies the body has made to gp120, Ab120. If an Ab120 has affinity to the gp120 site that binds with CD4, it may also have affinity to the CD4 receptor on the MHC II molecule, also blocking AP. Anti-idiotypic antibodies (AIAbs) may form against the Ab120s. These AIAbs may also have affinity for the CD4 site that binds to MHC II, again blocking normal AP. On the other hand, AP to HIV-infected Th cells appears to activate HIV virion production from these cells, leading to their deaths. If AP is blocked by sgp120, Ab120, or AIAb120, the rate of viral production will be slowed, but at the expense of reduced overall immune function. It has also been shown that HIV-infected Mφs can transmit HIV to Th during AP (Mann et al. 1990). Thus, the HIV infection process appears to be quite complex, including inherent negative feedback on HIV production, as well as positive feedback.

Still another scenario of immune system debilitation in HIV infection involves collateral damage of Th cells by NK cells. NKs combine with the Fc region of Ab120s on the surfaces of HIV-infected and uninfected CD4+ T-cells that have had the misfortune to have soluble gp120 (sgp120) bind with their CD4. The NK cells secrete perforin molecules that lyse the marked, infected, and normal T-cells. Another theory involving sgp120 has each arm of one Ab120 binding to gp120 bound to two adjacent CD4 molecules. This binding may induce the cell to produce the Fas molecule. If one Fas molecule encounters the Fas ligand on the cell or an adjacent cell, premature Th cell death by apoptosis may be induced. See Figure 8.19 for a schematic of this putative process.

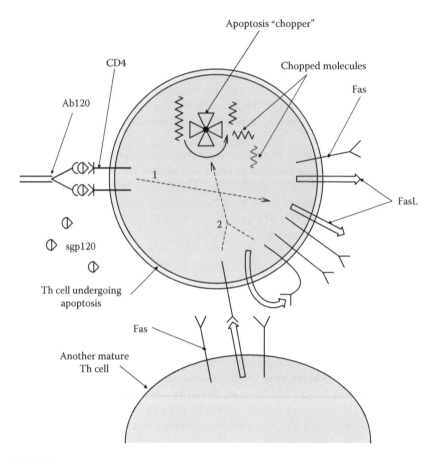

FIGURE 8.19 Schematic illustrating the hypothesis that apoptosis can be prematurely induced (1) in mature Th cells by antibodies binding to the HIV coat protein, gp120. The binding of the Th coat protein, Fas, to FasL is thought to trigger the production and activation (2) of internal proteolytic enzymes (the "chopper") that destroy the cell from within (PCD). See text for description. (From Northrop, R.B., *Endogenous and Exogenous Regulation and Control of Physiological Systems*, Chapman & Hall/CRC Press, Boca Raton, FL, 2000.)

Banki et al. (2005) examined the role of the human complement system (hCS) in HIV infection, and concluded that "…interactions of HIV with the complement system promote HIV infection facilitating its spread and infectivity" [my underline]. HIV infectivity is expedited by complement opsonization, which facilitates viral attachment to susceptible IS cells through complement receptors. This counter-intuitive behavior of the complement system is the result of a complicated series of interactions of complement molecules with both the complex adaptive IS and the innate (complement) IS. For example, activation of the hCS by HIV results in the cleavage of C3 and C5 molecules, leading to the production of the complement anaphylatoxins C3a and C5a. C5a and its derivative were shown to increase the susceptibility of Mφs to HIV infection through the induction of secretion of the cytokines, TNF-α and IL-6. (Mφs have C5a receptors.) Both IL-6 and TNF-α have been shown to trigger HIV replication in infected Mφ (Kacani et al. 2001). Thus, the innate hIS expedites HIV infection in the process of trying to fight it. This HIV–hCS interaction has not been modeled in the dynamic models of HIV infection given below (to do so would require modeling the hCS).

Because HIV infects key cells in the AP process using cell surface molecules used by the immune response, and the immune system responds to viral infection by killing infected cells, we see that the immune system can in fact attack itself. HIV-infected Mφs can pass on the HIV to uninfected Th during AP. Infected CD4+ T-cells are depleted because of viral eclosion, and their attack by NK and CTL cells. Ab120 and anti-Ab120 may jam AP, and premature apoptosis may be induced in uninfected Th. Fewer activated Th cells mean lower concentrations of key immunocytokines, hence there will be lower immune system activation in response to opportunistic infections; hence AIDS.

Another scenario that contributes to a weak immune defense against HIV involves the continual mutation of the noncritical structure of the gp120 molecule. This mutation rate can exceed the immune system's ability to mount an Ag-specific attack with CTLs and Ab120s. Thus, the normal lags in clonal expansion of CTLs and plasma B-cells specific for "flavor A epitope" of gp120 produce a response that is weakly effective against "flavor B," etc.

Several workers have sought to describe the dynamics of HIV infection with mathematical models. As we have seen above, HIV infection is a very complex process, and not all scientists working on the problem of HIV and AIDS agree on why the complex hIS gradually looses competence. Obviously there are many factors involved. One advantage of a good mathematical model is that one can explore possible new immunotherapies before one commits them to animal then human studies. It is possible to introduce viral or bacterial "opportunistic" pathogens into the model at various times to test the model immune system's competence.

Some simple mathematical models exploring the dynamics of HIV infection in different classes of immune system cells have appeared in the literature of mathematical biology over the past 20 years or so. For example, Bailey et al. (1992) described a relatively simple, 5-state model for HIV infection of CD4+ T-cells. Their model was based on the results of experimental, in vitro studies of HIV infection of CD4+ Th cells. The cells were cultured in the presence of IL2 for 50–60 days. The HIV-infected cells did not express virions or secrete extra IL2 unless they were stimulated with phytohemagglutanin (PHA). (PHA is a protein derived from beans that activates Th cells.) PHA-stimulated, infected CD4+ Th cells secreted IL2 for 2 days, eclosed virions transiently after 5–6 days, and then lysed and died (apoptosis?) after 7–10 days. Bailey et al. concluded that "...the critical series of events relevant to T4 cell depletion and disease progression appeared to be: infection, stimulation [as by Ag presentation], IL2 secretion, virus expression, and finally cell death."

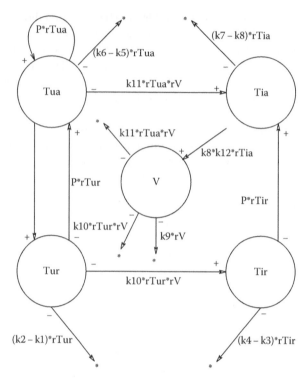

FIGURE 8.20 The five-compartment model for the HIV infection of CD4+ Th cells given by Bailey et al. (1992). The 5-state equations in the Bailey model are given by Equations 8.12A through E. (From Northrop, R.B., *Endogenous and Exogenous Regulation and Control of Physiological Systems*, Chapman & Hall/CRC Press, Boca Raton, FL, 2000.)

A compartmental diagram of the Bailey model is shown in Figure 8.20. Each compartmental concentration or pd is a state. The five Bailey state equations are written below in Simnon notation. The r preceding the cell pd symbols (e.g., rTir) means that rTxys are made nonnegative (i.e., they are half-wave rectified). (There can be no negative pds.) Note that dTur=dTur/dt, etc.

```
" Bailey Model 5 STATE EQUATIONS:
"
" Tur=Uninfected resting T-cell population density.
```

$$dTur = -P*rTur - k10*rTur*rV + (k1-k2)*rTur + k0*rTua$$

$$(8.12A)$$

```
"
" Tir=Infected resting T-cell p.d.
```

$$dTir = -P*rTir + k10*rTur*rV + (k3-k4)*rTir \quad (8.12B)$$

"

" Tua = Uninfected, activated T-cell p.d.

dTua = P*(rTur + rTua) - k11*rTua*rV + (k5 - k6 - k0)*rTua (8.12C)

"

" Tia = Infected, activated T-cell p.d.

dTia = P*rTir + k11*rTua*rV + (k7 - k8)*rTia   (8.12D)

"

" V = free HIV p.d.

dV = k12*k8*rTia - (k10*rTur + k11*rTua)*rV - k9*Rv (8.12E)

"The parametric input is

P = IF t > to THEN A*EXP(- (t-to)/tau) ELSE 0. (8.12F)

The constants and ICs are
" CONSTANTS:"
to:50, Poo:10, A:1, k00:0.01, k1:0.10, k2:0.05, k3:0.05,
k4:0.05, k5:0.05
k6:0.1, k7:0.1, k8:0.25, k9:2.64, k10:5E-7, k11:1E-7, k12:75,
TAU:5 days.
"

" INITIAL CONDITIONS:
Tur:1E6, V:1E6

Figure 8.21 illustrates the behavior of the model when two simulated boluses of PHA are given. There are rapid eclosions of HIV virions that decay exponentially. Also, note that in the steady-state following activation, there is a constant population of infected, resting Th that acts as a reservoir for future HIV eclosions.

8.4.3.3   Reibnegger hIS Model

A more complicated, 19-state mathematical model of HIV infection was formulated by Reibnegger et al. (1987). Reibnegger included a self-replicating pathogen, HIV virions, CD4+ Th cells with TCRs specific for pathogen Ag and also for HIV Ag. Monocyte/Mφs were also included, as well as CTLs specific for pathogen or HIV Ag, pathogen debris, HIV debris, and non-state functions of state variables for Mφs presenting Ag for pathogen (Mx) or HIV (My). They also used a non-state function for "factor" (**F**), which is evidently pooled, stimulatory immunocytokines secreted by active Th following AP by Mx and My. Reibnegger et al. did not include HIV-infected Mφs, B-cells, antibodies, NK cells, or specific immunocytokines in their model, a simplification that we have accepted considering the detail in the rest of their model. Figure 8.22A through C illustrate

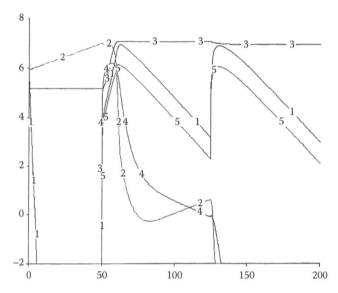

FIGURE 8.21 Simulation of the behavior of the four types of CD4+ Th cells and free HIV virions in a test volume using the 5-state model of Bailey et al. (1992). Boluses of PHA are given at t=50, and again at t=125 days. PHA is a mitogen for lymphocytes. Its effect is simulated by transiently increasing the arameter P in the model. Vertical axis: log(*) units. Horizontal axis: days. Trace 1=log(V), trace 2=log(Tur), trace 3=log(Tir), trace 4=log(Tua), trace 5=log(Tia). Note that initially, the HIV population density (pd) decays exponentially while the pd of Tur increases and there is a constant, steady-state pd of Tir. When the PHA is given, it stimulates the infected Th to make and eclose new HIV virions, hence the abrupt increase in V and Tir. Tur and Tua decrease because they are infected and thus become Tir and Tia. Tia follows V(t). At the second bolus of PHA, the pds of Tur and Tua plummet. (From Northrop, R.B., *Endogenous and Exogenous Regulation and Control of Physiological Systems*, Chapman & Hall/CRC Press, Boca Raton, FL, 2000.)

the 19 compartments (state ODEs) used by Reibnegger et al. and the transfer rates between them. We wrote the Simnon model, **reibsys2.t**, in order to simulate Reibnegger's model; **reibsys2.t** is given in Appendix A.5. (Note that in the program, any line preceded by quotes is a comment [non-executable code].) A great deal of line space in this program is taken up with the rectification of states and logging of states and variables. Rectification is required because in the simulation of large, nonlinear systems with poorly known rate constants, states can go negative under certain conditions of initial conditions and inputs. Negative states are meaningless when dealing with population densities and concentrations; hence, rectification is used to preserve nonnegativity on the ODEs, and hence physical reality. Logging states allows us to use one set of axes to display all states, because in sets of stiff equations, which we have here, some states can be very large while others approach zero.

Our program, **reibsys2.t**, follows the Reibnegger et al. model faithfully with the exception of the quasi-steady-state variable **F** for "lymphoid factors." In their original paper, Reibnegger et al. stated, "These factors, **F**, are assumed to (1) be identical kinetically, (2) be produced by helper T-cells Hx upon presentation of antigen X [X is P, the pathogen], (3) be produced instantaneously, and (4) decay rapidly. Thus, the different factors (e.g., IL2, 3, 4, 5, 6, 7, 10, 13, γIFN, αIFN, GM-CSF, and TNFβ) are simply represented by one quasi-steady-state variable: $F = H_x \cdot X/(k_f + X)$." Here, $H_x = x5$ and $X = x1$ above.

AP by Mϕs, as we have seen, initially involves an MHC II molecule containing the Ag binding with a helper T-cell receptor (TCR) specific for that antigen. Following the binding of the Th's CD4 to the MHC II molecule, and activation

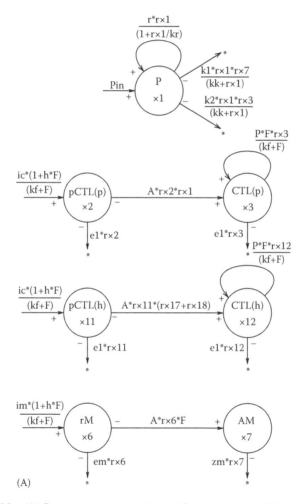

(A)

FIGURE 8.22   (A) Seven compartments in the 19-state model of HIV infection of the hIS described by Reibnegger et al. (1987). The self-feedback on states x1, x3, x5, x12, x14, x17, and x18 represent cytokine-induced clonal expansion. The ODEs describing this model are given in the model, **reibsys2.t** in Appendix A.5.

(continued)

of CD3, the Ths are activated, and among other things, they release the many "factors" comprising **F**. Reibnegger's expression for **F** implies, by mass action kinetics, that the Th combines directly with pathogen. This is not the case, biologically. The Th combines with an antigen-presenting Mφ or dendritic cell bearing the pathogen antigen. Reibnegger et al. gave an algebraic expression for the population concentration of antigen presenting Mφs presenting pathogen antigen, M_x. In their notation, $M_x = D_x(M + M^*)/(k_d + D_x)$. Here, D_x is the population concentration (pc) of pathogen debris (after lysing of pathogen by Mφs and CTLs). $D_x = x8$. $M = $ pc of resting Mφs $= x6$. $M^* = $ pc of activated, cytotoxic Mφs $= x7$.

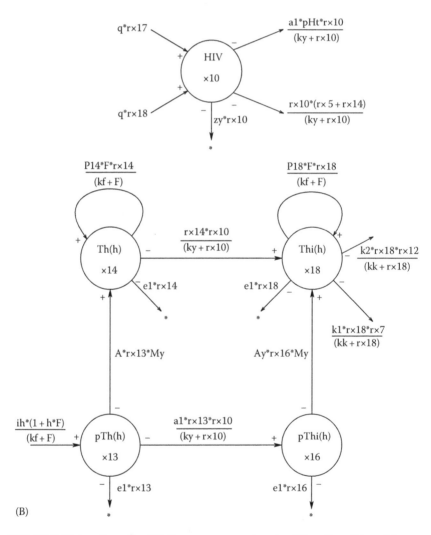

FIGURE 8.22 (continued)   (B) Five more compartments of the **reibsys2.t** model.

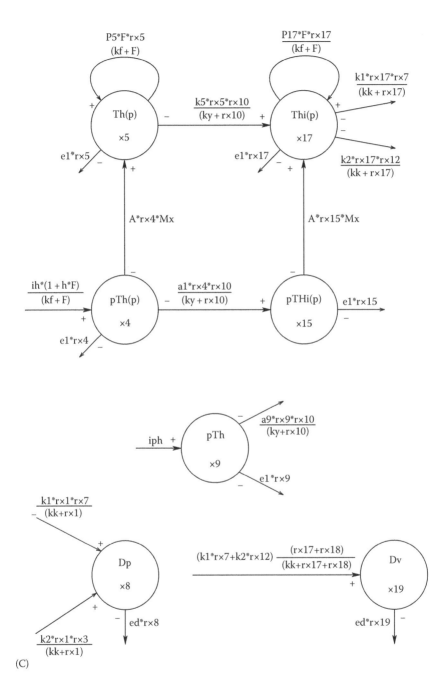

FIGURE 8.22 (continued)   (C) Seven more compartments of the **reibsys2.t** model. (From Northrop, R.B., *Signals and Systems in Biomedical Engineering*, 2nd edn., CRC Press, Boca Raton, FL, 2010.)

Succumbing to the temptation to tinker, we wrote a more realistic expression for **F** involving the presentation of pathogen and HIV antigen to appropriate Th cells. It is

$$\mathbf{F} = \frac{(rx5 + rx17) * Mx}{(kf + Mx * (rx5 + rx17))} + \frac{(rx14 + rx18) * My}{(kf + My * (rx14 + rx18))} \tag{8.13}$$

Here, we have assumed that antigen-presenting Mϕs with pathogen antigen (Ag) combine with Ths with TCRs specific for pathogen Ag. Both normal Th and HIV-infected Ths are considered. The factors so produced are given saturation by use of a Hill function. The second term mirrors the first term except that it is the HIV antigen, gp120, that is being presented to Ths with TCRs specific for HIV Ag. The immunocytokines produced by the AP-activated Th are assumed to be identical and independent of the Ag presented. In the model they stimulate Th and CTL cell clonal expansion and the input of precursor cells. Again, the "r" prefix on the variables signifies they were made nonnegative.

Figure 8.23A and B illustrates how the modified Reibnegger model responded to an HIV input at $t = 0$, then a pathogen inoculation at $t = 75$ weeks, and a further infection by pathogen at week 300. Note that the pathogen when first introduced is fought to an innocuous, chronic population density (pd). The parameters used in the simulation are given in Appendix A.5.

In Figure 8.23B, we see that as a result of immune system activation, HIV virions are released by infected Th cells. When the pathogen is introduced for the second time, the immune system is weakened, and it grows rapidly, in effect "killing" the model host. Note that it is possible to introduce into the model exogenous immunotherapies, such as monoclonally grown CTL specific for HIV antigen.

8.4.3.4   hIS Model of Northrop (2000)

In order to explore more detailed HIV immunotherapies, including specific, exogenous immunocytokines, the author and certain graduate students in 1989 developed a more detailed mathematical model based on Reibnegger's approach, but including, in addition to the cells modeled by Reibnegger et al., a viral pathogen that infects certain somatic cells in the body (SCp), NK cells, B-cells specific for HIV Ag and viral pathogen Ag, plasma B-cells secreting Abs for pathogen and HIV, HIV infection of Mϕs, as well as CD4+ Th cells, states for IL2 and γIFN, and quasi-steady-state variables for IL1, antigen-presenting Mϕs (not B-cells), and B-cell factors [e.g., IL6] causing Ab release. Our model used 32 states. While complicated to set up and interpret, it allowed more depth in simulations than did previous models. Like all of the models described in this section, it necessarily is a gross oversimplification of nature. Many rate constants are taken from the literature, others are "reasonable estimates." This model allowed us to investigate, in silico, mono- and multimodal immunotherapies such as exogenous inputs of IL2, γIFN, AZT, monoclonal CTL specific for HIV, etc. Our HIV Simnon program, **AIDS12.t** is given in Appendix A.6.

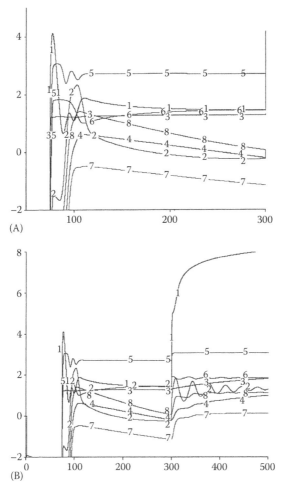

FIGURE 8.23 (A) Simulation of HIV infection and pathogen growth using the **reibsys2.t** model, with the REIBPAR8 parameters. At t=0 days, the model is "infected" with 100 HIV virions. At t=75, a pathogen pd of 1.E5 is inoculated. The pathogen pd expands rapidly, then is quickly brought down to about 1.5 log units by immune system action, giving a chronic infection. Following the pathogen peak, there are peaks in the pds of HIV, and Hit. Horizontal axis: days. Vertical axis: log(*) units. Trace 1=x1=pathogen, trace 2=x10=HIV, trace 3=x3=CTL(p), trace 4=x12=CTL(v), trace 5=x7=AM, trace 6=x5=Th(p), trace 7=x14=Th(v), trace 8=Hit. (Hit is total HIV infected Th.) (B) The same simulation as in (A), except an expanded time scale is used. A second inoculation of pathogen (1.E5 pd) is given at t=300 days. The pathogen grows rapidly and "overcomes" the "weakened" model hIS, leading to AIDS. Note that the strong growth of the free HIV pd and Hit in the AIDS condition for t>400. Trace 1=x1=pathogen, trace 2=x10=HIV, trace 3=x3=CTL(p), trace 4=x12=CTL(v), trace 5=x7=AM, trace 6=x5=Th(p), trace 7=x14=Th(v), trace 8=Hit. (From Northrop, R.B., *Endogenous and Exogenous Regulation and Control of Physiological Systems*, Chapman & Hall/CRC Press, Boca Raton, FL, 2000.)

All of the simulations described below were done using Euler integration with $\Delta t = 0.001$ over a model time scale of 0–600 days. The vertical axis is calibrated in $\log_{10}(pd)$ of the states and variables.

In Figure 8.24, we show the response of the system when the model immune system is free of HIV virions. Pathogen (P) is introduced as an injection of P at $t = 100$ days. P initially grows rapidly, then is quickly defeated by the immune system. A second pathogen injection is given at $t = 300$ with the same effect. In Figure 8.25, an initial inoculation of pathogen is given at $t = 100$; the pathogen is quickly destroyed by the "normal" immune system model. An inoculation of 10^3 HIV virions is given at $t = 200$. A second pathogen inoculation is given at $t = 300$. However, this time the immune system is weakened by the HIV infection, and there is pathogen regrowth to a steady-state level, representing AIDS. Figure 8.26 illustrates what happens in the model when 10^4 "units" of AZT are infused from $t = 250$ to $t = 450$. Up to $t = 250$, the system's response is the same as in Figure 8.25. Note that at $t = 250$, the pd of HIV begins decreasing; the free HIV pd [trace 2] crashes at about $t = 390$ but the pd of total HIV infected Th cells, $[H_{itot}]$ [trace 8], remains high. Curiously, the pathogen pd regrows rapidly to a third peak at $t = 430$ as a result of the AZT infusion, and then abruptly decreases when AZT infusion is stopped at $t = 450$. This behavior is counter-intuitive, not unexpected in a complex dynamic system.

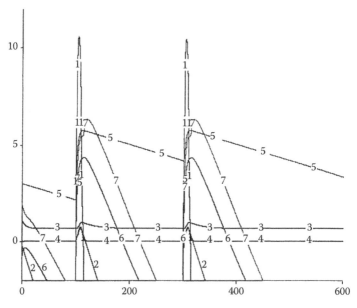

FIGURE 8.24 Simulated response of our 32-state model of the hIS, **AIDS12.t,** given in Appendix A.6. This hIS is not infected with HIV. Two inoculations of pathogen P are given at $t = 100$ (1.E4 pd) and again at $t = 300$ (1.E4 pd). Vertical scale: log(*) units. Horizontal scale: days. Trace $1 = \log(P)$, trace $2 = \log(Cx)$, trace $3 = \log(AM)$, trace $4 = \log(NK)$, trace $5 = \log(Hx)$, trace $6 = \log(PBx)$, trace $7 = \log(Abx)$. (From Northrop, R.B., *Endogenous and Exogenous Regulation and Control of Physiological Systems*, Chapman & Hall/CRC Press, Boca Raton, FL, 2000.)

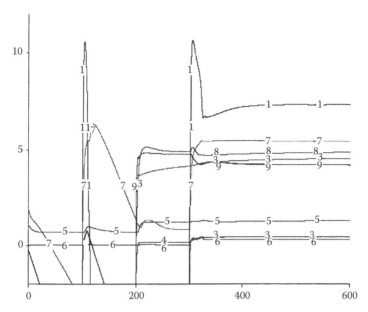

FIGURE 8.25 Response of the **AIDS12.t** model when 10^3 HIV virions are inoculated at $t = 100$ days. Same axes as in Figure 8.24. Trace $1 = \log(P)$, trace $2 = \log(Y)$ [Y = free HIV pd], trace $3 = \log(Cx)$, trace $4 = \log(Cy)$, trace $5 = \log(AM)$, trace $6 = \log(NK)$, trace $7 = \log(AbX)$, trace $8 = \log(PBy)$, trace $9 = \log(Hitot)$. After the second pathogen spike, P reaches a chronic, steady-state pd corresponding to AIDS. (From Northrop, R.B., *Endogenous and Exogenous Regulation and Control of Physiological Systems*, Chapman & Hall/CRC Press, Boca Raton, FL, 2000.)

Following the cessation of AZT infusion, the pds of pathogen and HIV again increase. In Figure 8.26, we illustrate the model's response to an infusion of IL2 of 10^3 "units"/time from $t = 250$ to $t = 450$. Again, the normal immune model conquers the pathogen inoculated at $t = 100$. HIV is introduced at $t = 200$, as in the previous simulations. Now, a simulated infusion of IL2 of 10^3 "units"/day is begun at $t = 250$ and continued until $t = 450$. Note that there is an increase in the slope of the log(HIV) concentration and a jump in the log(pd) of antibodies (AbX) for the viral pathogen when IL2 infusion is begun. At the second pathogen inoculation, its pd spikes and then remains flat at a log(pd) of about 6.3. When the IL2 infusion is stopped, the pathogen again grows rapidly.

To investigate what happens if exogenous IL2 is given along with AZT (bimodal therapy), we examined the simulation of Figure 8.27. Here, an IL2 infusion of 100 u/t, and an AZT infusion of 250 u/t are given from $t = 250$ to $t = 450$. (These are the minimum simulated doses to cause the pathogen pd to $\to 0$.) After the IL2 and AZT infusions are stopped, the pathogen tries to regrow, but the system has regained enough strength to cause the pathogen to $\to 0$ at $t = 570$. Stronger doses of AZT and IL2 cause the pathogen pd to crash at about $t = 450$. This simulation illustrates the putative advantage of multimodal therapies to treat HIV infection.

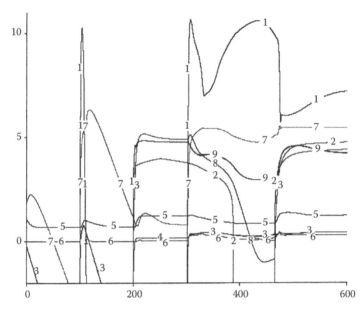

FIGURE 8.26 Response of the **AIDS12.t** model to pathogen inoculation at t = 100, HIV infection at t = 200, and a second pathogen inoculation at t = 300. An infusion of 1.E4 "units"/day of AZT is begun at t = 250 and ends at t = 450. Trace 1 = log(P), trace 2 = log(Y), trace 3 = log(Cx), trace 4 = log(Cy), trace 5 = log(AM), trace 6 = log(NK), trace 7 = log(AbX), trace 8 = log(PBy), trace 9 = log(Hitot). Note bizarre behavior of pathogen pd during AZT infusion. Also note that log(HIV) < −2 at about t = 390. Free HIV increases once the AZT infusion has stopped. (From Northrop, R.B., *Endogenous and Exogenous Regulation and Control of Physiological Systems*, Chapman & Hall/CRC Press, Boca Raton, FL, 2000.)

Other possible therapies for HIV can also be explored using the model. For example, exogenous inputs of γIFN, externally cultured and propagated CTL specific for HIV (Cy), and antiviral drug for pathogen can be tried. Also, various scenarios for sCD4 infusion can be investigated. As we have discussed previously, sCD4 will not only decrease the number of active gp120 sites for HIV attachment to CD4+ T-cells, thus reducing the rate of HIV infection, but it may also decrease the effective number of MHC II antigen-presenting cells by binding to MHC II and preventing normal AP. (These possible scenarios can be inserted in the appropriate ODEs and algebraic equations.)

8.4.4 Human Immune System vs. Cancer

8.4.4.1 Introduction

There are many theories why normal human cells in organs and tissues become cancerous. The mutation from normal cell function to cancer (oncogenesis) is often correlated with multiple gene mutations (in breast cancer, intestinal cancer), diet, environmental factors including the inhalation of tobacco smoke,

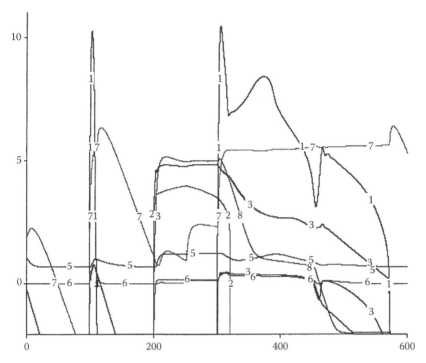

FIGURE 8.27 Response of the **AIDS12.t** model to pathogen inoculation at t = 100, HIV infection at t = 200, and a second pathogen inoculation at t = 300. In this scenario, a multi-modal therapy is explored. Both exogenous IL2 at 100 and 250 u/day of AZT are infused from t = 250–450. Trace 1 = log(P), trace 2 = log(Y), trace 3 = log(Cx), trace 4 = log(Cy), trace 5 = log(AM), trace 6 = log(NK), trace 7 = log(AbX), trace 8 = log(PBy), trace 9 = log(Hitot). Note that the pathogen regrows at the cessation of infusions, but eventually crashes at about t = 570 as a result of this minimal dose therapy. (From Northrop, R.B., *Endogenous and Exogenous Regulation and Control of Physiological Systems*, Chapman & Hall/CRC Press, Boca Raton, FL, 2000.)

excess sunlight (UV) on the skin, drinking water contaminated with chemical carcinogens, exposure to ionizing radiation (from radon gas, x-rays), etc. The genetic machinery of one cell, perhaps a stem cell, becomes mutated, either because its DNA is damaged, or because foreign DNA is inserted into its normal DNA by a viral infection. (In some types of cancer, viruses are known to be the oncogenic agent [e.g., the human papiloma virus causes cervical cancer]). In any case, the malignant progeny of the mutated cell produce altered surface proteins; their adhesion to other cells often decreases. Cancer cells' apoptosis mechanisms may also be disabled so they can reproduce clonally with altered DNA and proteins.

Most cancers are characterized by rapid growth, even though the normal cells in the tissue in which they lie have a slower, regulated growth. At first, the cancer-ous cells grow exponentially in time due to clonal "twinning." The time between divisions can be as short as 10 s of hours in a fast-growing cancer. After the tumor reaches a certain size, its growth slows. Then autacoids secreted by the cancer cells cause blood vessels from the host to invade the mass of the tumor to provide nourishment. This vascularization is called angiogenesis. Once the tumor is vascularized, its growth rate again increases. A 1 g tumor may contain about 10^9 cells. If untreated, cancers kill their hosts by crowding out normal cells and disrupting organ function. Malignant cancers can metastasize, or spread to other sites in the body. Certain lung and breast cancers, for example, can metastasize to the bone marrow and the brain.

As we have mentioned above, certain surface proteins on cancer cells are unique, or are altered normal proteins, providing in theory, a means whereby the immune system can recognize their difference from normal cells, and attack and destroy them. Obviously, in some cases, the immune system for whatever reason does not do this, and the cancer grows until it is diagnosed. The immune system may be suppressed by the cancer cells themselves, or the IS "sees" normal cell-surface proteins on the cancer cells and is inhibited from attacking them. Kochar (2006) lists a number of means that tumor cells use to foil the hIS. These include the production of cyclooxenase-2 (COX-2), an enzyme produced by lung cancer cells that causes immunosuppression due to decreased IL-10, and an increase of IL-12. Increased IL-10 reduces the expression of dendritic cell (DC) surface antigen molecules (used in AP), also proteins CD80 and CD86 required for T-cell activation in AP. Still another attack on DCs is by tumor cells that secrete nitric oxide (NO) and H_2O_2 that trigger DC apoptosis. Some tumor cells secrete the enzyme arginase that breaks down the AA arginine. Arginine is a key AA in the T-cell receptor. Thus arginine depletion leads to the weakening of AP to the CTL, hence decreased CTL function against tumor cells. Kochar also cited other clever ruses that have evolved in tumor cells in their "Red Queen" battles with the hIS to ensure their survival.

The rate of cancer growth appears to be important in eliciting immune response. A rapidly growing cancer is generally more likely to excite the host's immune defenses than is a slowly growing tumor. One reason for this phenomenon may be the half-life of helper T-cells. If the half-life is too short, and the tumor grows slowly, the density of activated Th does not reach a critical threshold, and the tumor "sneaks though," avoiding immune rejection (DeBoer et al. 1986).

An interesting phenomenon to contemplate is the spontaneous regression of tumors. Anecdotal, circumstantial evidence of spontaneous regression (Lewison 1976) leads us to speculate that the immune system, having once missed the cancer, can return in full force and destroy it, even though it has reached a mass large enough to permit diagnosis. This very rare reawakening of immune control has been found to generally accompany a simultaneous massive stimulation of the immune system by a severe viral or bacterial infection (Kochar 2006).

8.4.4.2 Cancer Vaccines and Immunotherapy

Recently, considerable attention has been drawn to the use of therapeutic vaccines to sensitize the hIS to unique surface antigens on cancer cells (Zhou 2005,

Kochar 2006, Acres and Bonnefoy 2008). The advantage of using the hIS to fight cancer is that its targets are unique. Immunotherapy avoids the broad, adverse side-effects of cancer chemotherapy drugs and radiation treatments. In February of 2009, Dendreon Corp. of Seattle announced a successful Phase III trial of their anti-prostate cancer vaccine, Provenge (Moss 2009). After 3 years, 34% of the men who received Provenge were still alive versus 8.9% of patients in the placebo group.

Provenge is an autologous vaccine. It is produced in vitro from the patient's own blood and prostate tumor cells, only for that patient. First, antigen presenting cells (APCs) (Mϕs, dendritic cells) are extracted extracorporally from the patient's blood. The APCs are mixed with a signature protein called prostatic acid phosphatase (PAP) (found on the surface of the patient's prostate cancer cells), then the PAP is fused with another immune-stimulating substance, granulocyte/monocyte-colony stimulating factor (GM-CSF). The treated APCs are then injected into the patient in a 1 h infusion. This process is repeated three times over the course of a month. The injected dendritic cells and Mϕs present the PAP antigen to the patient's cytotoxic T-cells (CTLs). Presumably the CTLs then attack the prostate tumor cells that are coated with the PAP epitope. The results of the Provenge Phase III trials are unusual, because tumors, in general, are weakly immunogenic; most tumor Ags are self-antigens and the hIS has been trained not to attack normal cells, and tumor cells have various immunosuppressive mechanisms. Thus development of a successful therapeutic vaccine requires some skill.

Those readers interested in pursuing the active area of cancer vaccine development R&D and trials should begin by reading the review article by Kochar (2006). Kochar categorized several types of cancer vaccines:

1. Antigen-based vaccines
   a. Tumor Ag DNA vaccines
   b. Tumor Ag synthetic peptide vaccines
   c. Ab-inducing vaccines against carbohydrate Ags
2. Monoclonal Abs
   a. Anti-idiotype vaccines
   b. Use of monoclonal Abs against a costimulator for CTL activation
3. Cell-based immunotherapy
   a. Tumor-cell-based vaccine
   b. Dendritic-cell-based vaccines
   c. T-cell-based vaccines
   d. NK-cell-based therapy

Also see the paper by Zhou (2005) on the complex immunosuppressive networks in the tumor environment and their impacts on vaccine therapy.

Another approach to cancer immunotherapy was described by Acres and Bonnefoy (2008). They reported on the use of a modified, recombinant vaccinia virus (VV) to introduce tumor-associated antigens (TAAs) into the body to fight specific classes of cancers. Attenuated VVs had genes for human breast cancer HER-2, prostate-specific antigen (PSA), or prostatic acid phosphatase (PAP) inserted in their genomes. When the patient is infected with the GM virions, they do not reproduce themselves, but do make large quantities of the recombinant protein before their host cells lyse. Presumably Mφs and/or dendritic cells clean up the debris, including the recombinant protein, then present it to CTLs who go on to fight the specific cancer cells.

Still another promising experimental approach to cancer therapy is by using catalytic antibodies (CAbs) (Ali et al. 2009). This is a bioengineered approach in which special monoclonal CAbs that have two paratopic sites are grown: one binds with a unique epitopic molecule on the surface of cancer cells; the other contains a catalyst that activates a specific anticancer, chemotherapeutic, prodrug. The monoclonal CAb is injected first and allowed to bind to the cancer cells, then the prodrug is injected. At the cancer cell surface, the CAbs cleave the prodrug to form a locally concentrated, active chemotherapy agent. This minimizes side effects of the chemotherapy agent on other tissues. The uncleaved prodrug is not toxic.

Certain cancers are "inoperable," and chemotherapy and/or radiation may be unsuitable because of the patient's medical condition. Needless to say, using exogenous stimulation of the patient's immune system to fight cancer is a current, active area of research. Certain aspects of immunotherapy for cancer can be modeled mathematically.

8.4.4.3 DeBoer's Model of the hIS vs. Cancer

One of the first, detailed mathematical models of the immune system versus cancer was described by DeBoer et al. (1985). DeBoer's model pitted the cellular arm of the immune system versus cancer. Their model was based on the experimental behavior of an ascitic lymphoma injected into the peritoneal cavity of mice.

DeBoer et al.'s (1985) model used eight states: the population densities (pds) of cytotoxic T-lymphocyte precursor cells (CTLP), cytotoxic T-lymphocytes (CTL), "normal" Mφs (MPH), activated, cytotoxic Mφs (ANGRY), unprimed, precursor, helper T-lymphocytes (HTLP), activated helper T-lymphocytes (HTL), tumor cells (TUMOR), and debris (DEBRIS) from tumor cells killed by CTL and ANGRY. B-cells, antibodies, and NK cells were not considered in the DeBoer model. The only specific immunocytokine used in DeBoer's model was IL1. (Elevated [IL1] causes B-cell proliferation, and leads to production of other cytokines by T-cells, fever, and cachexia.) In order to keep poorly estimated or unknown rate constants to a minimum, DeBoer defined algebraic parameters that pool the effects of immunocytokines known to be active in the cellular (adaptive) arm of the immune system. Their model included AP by Mφs to helper CD4+ T-cells using the Hill parameter, **APC**:

$$APC = \frac{(MPH + ANGRY){*}DEBRIS}{KMD + DEBRIS} \qquad (8.14)$$

The constant $KMD = 10^7$. The parameter FACTOR acts to activate Mϕs to ANGRY, and also stimulates clonal expansion of CTL and HTL (Th). FACTOR may have an analog in IL2. IL2, however is produced by CD4+ Th that have been activated by AP by Mϕs. (It would make more sense to have FACTOR as a function of **APC***HTL rather than TUMOR*HTL.) The constant $KMT = 50$.

$$FACTOR = \frac{HTL*TUMOR}{KMT + TUMOR} \tag{8.15}$$

The role of INFLAM in the model is to increase the production of hematopoietic stem cells and cause their differentiation into Mϕs, CTL, and Th. INFLAM is a saturating Hill function of FACTOR. IL3 and IL12 are now known to promote the proliferation of stem cells and their differentiation and maturation. A number of other immunocytokines also promote hematopoiesis. DeBoer et al. gave

$$INFLAM = \frac{H*FACTOR}{KMF + FACTOR} \tag{8.16}$$

where $H = 9$ and $KMF = 50$.

We put De Boer's model's 8-state equations (ODEs) into the Simnon form:

$$dCTLP = I1*(1 + INFLAM) - A*CTLP*TUMOR - EL*CTLP \tag{8.17A}$$

$$dCTL = A*CTLP*TUMOR - EL*CTL + R*CTL$$
$$*FACTOR/(KMF + FACTOR) \tag{8.17B}$$

$$dHTLP = I2*(1 + INFLAM) - A*HTLP*\mathbf{APC} - EL*HTLP \tag{8.17C}$$

$$dHTL = A*HTLP*\mathbf{APC} - EL*HTL + R*HTL$$
$$*FACTOR/(KMF + FACTOR) \tag{8.17D}$$

$$dMPH = I3*(1 + INFLAM) - A*MPH*FACTOR - EM*MPH \tag{8.17E}$$

$$dANGRY = A*MPH*FACTOR - DM*ANGRY \tag{8.17F}$$

$$dTUMOR = R*TUMOR/(1 + TUMOR/KR) - KILL$$
$$*(ANGRY + CTL)*TUMOR/(KMK + TUMOR) \tag{8.17G}$$

$$dDEBRIS = KILL*(ANGRY + CTL)*TUMOR/(KMK + TUMOR)$$
$$-ED*DEBRIS \tag{8.17H}$$

The constants that DeBoer et al. used initially were $A=0.001$, $EL=0.02$/day, $I1=0.5$ or 10, $I2=0.01$–100, $I3=1.25\times10^5$, $EM=0.05$/day, $R=1$, $KMF=50$, $DM=1$/day, $KILL=10$, $KR=10^9$, $KMK=10^5$, $ED=2$/day, $KMD=10^7$, $KMT=10^3$, $H=9$, $KMF=50$.

DeBoer et al. stated, "We consider one compartment in which a tumor grows autonomously and in which tumor cells are killed upon contact with cytotoxic effector cells. Effector cells are generated upon local (i.e., within the compartment) activation of precursor cells. Cells do not recirculate: precursors immigrate into the compartment, leave (decay), or become effector cells: effector cells, on the other hand, only leave the compartment (or decay locally)."

Simulations of the DeBoer et al. model equations above were done by the author using Simnon (Northrop 2000). The equations are stiff; this was seen in the computational "noise" on some of the traces when using the default, order 4/5 Runge–Kutta/Fehlberg integrator at the default (1.E−3) error tolerance. As might be expected, this noise is much reduced when the error tolerance was reduced to 1.E−6. Figure 8.28 illustrates the time course of the DeBoer model's states when the scenario does not include enough aggressive immune behavior toward the growing cancer cells. The logarithms of states are plotted, since an active tumor can reach millions of cells in a month.

Figure 8.29 illustrates an interesting, nonintuitive result of the DeBoer model: here, the tumor growth rate parameter is low ($R=0.9$, instead of 1), and the kill parameter $=17$, instead of 15. Note that the tumor regrows to a low cell count after apparently being destroyed. In Figure 8.30, the growth rate $R=1$, kill $=17$, and the tumor does not regrow.

In their papers, DeBoer et al. used their models to demonstrate the effect of antigenic modulation of the tumor cells. In this scenario, the tumor cells progressively loose surface antigen proteins as the tumor grows. This loss causes the tumor to appear less "visible" to the immune system, which may never reach a level of stimulation necessary to successfully fight the tumor.

8.4.4.4   hIS vs. Cancer Model of Clapp et al. (1988)

Clapp et al. (1988) developed an 18-state model of the immune system versus cancer. Their model expanded the model of DeBoer et al., and included the then more recent knowledge about the action of certain immunocytokines. It included B-cells, antibodies, and NK cells as well as CTL, Mφs, and Th. It also used certain, specific immunocytokines. This model has been slightly revised and is presented below. The 18 states are

   pNK = Precursor natural killer cells
   NK = Natural killer cells
   pTh = Precursor T-helper cells
   Th = T-helper cells
   pTc = Precursor cytotoxic T-cells (CTL)
   aTc = First-stage activated CTL
   Tc = Fully activated CTL

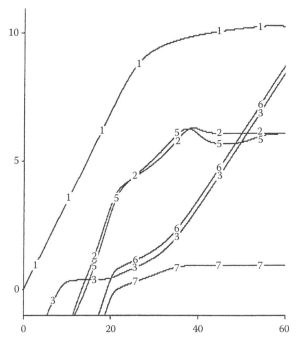

FIGURE 8.28 A simulated runaway growth of cancer using the 8-state model of DeBoer et al. (1985) (see Equations 8.17A through H). Vertical axis: $\log_{10}(*)$ units unless noted. Horizontal axis: time in days. Trace 1 = log(TUMOR) [cancer cell pd], trace 2 = log(ANGRY pd), trace 3 = log(CTL pd), trace 4 = log(HTL pd), trace 5 = log(APC pd), trace 6 = FACTOR, trace 7 = INFLAM. Parameters used are R = 0.8, KILL = 10, I1 = 0.5, A = 1.E–3, EL = 0.2, KMF = 50, I2 = 1, I3 = 1.25E5, EM = 5.E–2, DM = 1, KR = 1.E9, KMK = 1.E5, ED = 2, KMD = 1.E7, KMT = 1.E3, H = 9. (From Northrop, R.B., *Endogenous and Exogenous Regulation and Control of Physiological Systems*, Chapman & Hall/CRC Press, Boca Raton, FL, 2000.)

> pMp = Precursor Mφs
> aMp = First-stage activated Mφs
> Mp = Fully active Mφs
> Lys = Cellular debris
> CA = Cancer/tumor
> IL2 = Interleukin-2
> nB = Nonactivated B-cells
> aB = Antigen-activated B-cells
> pB = Plasma B-cells specific for CA
> AbCA = Antibodies (free and cellular) for CA
> mBCA = Memory cells for CA

(The complete Simnon program, **TUMOR5.t**, is given in Appendix A.7.)

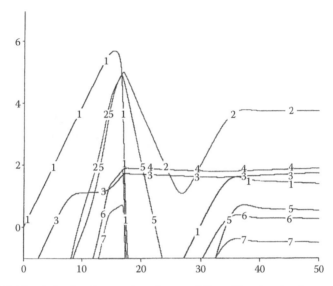

FIGURE 8.29   A simulated regression of cancer followed by a low-level regrowth, using the 8-state DeBoer model. Same parameters as in Figure 8.28 except R = 0.4, KILL = 17, and KR = 1.E7. Trace 1 = log(TUMOR) [cancer cell pd], trace 2 = log(ANGRY pd), trace 3 = log(CTL pd), trace 4 = log(HTL pd), trace 5 = log(APC pd), trace 6 = FACTOR, trace 7 = INFLAM. Same axes. (From Northrop, R.B., *Signals and Systems in Biomedical Engineering*, 2nd edn., CRC Press, Boca Raton, FL, 2010.)

The model of Clapp et al. is useful because it can be used to explore the effects of putative immunotherapies, such as the infusion of IL2, γ-IFN, activated NK cells, or monoclonally grown, activated CTLs on model cancer growth (see the discussion on cancer vaccines above). (IL2 stimulates proliferation of Ab-producing B-cells and activated T-cells, and increases NK cell functions. It also activates CTL antitumor responses. γ-Interferon stimulates MHC molecule expression on antigen-presenting cells (APCs) and somatic cells; it also activates Mφs, neutrophils, NK cells, and antitumor CTL actions.

Figure 8.31A and B illustrates the nonlinear signal flow graphs constructed from the 18-state equations of the **TUMOR5.t** model. Note that many of the branch gains are parametric functions of system states. While certain states may appear not to be directly coupled by branches, they are coupled parametrically.

Figure 8.32 illustrates the response of the Clapp model to one cancer cell arising at time t = 0. The model tumor grows exponentially until it reaches a critical size around day 20, when growth abruptly slows. No immune rejection is seen in the model over 45 days. In Figure 8.33, 15 u of IL2 is given/day as an infusion from day 14 to day 28. Here, the model tumor grows exponentially until about day 20, when the tumor cell population begins to decay. The population crashes around day 33; there is no tumor regrowth. Figure 8.34 summarizes the effect of IL2 infusions on cancer growth rate. Infusion rates of over 9 u/day cause the

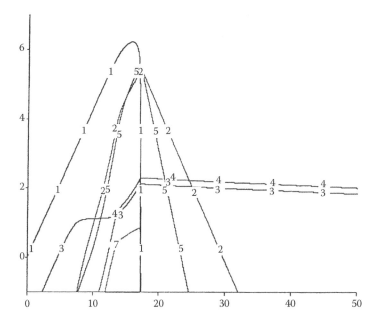

FIGURE 8.30 Simulated regression of TUMOR with no regrowth. DeBoer model. Trace 1 = log(TUMOR) [cancer cell pd], trace 2 = log(ANGRY pd), trace 3 = log(CTL pd), trace 4 = log(HTL pd), trace 5 = log(APC pd), trace 6 = FACTOR, trace 7 = INFLAM. Same parameters as in the simulation of Figure 8.29 except R = 1.0, KILL = 17. Same axes. (From Northrop, R.B., *Signals and Systems in Biomedical Engineering*, 2nd edn., CRC Press, Boca Raton, FL, 2010.)

model cancer cell density to crash to zero inside the 0–45 day time window, that is, there is complete immune rejection of CA. The time at which the CA population density crashes is dose dependent.

Other adaptive immunotherapy scenarios can also be investigated with the model. For example, we consider the bolus infusion of in vitro grown, monoclonal, CTL-specific for the cancer antigen, given at day 10. Figure 8.35 illustrates the model's response to 100 CTLs at time 10. Note that there is a temporary regression of cancer cells until the population of CTL (Tc) dies out, then the tumor regrows. In order to see if a critical number of CTL cells will cause a permanent regression, a series of escalating mTcinj values was given, and the log(Ca) values plotted. In Figure 8.36, we see that regrowth appears to cease for mTcinj ≥ 400 cells for t < 35 days. On a 0–100 day scale, we see that CA regrows at t = 50 for mTcinj = 400 cells. An mTcinj = 500 gives [Ca] = 7.E–3 at t = 100. mTcinj = 1000 cells gives [Ca] = 0 at t = 100.

8.4.4.5 Discussion

The immune system versus cancer modeling adventures described above illustrate, or suggest, general trends in immunotherapies. They should be viewed skeptically because they are gross simplifications of what are very complex situations. Like all PSs, the hIS is not a "stand-alone" system. It also receives signals indirectly from the CNS, and it is known to be affected by stress and the hormones stress produces, as well as endorphins resulting from pleasure. Even the metabolic hormone leptin modulates cells and cytokines in the hIS (Faggioni et al. 2001).

Rate constants and half lives used in hIS models are often estimates, or even educated guesses. Some are taken from in vitro data, or animal in vivo data, and

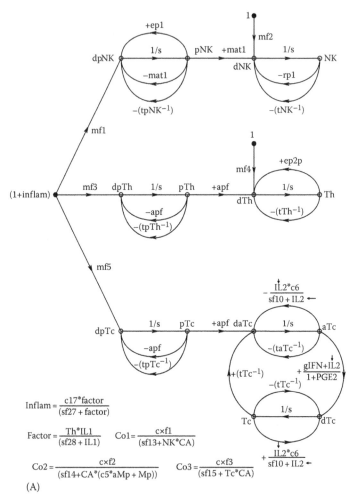

$$\text{Inflam} = \frac{c17^*\text{factor}}{(sf27 + \text{factor})}$$

$$\text{Factor} = \frac{Th^*IL1}{(sf28 + IL1)} \qquad \text{Co1} = \frac{c \times f1}{(sf13 + NK^*CA)}$$

$$\text{Co2} = \frac{c \times f2}{(sf14 + CA^*(c5^*aMp + Mp))} \qquad \text{Co3} = \frac{c \times f3}{(sf15 + Tc^*CA)}$$

(A)

FIGURE 8.31 (A) and (B) Nonlinear signal flow graphs constructed from the 18-state equations of the **TUMOR5.t** model of Clapp et al. (1988) (see Appendix A.7). Note the parametric coupling between the NLSFG in (A) and in (B). (From Northrop, R.B., *Signals and Systems in Biomedical Engineering*, 2nd edn., CRC Press, Boca Raton, FL, 2010.)

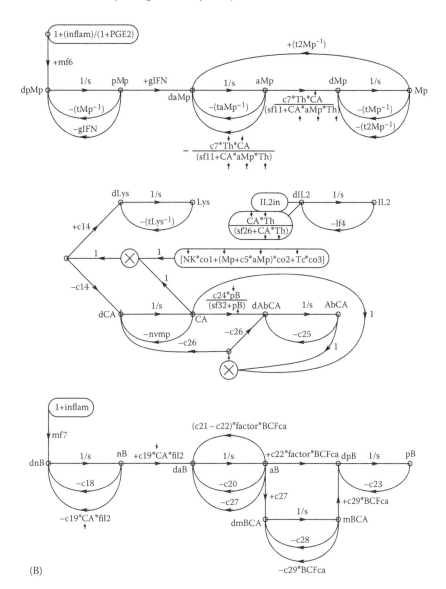

(B)

FIGURE 8.31 (continued)

may be far from human numbers. Also, different kinds of tumors vary in their antigenicity and growth rates. Tumors with weak antigenicity may avoid strong attack by antibodies and CTLs, but still may fall prey to NK cells under circumstances where NK cells are attracted to the tumor cells, and are not inhibited from killing them. Also, we do not appreciate the role of the complement system in tumor surveillance; this is a largely uninvestigated topic.

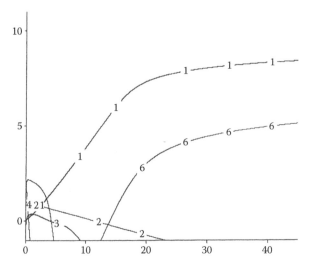

FIGURE 8.32 Simulation of a cancer growth that avoids immune rejection. 18-state Clapp et al. model, **TUMOR5.t**. Trace 1 = log(Ca), trace 2 = log(Th), trace 3 = log(CTL), trace 4 = log(NK), trace 5 = log(aMp), trace 6 = log(AbCa). See text and Appendix A.7 for parameters. (From Northrop, R.B., *Signals and Systems in Biomedical Engineering*, 2nd edn., CRC Press, Boca Raton, FL, 2010.)

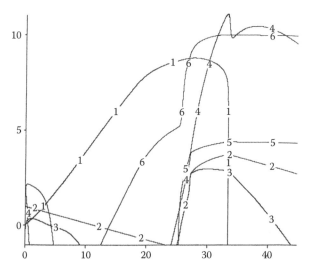

FIGURE 8.33 Simulation of the effect of an infusion of IL2 to the **TUMOR5.t** model of Figure 8.31. 15 u/day of IL2 are given from day 14 to day 28. Trace 1 = log(Ca), trace 2 = log(Th), trace 3 = log(CTL), trace 4 = log(NK), trace 5 = log(aMp), trace 6 = log(AbCa). See text for parameters. Note that the pd of Ca cells crashes at day 33. Note general immune activation occurring around day 25. (From Northrop, R.B., *Endogenous and Exogenous Regulation and Control of Physiological Systems*, Chapman & Hall/CRC Press, Boca Raton, FL, 2000.)

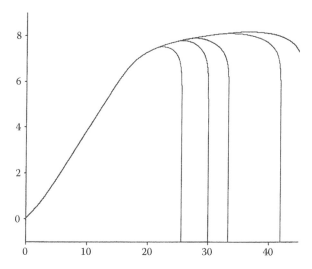

FIGURE 8.34 Same scenario as in Figure 8.33, except just log(Ca) plotted for various dose rates of IL2 infusion given from day 14 to day 28. From left to right, the IL2 rate is 40, 20, 15, 10, and 9 u/day. No effect was seen on Ca growth for IL2 rate < 9 u/day. (From Northrop, R.B., *Signals and Systems in Biomedical Engineering*, 2nd edn., CRC Press, Boca Raton, FL, 2010.)

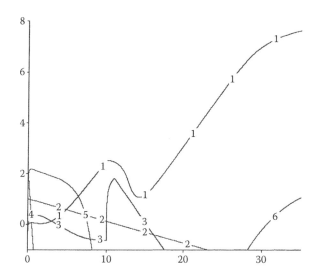

FIGURE 8.35 Simulation of the effect of another possible immunotherapy on the growth of Ca using the **TUMOR5.t** model. 100 monoclonally grown CTL specific for the Ca were injected at t = 100 days. Note the jump in CTL pd and its decay. The CA pd dips the regrows robustly. Trace 1 = log(Ca), trace 2 = log(Th), trace 3 = log(CTL), trace 4 = log(NK), trace 5 = log(aMp), trace 6 = log(AbCa). (From Northrop, R.B., *Signals and Systems in Biomedical Engineering*, 2nd edn., CRC Press, Boca Raton, FL, 2010.)

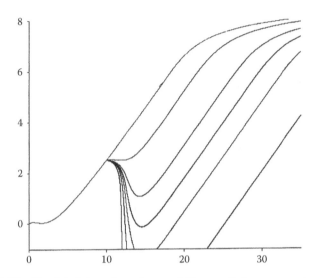

FIGURE 8.36 Simulated dose-dependency of Ca growth in response to exogenous, monoclonally grown CTLs (mCTL) specific for the Ca, injected at $t = 10$. Only log(Ca) is plotted. Top curve, 0 mCTL; second curve down, 20 mCTLs; third down, 100 mCTLs; fourth down, 150 mCTLs; fifth down, 200 mCTLs, sixth down, 300 mCTLs; seventh down, 400 mCTLs (at time $t = 50$, the Ca does not regrow for 400 mCTLs). See text for details. (From Northrop, R.B., *Endogenous and Exogenous Regulation and Control of Physiological Systems*, Chapman & Hall/CRC Press, Boca Raton, FL, 2000.)

8.5 EXAMPLES OF WHEN COMPLEX PHYSIOLOGICAL SYSTEMS FAIL

8.5.1 INTRODUCTION

Much of modern medical therapy is directed at curing or minimizing the untoward effects of component and regulatory failure in complex PSs. Some examples are hypertension (excessively high blood pressure); blood clotting disorders; mental disorders such as bipolar disease, schizophrenia, paranoia, and phobias; disorders of the bone structure (osteoporosis); and disorders of the human immune system (autoimmune diseases such as Type I diabetes, celiac disease, and myasthenia gravis).

In the following sections, we describe in more detail (1) the effects of dysregulation in general organismic homeostasis caused by organ failure, (2) certain autoimmune diseases where the hIS attacks normal body cells, and (3) circulatory shock as an example of a tipping point in a CNLS.

8.5.2 GENERAL ORGAN FAILURE

There are a number of ways a complex PS can malfunction or fail. As a first example, consider the failure of an organ from disease or an accident that simultaneously mediates regulation of several critical physiological parameters. Such failure can lead to organismic death. The kidneys are one example of such critical, multifunction organs. They are necessary for regulation of the blood concentrations of Na^+, K^+, H^+, Cl^-, HCO_3^-, Ca^{++}, Mg^{++}, glucose, urea, uric acid, certain plasma proteins, osmotic pressure, total water, as well as mean arterial pressure (MAP). Much of renal regulation is accomplished by active reabsorption of ions and molecules from the glomerular filtrate by transmembrane proteins imbedded in the outer membranes of epithelial cells in the tubules and the collecting ducts associated with each glomerulus in a kidney.

Active reabsorption is generally against concentration gradients, and often electric field gradients. Active reabsorption of substances is generally under hormonal regulation. A good example is the regulation of plasma $[K^+]$ ($[*]$ = concentration) by the adrenal cortical hormone, aldosterone (ALDO). The action of the K^+ aldosterone regulator is to maintain $[K^+]$ in plasma and interstitial fluid (IF) within narrow, physiological bounds, in spite of dietary inputs and other renal events, such as changes in the glomerular filtration rate (GFR). The normal (regulated) range of $[K^+]$ in IF is 3.5–5.0 mEq/L; the nonlethal range is 1.5–9.0 mEq/L. There are about 3500 mEq K^+ in the body, of which only 60 mEq is in the interstitial and extracellular fluids. Potassium is concentrated in the cells by ATP-driven, active transport "pumping" in which Na^+ is expelled from inside cells and K^+ is pumped in, both ions against concentration gradients, and Na^+ also against an electrical potential gradient.

ALDO is involved in a complex manner in both the regulation of blood volume and IF $[K^+]$. Even though Na^+ ions figure in these coupled regulatory systems, aldosterone concentration has little effect on IF $[Na^+]$, which is, in turn, regulated by the posterior pituitary hormone vasopressin, aka antidiuretic hormone (ADH) (Northrop 2000).

Conversely, [ADH] has little effect on IF $[K^+]$ if the ALDO system is intact. The cells in the zona glomerulosa of the adrenal cortex secrete ALDO in primarily in response to elevated IF $[K^+]$. (The mechanism by which IF $[K^+]$ affects ALDO secretion rate is not known.) Adrenocorticotropic hormone (ACTH) is important for ALDO secretion in that if it is absent, no ALDO will be secreted. Some [ACTH] is required for normal IF $[K^+]$-controlled secretion of ALDO. Guyton (1991) stated that a 10%–20% decrease in IF $[Na^+]$ can, on rare occasions, double $\dot{Q}_{ALDO}$. It is generally agreed that IF $[Na^+]$ and [ACTH] have minor effects on the secretion of ALDO by the adrenal cortex. The "normal" $\dot{Q}_{ALDO} = 150\,\mu g/day$ or 104.2 ng/min (average). The mean IF concentration of aldosterone is 100 ng/L. Aldosterone is broken down in the liver, and byproducts are secreted in the feces and urine. The washout (loss) time constant for ALDO is 17.3 min (Northrop 2000).

The hormone aldosterone has pleiotropic actions in the body. While its principal effect is in on the kidneys' epithelial cells, it also targets cells in the salivary glands, sweat glands, and the mucosal cells in the colon. The action of ALDO on these cells causes them to absorb Na^+ back into the body. Water generally follows the absorption of Na^+ in the intestines, so P_{osm} is preserved, and blood volume increases. The effect of ALDO in the kidneys is on the epithelial cells of the collecting ducts. Here, elevated [ALDO] causes an increase in the rate of active pumping in which potassium is exchanged for sodium. Specifically, high [ALDO] increases the rate at which K^+ is transported into, and Na^+ is transported out of the collecting duct's epithelial cells outer borders. K^+ then diffuses out of the inner borders of the duct's epithelial cells into the duct's lumen, and sodium diffuses out of the duct's fluid into the epithelial cells. The block diagram of Figure 8.37 illustrates the effect of IF [K^+] on $\dot{Q}_{ALDO}$ (Note that the $\dot{Q}_{ALDO}$ versus [K^+] curve is really a hypersurface, because $\dot{Q}_{ALDO}$ is affected by Angiotensin II and ACTH as well as [K^+].) The IF concentration of potassium increases the loss rate of K^+ in the urine and decreases the loss rate of Na^+ in the urine. This means more Na^+ is retained in the body when [ALDO] is high. Paradoxically, the aldosterone-induced retention of Na^+ does not significantly raise IF [Na^+] or P_{osm}. Apparently, water is reabsorbed through aquaporin proteins following the sodium ions, keeping P_{osm} constant. Thus, nature has to a large degree decoupled the K^+ and Na^+ regulatory systems, even though they share a common pump.

A comprehensive, nonlinear signal flow graph describing for the regulation of IF [Na^+], [K^+], and extracellular fluid volume is shown in Figure 8.38 (Northrop 2000). At the top of the diagram, we see the [K^+] regulator. Note that there is an inverse relationship between potassium and sodium loss rates. Total water volume enters the system parametrically, determining K^+ and Na^+ concentrations and P_{osm}. Note that P_{osm}, derived from [Na^+], also activates the subject's drinking behavior, here described by a hysteresis function. Nature has provided man and other mammals with a drinking satiety behavior mode in which just enough water

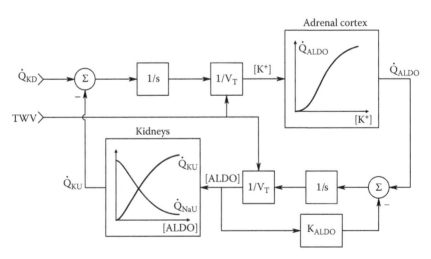

FIGURE 8.37 Block diagram model for the [K^+]-ALDO system. Note that the total water volume (TWV) enters parametrically. First-order loss dynamics are assumed for aldosterone. (From Northrop, R.B., *Endogenous and Exogenous Regulation and Control of Physiological Systems*, Chapman & Hall/CRC Press, Boca Raton, FL, 2000.)

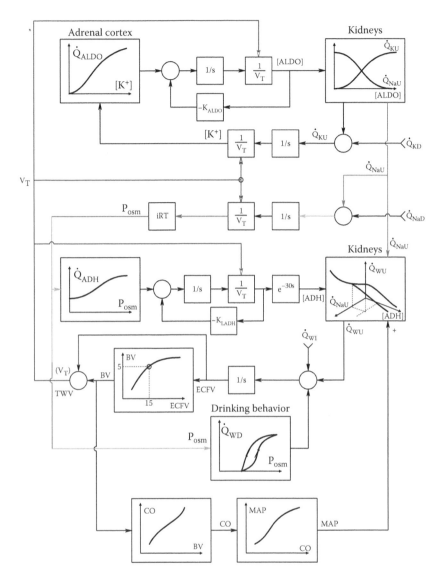

FIGURE 8.38 Nonlinear SFG modeling the regulation of interstitial fluid (IF) [Na+], [K+], and extracellular fluid volume. The parameter V_T (TWV) is critical in determining concentrations. (From Northrop, R.B., *Endogenous and Exogenous Regulation and Control of Physiological Systems*, Chapman & Hall/CRC Press, Boca Raton, FL, 2000.)

is drunk to restore a normal P_{osm}, and then the urge to drink disappears. There is a ½ to 1 h delay for all of the water drunk to become uniformly distributed in the body and give a steady-state, lowered P_{osm}. If this behavioral hysteresis were not present, the lag would cause a tendency to over-drink, yielding a hypotonic P_{osm}. Of particular interest in the diagram is the relation between water loss rate and the sodium loss rate and [ADH]. It is clear that high [ADH] increases the rate of water absorption back into the IF, and thus reduces water loss in the urine.

Aquaporins are a recently discovered class of transmembrane proteins (TMPs) that have the unique property of regulating the passage of water molecules through the plasma membranes of cells in animals, plants, yeasts, and bacteria. These complex TMPs have the property of excluding all large molecules, and ions such as Cl^-, Na^+, K^+, Ca^{++}, and Mg^{++} from passing through the membrane, that is, they are H_2O specific. The aquaporin channels also repel and exclude protons in the form, H_3O^+. Water molecules pass though aquaporin channels serially (i.e., in single file) at rates up to ca. 10^9 molecules/s (Verkman and Mitra 2000, Agre et al. 2002). Chavda and Patel (2008) cited 13 known mammalian aquaporins (AQPs), with eight types located in human kidneys. Calamita (2005) stated, "We now know of at least 300 aquaporins, functionally divided into orthodox aquaporins (permeable only to water) and aquaglyceroporins (permeable to small neutral solutes such as glycerol, in addition to water), variously expressed in vertebrates, invertebrates, microbes and plants." There are more than 150 AQPs in the Plantae (Zhao et al. 2008).

A fraction of the water entering certain cells (oocytes) is attributable to an active, transmembrane, sodium ion-driven, glucose-Na^+-H_2O-coupled water pump. The electric and concentration gradient of Na^+ ions provides the potential energy to drive this curious pump that requires glucose to work (Loo et al. 2002, Zeuthen 2007).

Binding of ADH to receptors on the surface of kidney collecting duct epithelial cells triggers the second-messenger, cyclic-AMP mechanism to make cAMP within the cells. The cAMP in turn stimulates transcription of the aquaporin-2 (AQP-2) gene and also causes the cytoplasmic store of AQP-2 molecules to be inserted into the cell membrane of the collecting duct cells facing the lumen of the collecting duct. Thus an increased density of AQP-2 allows more water from the collecting duct filtrate to enter the epithelial cells surrounding the collecting duct. This serves to concentrate the filtrate in the collecting duct (eventually urine), as well as adding water to the cytosol of the epithelial cells. Another aquaporin, AQP-3, is located on the basolateral membranes of collecting duct epithelial cells. The AQP-3 TMPs transport intracellular water passed through the ADH-regulated AQP-2 units into the extracellular fluid, thence to the blood, increasing blood volume.

At high IF $[K^+]$ levels, IF [ALDO] is elevated and ATPase pumping is stimulated so there is an increased K^+ loss rate in the urine, and as a result of the K^+–Na^+ coupling in the pump, there is a reduced rate of Na^+ loss in the urine, that is, Na^+ ions from the collecting duct fluid are pumped back into the IF at an elevated rate. Water follows these sodium ions through the aquaporins, keeping P_{osm} relatively constant in the IF around the collecting ducts. Thus, high IF $[K^+]$ leads to reduced water loss rate in the urine. These mechanisms make the $[Na^+]/P_{osm}$–ADH system relatively insensitive to the operation of the $[K^+]$–ALDO regulator. The net water balance includes insensible water loss rate, $\dot{Q}_{WI}$, water input from drinking, $\dot{Q}_{WD}$, and water loss rate through the urine affected by ADH and ALDO-induced sodium pumping, $\dot{Q}_{WU}$.

In summary, the regulators for IF [K⁺], IF [Na⁺] (Posm), and blood volume are intimately coupled through the mechanisms of the kidneys' excretory and absorption mechanisms. Nature has effectively decoupled the [Na⁺]–ADH system from the [K⁺]–aldosterone system so that if the ALDO system is blocked, [Na⁺] is still regulated (Northrop 2000).

Other multifunction organs whose failures are critical include the liver, the anterior and posterior pituitary, and the pancreas. It is hard to live without a pancreas. For example, the pancreatic alpha, beta, and delta cells produce the important endocrine hormones GLC, insulin, and somatostatin, respectively. The pancreas also secretes the important digestive enzymes: trypsin, chymotrypsin, carboxypeptidase, elastase, nuclease, pancreatic amylase, pancreatic lipase, phospholipase, and cholesterol esterase. Secretion of pancreatic enzymes is regulated both by the autonomic nervous system and other hormones (Guyton 1991).

8.5.3 Dysfunction of the hIS

The human immune system (hIS) is another complex PS that can suffer multiple kinds of regulatory failure. Autoimmune (AI) diseases occur when the hIS attacks normal body tissues (Mackay and Rosen 2001). Some examples of autoimmune diseases include, but are not limited to, myasthenia gravis (antibodies [Abs] attack acetylcholine synaptic receptors in muscle innervation), systemic lupus erythematosis (SLE) (antibodies are produced to DNA and a variety of other self-constituents in cells), type I diabetes (pancreatic beta cells are attacked by Abs leading to insulin deficiency), rheumatoid arthritis (a chronic inflammatory disorder of synovial joints), Guillain–Barre syndrome (an AI attack on the myelin coating of nerve axons), and celiac disease (an AI attack on the cells of the villi lining the small intestine induced by gluten proteins in the diet that leads to malabsorption).

There are a number of hypotheses on the etiology of AI diseases: one hypothesis that has recently received recent experimental attention is that AI diseases are exacerbated by certain genetic defects that affect the hIS. Quoting Gregersen and Behrens (2006), … "autoimmune disorders have diverse phenotypes, even within a given disease category, and it is currently unclear to what degree genetic versus stochastic or environmental factors contribute to this diversity.… "the environmental risk factors for autoimmunity are not well-defined." These authors go on to comment that AI can develop slowly over time; often circulating AI Abs can be detected years before clinical AI disease is expressed. Such delays in the progress of AI diseases make it difficult to identify environmental causes (e.g., virus or bacterial infections and exogenous toxins). For example, an established risk factor for rheumatoid arthritis is known to be smoking (Gregersen and Behrens 2006).

Treatment of AI diseases usually involves blocking some AI regulatory pathway. For example, in rheumatoid arthritis, activated Mϕs contribute to joint inflammation, so therapeutic strategies that modulate the cytokines that activate Mϕs have been developed. For example, Infliximab, a monoclonal antibody against TNF-α (a Mϕ activator protein), is effective against joint erosions as well as against Crohn's disease (Mackay and Rosen 2001). Another emerging approach to blocking AI-induced inflammation is by blocking the signals that control T-cell trafficking to the inflammation site, that is, by blocking the T-cells' chemotaxic responses (von Andrian 2003).

Systemic anaphylaxis is another life-threatening manifestation of loss of regulation in the complex hIS. Systemic anaphylaxis is a rapidly spreading, hypersensitivity allergic reaction, initially mediated by excess immunoglobulin E (IgE) antibodies (Abs). A number of common substances can initially hyper-sensitize the hIS, including, but not limited to, foods (tree nuts, peanuts, seafood, berries, egg albumin, cotton-seed oil, sulfite-containing food additives), venoms (bees, wasps, hornets, fire ants, spiders, snakes, jellyfish), medicines (local anesthetics, aspirin, sulfa drugs, penicillin, other antibiotics, vaccines containing egg or other animal serum proteins, injected radiographic contrast media, fluorescein, etc.), latex gloves, and bandages.

For reasons poorly understood, exposure to a hypersensitizing antigen (Ag) causes an unusually high production of IgE Abs with paratope regions specific for the antigenic epitope. This IgE antibody (Ab) production generally takes place in lymph nodes and is enhanced by type II helper T-cells (Th2). The activated Th2 cells release interleukins 4 and 13 (IL-4 and IL-13), which promote B-cells to make excess IgE Abs. The large numbers of specific IgE Abs then bind to Fc membrane receptors on circulating basophils, and on the surfaces of fixed mast cells found throughout connective tissues and in the airways, often near small blood vessels. IgE Abs also bind to Fc receptors on human monocytes, Mϕs, and platelets. The IgE Abs have a unique, long-lived interaction with their high affinity receptor (FcϵR1) on mast cells and basophils so they become primed to release their cytokines when the Ag molecules bind to them.

The trouble begins upon re-introduction of the antigen (Ag) into the body. The Ag binds to the IgE Abs already present, activating a cascade of reactions that trigger the Mast cells to degranulate (release powerful IS cytokines that trigger an acute inflammatory reaction in surrounding tissues). The IS-activating cytokines released from the Mast cells and basophils include histamine, leukotrienes, certain interleukins, certain prostaglandins, and the powerful platelet activating factor (PAF). Immunoglobulins IgG or IgM can enter the reaction and activate the release of complement fractions, which add to the inflammation. Thus, a disproportionately large activation signal is sent from the adaptive IS to the innate IS.

The acute phase of the systemic anaphylactic response is swift and massive, occurring within minutes of re-exposure to the Ag. Activation of the mast cells is basically an autocatalytic process, and spreads as the cytokines initially released diffuse through the tissues and are carried by the circulation, triggering more cytokine release.

Anaphylactic shock can occur because of loss of circulating blood volume. The cytokines of anaphylaxis (including PAF) cause a sudden increase in vascular permeability, causing water and certain blood proteins to leave the blood vessels and go into the intracellular volume, causing angioedema, hypotension, then hypovolemic shock followed by cardiac dysfunction, and even death. Shock affects all body systems.

Other symptoms of systemic anaphylaxis can include hives (urticaria), swelling in the throat, severe bronchospasm (asthma), chest pains, stomach cramps, nausea, vomiting, diarrhea, tremors, etc.

Local and systemic anaphylaxis are basically due to a failure in the regulatory mechanisms for IgE production. This dysregulation evidently has a genetic etiology; the inherited propensity to over-produce IgE Abs is called atopy. Atopic individuals may have as high as $12\,\mu g/mL$ of circulating IgE (normal is ca. $0.3\,\mu g/mL$) and near 100% of the FcɛR1 receptors on Mast cells are occupied with IgE antibodies (20%–50% is the normal IgE density) (Kimball 2009c).

Regulation of IgE production by control of B-cell differentiation to Ab-secreting plasma cells is thought to involve the low affinity Fc receptor, FcɛR2 (aka CD23). CD23 may also permit facilitated Ag presentation, an IgE-dependent mechanism through which B-cells expressing CD23 are able to present Ag and stimulate specific Th cells, causing the enhancement of a Th2 response, producing yet more IgE Abs.

IgE over-production is also implicated in the etiology of asthma. Asthmatic bronchospasm may be due to IgE Abs (both circulating, and fixed on lung mast cells) blocking β2-adrenergic receptors on bronchiolar smooth muscle (Berger et al. 1998, Cruse et al. 2005).

Desensitization of the hIS by periodic, small injections of the offending Ag ("allergy shots") is believed to work by switching the hIS to make more Th1 helper T-cells (versus Th2 cells). The presence of more Th1 cells shifts the immune response from IgE antibody-mediated immunity to cell-mediated immunity, reducing the occurrence of anaphylactic reactions.

Still another symptom of IS dysregulation is seen in the growth of tumors that "escape" hIS "surveillance." There is evidence that the hIS can kill tumor cells under certain conditions. Altered cell surface proteins on tumor cells, including mutations of major histocompatibility complex (MHC) molecules, can lead to IS attack. It is clear that some types of tumor cells have robust resistance to the biochemical signals from the IS that trigger programmed cell death (PCD = apoptosis) (of the tumor cells). Some of these signals include tumor necrosis factors α and β (TNF-α, β) and Fas Ligand (Northrop and Connor 2008). Tumor cells may also be resistant to PCD because they have blocked the transcription of their *p53* gene. (Cell damage normally induces p53 expression; p53 is an intracellular apoptosis inducer molecule.)

8.5.4 CIRCULATORY SHOCK

We consider shock because it is an example of a PS failure with a tipping point. The circulatory system services all body tissues. If it fails, so do the other complex PSs and organs dependent on it, and death of the organism occurs rapidly. We are concerned here with what happens when the circulatory system fails gradually.

Shock is a physiological/medical term given to the process of circulatory system collapse, generally resulting from inadequate cardiac output (CO). Many factors can affect the performance of the heart, so there are many types of shock. Severely reduced arterial flow from reduced mean arterial pressure (MAP) and cardiac stroke volume can lead to tissue anoxia (including cardiac muscle) and the accumulation of metabolic wastes (leading to acidosis), and a tipping point occurs. Weakened cardiac muscle pumps even less oxygenated blood to tissues, and irreversible damage (e.g., tissue necrosis) is done to the heart, kidneys, CNS, etc., and death can follow. From a systems point of view, shock leading to death is an autocatalytic or positive feedback process.

One of the main causes of shock is hypovolemia due to uncontrolled hemorrhage. Figure 8.39 illustrates an example of the time course of fatal hemorrhagic shock. Note that blood pressure loss from uncontrolled bleeding is at first compensated for by the body (e.g., through vasoconstriction), then as blood volume (BV) reaches a critical low value, CO plummets. Stopping the hemorrhage allows a small, temporary recovery of CO, but the damage has already been done; CO begins to fall again. In this scenario, when the CO reaches about 15% of normal a massive transfusion is given, bringing CO quickly back to normal. However, the recovery is temporary, the damaged heart cannot carry the increased load caused by the increased BV, and CO falls back to zero and death occurs.

Another cause of hypovolemia is edema caused by severe infection and/or anaphylaxis. Cytokines released in the acute inflammatory process by immune system cells cause a massive increase in peripheral vascular permeability, and osmotic and hydraulic pressure forces blood plasma proteins (and water) to enter

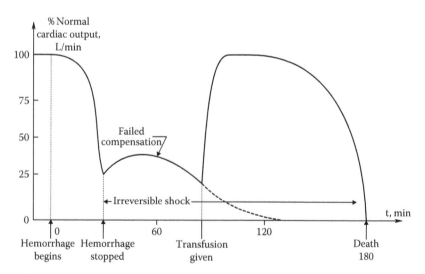

FIGURE 8.39 The time course of fatal hemorrhagic shock. (From Northrop, R.B., *Endogenous and Exogenous Regulation and Control of Physiological systems*, Chapman & Hall/CRC Press, Boca Raton, FL, 2000.)

the extracellular volume, lowering BV, hence cardiac venous return and CO drop. Anaphylactic cytokines also cause peripheral vasodilation, contributing to low blood pressure and low CO. Again, death can occur from hypovolemia, low CO, and the irreversible accumulation of cell damage.

Decreased venous return can also contribute to shock. Slow blood flow can cause clots to form in major body veins and the pulmonary blood vessels, exacerbating low venous return, hence low CO.

Treatment of shock generally involves restoring BV and CO as quickly as possible, before the tipping point is reached (irreversible damage to the heart, CNS, kidneys, etc.). Inhaled O_2 is given and also hyperbaric O_2 may aid the recovery of anoxic, damaged tissues. Inflatable, pneumatic shock pants can aid venous return, and injected glucocorticoids may control the apoptosis of anoxia-damaged cells.

8.6 SUMMARY

In this chapter, we have defined physiology and described it as a complex science of living systems. Also defined was homeostasis, the active maintenance of steady-state conditions in PSs.

Three complicated mathematical models of complex PSs, that is, basic glucoregulation, the hIS versus HIV/AIDS scenario, and the hIS versus cancer, were presented. The models were based on nonlinear ODEs describing mass-action reactions between chemicals, between chemicals and cells, and between cells and cells. Model results were consistent with observed physiological behavior.

Finally, we gave some examples of the results of PS component failure; general organ failures, dysfunctions in the hIS including autoimmune diseases, and circulatory shock were considered.

PROBLEMS

8.1 Make a systems block diagram describing the extensive metabolism and actions of the pleiotropic hormone, leptin. Show the effects of leptin on NPY, insulin secretion, GLC secretion, fat metabolism, the hIS, etc. Also show how leptin secretion is affected by the preceding substances. Assume simple, first-order kinetics for the disappearance of leptin.

8.2 Make a systems block diagram describing the extensive metabolism and actions of the pleiotropic hormone, NPY. Show its actions and the factors regulating its release. Assume simple, first-order kinetics for the disappearance of NPY.

8.3 Hill functions are widely used in modeling biochemical systems. They can be used to account for a finite number of receptors on a cell membrane, for example. Plot the saturation Hill function, $f_s([A]) = (k_{max}[A]^n/B^n + [A]^n)$ versus $[A]$, for $n = 1, 2, 4$, and 10: B, $k_{max} = 1$. $[A] \geq 0$.

8.4 Hill functions can also be used to model parametric inhibition. Plot the repressor Hill function

$$f_r([X]) = \frac{k_{max}^n}{(1+([X]/B)^n)}, \text{ vs. } [X], \quad \text{for } n = 1, 2, 4, \text{ and } 10 : B,$$
$$k_{max} = 1. [X] \geq 0$$

8.5 The hIS model of Northrop (2000) (see Section 8.4.3.4 and Appendix A.6) does not include a regulated population of dendritic cells. Show how the actions of these important immune cells can be added to the model by one or more new state equations. Assume they can ingest pathogens and do AP to Th cells, and are stimulated by the same cytokines that affect Mϕs.

8.6 Discuss how exogenous monoclonal catalytic antibodies might be used to fight specific cancers.

8.7 Consider the model for glucoregulation, `Glucose6.t,` given in Appendix A.6. Assume the pancreatic alpha cells are damaged by a viral infection so very little GLC can be produced. Investigate by simulation what happens during an IVGTT with a pancreas with very low GLC release rate. Also examine the system responses for an insulin IV bolus dose and reduced GLC. Now try an IV bolus of GLC. Simulate with MATLAB/Simulink®, or Simnon.

8.8 Consider HIV virions as a viral quasispecies (VQS). How many varieties of HIV-1 and HIV-2 have now been identified infecting humans?

8.9 Make an exhaustive list of human autoimmune diseases. How many can be considered to have a genetic etiology?

8.10 The Visna virus attacks sheep monocytes/Mϕs, not their Th or B-cells. (A) Modify the model for HIV infection given in Appendix A.6 to simulate Visna virus infection of a sheep's IS. (B) Using an infecting bacterial pathogen, simulate the sheep IS versus a pathogen, with and without the Visna virus infection. Try AZT and IL2 therapies.

8.11 It is reasonable to identify 12 PSs in humans: (1) cardiovascular (circulatory), (2) the CNS/PNS, (3) endocrine, (4) gastrointestinal (digestive), (5) glucoregulatory, (6) immune, (7) integumentary, (8) muscular (striated, smooth, cardiac), (9) renal/urinary, (10) reproductive, (11) respiratory (including O_2 and CO_2 transport in blood), and (12) skeletal. Distinctions between these systems can be blurry; they all interact to some degree (some more than others). Which do you consider the least complex human physiological regulatory system? Give reasons for your choice.

8.12 Describe what would happen (in as much detail as you can) what would happen to you if your complete, adaptive immune system suddenly ceased to exist.

9 Quest for Quantitative Measures of Complexity

9.1 INTRODUCTION

Because complexity is seen to exist in many diverse, natural, and man-made systems, it has been very difficult to agree on common, working definitions for complex systems (CSs) and, consequently, to agree on a universal set of quantitative descriptors for their complexity. In fact, universal descriptors may not exist, and probably should not exist. Complexity measures have thus evolved ad hoc for various classes of CSs (i.e., genomic, biochemical, physiological, ecological, economic, etc., as well as computer programs, data structures, and time series). Note that any quantitative measure of complexity must be graded, thus an arbitrary threshold criterion must be used to compare the complexity "score" of CS "A" with CS "B." Some CSs are called very complex by acclaim (e.g., the adaptive, complex human CNS), while the levels of complexity of others can be argued (e.g., the human complement system vs. the human cardiovascular system). It is the graphical models of CNLSs to which we generally apply complexity measures.

A common approach to describing CNLSs lies in *graph theory*. In molecular biological and many physiological systems modeled by sets of mass-action and diffusion ODEs, causal connections can be quantified by *nonlinear, directed signal flow graphs* in which states are characterized by *nodes* (aka vertices) and the causal connections between them by unidirectional (aka directed) gain *branches* (aka edges). The branch gains can contain algebraic nonlinearities and parametrically determined gains determined by one or more node state variables, or by integrators (a linear operation).

The quantification of the complexity of a nonlinear SFG can be a score or formula based on structural quantities such as: The *total number of state nodes*, the *number of branches*, the *number of cycles* (feedback loops), the *total number of branches* (linear and nonlinear) *leaving a node*, the *total number of branches summing into each node*, and the *cyclomatic number,* which is the number of *independent* (nontouching) closed loops (cycles) in a minimal digraph. Or, these measures can be expressed together in some combination as a multi-dimensional vector.

Many metrics of system complexity rely on, or are based on, information theory. Seth Lloyd (2001) gave an extensive list of measures based on the *Difficulty of Description* of CNLSs (typically measured in bits). These include but are not limited to: *Information (Shannon), Entropy, Algorithmic*

Complexity or Algorithmic Information Content (the *Kolmogorov complexity measure*, an upper bound to Shannon's entropy), *Minimum Description Length, Fisher Information, Renyi Information, Code Length, Dimension, Fractal Dimension*, and *Lempel-Ziv Complexity*. There are metric criteria based on the *Difficulty of Creation of a CNLS* (typically measured in dollars, time, energy, etc.): *Computational Complexity, Time Computational Complexity, Space Computational Complexity, Information-Based Complexity, Logical Depth, Thermodynamic Depth, Cost* and *Crypticity*. There are also metric criteria based on the Degree of Organization of a CNLS: *Metric Entropy, Fractal Dimension, Excess Entropy, Stochastic Complexity, Sophistication, Effective Measure Complexity, True Measure Complexity, Topological Epsilon-Machine Size, Conditional Information, Conditional Algorithmic Information Content, Schema Length, Ideal Complexity, Hierarchical Complexity, Tree Subgraph Diversity, Homogeneous Complexity, Grammatical Complexity, Algorithmic Mutual Information, Channel Capacity, Correlation, Stored Information and Organization*. Note that many of the above metrics are applicable only to computer programs, information transmission, and coding, and are not of general interest to us here, except perhaps for measuring mutations in genomes and in *RNA viral quasispecies* (VQS).

In an attempt to add detail to the exhaustive list of complexity measures given by Seth Lloyd (2001), the CSE group at UC Davis published an on-line list of complexity measures (CSE Group 2008) in which they elaborated mathematically on many entries on Lloyd's listing.

9.2 INTUITIVE MEASURES OF COMPLEXITY

9.2.1 INTRODUCTION

As a species, we are becoming more and more aware of CSs, complex issues, and complex behavior. We hear the adjective "complex" used more and more by persons in politics, the media, and in the sciences. On what criteria can one make the judgment that system A is more complex than system B? You might examine (1) The difficulty in understanding, describing, and modeling the system. (2) The number of known, identifiable input and output variables (states) in the system. (3) The number and degree of the quantitative interrelationships between the variables, i.e., the system's *sensitivity matrix*. (4) The number of connections to other systems. (5) Is the system in the steady-state, or are certain inputs continually changing? (6) Can the system be effectively modeled quantitatively and dynamically by sets of ODEs or DEs, or by agent-based mathematical models? (7) Look quickly at the digraph model of CS A, and that of CS B. Which is more complex? One is tempted to visually estimate the density of nodes (N) and branches (B) (vertices and edges) to arrive at an intuitive complexity comparison. Such a subjective estimate does not use quantitative data on N and B, as well as topological features, such as feedback loops, feed-forward paths, branch transfer functions, etc., and necessarily must lack precision.

Science has revealed to man the awesome complexity of life and living systems. All living systems follow natural physical and chemical laws. Economic systems differ from living systems in that most of the rules governing their behavior depend on human behavior, all of which are challenging to model (e.g., greed, generosity, fear, paranoia, hoarding, anger), although *agent-based models* are now being used (Macal and North 2006). The availability of goods, the cost of producing them, the cost of shipping, and distributing them all figure in the modeling of economic systems.

9.2.2 EXAMPLE OF INTUITIVE APPROACHES TO COMPLEXITY

Almost everyone will agree that the U.S. economy is a CNLS. Just try to describe it verbally, or mathematically.

Affecting everyone's life today is the specter of changes in lifestyles caused by the scarcity of jobs. There are no simple quick fixes to reducing ca. 10% unemployment (April 11, 2010) **u** on a national scale ($\mathbf{u} = 100 - \mathbf{e}$). Employment (**e** %), as a parameter in complex economic systems, depends on many economic variables and parameters. Many of the branches connecting **e** to variables in an economic system model contain transport lags (delays), as well as thresholds which create effective delays. The decision to hire workers is complicated. It involves facts as well as human feelings; humans procrastinate actions, often for irrational reasons.

Suppose you were given the job of reducing **u** to pre-2007 levels. How would you do it? Intuitively, you might use the law of supply and demand. Employed labor creates goods and services for which there must be markets (consumers). You might recommend giving low-interest government loans to small businesses and banks to "prime the pump," so that businesses can buy raw materials and hire labor. These newly hired workers now have cash to pay off their debts and buy basic goods and services (e.g., food, healthcare, transportation). These loans add to the national debt, and the tax burden of citizens in the future; formally, they are called deficit spending. Intuitively, employers (e.g., manufacturers) postpone hiring workers and stocking raw materials until they are sure there is a robust market for the goods they make. However, the robust market will only exist for high **e**. Thus, there appears to be a critical value of **me** required for **u** to decrease to some low steady-state level. (**m** is the number of manufacturers manufacturing goods.) The business–employment cycle is autocatalytic; it involves regenerative positive feedback loops and fuzzy social thresholds.

What other strategies might one use in lowering **u**? In the great depression of the late 1920s to late 1930s, the government of Franklin Delano Roosevelt passed many "New Deal," "shoot from the hip," intuitive, spending laws aimed at reducing **u** and reactivating the U.S. economy. They created many "make-work" agencies to hire unemployed workers; some were declared unconstitutional by the Supreme Court, some were successful in creating employment, goods, and services. Of the many "alphabet soup" agencies created in the New Deal, the three that stand out in my mind were the Civilian Conservation Corps (CCC, 1933–1942), the Work

Projects Administration (WPA, 1935–1943), and the Tennessee Valley Authority (TVA, 1933–…).

The CCC employed unmarried, unemployed young men who were paid ca. $30/month. CCC camps were set up in all 48 states as well as Alaska, Hawaii, Puerto Rico, and the U.S. Virgin Islands. CCC workers were given public service tasks; to improve our access to public lands, improve national parks, create fire roads, work on flood control dams and hydro-power plants, etc. "By design, the CCC worked on projects that were independent of other public relief programs" (CCC 2010). The U.S. Forest Service administered more than 50% of all CCC public works projects; others were supervised by the National Park Service and the Soil Conservation Service. That is, government money was spent for CCC jobs that performed public services. They built fire towers, fire roads, fought forest fires, planted over 3 billion trees in reforestation projects, built campgrounds and camping facilities in national parks, worked on flood control and land drainage, etc. The efforts of the CCC were an effective (temporary) pump-priming strategy because they reduced u and provided the workers with wages to purchase goods, food, healthcare, and transportation. Although government money was spent, useful services and goods were received, and the workers paid taxes.

The goal of the WPA was to employ most of the unemployed people on relief until the economy recovered. It employed a maximum of 3.3 million people in November 1938. Worker pay varied from $19 to $94/month, depending on skill and experience. WPA projects totaled ca. $11.4 billion through June 1941. These included highway construction (e.g., The Merritt Parkway in Connecticut with its unique bridges and arboretum-like medians), construction of public buildings, publicly owned or operated utilities, and various welfare projects.

The third New Deal mega-agency was the TVA (1933–present), created to produce cheap electric hydro-power (later nuclear power), make rivers navigable, flood control, malaria prevention, reforestation, and erosion control. TVA also developed fertilizers, taught farmers how to improve crop yields, did reforestation (with CCC), fought forest fires, improved habitats for fish (fish ladders at dams), and wildlife (TVA 2010).

Were New Deal programs effective at increasing employment and sparking economic recovery? There is great debate on these issues. Unemployment in the Great Depression peaked at over 21% around 1933, then started to slowly decline to 15% in 1940. Something was working. u sharply decreased in 1940, probably because of the lend-lease program to ship military supplies to Britain to help its war effort against Germany. (About 600,000 workers were employed building merchant vessels (liberty ships) in 1940.) u reached a low of ca. 2% around 1943 due to the massive U.S. war mobilization. Many New Deal programs such as the CCC and WPA were terminated. The U.S. war effort involved spending billions of dollars; the National Debt/GNP shot up from ca. 40% in 1941 to a peak of ca. 130% in 1946. One might argue that the New Deal alphabet agencies had the right approach, but were not large enough in scale, and it was the massive WW II spending that caused the economy to recover, but at the expense of a huge national debt.

Perhaps, future attempts to regulate economic systems will proceed from less intuitive approaches. We predict that the effective use of validated and verified, agent-based models of economic systems will lead to coordinated, multifactorial manipulations to achieve national economic "health" and low **u**.

9.3 STRUCTURAL COMPLEXITY MEASURES

Bonchev (2004) has given a review of quantitative complexity measures (also see Ay et al. 2006, CSE Group 2008). *Structural complexity* accounts for the manner in which the elements of a system's model graph are organized or connected. The number of graph nodes (vertices), **V**, characterizes the *structure size*. The total number of graph branches (edges), **E**, is one measure of the graph's connectivity. A cycle is a closed, directed path, or loop, with no other repeated vertices (nodes) than the starting and ending vertices (nodes). An engineer will recognize a cycle as a feedback loop. The number of cycles **C** in a graph tells us about the graph's topological complexity. The more connected the structure, the larger **E** and **C** will be. The *Euler equation* relates **C**, **E**, and **V**, it is a topological descriptor of a graph modeling a CNLS.

$$\mathbf{C} = \mathbf{E} - \mathbf{V} + 1 \tag{9.1}$$

Another topological measure of graph complexity is its *percent connectedness*:

$$\mathbf{Conn}\% \equiv 100 \times \frac{2\mathbf{E}}{\mathbf{V}(\mathbf{V}-1)} \, [\%] \tag{9.2}$$

Note that $\mathbf{V}(\mathbf{V}-1)/2$ is the *maximum number* of edges (branches) in a *simple graph* having **V** vertices (nodes). In *multigraphs*, more than one branch can exist between two nodes, and **Conn**% can exceed 100%! Another definition is the *connectivity of a graph*, **Cn**. This is the percentage of nodes in a graph having at least one connection with another graph's node.

The *local connectivity* of the kth node in a graph is described by the *node (vertex) degree*, a_k, which counts the number of the node-nearest *neighbors*. **A** is defined as the *total adjacency* of the graph:

$$\mathbf{A} \equiv \sum_{k=1}^{N} a_k \tag{9.3}$$

The *average vertex (node) degree*, $\langle a_k \rangle \equiv \mathbf{A}/\mathbf{V}$, is also used as an average measure for the graph's connectivity. In *directed graphs* (e.g., signal flow graphs), the branches have a direction (i.e., to, from a node). Thus one can specify *in-degree* and *out-degree* for each node (vertex).

Another descriptor of the topological character of a graph is the *distance*, d_{jk}, between the jth and kth nodes. d_{jk} is an integer quantity equal to the number of branches (edges) connecting the nodes along the shortest path between them. We also have the *vertex (node) distance (distance degree)* which is the sum of the distances between the node j and all other nodes in the graph. *Network distance*, known as the *Wiener number*, **W**, is defined as the sum of all the *distances* in the graph:

$$\mathbf{W} \equiv \frac{1}{2} \sum_{j,k=1}^{V} \mathbf{d}_{jk} = \frac{1}{2} \sum_{j=1}^{V} \mathbf{d}_j \tag{9.4}$$

The *average vertex distance*, $\langle \mathbf{d}_j \rangle \equiv 2\,\mathbf{W}/\mathbf{V}$, characterizes how easily a node can reach all other vertices in the graph. Yet another graph descriptor is the *average intersite distance* or *graph radius*, $\langle \mathbf{d} \rangle \equiv 2\mathbf{W}/[\mathbf{V}(\mathbf{V}-1)]$, which shows how easily (on an average) one node can reach another node.

Bonchev (2004) commented that biological networks have been characterized as "small-world networks," because the average distance in them is generally small. In graphs with *directed paths (edges)* (e.g., all biochemical pathways), there can be no paths connecting some pairs of nodes. Thus the distance between such a pair of nodes is infinite, which makes it impossible to define the *Wiener number*. Bonchev concluded that for practical purposes, distance-based descriptors can be calculated by counting only the finite distances along existing paths in the network.

An excellent discussion paper by Kim and Wilhelm (2008) treats various measures of graph complexity. However, they only considered undirected and unweighted networks without loops or multiple edges. They defined and evaluated eight new measures of graph complexity, including: three different subgraph measures, three product measures, and two entropy measures. They also evaluated a previously proposed graph entropy measure, the off-diagonal complexity, *OdC*. They compared different measures to characterize the complexity of 33 real-world, connected graphs of different biological, social, and economic systems with the six computationally more simple measures. All of their complexity measures were normalized to have values between zero and one.

As an example of the complexity in measuring graph complexity, we examine one of Kim and Wilhelm's measures, the *maximum medium articulation*, *MMA*. They defined *MMA* as the product, **R*I**. **R** is the *redundancy*, given by

$$\mathbf{R} = -\sum_{i,j} \mathbf{T}_{ij} \log \left(\frac{\mathbf{T}_{ij}^2}{\sum_k \mathbf{T}_{kj} \sum_l \mathbf{T}_{il}} \right) \tag{9.5}$$

The mutual information **I** is defined as:

$$I = \sum_{i,j} T_{ij} \log\left(\frac{T_{ij}}{\sum_{k} T_{kj} \sum_{1} T_{il}} \right) \tag{9.6}$$

where

T_{ij} is the normalized flux from node i to j: $T_{ij} = t_{ij} / \sum_{k,l} t_{kl}$

$R \rightarrow 0$ for a directed ring, but maximum for the fully connected graph

I is the opposite, so $MA \rightarrow 0$ in the extreme cases but has a maximum in between

Another complexity measure defined by Kim and Wilhelm is the *efficiency complexity measure*, **Ce**.

$$Ce = 4\left(\frac{E - E_{path}}{1 - E_{path}} \right)\left(\frac{1 - E}{1 - E_{path}} \right) \tag{9.7}$$

The *graph efficiency* **E** is defined as the arithmetic mean of all inverse shortest path lengths:

$$E \equiv \frac{1}{n(n-1)/2} \sum_{i} \sum_{j>i} \left(\frac{1}{d_{ij}} \right) \tag{9.8}$$

Perhaps, the most valuable insight into the graph complexity measures set forth by Kim and Wilhelm is their graphical comparison of their measures on various seven-node graphs and on other graphs with more nodes and diverse edge structures. The interested reader is urged to read their paper.

Piqueira (2008) summarized several useful measures of biological complexity, based on the assumption that the complex biological system could be modeled by a very large set of ODEs. He cited a general structural complexity measure in which a system of **m** parts has **Q** different physical arrangements, given by

$$Q = 2^{\frac{m(m-1)}{2}} \tag{9.9}$$

A probability, p_i, is attributed to each possible physical arrangement, $i = 1, 2, \ldots, Q$. Thus an informational entropy, H_e, can be associated with the system's arrangements:

$$H_e = \sum_{i=1}^{Q} p_i \log\left(\frac{1}{p_i} \right) \tag{9.10}$$

It is well known that the max H_e occurs when all p_i are equal. Sic:

$$\max H_e = \frac{m(m-1)}{2} \tag{9.11}$$

Piqueira now defines the system's *structural complexity*, C_e, as

$$C_e \equiv \left(1 - \frac{H_e}{\max H_e}\right)^{k_1} \left(\frac{H_e}{\max H_e}\right)^{k_2} \qquad (9.12)$$

where k_1 and k_2 are positive real constants, "chosen according to the particular object to be studied." The disorder functional is defined as $D = H_e / \max H_e$.

We see that Piqueira's structural complexity, C_e, is in reality, close to an informatic measure (see below). Equation 9.12 is compatible with the property that neither well-ordered nor completely random systems are seemingly complex.

Piqueira (2008) goes on to define a *functional (dynamic) complexity*, C_f, in a similar manner used for C_e, and uses this probabilistic measure to define an *overall biological system complexity*, C_b.

$$C_b \equiv N^\varepsilon \, C_e^\zeta C_f^\eta \qquad (9.13)$$

where
   N is the total number of state variables in the system
   ε, ζ, and η are real positive exponents chosen according to the weight to be attributed to the three factors
   C_b may be considered to be a function of time if the system is a CAS

Piqueira cautioned that C_b must be used comparatively; it should not be used as an absolute measure.

Quantifying the complexity of biological sequences (e.g., genes, RNAs), and biological waveforms (e.g., EEGs, animal songs) has attracted many applied mathematicians and quantitative biologists. Nan and Adjeroh (2004) wrote a review paper: *On complexity measures for biological sequences*. They described: *Shannon's entropy, T-Complexity, Linguistic complexity*, as well as performance measures including *Average of complexity profile, apparent periodicity*, and *relative entropy*.

Complexity in ecosystems modeling was treated in the paper by Lawrie and Hearne (2007). The focus of this research was to generate simpler models through the use of output parameter sensitivities (see Section 4.6.2) of an unreduced, mathematical model (a set of NL, nonstationary ODEs). They assumed an ecosystem model described by a large set of N nonlinear ODEs:

$$\dot{\mathbf{x}} = f(\mathbf{x}, \mathbf{u}, t : \mathbf{p}), \quad \mathbf{y} = y(\mathbf{x}) \qquad (9.14)$$

where
   $\mathbf{x}$ are the N unreduced model ecosystem's states
   $\mathbf{u}$ are its inputs
   $\mathbf{y}$ are its outputs
   $\mathbf{p}$ is a parameter vector

The objective was to reduce the **N**-model to those states and rate constants that are *biologically significant* in the model. (This model "pruning" had nothing to do with the technical aspects of running the simulation. It was to find what states were largely irrelevant to model behavior.) Their reduced model had **M < N** states and was given as

$$\dot{\hat{\mathbf{x}}} = \hat{\mathbf{f}}(\hat{\mathbf{x}}, \mathbf{u}, t; \hat{\mathbf{p}}), \quad \hat{\mathbf{y}} = \hat{\mathbf{y}}(\hat{\mathbf{x}}) \tag{9.15}$$

Lawrie and Hearne (2007) defined two approaches to model reduction: (1) the advanced rate elimination method (AREM) and (2) the variable simplification method (VSM). They devised methods for estimating the sensitivities, $s_{ijk} = \partial y_i / \partial f_{jk}$, for each output, y_i. f_{jk} is the kth rate of the jth derivative in the full model. In the AREM, if a sensitivity, s_{ijk}, exceeds a set threshold, φ, then the rate f_{jk} is retained, else it is set to zero. In the VSM, each dx_i/dt is compared to a threshold, and if its max $< \varphi$, that dx_i/dt ODE is removed from the model, reducing its order. The authors satisfactorily tested their two-model order reduction techniques on a 29-state ecosystem model of Port Philip Bay. Their study was interesting because it used model sensitivities and thresholding to reduce model order, while preserving the CNLS's complex behavior.

9.4   INFORMATIC COMPLEXITY MEASURES

Besides structural measures of graph complexity based on topology and connection, we also have many descriptors based on *information theory*. Informatic measures of complexity have been used to quantify complexity in computer programs and data sequences, including nonstationary waveforms (such as ECGs of cardiac arrhythmia), the DNA base sequences (*codons*) in genes (*exons*) and in *introns*, and the compositional complexity of the proteome by means of its normalized information content. A number of these measures are based on Shannon's Information Theory 1949, and Shannon's *entropy* measure. The information content, $I(\alpha)$, of a system in which N elements are organized into k sets having N_i elements each can be written as

$$I(\alpha) = H_{max}(\alpha) - H(\alpha) = \sum_{i=1}^{k} N_i \log_2 N_i \quad \text{bits} \tag{9.16}$$

where α is a certain *equivalence criterion* whereby the N elements are distributed into k sets.

$$H_{max}(\alpha) \equiv N \log_2 N, \quad \text{so } H(\alpha) = N \log_2 N - \sum N_i \log_2 N_i \tag{9.17}$$

The relative information content, $I_r(\alpha)$, is defined as

$$I_r(\alpha) \equiv \frac{1}{N \log_2 N} \sum_{i=1}^{k} N_i \log_2 N_i, \quad 0 \leq I_r(\alpha) \leq 1.0 \tag{9.18}$$

(In considering the information in a gene codon sequence, it may be more expedient to use logs to the base three ["trits"], than to the base two [bits].) Bonchev and Trinajstic (1977), cited by Bonchev (2004) has adapted the informatic approach to graph topology.

Tononi et al. (1999) used stochastic informatic modeling of small (12-node) networks to examine the role of degeneracy and redundancy in synthetic biological networks. They demonstrated that the networks that exhibit degeneracy also have high values for complexity, a measure of the average mutual information between the subsets of a system.

The LMC complexity measure of a thermodynamic, CNLS was described in detail by Solé and Luque (1999). Assume the CNLS has N accessible states belonging to a set $\Sigma_\mu = \{x_i(m); \quad i = 1, \ldots, N\}$, and have an associated probability distribution:

$$\Pi_\mu = \{p_i(\mu) = P[x = x_i(\mu)]; \quad i = 1, \ldots, N\} \tag{9.19}$$

where μ stands for a given parameter which allows the description of transition from the ordered (low μ) to the disordered (high μ) regimes. Also,

$$\sum_{i=1}^{N} P_i(\mu) = 1 \quad \text{and} \quad p_i(\mu) > 0 \ \forall i = 1, \ldots, N \tag{9.20}$$

The LMC measure is based on (1) The Boltzmann entropy, H_μ:

$$H_\mu(\mathbf{P}_\mu) = -\sum_{j=1}^{N} p_j(\mu) \log[p_i(\mu)] \tag{9.21}$$

where $\mathbf{P}_\mu = [p_1(\mu), \ldots, p_N(\mu)]$, and (2) the so-called disequilibrium function, $\mathbf{D}_\mu$, defined as

$$\mathbf{D}_\mu(\mathbf{P}_\mu) \equiv \sum_{j=1}^{N} (p_j(\mu) - N^{-1})^2 \tag{9.22}$$

Using Equations 9.12 and 9.13, Solé and Luque defined the LMC measure by the product functional (Equation 9.23)

$$C_\mu(\mathbf{P}_\mu) \equiv H_\mu(\mathbf{P}_\mu) * \mathbf{D}_\mu(\mathbf{P}_\mu) \tag{9.23}$$

Note that as a CNLS becomes more disordered, the entropy increases, while the disequilibrium function $\mathbf{D}_\mu$ decreases as the system approaches equiprobability/disorder. Solé and Luque gave thermodynamic examples where they plotted LMC complexity, H_μ, and $\mathbf{D}_\mu$ vs. Kelvin temperature. The LMC showed a sharp peak around a certain temperature. LMC complexity has application in physics and chemistry.

In a review paper by Ay et al. (2006), the authors developed a unifying approach for complexity measures based on the principle that complexity requires interactions at different scales of description. They based their criteria on probability, information theory, entropy, conditional entropy, and conditional mutual information. They describe in detail the TSE complexity measure (TSE stands for Tononi, Sporns, and Edelman 1994). While clearly presented, the paper of Ay et al. is abstract and lacks any practical applications or examples in biology.

9.5 SUMMARY

We have seen that there are a bewildering number of quantitative measures of complexity that can be applied to the graph model of a CS. There are also many complexity measures that can be applied to information (time series, computer programs, genomic and proteomic sequences).

The quantitative complexity measures we have reviewed are functionals; that is, they generally return a single, positive integer number as a measure of the complexity of the CNLS they are applied to. This is philosophically bothersome. One attribute of a CS is the difficulty one has in describing and modeling it. Should its complexity be distilled down to *one* positive number? Is not relative information about the CNLS's structure and function lost? It would be interesting to see someone develop a multi-dimensional complexity index, perhaps a 3-D vector surface based on three different complexity indices, or a 2-D plot, like a joint time-frequency (JTF) plot for a nonstationary time signal (Northrop 2003). Such complexity surfaces could then be examined for fits, intersections, and congruencies. In other words, make complexity measures more complicated.

It is clear that all functional complexity measures must be used comparatively, and applied to like CNLSs.

10 "Irreducible" and "Specified Complexity" in Living Systems

10.1 INTRODUCTION

Irreducible complexity (IC) is a viewpoint based on religious faith and dogma. It is a term coined by biochemist Behe (1996), who claimed that many physiological and biochemical systems are irreducibly complex. To be IC, such a system must be composed of several interacting "parts" (subsystems, components, modules) that contribute to its overall function, and the removal of any one of the parts, subsystems, components, or modules will cause the IC system to effectively cease functioning. Behe claimed that "An IC system cannot be produced gradually by slight, successive modifications of a precursor system, since any precursor to an IC system is by definition nonfunctional." He argued that IC systems cannot have evolved, and therefore must have been "designed," linking his definition of IC biological systems to the *intelligent design* (ID) religious movement.

The thrust of this chapter is not to argue theology, faith, or the existence of an intelligent designer (aka God), but to cite scientific evidence that the so-called IC systems Behe identified are, in fact, the products of gradual evolution. Yes, some systems may fail when components are removed, but this begs the question of how the components got there in the first place. We assert that failure of a system due to component removal is not a criterion for complexity. In fact, one is more likely to see system dysfunction in a simple system that loses a component or module! Note that a *complex adaptive system* (CAS) such as the human brain can compensate for the loss or damage to a subsystem and maintain nearly normal function. Focal damage to the retina of the complex human eye can produce degraded vision (blind spots, fuzzy vision) but not necessarily blindness. There is anatomical and functional redundancy in the neural networks making up the vertebrate retina.

In this section, we address some of the contemporary thoughts surrounding ID, and arguments for IC put forth in Behe's controversial book, *Darwin's Black Box: The Biochemical Challenge of Evolution* 1996. Behe's thoughts are of interest to us because, as we discussed in Section 5.2, evolution and speciation are complex processes involving several interacting, complex, biological systems. In his book, Behe claimed that many biological systems are "irreducibly complex," and that in order to have evolved, multiple systems would have to have arisen simultaneously, presumably by the "hand" of an (unknown) "intelligent designer."

Behe claimed that many examples of IC systems exist in biochemistry, and their existence argues for an "intelligent designer" (Robison 1996).

In other words, Behe's position is that an IC system must have been created or designed as an intact, whole entity; it could not have evolved gradually over time. This assertion flies in the face of contemporary thinking on evolution. Behe's position essentially denies the existence of adaptable functional modules in nature. Naturally, Behe's (1996) assertions created a whirlwind of controversy not only from evolutionary biologists, but also from biochemists, physiologists, ecologists, etc.

Theobald (2007) has called Behe's reasoning invalid: Behe's premise 1 is: Direct, gradual evolution proceeds only by stepwise addition of parts. Behe's premise 2 is: By definition, an IC system lacking a part is nonfunctional. Behe's conclusion: Therefore, all possible direct gradual evolutionary precursors to an IC system must be nonfunctional. Theobald claimed that Behe's first premise is invalid because gradual evolution can do much more than just add parts. For example, evolution can also modify (augment, diminish, alter, subtract) subsystems, as well as totally remove parts. Theobald pointed out that Behe's IC definition is restricted only to reversing the addition of *entire parts*. Note also that "parts" are not defined by Behe. (He implies that they are "whole proteins or even larger.") Nor does Behe define *complex* or *complexity*!

Dunkelberg (2003) argued that organisms do not come with *parts*, *functions*, and *systems* labeled. They are context-dependent, terms of convenience. Dunkelberg gave the example of the human leg. "We might say, for instance, that the function of a leg is to walk, and call legs walking systems. But what are the parts? If we divide a leg into three major parts [thigh, calf, foot], removal of any part results in loss of the function. Thus legs are IC …if we count each bone as a part, then several parts, even a whole toe, may be removed and we still have a walking system." By Behe's criteria, this leg is not IC. Clearly, it can't be both.

10.2 IRREDUCIBLE COMPLEXITY

10.2.1 Examples of IC in Nature

Dunkelberg (2003) offered two examples of complex systems in nature and discussed whether or not evolution can lead to IC. His first example was the Venus' flytrap plant, *Dionaea muscipula*. *Dionaea* is an insectivorous plant that grows naturally in acidic forest wetlands in the Carolinas. When an insect climbing the plant brushes against trigger hairs in the center of the clamshell-like trap, it closes rapidly, trapping the insect. Then over a period of days, it secretes proteolytic enzymes that "digest" the insect. The plant makes metabolic use of the protein nitrogen and minerals from the digested prey. The plant has the capacity to close fully about four times.

Clearly, this insectivorous plant has a complex structure and physiology. What Dunkelberg did was show a plausible means whereby this plant could have slowly evolved by comparing it with the insectivorous sundew plants (*Drosera* sp.).

Sundews use sticky-tipped tentacles to trap insects; when a tentacle is stimulated mechanically by an insect sticking to it, it transmits an electrical signal to its neighbors that cause them to all close in on the prey, and the leaves curl in, too. *Drosera* are genetically related to flytrap plants and the insectivorous waterwheel plant, *Aldrovanda* sp. Dunkelberg theorized that in evolution, *Dionaea* lost the sticky glue of *Drosera's* touch-sensitive tentacles from its touch-sensitive hairs, and also developed a modified pair of leaves that closed fast enough to trap insects. Note that the flytraps' leaves also evolved spines on their outer edges to facilitate prey capture. Thus *Dionaea*, by Behe's criterion, cannot be IC because it evolved from a related simpler form of insectivorous plant. Here, the evidence is genetic as well as morphological.

Dunkelberg also discussed the possible IC of swimming systems (flagella, cilia) of bacteria, *Archaea*, some eukaryotic protists, and some eukaryotic body cells (in the gut and airway). (Cilia are called flagella, especially when a cell has only one or two.) A swimming system requires what is called a molecular motor, a highly structured organelle located inside a cell's cytoplasm under its membrane, attached to the membrane and the flagellum. The molecular structure of cilia and their motors is delightfully intricate. Motors convert metabolic energy to mechanical work (force times distance). Many contain ca. 200 different proteins and various numbers of contractile microtubules. Dunkelberg stated that some structural proteins are always present, others vary from species to species.

In order to claim that a flagellum is an IC system, Behe divided it into only three "parts": the "motor," the "connector," and the "paddle" (external projection). It is clear that a flagellum will not work, lacking any one of Behe's major parts; therefore, by his criterion, all flagella and cilia must be IC. However, if we choose other, larger sets of molecular "parts," then flagella and cilia are not IC. Normal ("indirect") evolutionary changes can certainly modify, subtract, add to ciliary proteins, and select altered ciliary functions. Dunkelberg went on to compare ciliary structures and functions in various bacteria, *Archaea*, and eukaryotes. He noted that the Archaeal flagellum is analogous to a bacterial flagellum, but simpler in structure and different in molecular detail. He noted that it resembles a bacterial transmembrane projection called a type IV *pilus*, to which it is probably related. Bacteria use their *pili* for simpler locomotion called twitching. However, bacteria with flagella are diverse and complicated, even a single coccus bacterium may have different flagella at its ends and around its sides.

Flagellar function appears at first to be locomotion, but examples exist of bacteria using them for adhesion and for exporting proteins including those that cause disease. They participate in a number of bacterially caused diseases including: diarrhea, ulcers, urinary tract infections, and bubonic plague. An organelle lying below the cell wall called a type III secretion system (T3SS) has been shown to be responsible for assembling a flagellum, motor and all, and bacterial exotoxins are secreted by the T3SS (Young and Young 2002).

In conclusion, examination of the flagella of many organisms has shown that their structure and function are variable, the simplest designs being found in the simplest organisms (*Bacteria, Archaea*) and the most sophisticated designs being in eukaryotes. Flagella are usually associated with propulsion in a liquid medium. Sperm tails are flagella, and the junction of the outer segment and inner segment of the visual rods and cones in the vertebrate retina are connected by a ciliary structure (Kandel et al. 1991). Fixed, ciliated cells lining the bronchial tubes expedite the clearance of mucous and foreign particles from mammalian lungs. Clearly, over the past 3 billion years or so, flagella have had ample time to evolve many highly functional designs and applications, starting with the simplest of bacteria. Cilia are certainly not IC by biochemical criteria.

10.2.2 COMPLEX HUMAN COMPLEMENT SYSTEM AND IC

In a *Talk Origins* essay, Coon (2002) tackled the gnarly problem of whether the *complement system* (CS) of the hIS is in fact IC (we described the CS in Section 6.4.5). Coon adequately summarized the function and organization of the human compliment system, and we are left with the impression that this system is quite complicated in its function; but is it IC? To not be IC by Behe's definition we must show it evolved gradually in lower (non-vertebrate) animals and preserved its immune functions as it changed. In other words, we must examine the comparative biochemistry of the CS through the existing phyla, and show how evolutionary changes have eventually led to the *human CS* (hCS). (Reactions in the complex hCS are summarized in Figure 4.1.)

One of the claims Behe made is that enzyme cascades, such as the activation of the hCS, are IC because each new step in the cascade "would require both a proenzyme and also an activating enzyme to switch on the proenzyme at the correct time and place." Coon noted that all three initial complement cascade pathways rely on the ubiquitous activity of C3 convertase in order to activate the final complement pathways (MAC, opsonization, inflammation). Humans lacking the ability to express C3 lack a functional CS and are greatly more susceptible to bacterial infections. Coon observed that it at first appears that the vertebrate pathways could not have evolved in a Darwinian fashion from earlier lectin pathways, because they are all dependent on the activity of *C3 convertase*. Coon went on to review two lines of evidence that the hCS gradually evolved to its present status. That is, it was tinkered with by evolution while maintaining its immune system function.

In Coon's (2002) review of the comparative biochemistry of CSs, he examined the CSs of the lamprey eel (an agnathan), sea squirts (in the ascidian group of urochordates), sea urchins (echinoderms), and of course, humans. The relationships are too detailed to cover in detail here, but he concludes that: "It is thought that much of the vertebrate complement system arose from duplication of genes encoding complement protein molecules C3/C4/C5, Bf/C2, C1s/C1r/MASP-1/MASP-2, and C6/C7/C8/C9." In the lectin pathway, *mannan-binding protein* (MBP) has an affinity for *N*-acetlyglucosamine residues on surface proteins of pathogens, activating MASP. MASP is mannan-binding protein-associated serine protease, a proenzyme. MASP cleaves the complement proteins, C4 and C2 → C2b + C2a. C2b is also known as *C3 convertase*, an enzyme common to all three hCS pathways. Deletion of the C3 gene in knockout mice destroys their complement pathway. However, the sea squirt C3 protein does not have the C3 convertase cleavage site, so an enzyme not like vertebrate C3 convertase (aka C2b) activates the sea squirt's C3 protein; thus natural evolution has provided an alternative pathway. Coon observed that in mammals, MASP can act directly as a weak C3 activator, evidently another older, alternative pathway.

Coon stated that it has been recently shown that a urochordate has genes that correspond to the vertebrate genes for C3, C4, and C5, but not for C2. In the process of evolution, several important CS genes have been conserved, others have been modified (tinkered), and in the lamprey, the Classical pathway genes have been entirely lost [C1(q,r,s), C2 and C4], yet the lamprey's CS still functions with the alternative and lectin pathways. Evidently, it is not IC. Removal of "parts" (other than C3) still allow the CS to function, albeit at a reduced level.

By tracing the genes for complement proteins through phylogenetic space, it is clear that the hCS system has indeed evolved, and thus fails to meet Behe's defined criteria for an IC system. Still, it is part of the incredibly complex hIS.

10.2.3 Human Eye and IC

Anyone who has studied the anatomy and physiology of the human eye will agree that it is a complex optical/neural structure. However, anyone who makes an examination of the comparative anatomy of the eyes of vertebrates and invertebrates (cf., Bullock and Horridge 1965, Walls 1967) will be impressed at nature's ability to "tinker" with structure and function in the process of evolution. It is not surprising that the ID movement has dubbed the human eye IC. However, as in the cases cited above, comparative anatomy, physiology, and genomics have shown that eyes have evolved from the most primitive pigmented neural receptor cells to their present complex forms. The vertebrate and molluscan eyes have a heritage of ciliated photoreceptor cells, while the rhabdomeric photoreceptors in arthropod compound eyes took a divergent biochemical and anatomical pathway billions of years ago (Bullock and Horridge 1965, Northrop 2001). The transmembrane potentials of insect *retinula cells* depolarize in response to input photon energy. The relatively simple rhabdomeric compound eye of the modern horseshoe crab, *Limulus polyphemus*, evolved over 400 mya (Mittmann 2002), but the same anatomical patterns are largely preserved in today's insect and crustacean compound eyes. Note that modern *Limulus*, in addition to its two compound eyes, each having ca. 10^3 ommatidia, has two smaller visual sensors called ocelli located just posterior to each compound eye. Two more ocelli are found in the middle front of its carapace, yet another ocellus is located on the ventral medial line under the carapace, and two more ocelli are located ventrally in front of the mouth for a grand total of seven ocelli (plus the two compound eyes). *Limulus* is more closely related to spiders (which have eight eyes), ticks, and scorpions, than to crabs. Note that on the other hand, the cilia-based rods and cones in a vertebrate retina hyperpolarize in response to input photons.

It turns out that the *rag worm*, an aquatic relative of earthworms, has both rhabdomeric photoreceptors in its eyes, and ciliary (vertebrate-type) photoreceptors in its brain where they apparently sense light and synchronize the worm's master biological clock.

Most eyes have lenses. The insect compound eye focuses incoming light with a clear chitonous lens (that is part of its exoskeleton) over each ommatidium (photoreceptor bundle). In mollusks, plecypods have simple "pit eyes" with several photoreceptor cells and mucous-like "lenses" (LaCourse and Northrop 1983); cephalopod mollusks actually have sophisticated "camera" eyes with crystalline protein lenses capable of accommodation. The highly evolved cephalopod eyes also have a variable pupil. While vertebrate eyes have crystalline lenses capable of accommodation as well as pupils, they also have an inverted retina in which light must pass through the retina's neural layers to reach the rod and cone photoreceptor cells.

While the eyes of a hawk or a dragonfly are complex, there is ample phylogenetic evidence that these systems have evolved gradually, and they are certainly not irreducibly complex.

10.3 SPECIFIED COMPLEXITY

Dembski (1998, 1999) has taken a mathematical approach to define a property of biological systems called *specified complexity* (SC), and claimed its existence as support for the faith-based theory of ID. He claims that SC is a reliable marker of design by an "intelligent agent," a keystone concept for ID, which opposes modern evolutionary theory. We have seen in the previous sections of this chapter that by artfully choosing the definitions of terms for a system, Behe could support a variety of outcomes to make the desired point that certain biological/biochemical systems are irreducibly complex, and thus must have been "designed" by a supreme being.

Dembski's approach to SC, aka complex specified information (CSI), follows the same pattern except Dembski uses information theory and probability to examine "patterns" in biological systems. He asserts that SC/CSI is present in a "configuration" when it can be described by a pattern that contains a "large amount" of independently specified information and is also complex, which he describes as having a "low probability of occurrence." (If only it were that simple.) Dembski holds that SC/CSI exists in numerous features of living organisms (e.g., DNA, metabolic pathways, structures), and argues that they cannot have been created by the known mechanisms of physical laws and chance. He claims that this is so because "laws" can only shift around or lose information, but do not produce it. Also, chance can produce *complex unspecified information*, or *unspecified complex information*, but not CSI. Dembski provides, in our opinion, a shaky mathematical analysis that he claims demonstrates that natural law and chance together cannot generate CSI. The ultimate thrust of his arguments and proofs is the assertion that once a system with CSI has been defined, it must be the result of an ID by a "Creator."

Dembski (1999) gave the examples: "A single letter [of the alphabet] is specified without being complex. A long sentence of random letters is complex without being specified. A Shakespearean sonnet is both complex and specified." Arbitrarily, Dembski defined CSI initially as being present in a "specified event" whose probability did not exceed 1 in 10^{150} (or 10^{-150}, an exceedingly small number!). He called this infinitesimal number the "universal probability bound" (UPB), corresponding to the inverse of the upper limit of... "the total number of [possible] specified events throughout cosmic history." More recently, Dembski redefined the UPB as the inverse of the total number of bit (binary) operations that could possibly have been performed in the entire history of the universe.

Note that if a coin is tossed 1000 times and the outcome recorded, the probability of a particular outcome (e.g., H,T,T,H,H,T,H,T,T,...) is 2^{-1000} or ca. 10^{-301}, thus for any specific outcome of the coin-tossing, the a priori probability that this one pattern occurred is thus ca. 10^{-301}, which is incredibly smaller than Dembski's UPB of 10^{-150}! This example illustrates Dembski's specious use of arbitrary, very small numbers to demonstrate SC.

Dembski (1998) formulated his *Law of Conservation of Information*: "Natural causes can only transmit CSI but never originate it." He gave four corollaries to his *Law*: "1) The SC in a closed system of natural causes remains constant or decreases. "2) The SC cannot be generated spontaneously, originate endogenously or organize itself (as these terms are used in origins-of-life research). "3) The SC in a closed system of natural causes either has been in the system eternally or was at some time added exogenously (implying that the system, though now closed, was not always closed). "4) In particular, any closed system of natural causes that is also of finite duration received whatever SC it contains before it became a closed system."

Dembski's Law is in fact a claim, and not a law. Its actual validity and utility are uncertain; it is neither widely used by scientists nor has it been cited in mainstream scientific literature (Baldwin 2005).

Dembski defined *information* as $-\log_2(p)$, where p represents the probability of an event. This is based on Shannon's usage in classical information theory. Dembski goes on to use $-\log_2(p)$ as a complexity measure.

Dembski (2005) proposed to view design inference as a statistical test to reject a chance hypothesis P on a space of outcomes, Ω. This test is based on the Kolmogorov complexity (cf. Glossary) of a pattern T that is exhibited by an event E that has occurred. Mathematically, E is a subset of Ω, the pattern T specifies a set of outcomes in Ω and E is a subset of T. Given a pattern T, the number of other patterns that may have Kolmogorov complexity no larger than that of T is denoted by $\varphi(T)$. The number $\varphi(T)$ thus provides a ranking of patterns from the simplest to the most complex. Dembski claims to find the upper bound $\varphi(T) \leq 10^{20}$ for the pattern T describing the bacterial flagellum. (Exactly what is this pattern [genomic, anatomical, etc.]?) Dembski's use of large numbers is very artful and generally lacks justification.

Dembski (2005) devised an original numerical definition for SC:

$$\sigma \equiv -\log_2[R \times \varphi(T) \times P(T)] = \log_2 \left\{ \frac{1}{[R \times \varphi(T) \times P(T)]} \right\} \qquad (10.1)$$

where σ is the numerical value of the SC of T. "R corresponds roughly [sic] to repeated attempts to create and discern a pattern, T." Dembski asserts that R can be bounded by 10^{120}. P(T) is the probability of observing the pattern T. (Note that if T is unique, there is no way of knowing P(T).)

Dembski claimed that the $\log_2$ argument can be used to infer design for a configuration: If there is a target pattern T that applies to the configuration and whose SC exceeds 1, it implies ID. This condition gives the inequality:

$$\left[10^{120} \times \varphi(T) \times P(T) \right] < \frac{1}{2} \qquad (10.2)$$

Dembski's mathematical approach to SC has been judged to be unusual, to say the least. In a review, Elsberry and Shallit (2003) commented that SC "...has not been defined formally in any reputable peer-reviewed mathematical journal, nor (to the best of our knowledge) adopted by any researcher in information theory." Dembski uses mathematics in describing SC in a nonrigorous, ad hoc manner. Dembski (2002b) himself has stated that he is not "...in the business of offering a strict mathematical proof for the inability of material mechanisms to generate specified complexity." His SC and its attendant mathematical formulae and criteria were created to argue for ID and an all-powerful designer (God). Perhaps what Dembski has not realized is that it is impossible to prove items of faith using science and mathematics. The converse is true, as well.

Baldwin (2003) wrote a critique of Dembski's concept, definition, and use of *complex specified information* (CSI). Baldwin stated that "the mathematical rigor supporting Dembski's case is lacking." He also points out that Dembski's assumption "...that if a pattern exists prior to a possibility being actualized, it must be causal" is flawed. Baldwin goes on to note that Dembski's argument that CSI cannot be created by a combination of chance and necessity (or mutation and natural selection) is an argument from ignorance. He points out that Dembski asserts that it cannot be done, but fails to demonstrate why. Finally, Baldwin considers Dembski's "Law of Conservation of Information," which Dembski calls a "strong proscriptive claim." Baldwin stated: "Since it is a claim and not a law, any arguments based on it can and should be rejected as pseudo-science."

10.4　SUMMARY

In this chapter, we have examined and using scientific evidence, refuted the faith-based theories of IC and SC. To believe in evolutionary theories is not incompatible with having faith. If God created life, He also created the genomic machinery to support evolution, complexity, and the human mind to try to understand it all. Darwin (1859) believed this.

Using modern genomics, it is now possible to create phylogenetic trees for species as well as their organs and organ-systems, allowing us to track evolution on all levels of scale: genomic, biochemical, cellular, physiological, and organismic.

In closing, we note that scientific theories are based on objective evidence, collected over time, and they are subject to change over time as new evidence accrues. The "theories" of IC and SC, ID, and creationism are based on a religious belief system that does not propose an evidence-based and genetically sound mechanism through which all living things (and viruses) change over time. "The scientific theory of evolution and the mechanism it proposes (natural selection) is 'scientific' because it is based on substantial and varied sources of evidence and because it is subject to evidence-based change. There are fundamental differences between scientific theories and religious theories in the way they ask and answer questions" (Rollins 2010).

11 Introduction to Complexity in Economic Systems

11.1 INTRODUCTION

While this text has focused mostly on the general properties of complexity and complex living systems, we would be remiss in not examining the very important CNLSs that probably receive the most public attention today, i.e., economic systems (ESs). You will see that three reasons why ESs must be viewed as complex are (1) their models contain many nodes and directed edges, and have nonlinear, time-variable, and parametric gains; (2) they are noisy; and (3) human behavior is an integral component of their dynamics. Human behavior on the individual level, in small groups, and *en masse* is modulated by communications and information. Humans make the decisions to invest, buy, sell, adjust prices, raise and lower taxes, grow, manufacture, catch, plant, harvest, mine, refine, stockpile and hoard goods based on information from surveys, in the media, on the Internet, and from interpersonal contacts, all of which influence individual, group-, or mass-behavior. The quantification of human attitudes and behavior on economic issues presents a daunting challenge to ES modelers.

ESs have taken the forefront of public (and media) attention in 2008–2009–2010 in the United States because of the spike in energy costs, and their linkage with worldwide increases in population, prices for food, durable manufactured goods, transportation, real estate, etc. Other factors affecting ESs include the rise in U.S. unemployment, the steady loss of manufacturing jobs to overseas facilities, and, more recently, the October 2008 lending crisis, which morphed into the 2009 U.S. recession going into 2010, as well as government attempts to break the recession by targeted stimulatory spending and lending. We already (3/10) have heard complaints (largely political) that the stimulus spending has been in vain because unemployment is still high. Do you think this is because not enough government stimulus money was spent, was spent in the wrong places, or because there are built-in time delays in any ES (in outputs resulting from various inputs), all of the preceding, or some other set of reasons? ESs are every bit as complicated as biological systems.

Several time-variable input parameters make ESs of all kinds nonstationary and exacerbate our difficulty in modeling them. These time-variable input parameters include but are not limited to the following: (1) A steadily increasing world population, which puts strains on markets, resources, and job availability.

(2) Limits on resources (e.g., water, food, energy) that fluctuate with planetary weather, which in turn is affected by such factors as *el niño*, sunspots, global warming, etc. Finite, nonrenewable energy resources are becoming exhausted and more expensive to extract, e.g., oil, natural gas, and coal. (3) The gradual exhaustion of easily mined metal ores and mineral deposits, and the increased energy cost of extracting these materials. (4) Human behavior in attempting to manage ESs. In summary, we view ESs as nonstationary, noisy CNLSs.

11.2 INTRODUCTION TO ECONOMIC SYSTEMS

11.2.1 INTRODUCTION

Anyone who watched CNN news and commentary on television in the last half of 2008 will recall that the U.S.'s and the world's economic problems were their "Issue #1." Unquestionably, ESs are complex. To understand them and predict their behavior, we need to be able to construct valid mathematical models of them. To do this, we must define the nodes (input variables, states and output variables), and unidirectional branches or edges (describing quantitative causal relationships between the variables).

Why are ESs complex? Humans run economies. ES complexity comes largely from human behavior in response to information (Arthur et al. 1997, Arthur 1999, Bowles and Gintis 2002). Human psychology, group dynamics and communication figure in economic decisions. The human emotions of greed, fear, anxiety, and panic are hard to quantify, but their effects are seen reflected in the stock market, commodities trading, and the loan industry. As we remarked above, economic complexity also arises from the nonstationary and noisy behavior of economic variables, including population growth and weather.

Pryor (1996) described the *structural complexity* of ESs in three ways: (1) An economic process or system is more complex if it requires an increase in information for its effective operation. (2) ES complexity increases if, in its model, there is an increase in the density of paths between nodes. (3) Increased complexity also occurs when there is an increase in the number of heterogeneous *modules* in it (see Chapter 4). Thus a metropolitan area's economy might be more complex by population growth, the inclusion of more heterogeneous ethnic groups, and by the spatial segregation of ethnic groups. Because the exchange of financial information is an integral part of both macro- and micro-ESs, the increased use of cell phones, PDAs, and the Internet has contributed to the increase of structural economic complexity. More sophisticated, targeted TV and Internet advertising has also affected demand-side economics.

Like many other classes of CNLSs, macro-ESs can be partitioned into interconnected modules (subsystems). We have the world economic system (WES), which can be subdivided into regional ESs: the U.S. ESs, the E.U. ESs, the Chinese ESs, etc. Clearly, these regional macro-ES models are all functionally linked, increasingly so with globalization and the Internet. Financial information and money can be exchanged at the click of a mouse. Each regional ES can be further subdivided into modules in order to facilitate analysis. Perhaps, one of the more important economic modules or subsystems in any national economy concerns energy (sources, sinks, reserves, costs, delays in production, etc.); another module deals with agriculture (food production, marketing, and distribution); others include health care, manufacturing, recycling, and stock markets. The energy module appears to be more highly connected to the agriculture (growing, transportation), manufacturing, and mining modules. Weather certainly affects food production (hence prices) and also alternate electrical energy production (wind turbines, solar cells, hydropower). It is clear that energy is required for mining, structural metal production (steel, aluminum, titanium, etc.), agriculture, manufacturing, precious metal (gold, platinum, silver, etc.) refining, transportation (personal automobiles, trucks, trains, ships, aircraft), and obviously, personal comfort. While a new nuclear energy plant is safer than its predecessors, and has a near-zero carbon footprint to operate, it is expensive, and its construction has a huge carbon footprint.

Economics as an academic discipline has been largely balkanized; i.e., a number of "schools of economic thought" and approaches to economic analysis have arisen since the seminal work of Adam Smith in 1776, *The Wealth of Nations*. In addition, economics is generally studied on two levels of scale; *macroeconomics* and *microeconomics*. Both levels are complex systems, and their behaviors are quite interdependent. Foster (2004) cited *neo-classical, evolutionary, post-Keynesian, Chicago* and *neo-Austrian* economic schools of thought. There is also a *thermodynamic* (second law) approach to economic analysis (aka *econophysics*) (Georgescu-Roegen 1971, Saslow 1999, Raine et al. 2006, Kafri 2008). Over 29 schools of economic thought were described in the Wikipedia (2009a) essay on Schools of Economics!

Foster (2004) made a strong case for viewing ESs as complex, dissipative structures that import free energy (in a thermodynamic sense) and export *entropy* in a way that enables them to self-organize their structures and functions. These systems are open systems that absorb information from their environments and create stores of knowledge that facilitate their growth. They are in fact, *complex adaptive systems* (CASs). Foster makes the point that adaptation, hence evolution, of an ES cannot occur if there is a very high degree of connectivity (edges) between its nodes, and there is no room for learned plasticity in the edge structure, which can affect behavior. Foster viewed complexity in his paper simplistically as the "...connective structure of a system."

Why are there so many approaches to modeling and analysis of ESs? The answer is simple; ESs are very complex, and most economists treat the facets of economics they are attracted to and are familiar with. By this, I do not mean their analyses are necessarily invalid, only limited in scope and effectiveness. The "big picture" must be considered in any CNLS. For example, in the nineteenth century, economists developed the *Quantity Theory of Money* (QTM) (inflation or deflation could be controlled by varying the quantity of money in circulation inversely with the level of prices). In 1936, economist John Maynard Keynes argued that the effect of circulating money on the prices of goods was virtually nil, illustrating that the sensitivity of prices to circulating money was small, invalidating the QTM for most economists. Keynes maintained that government budgetary and tax policy, and the direct control of investment had higher sensitivities with respect to the prices of goods. Then in the 1960s, economists Milton Friedman and Anna Schwartz refuted Keynes' approach and reestablished the validity of the QTM. Their ship floated until the 1990s, when other economic models focusing on growth and development surged. And so contemporary economists have also included the areas of public finance (taxation), labor, industrial organization, international economics, agriculture, information, and law in their modeling and analyses. In other words, economists have now justifiably expanded the scope of their models to include more variables and parameters.

Horgan (2008) commented that: "Economics [economic analysis] keeps lurching faddishly from one approach to another rather than converging on a single paradigm the way that more successful scientific fields such as nuclear physics or molecular biology do. The obvious reason is that economies are fantastically complicated in comparison to atomic nuclei or galaxies or *E. coli*." In our opinion, to be successful studies of ESs must be interdisciplinary, drawing on complex systems theory, chaos theory (Horgan's *Chaoplexy*), dynamic modeling, as well as mathematical approaches adapted from ecology and evolutionary biology, and very importantly, psychology (after all, fallible humans operate economies). Perhaps economics should be viewed more from a social science viewpoint. This approach is, in fact, taken in the interesting text, *Complex Adaptive Systems: An Introduction to Computational Models of Social Life*, by Miller and Page (2007). The role of CASs in economics is treated in this interesting text, but unfortunately there is a dearth of mathematical detail in describing various models.

11.2.2 BASIC ECONOMICS: STEADY-STATE SUPPLY AND DEMAND

One of the most fundamental concepts in *neo-classical economic* (NCE) *theory* are the "laws" of *supply*, *demand*, and *supply and demand* (S&D). These "laws" describe steady-state (SS) or equilibrium relations, and are usually analyzed in the framework of *microeconomics*. NCE models rely on the assumption that there is no trading at all unless and until all prices reach equilibrium, at which point, all sellers and buyers simply exchange a good at a certain price. In an NCE model, there is no excess demand or shortage of goods, labor, or services. Also, unemployment is not considered. Probably, the most severe criticism of NCE models is that they do not consider money; it does not and cannot appear. There is no capital accumulation, earnings, savings, taxation, etc. The price P in a S&D model is only a label (McCauley and Kuffner 2004).

We shall review the generalities of NCE systems, and discuss how their principles might guide us to formulating an SS mathematical model for an ES. First, some terms we will use:

> *Good* (*n.*) (*in economics*): A good is any object or service that increases the utility, directly or indirectly, of the consumer. A good is manufactured, grown, crafted, etc. and generally sold for profit. A good is *supplied* to meet a general *demand*.

Price: The cost a consumer must pay for goods or services.

Demand: The (average) quantity of a certain good or service desired by consumers.

Service: Work done for others as an occupation or business. E.g., income tax preparation, mowing grass, etc.

Supply: How much of a good or service the market can supply at a given price?

Production Function: An equation that expresses the fact that a producer's output depends on the quantity of raw material inputs it employs. Inputs can be combined in different proportions to produce a given level of output.

Utility Function: An equation that attempts to model the pleasure or satisfaction households (basic microeconomic consuming units) derive from consumption. It depends on the products purchased and how they are consumed. Utility functions provide a general description of the household's preferences between all of the paired alternatives it might be presented with.

These terms are related functionally, and can vary parametrically. S&D *curves* are used in *microeconomics* to illustrate how the supply of a *good*, $P_S(Q)$, has a common, SS solution (price P_o) with the consumer demand for that good, $P_D(Q)$. Various production and *utility functions* taken with certain assumptions about human behavior lead to S&D curves such as those illustrated in Figure 11.1. Note that not all S&D curves look alike. In general, most *supply curves* (S) have positive slopes, while most *demand curves* have negative slopes. For example, the

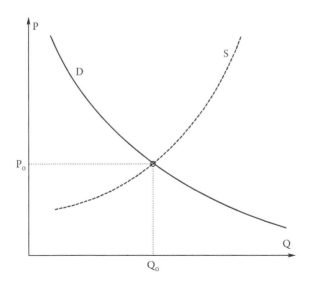

FIGURE 11.1   Representative supply and demand curves for a good or service. Horizontal axis: Q = *quantity of a good or service*. Vertical axis: P = *price of a good or service*.

demand curve, $P_D(Q)$, represents the *quantity of a good or service* (Q) consumers are willing and able to purchase at various *prices* (P). $P_D(Q)$ is also a function of geographic location (not much demand for air conditioners in the Antarctic), the consumer population density, demographics, seasonal need (e.g., snow shovels, seed corn), etc. The supply curve, $P_S(Q)$, illustrates the SS relationship between the market price and the amount of a good produced (Q). (Suppliers are likely to produce more of a good at higher prices; they try to maximize their profit.)

In the simple, SS demand curve illustrated in Figure 11.1, the price (P) of a good is seen to be a decreasing function of the quantity (Q) willing to be purchased by consumers, *all other factors remaining constant (ceteris paribus)*. As the price increases, so does the opportunity cost for that good. The SS supply curve in Figure 11.1 generally has a positive slope, meaning that producers will supply more goods at a higher price because selling a higher quantity at a higher price increases their profit, *ceteris paribus*.

The *SS S&D equilibrium* is at (P_o, Q_o) in Figure 11.1. This intersection illustrates the *law of S&D* for a good. The $P_S(Q)$ and $P_D(Q)$ curves have a unique intersection where supply price of a good, P, equals the demand price (P_o). At this point, the amount of a good being supplied equals, in theory, the amount demanded by the consumers. S&D curves represent SS analysis, and are only valid in the short-term. Note again that ESs are nonstationary and noisy and, as we argue below, are best modeled and studied using dynamic models based on large sets of nonlinear ODEs.

A challenge in understanding SS S&D is to consider what happens to prices when other factors cause a shift in the SS demand curve, $P_D = f(Q)$, while the supply curve, $P_S = g(Q)$, remains fixed. Such a scenario is illustrated generally in Figure 11.2. Here, the demand curve shifts to the right, from **D** to **D'**. As a result of generally increasing consumer demand, a new equilibrium, Q', P' is established. Along with the increased demand, the price increases by $\Delta P = (P' - P_o) > 0$, and the number of goods sold also increases by $\Delta Q = (Q' - Q_o) > 0$. The converse relation holds as well, if **D'** shifts to the left, then both ΔP and ΔQ are negative.

In some cases, the supply curve can be vertical; that is, the quantity supplied to the market is fixed (or nearly fixed) by the manufacturer, regardless of market price. (A short-term example of a nearly horizontal supply curve has been oil production regulated by OPEC, land availability in a region, or diamond production regulated by De Beers.) A nearly vertical supply curve, S_L, is shown in Figure 11.3. When demand increases (shifts to the right), $\Delta P > 0$.

A supply curve can also be double-valued. Figure 11.3 shows such a curve (S_C) that has been observed in the labor market and the crude oil market (Samuelson and Nordhaus 2001). After the 1973 oil crisis, many OPEC countries *decreased* their production of oil even though prices increased. This was due to *human behavior*: Why sell a valuable asset? Save it to market later at a higher price (here, the seller is hoarding or warehousing a commodity).

The *price elasticity of S&D* is an important concept in static S&D theory. For example, if a vendor decides to increase the price of his good, how will this affect his sales revenue? Will the increased unit price offset the likely decrease in

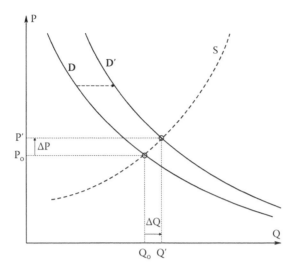

FIGURE 11.2 Supply and demand curves illustrating the fractional increases in P and Q when the demand curve **D** shifts to the right to **D'** (increasing Q at constant P). Note that $\Delta P/\Delta Q > 0$.

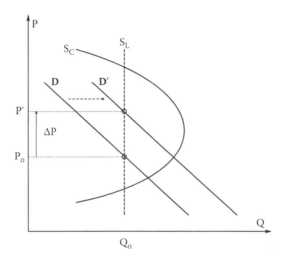

FIGURE 11.3 Illustration of supply and demand when the supply curve is vertical; that is, the quantity supplied to the market is fixed (or nearly fixed) by the manufacturer, regardless of market price. When the demand curve shifts to the right (to **D'**), Q remains at Q_o and $\Delta P > 0$. A double-valued supply curve is shown in S_C. See text for discussion.

sales volume? Or, if a government imposes the tax on a good, thereby effectively increasing the price to consumers, how will this affect the quantity demanded?

One way to define *elasticity* is the percentage change in one variable divided by the percentage change in another variable. Thus, elasticity is basically a system sensitivity as defined in systems engineering (see *Glossary*). For example,

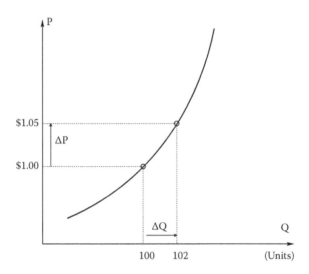

FIGURE 11.4 Illustration of supply elasticity. See text for discussion.

if the price of a widget is raised from \$1.00 to \$1.05, and the quantity supplied rises from 100 to 102 u, the supply slope at a point is $2.00/0.05 = 40$ u/dollar. (See Figure 11.4.) Because the elasticity is defined in terms of percentages, the quantity of goods sold increased by 2%, and the price increase was 5%, so the price elasticity is $2/5 = 0.4$. Note that for a perfectly inelastic supply, the supply curve is vertical, as shown by S_L in Figure 11.3.

The income elasticity of demand (IED) tells us how the demand for a good will change if purchaser income is increased (or decreased). (Figure 11.1 illustrates a generic demand curve.) For example, how much would the demand for a giant plasma TV screen change if the average purchaser income increased by 10%? Assuming it is positive, the increase in demand would be reflected by a positive (up and right) shift of the demand curve, resulting in a positive ΔP and ΔQ. This is shown in Figure 11.2. Thus, $IED \equiv (\Delta Q/Q_o)/(\Delta P/P_o)$ is seen to be sensitivity.

The *cross elasticity of demand* (CED) is another important *sensitivity* used in economic models. Here, we calculate the percent Q of a good demanded in response to a percent change in the price of another competing or substitute (alternative) good, or the price of a complement good that must be used with the good in question (e.g., a set of loudspeakers used with a sound system). Thus the sensitivity, $CED \equiv (\Delta Q/Q_1)/(\Delta P/P_c)$. An example given of CED is the decrease in the demand for a certain brand of SUV by 20% when the price of gasoline increases by 30%. Here, the $CED = -0.667$.

11.2.3 FORRESTER'S VIEWS

The challenge of formulating quantitative, dynamic models of complex ESs (CESs) was recognized over 54 years ago by economist/mathematician Forrester (2003) at MIT. In 1956, Forrester commented on the factors governing the time-dependent interplay of money, materials, and information flow in ESs and industrial organizations. The first point he made was that CESs contain causal feedback loops. He wrote: "The flows of money, materials, and information feed one another around closed re-entering paths." He went on to state that such systems often develop bounded oscillations (limit cycles), a well-known property of CNLSs having feedback paths and time delays (see Chapter 3). Forrester noted some other factors that operate in CESs:

The human factor of resistance to change, including habit, inertia, prejudices, traditions, etc. (Quantifying these properties lies more in the realm of psychology than economics because they involve human behavior.)

Accumulation (of materials or cash): Accumulation (saving, investing, hoarding, stockpiling, warehousing) also involves human decision making.

Delays (transport lags). (We have illustrated how transport lags can destabilize even simple, nonlinear systems in Section 3.5.3.) In 1956, Forrester cited transport lags in the following ESs: (1) The delay between actual sales and accounting reports submitted to corporate decision-makers. (2) The time required to process orders for goods, including building inventories of components. (3) The delay between the decision to make a good, and its production (i.e., manufacturing throughput time). (4) The length of time from the decision to plant a crop to the harvest of the mature crop. (5) The time between collection of taxes and disbursement of governmental funds. (6) The length of time from the decision to build new: factories, oil wells, refineries, pipelines, wind turbine "farms," solar cell farms, nuclear plants, power distribution lines, etc. (7) Other delay generators in a model ES include: mail delays, freight shipment transit times, time required for changes in human behavior in response to social or economic changes (generally, a tipping point is seen). (8) The $800B economic stimulus package appropriated by Congress in early 2009 appears to have built-in bureaucratic delays in its distribution to target agencies, causing lags in job creation and slowing economic recovery.

Quantizing, in which the periodic availability of financial information (e.g., monthly and quarterly economic reports) builds lags into decision making. This is actually a discrete, sampled input.

Policy and decision-making criteria. Forrester argues that these criteria have first-order effects on "amplification characteristics of the system" [i.e., path gains].

In commenting on the interrelationship of goods, money, information and labor, Forrester argued that previous models (≤1956) had generally neglected to adequately interrelate these variables. At this writing, evidently this is still largely true, or we would be able to understand and manage the U.S. economy more competently than insisting on tax cuts on one hand while attempting to implement expensive wars, health care programs, and multibillion dollar corporate bail-out loans on the other. Giving tax rebates in 2008 and arguing for a reduction on

Federal fuel taxes was a short-term palliative for a currently sinking U.S. economy, as well as a political gesture.

So much of ES behavior depends on human behavior in response to information. It is evident that such behaviors must be included in any comprehensive dynamic model of the U.S. economy. As we stated above, the challenge remains: How do you model human emotionally driven behavior such as: fear, greed, confidence/pessimism, dissatisfaction, satisfaction, happiness/paranoia, etc. in ESs, and how they spread among a population? Forrester commented: "I believe that many of the characteristics of a proper model of the national economy depend on deeply ingrained mental attitudes, which may change with time constants no shorter than one or two generations of the population." That is, one's (micro) economic attitudes and behavior can be influenced by one's parent's attitudes, or the attitudes of one's peers or ethnic group. Modern attitudes include factors such as the encouragement of personal deficit spending using credit cards. Buy and enjoy now, pay later (if you can). Remember the parable of the *Grasshopper and the Ant*?

Forrester concluded his prescient paper by discussing the factors he felt should be considered in ESs dynamic modeling. He noted: "Almost every characteristic that one examines in the economic system is highly nonlinear." (We should add, "...and time-variable," as well.) Without saying so explicitly, he argued that ESs are CNLSs. He called for the use of sets of nonlinear ODEs as modeling tools. All this over 53 years ago, on the threshold of the computer revolution, and at the dawn of organized complex system thinking.

Four texts that have appeared since 1990 deal specifically with the dynamic modeling of ESs: Brock and Malliaris (1992) *Differential Equations, Stability and Chaos in Dynamic Economics*, Ruth and Hannon (1997) *Modeling Dynamic Economic Systems*, Neck (2003) *Modeling and Control of Economic Systems*, Barnett et al. (2004) *Economic Complexity: Non-Linear Dynamics, Multi-Agents Economies, and Learning*, and Zhang (2005) *Differential Equations, Bifurcations and Chaos in Economics*. Forrester's call is finally being heeded. Hopefully, a few economists have the mathematical background to effectively extract information from such texts.

11.2.4 DYNAMIC MODELS OF ECONOMIC SYSTEMS

Forrester's seminal (1956) paper called for the dynamic modeling of ESs. It has been noted that that in general, neo-classical economic (NCE) theory ignores processes that take time to occur. This includes for example, delays on the supply side, including for example, the time for a crop to mature, the time to build a new wind farm and put it on-line, or the time to develop a new fuel-efficient automobile engine. It is also clear that the dynamic path of the economy cannot be ignored. Economists generally do not consider time when analyzing S&D, or any other key variables. That is, any economist who assumes an SS into the future, walks a stupid and dangerous path when dealing with a nonstationary, CNLS.

NCE does use mathematical models in static analysis; these models use simultaneous, linear, *algebraic equations* and linear algebra (matrix methods) to reach solutions. Forrester argued that ESs must be modeled using sets of *nonlinear differential equations*, and that such dynamic models will behave chaotically, ideally following limit cycle attractors similar to real-world ES behavior. We note that in extrapolating from models to the real world, economic variables are likely to be in disequilibrium—even in the absence of external transient inputs. The conditions that NCEs have proven to apply at equilibrium will thus be irrelevant in actual ESs. It is clear that SS economic analysis cannot be used as a simplified proxy for dynamic analysis. The real question is whether we can sufficiently control such unstable, chaotic systems. Can we constrain their instability (chaos) within acceptable bounds while producing desired outcomes?

Some of the time-dependent processes that ought to be included in dynamic microeconomic models include (1) The *functional lifetime* (FL) of a good. (Does FL follow Poisson statistics or some other distribution? Think of automobile batteries as an example.) What affects the human motivation to replace a failed good? Or do we buy a new, improved model? (2) Planned obsolescence in design. (We see this in the automobile and appliance industries.) (3) The role of an advertising campaign in shifting the demand curve. The "new model is better" marketing strategy is applied to all kinds of durable goods ranging from air conditioners, automobiles, to computers, furniture, and windows. A twenty-first century advertising approach is "greeness"; buyers must be convinced that the improved energy efficiency of a good is worth the capital outlay for future operational savings. (4) Population growth is important because it presents an ever-increasing need for tax-based resources such as health care, unemployment compensation, municipal services (water, sewer, power, refuse collection, schools, etc.), as well as providing an increasing source of labor and consumers.

There has been a slow trend toward dynamic modeling of ESs in the past 50 years. Wisely, some modelers have started with simpler model structures (three ODEs) (Abta et al. 2008), while others have constructed more complicated economic models (ca. 100 ODEs with more than 80 parameters), and then have had to reduce their model to a more realistic architecture (Olenev 2007). Olenev's model was faced with many unknown path parameters that had to be estimated, and there were unknown initial conditions on many model (node) variables that also had to be determined. Eight nonlinear ODEs were described in his paper (volume restrictions prevented inclusion of more detail). Both Olenev's and Abta's models were interesting because they described the flow of capital.

The relatively simple model of Abta et al. (2008) was used to study limit cycles in a business cycle model. Their three ODEs are summarized below:

$$\dot{Y} = \alpha\left[I(Y) - \delta_1 K - (\beta_1 + \beta_2)R - l_1 Y\right] \tag{11.1A}$$

$$\dot{K} = I\left[Y(t - \tau)\right] - (\delta + \delta_1)K - \beta_1 R \tag{11.1B}$$

$$\dot{R} = \beta\left[I_2 Y - \beta_3 R - M\right] \tag{11.1C}$$

where

  Y is the gross product
  K is the capital stock
  R is the interest rate
  τ is the delay for new capital to be installed
  α is the adjustment coefficient in goods market
  β is the adjustment coefficient in money market
  I(*) is the linear investment function
  S(*) is the linear savings function
  M is the constant money supply

[We have used the notation from the Abta et al. paper.]

Representation of this simple economic system model by a SFG (cf. Figure 11.5) shows that it has three integrators and eight nodes. Abta et al. used the "Kaldor-type investment function":

$$I(Y) = \frac{\exp(Y)}{1+\exp(Y)} \tag{11.2}$$

Notice that if $0 \le Y \ll 1$, $I(Y) \cong \frac{1}{2}(1+Y) \cong \frac{1}{2}$, and for $Y \gg 1$, $I(Y) \rightarrow 1$. Thus, the Kaldor function saturates at 1 for $Y \gg 1$. If $Y \ll 1$, $I(Y)$ is linear, and the model is in fact, a linear, cubic system. Inspection shows that the complicated linear SFG has five negative feedback loops; one has a delay. Their linear loop gains are: $A_{Ln}(s) = -(\delta + \delta_1)/s$, $-\alpha\delta_1 (1/4) e^{-s\tau}/s^2$, $-\beta\beta_3/s$, $-\alpha l_1/s$, and $-\beta(\beta_1 + \beta_2)l_2/s^2$. There are two positive feedback loops with linear loop gains: $A_{Lp}(s) = +\alpha(1/4)/s$ and $+\alpha\beta(\beta_1 + \beta_2)l_2/s^2$.

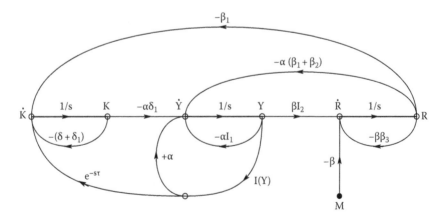

FIGURE 11.5 Three-state, nonlinear SFG based on the ODEs of Abta et al. (2008) used to model oscillations in the business cycle. Although all feedback loop gains are negative, instability is the result of the transport lag in the feedback gain between Y and $\dot{K}$.

Abta et al. found that when $\alpha = 3$, $\beta = 2$, $\delta = 0.1$, $\delta_1 = 0.5$, $M = 0.05$, $l_1 = 0.2$, $l_2 = 0.1$, $\beta_1 = \beta_2 = \beta_3 = 0.2$, the system exhibited SS, bounded, limit-cycle oscillations for $\tau \geq 1.7975$. (ICs and parameter units were not specified.) The results of this paper were not remarkable (otherwise linear, cubic, feedback systems with delay(s) can easily be unstable); what was of interest, however, is the fact the authors examined the stability of a simple ES model known to have a closed attractor for a sufficiently large delay in capital investment, Y.

Olenev's dynamic model for a regional economy used ODEs on the following eight variables (only the eight X variables are shown in his paper): $Q_Y^X(t)$ = shadow products in sector X, $W_X(t)$ = stock of open ("white") money of economic agent X, B^X = shadow incomes from sector X, Z^X = debts of agent X to bank system B, Q_X^L = stock for final product X of timber industry complex directed to household markets, L, p_X^L = consumer price index on product X, s_L^X = open wage sector X, W^G = stock of money in the regional consolidated budget. Production sectors X, Y, and Z were assumed. Additional variables are obtained by substituting Y or Z for X in the preceding list of variables. Space prevents us from writing the Olenev system's ODEs in detail. Olenev stated that "The main purpose of the paper is to illustrate the method, rather than present a state-of-the-art analysis."

Another approach to dynamic modeling of ESs is through the use on *agent-based models* (ABMs). (See the excellent review on ABMs by Macal and North [2006].) Instead of using NL ODEs, as in Equations 11.1, the ABM uses interconnected *agent models* to mimic human behaviors in the ES. Many different kinds of agents are incorporated into the economic ABM. For example, some agents are *consumers* (they decide whether to buy, hoard, or save their money). Others are *manufacturers* (they decide whether to manufacture and/or inventory goods, lay off workers). Still others can be *marketers/distributors*, others can be *suppliers of raw materials*, *investors*, etc. Each agent is a self-contained, discrete entity with a designed set of characteristics and rules governing its I/O behaviors and decision-making capability. In the ABM, agents interact with other agents, and have the ability to recognize the traits of other agents. An agent may be goal-directed, having goals to achieve in its behaviors (e.g., maximize manufacturing profit). It is also autonomous, self-directed, and has the ability to learn and adapt its behaviors, based on experience. This is quite a programming challenge for the designers of economic ABMs. Because of the complexity of agent-based, dynamic economic models, they are generally simulated on enhanced, multi-processor computers, not PCs. (See Tesfatsion [2005a,b, 2002], and Holland and Miller [1991] to explore more material on economic agent-based modeling.) How would you craft the attributes of a stock trader agent? Agent-based modeling is treated in more detail in Section 11.2.8.

11.2.5   What We Should Know about Economic Complexity

To be a competent economist in the twenty-first century, one should have interdisciplinary skills that embrace classical economic theory, mathematics (including complex systems theory, differential calculus, chaos theory, and the dynamics of nonlinear systems), and also psychology. Quite a requirement, but then competent biomedical engineers should know differential and integral calculus, ODEs, engineering systems analysis, physiology, as well as introductory anatomy, cell biology, biochemistry, and complex systems theory, etc.

To form meaningful dynamic models of ESs, one must be able to identify (define) input, output, and internal variables (node parameters), and equally important, the branch (edge) functions relating the variables. One should also have measured relevant system sensitivities. In short, one should be able to build a valid dynamic model of a selected ES and then run simulations. All models must be verified from known data before they can be used predictively.

Durlauf (1997) addressed the need for economists to have a background in complex systems theory. He used the example of a stock market having many traders with idiosyncratic beliefs about the future behavior of stock prices. These traders react to common information and exchange information rapidly through electronic media, but ultimately, stock prices are determined by a large number of decentralized buy-and-sell decisions (a scenario that can be simulated by ABMs). Durlauf commented that there is a feedback between the aggregate characteristic of the ES under analysis and the individual (microeconomic) effectors which comprise that environment. "Movements in stock prices influence the beliefs of individual traders and in turn influence their subsequent decisions." Information flow was seen to be critical.

Durlauf noted that *conformity effects*, in which *an individual's perceived benefit from a choice increases with the percentage of his or her friends who make the same choice*, must figure in economic dynamic models. Conformity effects are an example of autocatalytic, positive feedbacks between (human) economic actors. Another example of positive feedback, first noted by Forrester, is *economic attitude* passed from parents to offspring in a family. Such familial (group) feedback can determine education and job choices, as well as spending/saving patterns. Such mundane things as the choice of automobile make to purchase (e.g., Ford or Chevy) can also be influenced by family.

Durlauf offered two messages for economic policymakers: First, interdependence among various component subsystems can create multiple types of internally consistent overall behavior. As a result, ES behavior can become stuck in undesirable SSs. "Such undesirable steady-states may include high levels of social pathologies or inferior technology choices." Second, "…the consequences of policies will depend critically on the nature of the interdependencies." He pointed out that the effects of different policies may be highly nonlinear, producing unexpected, chaotic results. He stated: "At a minimum, detailed empirical studies which underlie conventional policy analysis should prove to be even more valuable in complex environments." Which is to suggest, to manipulate CESs, one must know estimates of the system's relevant sensitivities.

Lansing (2003) was of the opinion that complexity theory now has an impact on economics: there is "a shift from equilibrium models constructed with differential equations to nonlinear dynamics, as researchers recognize that economies, like ecosystems, may never settle down into an equilibrium." This trend was noted by Arthur (1999), who argued that: "complexity economics is not a temporary adjunct to static economic theory, but theory at a more general level, out-of-equilibrium level. The approach is making itself felt in every area of economics: game theory, the theory of money and finance, learning in the economy, economic history, the evolution of trading networks, the stability of the economy, and political economy." (See also Arthur et al. 1997 and Kauffman 1995.)

Krugman (1994) urged the use of the metaphoric phrase "complex landscapes" in dynamic economic analysis. *Landscape* in this context refers to the N-dimensional phase plane behavior of an N-state ES as it seeks equilibrium, given a transient input, or as an IVP. (Note that a three-state system has a 3-D landscape surface—similar to a contour map.) Krugman pointed out that CNLS ESs can have *many* point attractors, "basins of attraction," and closed attractors for each variable. Hence, a complex landscape can exist for a complex ES. Krugman also remarked: "The most provocative claim of the prophets of complexity is that complex systems often exhibit spontaneous properties of self-organization, in at least two senses: starting from disordered initial conditions they tend to move to highly ordered behavior, and at least in a statistical sense this behavior exhibits surprisingly simple regularities." Self-organization or learning behavior in CESs is not unexpected. After all, humans with brains operate them, and brains are adaptive CNLSs that can learn.

An assessment of economic complexity was treated in terms of complexity theory and measures in a paper by Lopes et al. (2008). Lopes et al. gave 12 algebraic indicators of connectedness (hence, complexity) based on the large $\mathbf{A}$ matrices describing the sets of N *linear, simultaneous equations* describing SS, input–output relations in ESs associated with OECD countries. For example, one of the simpler measures given to quantify complexity in ESs is the *inverse determinant measure*, IDET: $IDET = 1/|\mathbf{I} - \mathbf{A}|$. *Where* $\mathbf{I}$ is the $(N \times N)$ unit matrix. Another simple measure they applied is the *mean intermediate coefficients total per sector*, MIPS: $MIPS \equiv \mathbf{i}^T \mathbf{A} \mathbf{i}/n$. *Where*: $\mathbf{i}^T$ is the transpose of a unit vector, $\mathbf{i}$, of appropriate dimension, and n is the number of sectors.

They compared the results of their 12 measures applied to standard sets of economic data from 9 countries in the 1970s and the 1990s. Not unexpectedly, they found that large economies (United States and Japan) are more "intensely connected" and thus more "complex" than small ones (Netherlands, Denmark). The Lopes et al. paper is significant because it used quantitative measures of assessing complexity in ESs based on linear algebraic models of the economies. (Space prevents us from describing the Lopes criteria in detail.)

Raine et al. (2006) wrote an interesting paper in which they speculated on why biological and socio-ESs expand their structures (and populations) with the result that they use increasing amounts of "free energy" and associated materials. In biological systems, *free energy* is the thermodynamic energy released by breaking (or making) chemical bonds plus the energy from solar radiation. In socio-ESs, energy can be from exothermic chemical reactions (combustion), nuclear reactors, wind, solar, tides, etc. This energy is primarily converted to and distributed as transmitted electricity. (It is certainly not "free.") They address this natural increase in complexity from the viewpoint of a modified *Second Law of Thermodynamics*.

The traditional *second law of thermodynamics* states that any real process in an isolated system can proceed only in a direction that results in an *entropy increase*. Schneider and Kay (1994) (cited by Raine et al. 2006) reformulated the second law to apply to living systems: "The thermodynamic principle, which governs the behavior of systems is that, as they are moved away from equilibrium, they will utilize all avenues to counter the applied gradients. As the applied gradients increase, so does the system's ability to oppose further degradation." They asserted that ecosystems develop in ways that systematically increase their ability to degrade incoming solar and chemical free energy. According to Raine et al., "This reformulated second law is appropriate for the analysis of spatially fixed and open systems such as plants and forests [e.g., ecosystems]. Such systems are subject to two opposing gradients: the thermodynamic degradation gradient and the incoming solar radiation gradient." Solar radiation provides the free energy required for these systems to oppose thermodynamic degradation inherent in the entropy law.

Raine et al. (2006) observed that biological, ecological, and ESs are never near classical thermodynamic equilibrium. The ability to do self-organization requires structures (and algorithms) that throughput free energy in a way that resists the thermodynamic gradient. These self-organized processes in complex systems emerge and adapt through the creation of endogenous feedback paths, subsystems or modules in order to maximize the efficiency of energy utilization and resist the degradation process. These authors maintained that it is the role of knowledge (information) that differentiates economic evolution (development) from biological evolution. They cited three ways that CESs differ from their biological counterparts: (1) By sharing knowledge (information transmission). (2) Knowledge interactions imply that experimental mechanisms (R&D) may contribute to ES evolution (applied science can create information). (3) Growth limits are not reached as easily as in biosystems because of the continual innovation created by new knowledge (e.g., the better use of materials in manufacturing, the use of new materials to make goods). Economic growth also results from technical improvements in the physical means to process energy. The accumulation of knowledge in ESs is part of their evolutionary process.

Raine et al. concluded with the thoughts that: "Economic systems are characterized by the explicit use of knowledge [information] in harnessing energy, and consequently creating value [of goods and services]." "The reformatted second law suggests that knowledge structures be considered as unique complements that allow socio-ESs to utilize more energy than other biological and ecological species. Knowledge coevolves with energy using structures, facilitating economic growth through the use of functional and organizational rules under the governance of social institutions."

While the modified second law of thermodynamics provides an interesting viewpoint on socio-ESs, we view the major contribution of Raine et al. to be the viewpoint that knowledge and information are critical parameters in ESs. Only humans can create, process and store abstract information, and humans run ESs.

We submit that the viewpoint of Raine et al. is valid for ESs, but misses the mark in their consideration of biological systems. Genomic information is critical in directing life processes. It is the equivalent of their "knowledge" in CESs. Genomic information is passed on (or communicated) generationally, and its alteration is necessary for evolution to occur.

11.2.6 INFORMATION, ENTROPY, AND UNCERTAINTY IN ECONOMIC SYSTEMS

At this writing, *information theory* (IT) is now enjoying its 62nd birthday. Originally conceived by Claude Shannon at Bell Labs in 1948, it rapidly evolved into a mature branch of applied mathematics that, among other things, has been applied to the design of digital data transmission codes. It has also been applied to the disciplines of biology (Quastler 1958) and economics (Raine et al. 2006).

Let us examine the basic Shannon–Weaver definition of *information content*:

$$H(x) \equiv -\sum_{i=1}^{r} p(i) \log_2 p(i) \qquad (11.3)$$

where
  x is a *classification* with r categories i, and associated probabilities $p(i)$
  $H(x)$ is defined as the *information content* of x
  $H(x)$ is also called "uncertainty" in IT-speak

Quoting Quastler (1958): "The information function looks (except for a scale factor) like Boltzmann's entropy function; this is not a mere coincidence. The physical entropy is the amount of uncertainty associated with a state of a system, provided all states which are physically distinguishable are considered as different, that is, if the categorization is taken with the finest grain possible." He continued to state that "...physical entropy is an upper bound of the information functions which can be associated with a given situation, but it is a very high upper bound, usually very far from the actual value." Quastler preferred not to use the word "entropy" as synonymous with "information."

Some of the properties of H(x) are

  1. *Independence*: Let i be one of the possible categories of an event, x, $p(i)$ the associated probability, and $F(i)$ the contribution of the ith category to the overall uncertainty. $F(i)$ should be a function only of $p(i)$: The function:

$$F(i) = -p(i) \log_2 p(i) \qquad (11.4)$$

  meets this requirement.

2. *Continuity*: $F(i)$ is a continuous function of $p(i)$. That is, a small change in $p(i)$ should result in a small change in $F(i)$.
3. *Superposition*: It can be shown that the total information derived from two *statistically independent sources* is the sum of the individual uncertainties, i.e., $H(x, y) = H(x) + H(y)$.
4. *Natural scale*: If one plots $F(p)$ vs. p, $0 \le p \le 1$, one observes that $F(0) = F(1) = 0$, and $F(p)$ has a peak of 0.53 for $p = 0.37$.
5. *Averaging*: $F((p_1 + p_2)/2) \ge (1/2)[F(p_1) + F(p_2)]$. A special case occurs when all $p_i = 1/r$. Then

$$\max H(x) = \sum_{i=1}^{r} \left(\frac{1}{r} \right) \log_2 \left(\frac{1}{r} \right) = -r \left(\frac{1}{r} \right) \log_2 \left(\frac{1}{r} \right) = \log_2(r) \qquad (11.5)$$

If there is a binary classification, $r = 2$ and $\max H(x) = 1$.
6. *The effect of pooling*: $F(p_1 + p_2) \le F(p_1) + F(p_2)$. This means that pooling of two classes in one equivalent class reduces uncertainty.

Entropy described from a statistical mechanics viewpoint was first defined by Boltzmann in 1896 as being proportional to the logarithm of the number of microscopic configurations that result in the observed macroscopic description of the thermodynamic system. That is: $S \equiv -k \ln(\Psi)$. Where: k is Boltzmann's constant (SI units: $k = 1.381 \times 10^{-23}$ J/K), and Ψ is the number of microstates corresponding to the observed thermodynamic macrostate. This (microscopic) definition of entropy is considered to be the fundamental definition of entropy because all other definitions of entropy can be derived from it, but not *vice versa*. Statistical mechanics explains entropy as the amount of uncertainty about a system after its observable macroscopic properties (e.g., temperature, volume, phase) have been quantified. S measures the degree to which the probability of the system is distributed over different possible quantum states. The more states available to the system with higher probability, the greater the entropy. A closed system will tend to the macrostate that has the largest number of corresponding accessible microstates (Schneider and Kay 1994).

Entropy measures have been reformulated and applied to ESs, ecological, biological, and social systems. Schneider and Kay (1994) developed a "restated second law of thermodynamics" that they applied to living systems, and which Raine et al. (2006) extended to ESs. The Schneider and Kay law is: "The thermodynamic principle which governs the behavior of [complex] systems is that, as they are moved away from equilibrium, they will utilize all avenues available to counter the applied gradients. As the applied gradients increase, so does the system's ability to oppose further movement from equilibrium."

Raine et al. stated insightfully: "Economic systems are characterized by the explicit use of knowledge in harnessing energy, and consequently creating value. Traditional economic analysis has been reserved in its incorporation of energy, usually only as a factor of production. The reformulated second law suggests that knowledge structures be considered as unique complements that allow socio-economic systems to utilize more energy than other biological and ecological species [ecosystems]." These authors have argued that the growth of knowledge as a part of the process of increasing structural complexity in ESs may be an outcome of the second law.

And so we should ask, does information have a thermodynamic equivalency to energy? It is a big conceptual jump from PV = nRT thermodynamics, Carnot cycles and entropy, to information theory; entropy (uncertainty) derived for noisy binary communications channels, and information used in economic system development, yet they are all linked.

11.2.7   TIPPING POINTS IN ECONOMIC SYSTEMS:
RECESSION, INFLATION, AND STAGFLATION

There are several examples of tipping points in ESs: One is the *onset of a recession/depression*. Human behavior figures large in these scenarios. In a recession, there is a contraction of the business cycle, the economy stagnates and there is high unemployment, little circulating cash, factories close, the stock market plummets, etc. A sustained, severe recession may morph into a depression with massive unemployment and collapse of markets. A rule of thumb is that a recession occurs when the nation's gross domestic product (GDP) growth is negative for two or more consecutive quarters. Recessions are the result of the falling demand for goods; buyers lack the available funds and motivation to purchase certain goods. Prices generally fall, but not enough to stimulate turn-over. Lenders lack cash to loan, and are unwilling to extend poorly secured credit. U.S. unemployment soars as manufacturers cutback domestic operations due to a lack of demand for goods, or move overseas to find cheaper labor. The morphing of a recession into a depression involves autocatalytic or positive feedback behavior. (For an in-depth analysis of many of the factors that contribute to stock market crashes, see the text by Sornette [2003], *Why Stock Markets Crash.*)

To counteract recession, Keynesian economists may advocate government deficit spending (e.g., construction and research on alternate energy sources creating jobs) to catalyze economic growth. Supply-side economists may suggest tax cuts to promote business capital investment, and laissez-faire economists favor a Darwinian, do-nothing approach by government in which the markets sort themselves out; the weak fail (a capitalist survival-of-the-fittest) and the workers suffer. The populist economic approach is for government to implement lower- and mid-bracket tax relief and simultaneous subsidies for manufacturing, banks, and agriculture to stimulate the economy. Such subsidies are ultimately paid for by all taxpayers. Their object is to "prime the economic pump." The built-in delays of government fund allocation and setting up spending programs can make it appear that government efforts are ineffective; people expect instant results.

One tipping point for a recessionary (autocatalytic) spiral may be excessive debt, private and public. Nearly everyone uses credit cards, and credit card interest rates are usurious. Information in the form of advertising, "keeping up with the Joneses," variable-rate mortgages, "no cash down" payment plans, etc. encourages buyers to accrue debt. At some point, some individuals no longer can meet their combined tax, interest, health care and cost-of-living obligations and have to declare bankruptcy or default on their mortgages. Few have been willing to "live within their means," after all, the government doesn't. "I want it, and I want it now!"

Another symptom of economic pathology is *inflation*. Inflation is seen as a rapid rise in the level of prices for goods and services, as well as a decline in the real value of money (i.e., a loss in its purchasing power). The adverse effects on the economy from inflation can include the hoarding of consumer durables by households in the form of cash, canned goods, flour, etc., as stores of wealth. Investment and saving are discouraged by uncertainty and fear about the future. One trigger for inflation is known to be a high growth rate of the money supply, exceeding that of the economy. This happened in pre-WWII Germany. Typically, inflation is quantified by calculating a *price index*, such as the U.S. consumer price index (CPI). The annual U.S. CPI fell from 3.8% in 2008 to −0.4% in 2009, a sign of recession (BLS 2010). The CPI was 7.6% in 1978, and jumped to 11.3, 13.5, and 10.3 in 1979, 1980, and 1981, respectively, then fell to 6.2 in 1982, a brief inflationary surge. Observed from a microeconomic viewpoint, inflation has affected certain foodstuffs more than other durable goods. For example, the rise in the prices of beef, chicken, certain cereals, and corn syrup can be traced in part to the decrease in the supply of corn grown for food use. A significant fraction of corn grown now goes into the production of ethanol for a gasoline additive, rather than feeding farm animals or humans.

Many of the theories on the causes of inflation seem to converge on it being a monetary phenomenon. In the Keynesian view of the cause of inflation, money is considered to be transparent to real forces in the economy, and economic pressures express themselves as increases of prices seen as visible inflation. The monetarist viewpoint considers inflation to be a purely monetary phenomenon; the total amount of spending in an economy is primarily determined by the total amount of money in existence. According to the Austrian school of economic theory, inflation is a state-induced increase in the money supply. Rising prices are the consequences of this increase. Other economist splinter groups have devised other theories and explanations for inflation.

Because the causes of inflation are varied and debatable, a variety of approaches have been used to control it. The U.S. Federal Reserve Bank manipulates the prime interest rate. A raised interest rate and a slow growth of the money supply are the traditional tools to dampen inflation. Ideally, a 2%–3% per year rate is sought. Keynesian economists stress reducing demand in general, by increased sales taxes or reduced government spending to reduce demand. Government wage and price controls were successful in WWII in conjunction with rationing (to prevent hoarding). Supply-side economists propose fighting inflation by fixing the exchange rate between the domestic currency and some stable, reference currency or commodity (e.g., the Euro, the Swiss franc or gold).

So is there a crisp tipping point for inflation, or does it tend to increase monotonically if left uncorrected? The "wage-price spiral" is a well-known twentieth-century phenomenon. At present, if wages become too high, the manufacturer or service provider can move overseas to find a pool of cheaper labor, leaving behind unemployment that tends to dampen the inflationary spiral.

Stagflation is yet another pathological macroeconomic condition. Stagflation is simultaneous inflation and economic stagnation. (Stagnation is defined as low economic growth coupled with high unemployment.) The stagnation can result from an unfavorable *supply shock*, such as a sharp increase in the price/bbl. of light, sweet crude oil. Increased energy or material cost slows production of goods, and causes their prices to be raised to meet continued demand. Other commodities can trigger supply shock, e.g., a corn crop failure (due to weather), the failure of the Peruvian anchovy fishery (due to over-fishing, or global warming affecting ocean currents). (Anchovies are a major source of fertilizer and animal food protein.) If at the same time, central banks use an excessively stimulatory monetary policy to counteract a perceived recession (low interest rates), the money supply increases and a runaway, wage-price spiral can occur.

The cure for stagflation is to restore, or find alternatives to the interrupted supply while raising lending interest rates to dampen the inflation. Growth can also be stimulated by the government reducing taxes. Note that steady world population growth means increased competition for key commodities such as crude oil, hence a steady increase in its price. While not strictly a supply shock, this limited resource will have a dampening effect on production.

11.2.8 INTRODUCTION TO AGENT-BASED MODELS AND SIMULATIONS OF ECONOMIC AND OTHER COMPLEX SYSTEMS

An *agent*, as used in an ABM or *multi-agent simulation* (MAS) is generally a software module that is a component in an ABM. An agent inputs information from adjacent agents and the environment, and according to simple designed-in rules, it outputs actions and/or information. An ABM consists of dynamically interacting, rule-based agents. They can respond to externally introduced information as well as information from other nearby agents. An ABM can often exhibit complex, emergent behavior.

The agents in a MAS can have several important characteristics: (1) *Autonomy*: The agents are at least partially autonomous. (2) *Local connection*: No one agent has a full global view of the entire system; it is generally connected locally. (3) An agent can be programmed to exhibit *proactive behavior*, directed at achieving a goal. (For example, in an economic MAS, an agent can be made to be a hoarder.) (4) *Decentralization*: There is no designated controlling agent; every agent is its own "decider."

Gilbert and Terna (1999) commented: "There is no one best way of building agents for a multi-agent system. Different architectures (that is, designs) have merits depending on the purpose of the simulation." They go on to say that one of the simplest, effective designs for an agent is in the *production system* (PS).

The three basic components of a PS agent are: (1) A set of rules, (2) A rule interpreter, and (3) A working memory. The rules have two parts; a condition that specifies when (and if) a rule is executed, and an action part that determines the consequences of the rule's execution. The interpreter considers each rule in turn, executes those for which the conditions are met, and repeats the cycle indefinitely. Obviously, not every rule fires on each cycle of the interpreter because the agent's inputs may have changed. The MAS's agents' memories can change, but the agents' rules do not. In an adaptive MAS, the agent rules can be made to change or evolve, to pursue some goal(s).

The reader interested in pursuing agent-based modeling and MAS in-depth should consult the tutorial papers by Axelrod and Tesfatsion (2009), Macal and North (2006), Bonabeau (2002), and Gilbert and Terna (1999). Also see the texts by Gilbert (2007), Tesfatsion and Judd (2006) and Wooldridge (2002).

ABMs are used in MASs of a broad selection of CNLSs. There are many applications for MASs; for example: to investigate the behavior of micro- and macro-ESs, terrorist network structures, animal group behavior, behavior of immune system cell trafficking, behavior of the Internet, vehicular traffic flows, air traffic control, wars, social segregation, stock market crashes, disaster response, GIS, etc. For example, see the 2005 paper by Heppenstall et al. on their ABM for petrol price setting in West Yorkshire in the United Kingdom. This simple, economic ABM used interconnected, like-agent models for petrol station behavior. Ormerod et al. (2001) wrote an interesting paper entitled, *An agent-based model of the extinction patterns of capitalism's largest firms.* Their ABM contained N agents, all interconnected. Model rules specified: how the interconnections are updated, how the fitness of each agent is measured, how an agent becomes extinct, and how extinct agents are replaced. The overall properties of the ABM emerged from the interactions between the agents. They stated: "The empirical relationship between the frequency and size of extinctions of capitalism's largest firms is described well by a power law. This power law is very similar to that which describes the extinction of biological species."

See also the papers by Farmer and Foley (2009) and Buchanan (2009) on the need for ABMs in economics. *Agent-based computational economics* (ACE) is an officially designated special interest group of the *Society for Computational Economics.* See for example, ACE at the Iowa State University website: http://www. econ.iastate.edu/tesfatsi/ace.htm (accessed on December 10, 2009). Altreva™ offers *Adaptive Modeler™* simulation software for ABM applied to stock trading (http:// altreva.com [Accessed on December 10, 2009]). They stated: "Instead of optimizing one or a few trading rules by back-testing them over and over on the same historical data, *Adaptive Modeler* lets a multitude of trading strategies compete [through agents] and evolve on a virtual market in real time. "*Adaptive Modeler* automates the process of creating new trading rules to adapt to market changes...."

While ABM has found many applications in modeling social and ESs, it has also been applied to complex cell scenarios in cell biology and therapeutic medicine. For example, Bailey et al. (2009) devised a large, innovative, multi-cell, ABM of therapeutic *human adipose-derived stromal cells* (hASC) (a type of human adult stem cell) behavior. hASCs are injected to counteract the effects of local ischemia on tissues such as muscle. (Two ischemic scenarios that should be familiar are acute myocardial infarction and peripheral vascular disease.) The authors included such factors as adhesion molecule expression, chemokine secretion, integrin affinity states, hemodynamics, and microvascular network architectures in their elaborate ABM. Their model was verified and validated from known data; it successfully reproduced key aspects of ischemia and cell movements (trafficking) responding to cytokines. Their validated model was then challenged with systematic knockouts aimed at identifying the critical parameters mediating hASC trafficking. Simulations predicted the necessity of a yet-unknown selectin-binding molecule to achieve hASC extravasation, as well as any *rolling behavior* mediated by hASC surface expression of CD15s, CD34, CD62e, CD62p, or CD65. (Rolling behavior is just that, a circulating cell [e.g., an hASC] comes in contact with the vascular endothelium and can partially adhere to it, rolling along downstream with the blood flow. When a firm adhesion forms, the once-rolling cell binds firmly to the vascular endothelium and can then undergo extravasation, entering the tissue surrounding the capillary.) Bailey et al. commented that: "Our work led to a fundamentally new understanding of hASC biology, which may have important therapeutic implications." To further explore the trafficking of immune system cells, see the reviews by Luster et al. (2005) and Medoff et al. (2008), and the paper by Cook and Bottomly (2007). There appear to be many hIS cell scenarios amenable to ABM.

There are a large and growing number of software applications for ABM/MAS, some are proprietary, some open-source, some free. Wikipedia has an extensive *Comparison of Agent-Based Modeling Software* (ABM Software 2010, Tools 2010) (too many programs and toolkits are given to cite in detail here). Also see the review and evaluation of ABS platforms by Railsback et al. (2006).

In any modeling formalism, model verification and validation is imperative. A model's ability to correctly simulate system behavior for novel inputs and ICs can only be based on its competence in simulating known, past scenarios. This is true for ABMs, as well as large dynamic models based on sets of nonlinear ODEs. We note that Bailey et al. (2009) validated their model before trying cytokine knockouts, etc.

11.3 SUMMARY

This chapter has introduced and described classical static-(S&D) and dynamic-modeling of CESs. In 1956, Forrester (2003) urged that the dynamics of CESs be modeled by sets of nonlinear ODEs. The formulation of these ODEs promises to be every bit as challenging as writing the differential equations describing complex physiological systems such as the human immune system. They will contain transport lags, thresholds, and saturation terms, as well as parametric branch gain determination.

However, the largest problem in formulating accurate dynamic models of CESs is seen to be how to model human economic behavior, on both micro- and macroeconomic scales. We live in a new era of rapid communications: cell phones with cameras and GPS systems, the Internet, PDAs, twitter, cable news, on-line financial analysis, etc. Thus the transport delays in the communication branches are largely removed. Attitudes, motivations, decisions (buy, sell, manufacture, do nothing, etc.), fear, anger (e.g., at subprime lenders) must be cobbled into mathematical relationships in new dynamic models of CESs. Delays still exist in the execution of government programs designed to fight the recession, however.

The use of ABMs was introduced as a means of incorporating information-modulated, decision-making behavior into ES modeling. ABMs were shown have application in biology as well, such as modeling hIS cell trafficking.

Simulations of large sets of nonlinear ODEs (model equations) can easily be done with applications, such as MATLAB®/Simulink™, STELLA®, and Simnon™. These applications have been successful in modeling large, complex, nonlinear, ecological, biochemical, and physiological systems, etc., and are applicable in the study of dynamic models of CESs. A grand list of agent-based modeling software can be found on the Wikipedia web site, http://en.wikipedia.org/wiki/Comparison_of_agent-based_modeling_software (accessed 2/11/10).

12 Dealing with Complexity

12.1 INTRODUCTION

To be able to anticipate and predict quantitatively (or even just qualitatively) the behavior of a complex, nonlinear system (CNLS) that we have introduced to for the first time, we need to be able to model it, i.e., describe and understand the relationships between its states in terms of algebraic and differential (or difference) equations. This mathematical description allows us to model it in terms of a directed graph structure (relevant nodes and branches) and its dynamics (ODEs relating node parameters and branch transmissions), and then run computer simulations under different conditions. We need to be able to estimate the nonlinear relationships between parameters, and their initial conditions. One of the universal attributes of CNLSs is that one generally does not know all of the relevant signals (variables) and the relationships between them—we can only approximate or estimate this data. An important purpose in formulating a detailed model of a CNLS is to be able to simulate (and verify) its behavior under a variety of initial conditions and inputs in order to anticipate unintended behaviors.

In many examples of CNLSs, we can identify the system's major *independent variables* (inputs) and define *dependent variables* (outputs), but we have little idea as to how a given input will functionally affect the set of outputs. Inputs can interact in a nonlinear manner (superposition is not present), and tipping points and chaotic behavior may occur. More challenging is the need to establish branch gains relating interior (unobservable) states. Often, these gains have to be estimated and guessed. Branch gains often are functions of node variables, which further contribute to system complexity. This property in certain physiological and biochemical systems is called parametric regulation (Northrop 2000). The measurement of some real-world, MIMO CNLS sensitivities to parameter changes, and its cross-coupling gains, is also helpful in constructing a valid, verified model that will enable the *in silico* exploration of its properties.

Below, we examine two little-known but nevertheless effective approaches to dealing with general CNLSs.

12.2 DÖRNER'S APPROACHES TO TACKLING COMPLEX PROBLEMS

Dörner's (1997) book, *The Logic of Failure: Recognizing and Avoiding Error in Complex Situations*, stressed the importance of complex systems thinking in problem solving. Dörner is a cognitive psychology professor at the University of Bamberg who addressed the dark side of the LUC and its causes.

Things can go wrong (the LUC), according to Dörner, because we focus on just one element of a CNLS. We tend to apply corrective measures too aggressively or too timidly, we ignore basic premises, over-generalize, follow blind alleys, overlook potential "side effects," and narrowly extrapolate from the moment, basing our predictions on the future of those variables and parameters that most attract our attention. We also tend not to compensate correctly for system time lags between inputs and reactions. In short, we tend to oversimplify problems. Dörner stated: "An individual's reality model can be right or wrong, complete or incomplete. As a rule it will be both incomplete and wrong, and one would do well to keep that probability in mind."

Dörner identified four human behavioral trends that contribute to failures when working with complex systems: (1) The slowness of our thinking: We streamline the process of problem-solving to save time and energy. (2) We wish to feel confident and competent in our problem-solving abilities and hence we try to repeat past successes using the same methods, even though the system has changed. (3) We have inability to absorb and retain large amounts of information (in our heads), quickly. Dörner was of the opinion that we prefer static mental models, which cannot capture a dynamic, ever-changing process. (4) We have a tendency to focus on immediately pressing problems. We are captives of the moment. We ignore the future problems our solutions may create. "They may not happen." I will add my fifth and sixth reasons for getting adverse, unintended consequences: (5) Decision-makers (often of the political persuasion) try to look good and appear competent; they make this a priority over actually accomplishing the goal. Then when things go wrong, a scapegoat is found, or a specious, single-cause hypothesis is created to explain the failure. (6) Our responses are rate-sensitive to our informational inputs. If something adverse happens slowly enough, we tend to ignore it (perhaps, it will go away). Things that happen suddenly (e.g., 9/11, throughway bridge collapses) trigger massive responses, even over-responses, which often have not been thought through.

It is clear that fact-based, critical thinking leads to better decision outcomes than hunch-based, guess-based, or faith-based thinking. The system's steersman should also "look outside the window" to constantly observe how his actions affect the system's states. (In economic systems, this view should include marketing research, including a review of the conditions that shape consumer preferences, something the big three auto makers in Detroit have clearly failed at, at least until recently. Clearly, system sensitivities should be constantly monitored to generate feedback to the inputs, in order to prevent the LUC from exerting itself.

Dörner's book used a real situation (i.e., Chernobyl), and two, simulated research scenarios using human subjects: A fictitious African country, *Tanaland*,

and a fictitious community, *Greenvale*, were used to verify his thesis on complex systems operation. Dörner showed the various behaviors individuals exhibit when challenged with ambiguity in decision-making scenarios, and when facing overwhelming complexity. In some instances to avoid coping with complexity, subjects focused tightly in a small area in which they were comfortable, or allowed themselves to become distracted by small items. Some individuals were willing to change, others jumped right in without any situational analysis and became confused by unintended consequences and started creating bogus hypotheses on why they were experiencing problems.

Following his research, Dörner identified seven common problems people have when dealing with CNLSs: (1) Failure to state and prioritize specific goals. (2) Failure to reprioritize as events change. (3) Failure to anticipate "side effects" and long-term consequences. (4) Failure to gather the right amount of system information (neither to rush ahead with no detailed plan, nor to excessively overplan). (5) Failure to realize that actions often have *delayed consequences*, leading to overcorrection (and possible instability) when the results of an action (input) do not occur at once. (6) Failure to construct suitably complex models of the system/situation. (7) Failure to monitor progress and reevaluate input actions.

Clearly, one needs a detailed foreknowledge about the organization of a CNLS and its sensitivities before one attempts to manipulate it. Toward these ends, another German scientist, Frederic Vester (b. 1925, d. 2003), developed a set of procedures for characterizing CNLSs, including a software analysis package, *Sensitivity Model Prof. Vester®* (Vester 2004, Malik mzsg 2006). Vester's approach is summarized below beginning with his "paper computer."

12.3 FREDERIC VESTER'S "PAPER COMPUTER"

As part of his integrated approach to dealing with certain complex systems and situations, Vester developed a paradigm that he called the "paper computer." This procedure is a heuristic method of making estimates of the sensitivities between the variables of a CNLS (Sustainable 2008). Vester listed five steps in the paper computer process: (1) The systems analyst first picks a suitable CNLS to analyze. (2) Next, he or she makes a list of the relevant variables between which he or she wishes to establish causal relations. These can be obvious system inputs and outputs, as well as known "internal" variables. (Vester suggested picking between $N = 15$ and 30 variables, but 50 might be a more appropriate upper bound when the analysis is done on a computer.) (3) Now, an *impact matrix* is made on a piece of graph paper (or, on a computer). The formulation of the impact matrix is generally quite subjective, it depends on the interpretations and judgments of the person(s) evaluating the system. To make the impact matrix, one lists the N variables along the X-axis and the same N variables along the Y-axis. Then a 2-bit scale (0, 1, 2, 3) is used to estimate the impact (influence) of each variable (x_k) on each other variable (x_j, $j \neq k$). Obviously, $0 =$ no impact, $1 =$ small impact (a big change in the chosen variable makes a small change in the target variable, i.e., low sensitivity), $2 =$ medium impact (a small change causes a small change, but a large change causes a medium to big change in the target variable), 3=high impact (a small change in the variable causes a large change in the target variable). The impact matrix diagonal is all zeros. Note that there are $N(N-1)$ boxes to hold the impact numbers (sensitivity estimates). Table 12.1 illustrates an impact matrix for a hypothetical, relatively simple, $N=8$ system I created. Figure 12.1A illustrates the hypothetical system's connectivity in a directed graph with eight nodes, and the Vester diagraph for this hypothetical system. (In this example, the directed graph was made up before the impact matrix was constructed.)

(4) Having established all the obvious cause-and-effect relationships for the system in the impact matrix, for each of the N variables, one adds up the impact numbers in their vertical columns and also in their horizontal rows. Vester calls each row sum an *active sum*, and each column sum a *passive sum*. Thus, each variable is characterized by a pair of numbers that are an indicator of the strength of a given variable in influencing the whole system performance and also the degree to which that variable responds to the whole system's $N-1$ variables.

(5) The final operation in the paper computer paradigm is to plot the 2-D descriptors for each variable on an X–Y graph that Vester calls a *diagraph*. The passive sum numbers are plotted on the positive X-axis, the active sum numbers along the positive Y-axis. A pair of active and passive numbers determines the location of that variable in 2-D, impact vector space in the first quadrant.

Vester identified eight regions in the diagraph's 2-D space (see Figure 12.2). Points lying in *region1* (the top left corner) belong to *active variables* that exert a lot of influence on the system as a whole, but do not receive much impact from the

---

TABLE 12.1

Impact Matrix for the Hypothetical 8-State System

State	x_1	x_2	x_3	x_4	x_5	x_6	x_7	x_8	AS	PS	AS/PS	L_k	θ_k
x_1	**0**	1	0	2	3	2	0	0	**8**	**3**	2.67	8.54	69.4°
x_2	0	**0**	0	0	1	0	0	0	**1**	**1**	1	1.41	45°
x_3	1	0	**0**	0	0	0	0	3	**4**	**3**	1.333	5.0	53.1°
x_4	0	0	0	**0**	0	0	0	1	**1**	**3**	0.333	3.16	18.4°
x_5	0	0	1	1	**0**	0	2	0	**4**	**6**	0.667	7.21	33.7°
x_6	0	0	2	0	0	**0**	0	0	**2**	**5**	0.40	5.39	21.8°
x_7	2	0	0	0	0	0	**0**	1	**3**	**2**	1.50	3.51	56.3°
x_8	0	0	0	0	3	3	3	**0**	**6**	**5**	1.20	7.81	50.2°
PS	**3**	**1**	**3**	**3**	**6**	**5**	**2**	**5**	**0**	—	—	—	—

Notes: AS, active sum; PS, passive sum; length L_k is the impact vector magnitude, $L_k = \sqrt{AS_k{}^2 + PS_k{}^2}$. θ_k=impact vector angle=$\tan^{-1}(AS_k/PS_k)$. See the eight impact vectors ($L_k \angle \theta_k$) plotted on the diagraph in Figure 12.1B. We consider x_5 and x_6 to be system outputs, x_1 and x_8 system inputs, and the internal nodes are x_2, x_3, x_4, and x_7 in this example. Bold numbers are row and column sums, or functions of sums. Zeros in matrix diagonal are bold for emphasis.

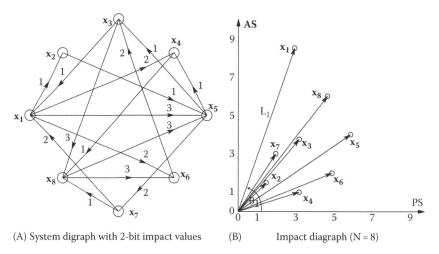

(A) System digraph with 2-bit impact values   (B)   Impact diagram (N = 8)

FIGURE 12.1 (A) An 8-node digraph with weighted branches used to illustrate Vester's sensitivity model. (B) The Impact Diagram derived from the example digraph of (A), and use of the Impact Matrix, Table 12.1. See text for description.

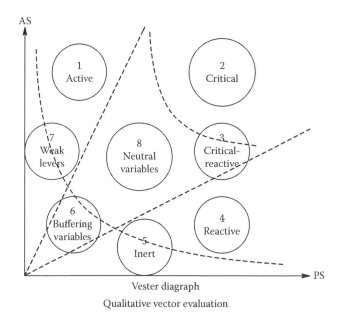

Vester diagraph

Qualitative vector evaluation

FIGURE 12.2 Eight "influence regions" in a diagraph's 2-D vector space. Region boundaries are fuzzy due to quantization noise (From Northrop, R.B., *Exogenous and Endogenous Regulation and Control of Physiological Systems*, Chapman & Hall/CRC Press, Boca Raton, FL, 2004).

N – 1 other variables. The *Region 1* variables are helpful levers for change in the whole system. They are, in effect, inputs. *Region 2* (high active and passive numbers) variables are called *critical variables*. They receive much impact and also exert much influence on the other variables (on the average). They may behave like switches that can shift the whole system up and down the performance scale. They are stronger than active variables. In *Region 3* are located the vectors for *critical reactive* variables that exert some influence on the system and are highly sensitive to the system's status. Reactive variables are located in *Region 4*; they can serve as indicators, but not for steering the system. *Inert variables* are found in *Region 5*. In *Region 6* are the *buffering variables* that are weakly connected to the system. Vester calls the variables in *Region 7*, *weak levers*. Finally, *Region 8* is in the center of the diagraph. In it are *neutral variables* that may serve for the self-regulation of the system. Note that the numbered regions in the diagraph have fuzzy boundaries. I have found by working several examples that it is possible for system outputs to appear in Vester's regions 8 and 6. This is not surprising, however, with 2-bit impact numbers and eight regions (3-bit) defined in the diagraph.

The entire diagraph with its N points can provide an insight on how to manipulate the complex system. Its effectiveness depends entirely on your skill in assigning 2-bit impact numbers between the variables. The paper computer process is used as one component of an iterative systems analysis paradigm developed by Vester called *the sensitivity model*. More about this below.

12.4   SENSITIVITY MODEL OF VESTER

The purpose of Vester's sensitivity model software is to enable effective management of certain CNLSs while minimizing unintended consequences. He recommends it for application in corporate strategic planning, technology assessment, developmental aid projects, examination of economic sectors, city, regional, and environmental planning, traffic planning, insurance and risk management, financial services and research and training. I suggest that the areas of antiterror strategy, ecosystems, energy planning, epidemiology, flight control, freight management (shipping, rail, truck, and air), waste management, and resource recovery be added to this applications list. On the Vester web site (accessed 12/08/09), some 66 licensees and users of his software are listed. Not surprisingly, 62 are in German-speaking countries (Germany, Austria, Switzerland); however, one licensee is Danish, one Namibian, one is in Taiwan and one is the European Air Control. No licensees were cited in English-speaking countries. Perhaps one reason Vester's scholarly works and software are relatively unknown in the United States is that he did not write in English, and his writings do not refer to the methodological developments of systems thinking in the past 30 years in the English literature. In our opinion, Vester's most significant work was his 1999 book: *Die Kunst vernetzt zu Denken: Ideen und Werkzeuge für einen neuen Umgang mit Komplexität* [The Art of Network Thinking: Ideas and Tools for a New Way of Dealing with Complexity]. This text was reviewed in English by Ulrich (2005) and Business Bestseller (2000).

The central theme of the book was described in its preface (translated from German by Ulrich):

> Do we have the right approach to complexity: do we really understand what it is? Man's attempt to learn how to deal with complexity more efficiently by means of storing and evaluating ever more information with the help of electronic data processing is proving increasingly to be the wrong approach. We are certainly able to accumulate an immense amount of knowledge, yet this does not help us to understand better the world we are living in; quite the contrary, this flood of information merely exacerbates our lack of understanding and serves to make us feel insecure... Man should not become the slave of complexity but its master.

One reason Vester's book is important is that it gives ordinary researchers, professionals, and decision-makers (as opposed to trained complex systems scientists) a new sense of competence in dealing with the complex systems issues in the twenty-first century. The other reason is that it describes his unique, sensitivity model software.

Figure 12.3 illustrates the recursive structure of Vester's sensitivity model paradigm. There are nine modules or software tools in which information is processed, presumably replicating how evolutionary management works in nature.

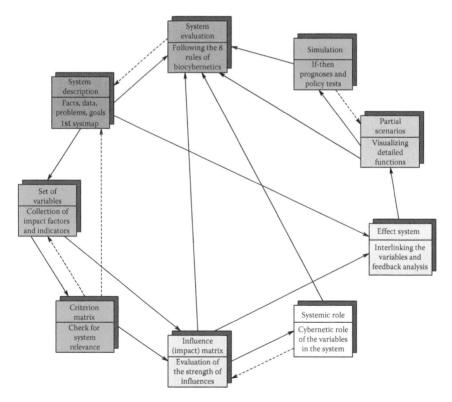

FIGURE 12.3 Recursive structure of Vester's sensitivity model paradigm. There are nine modules or software tools in which information is processed, presumably replicating how evolutionary management works in nature. See text for description.

The *Sensitivity Model Prof. Vester* runs on Windows PCs with *95, 08, NT, 2000, XP & Vista* operating systems, and requires only ca. 30 MB memory.

12.5 LEARNING FROM OUR MISTAKES

A foolish consistency is the hobgoblin of little minds.

Emerson

As we have noted above, in dealing with CNLSs with the best intentions, one is liable to make mistakes; i.e., take an action that leads to poor or inappropriate results. When one repeats the same action and expects a different result, one might be accused of insanity, or at best, naive behavior. Clearly, the manipulator(s) of the CNLS needs to "take the lid off" the system and reexamine in detail its behavior under a variety of branch parameters, inputs, and initial conditions. Generally, a more detailed model must be constructed—more variables, more sensitivities need to be measured or estimated in order to generate valid predictions. Above all, the manipulator(s) must not fall victim to the "single-cause mentality," or be of the "my box" mindset.

Nowhere is this need more obvious than in economics. Certain parameters of the U.S. economy have been manipulated by the U.S. government and the private sector to try to prevent the recession of 11/08 from morphing into a full-blown depression with soaring unemployment and market stagnation. In early 2010, there have been indicators that the hundreds of billion dollars the government has spent on bank bailouts and rescue for certain industries have done some good toward restoring U.S. economic health; however, high average national unemployment continues to lag these good indicators because of built-in delays in the complex, nonlinear, nonstationary U.S. economic system (at this writing, 4/11/10, the mean U.S. unemployment fraction is ca. 10%).

Have economists learned enough from dealing with the "Great Depression" and other more recent "mini-recessions" to advise appropriate government actions that will speed U.S. (and global) economic recoveries? Time will tell.

Smith (2004) published online a very insightful essay on *Systems Thinking: The Knowledge Structures and the Cognitive Process.* I recommend its reading. I was impressed with his "5th Law" (a systems equivalent to the "Peter Principle"): "Systems controllers rapidly advance to the level of systems complexity at which their systems competence starts to break down." Smith's paper echoes many of the thoughts found in Dörner's book; Smith gives many interesting examples of problems arising in complex systems brought about by human incompetence.

12.6 SUMMARY

Picture a 3-D, glass spherical volume filled with a large, directed graph modeling a CNLS. The graph consists of a very large number of randomly spaced nodes (vertices). Some of the nodes are found in densely connected clusters because they relate to subsystems or modules in the CNLS. From outside the sphere, you can see the nodes and the branches connecting them very well on the periphery of the volume and identify them quantitatively. Deeper into the volume, the nodes and branches are not so visible, but one can estimate their values. However, at the center of the 3-D graph, one is hard-pressed to even count the nodes, and the connections of the branches are mostly unknown. The central structure of the graph is hidden by the outer layers.

If the complex system could be completely described, all of its nodes and most of its branches could be mapped on the inner surface of the glass sphere, and would be completely visible (and could be characterized) by rotating the sphere. Unfortunately, many CNLSs cannot be completely described.

In complex systems analysis, we try to develop scientific methods to improve our resolution of the system and build reliable mathematical models of it. They allow us to effectively rotate the glass sphere and shine a focused, coherent light into it in order to see different views of the 3-D graph within as a hologram, clarifying its structure. The insightful analytical approaches of Dörner and Vester described above are a beginning at describing and understanding a new complex system we are presented with, having no prior knowledge of it. Vester offered a heuristic, first-step approach at clarifying the inner structures of a complex system's graphical model.

PROBLEMS

12.1 Make a Vester Diagraph for the 11-node digraph shown in Figure P12.1. 2-bit branch weights have been determined and are shown in the graph. Comment on how well the Diagraph vectors agree with the graph structure.

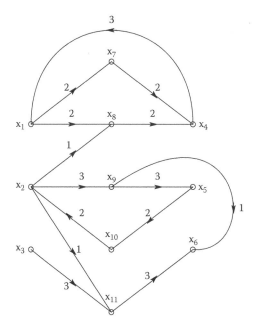

FIGURE P12.1

12.2 An 8-node system has a signed, 3-bit weighting scale for its directed branches.

(A) Construct an 8-node SFG for the system.

(B) From the impact matrix given below, draw the eight Diagraph vectors. Note that some may lie in the second, third, and fourth quadrants. Comment on how well the Diagraph's vectors agree with the SFG's structure. What is the significance of the vectors in the 2nd, 3rd, and 4th quadrants?

Impact Matrix for P12.2.

State	x_1	x_2	x_3	x_4	x_5	x_6	x_7	x_8	AS(y)	PS(x)	AS/PS	L_k	θ_k
x_1	0	0	0	0	0	3	0	0	3	−3	−1.0	4.24	−225°
x_2	0	0	0	0	0	0	3	0	3	−1.0	−3.0	3.16	−252°
x_3	0	0	0	0	1	0	0	0	1	2	0.5	2.24	26.6°
x_4	0	0	0	0	0	0	0	−2	−2	4	0.5	4.47	−26.6°
x_5	−3	0	0	0	0	0	0	0	−3	1	−3.0	3.16	−71.6°
x_6	0	1	2	1	0	0	0	0	4	3	1.333	5.0	53.1°
x_7	0	0	0	3	0	0	0	0	3	3	1.0	4.24	45°
x_8	0	−2	0	0	0	0	0	0	−2	−2	1.0	2.83	−135°
PS	−3	−1	2	4	1	3	3	−2	—	—	—	—	—

12.3 What emerging perturbations to ecosystems will present challenges to man in the twenty-first century?

12.4 What changes in the inputs and parameters of existing complex systems will present challenges to man in the twenty-first century? How will they do this? What may happen?

12.5 You are an internationally known systems biologist/ecologist with extensive experience in mathematically modeling complex systems. You have been asked by the federal government to serve as a consultant to evaluate any adverse unintended consequences resulting from the proposed adoption of a federal Internet use tax (on all e-mail messages sent, and on music, pictures, and videos downloaded). For purposes of this problem, let the tax be 3¢/transaction. List the unintended consequences of this proposed tax. Estimate how much the government would make per year from this tax if Internet use remained unchanged. [*Note*: Human behavior is involved.]

12.6 Repeat Problem 12.5 for cellular telephone calls made (voice and text), also, Twitter messages sent. Again, assume 3¢/transaction.

12.7 What would be the adverse unintended consequences if the federal government legalized the use, importation, and growing of marijuana? It could only be purchased from government-licensed dealers, and would have federal and state excise taxes on it.

Appendix A: *Simnon*™ Programs Used in Chapters 3 and 8

A.1 ENZYMATIC OSCILLATOR

```
CONTINUOUS SYSTEM snell2 " 8/24/98 Cross-coupled
enzymatic oscillator.*
STATE a b p i ap bp pp ip "8 states.
DER da db dp di dap dbp dpp dip "State derivatives.
TIME t
"
da=- k1*ra*re+k1r*rb+Ja " Ja is input rate of A.
"
db=k1*ra*re -(k1r+k2)*rb
"
dp=k2*rb-k3*rp-k4p*rp^n*rep+k4pr*rip*n
"
dip=k4p*rp^n*rep-k4pr*rip
"
re=e0-ri-rb " Initial, constant amount of E=e0.
"
rep=ep0-rip-rbp " Initial, constant amount of
E'=ep0.
"
dap=- k1p*rap*rep+k1pr*rbp+Jap
"
dbp=k1p*rap*rep-k1pr*rbp-k2p*rbp
"
dpp=-k3p*rpp+k2p*rbp-k4*rpp^n*re+k4r*ri*n
"
di=k4*re*rpp^n-k4r*ri
"
ra=IF a>0 THEN a ELSE 0 "Rectification of states.
rb=IF b>0 THEN b ELSE 0
rp=IF p>0 THEN p ELSE 0
ri=IF i>0 THEN i ELSE 0
rap=IF ap>0 THEN ap ELSE 0
rbp=IF bp>0 THEN bp ELSE 0
rpp=IF pp>0 THEN pp ELSE 0
```

```
rip=IF ip>0 THEN ip ELSE 0
"
Ja:3 "Constants used in the model.
Jap:1
e0:5
ep0:5
k1:0.1
k1r:0.05
k1p:0.1
k1pr:0.05
k2:1
k2p:1
k3:0.3
k3p:0.3
k4:0.5
k4p:0.5
k4r:0.5
k4pr:0.5
n:2
"
b:4 "ICs. All others zero by default.
bp:4
END
```

A.2   SUBHARMONIC-GENERATING NLS

```
continuous system tomovic1
STATE x1 x2
DER dx1 dx2
TIME t
"
dx1=A*cos(6*t)-x2-0.2*x2^3
dx2=x1
in=0.3*cos(6*t)
"
A:1
zero:0
"
END
```

A.3   GLUCOREGULATION MODEL

CONTINUOUS SYSTEM GLUCOSE5 " Creation date 11/01/94.
Mod. 8/25/97.

" Normal glucoregulatory model of 70 kg human. 12
States. 1:43 pm
"
STATE g nisG iscG lG gn gln bI ifI pI x1 x3 x2
DER dg dnisG discG dlG dgn dgln dbI difI dpI dx1 dx3
dx2
TIME t
"
" GLUCOSE PLANT
"
dg = -dgu - dgnis - dgisc + nhgb + dGin " dg is rate of
glucose
" mass input, g/min.
" dGin is dietary, gluconeogenesis, and gtt sources.
"
" Utilization Uptake by Non-Insulin-Sensitive Cells.
"
dgnis = Kdnis*(rpG - cnisG) " dgnis is rate of pG uptake
by non-ins-
" ulin sens. cells, g/min.
dnisG = -nisG*Kmnis + dgnis " nisG is mass of G inside
NIS cells.
"
cnisG = nisG/Vnis " Conc. Glucose in NIS cells, g/l
"
" Utilization Uptake by Insulin-Sensitive Cells.
"
discG = -iscG*Kmisc + dGisc " iscG is total mass of
glucose inside
" insulin-sensitive cells.
dGisc = Hi*Kdisc*(rpG - ciscG)" rate of mass diffusion
into iscs, g/min.
" from blood.
ciscG = iscG/Visc " Conc. isc Glucose, g/l
"
" OTHER EQUATIONS:
"
Hi = K6 + (K7*Ki*bI)/(1 + Ki*bI) " Hill function modelling
effect of ins-
" ulin on diffusion constant.
R = rpG*Hi " R function.
"
pG = g/Vol " Plasma glucose conc., g/l.
"

```
rpG = IF pG < 0 THEN 0 ELSE pG
"
dpG = dg/Vol " Rate of change of plasma glucose conc.
"
dgu = IF pG < 1.8 then 0 ELSE Kk*(-1.8+pG) " Kidney
spillover
" nonlinearity.
" NHGB EQUATIONS:
"
NHGB = Hi*Kld*(clG - pG) " NHGB in g/min.
"
dlG = -NHGB - Ks*Hi*clG + fgn*Hgln " lG is hepatic glucose
mass.
"
dgn = Ks*Hi*clG - fgn*Hgln " Glycogen stored in liver,
gms.
"
clG = lG/Vl " Conc. glucose in liver.
"
Hgln = bo + bm*Kgl4*gln/(1 + gln*Kgl4) " Hill fctn for
glucagon action.
"
fgn = IF gn < 0 THEN 0 ELSE K16 " Glycogen switch.
"
" GLUCAGON EQUATIONS:
"
dgln = - h*gln + Kgl3*fgln " Glucagon release from
pancreatic alpha cells.
"
fgln = Kgl1/(1 + Kgl2*R) " glucagon release rate inverse
to R parameter.
"
" PANCREATIC INSULIN DYNAMICS:
"
dpaI = x1 + x3 " Insulin output (fast + slow), microg/min.
"
rdpG = IF dpG < 0 THEN 0 ELSE dpG
"
dx1 = - beta*x1 + Kf*rdpG " Fast insulin
"
dx2 = -a*x2 + Ksi*a*HpGI " Slow insulin, eq. 1.
"
dx3 = -a*x3 + Ksi*a*x2 " Slow insulin, eq. 2.
"
rdpaI = IF dpaI < 0 THEN 0 ELSE dpaI
```

```
"
HpGI = K14*pG*pG/(K15 + pG*pG)  " Hill function for
insulin release
" as a function of pG.
"
" 3 INSULIN COMPARTMENTS: (K8, K9, K10, K11, KLpi,
KLifi)
"
dbI = - (K10 + K11)*bI + K9*ifI + K8*pI + dUxin  " Blood
insulin
"
difI = - (KLifi + K9)*ifI + K11*bI  " Interstitial fluid
compartment.
"
dpI = - (KLpi + K8)*pI + K10*bI + rdpaI  " Portal compartment
"
" EXOGENOUS INSULIN INPUT:
"
dUxin = IF t > tins THEN Iin*EXP(-(t-tins)) ELSE 0
"
" GLUCOSE IVGTT:
"
dGin = Gin1 + Gin2 + Gneo
Gin1 = IF t > tgluc THEN Gin ELSE 0
Gin2 = IF t > (tgluc + 5) THEN -Gin ELSE 0
"
" CONSTANTS
"
tins:800
tgluc:400
beta:0.1
Kf:.333
Ksi:1.7
K4:.001
K5:0.333
K6:.05
K7:15
K8:.7
K9:.7
K10:.4
K11:.75
KLpi:.2
KLifi:.2
K14:.3
K15:2.
```

```
K16:12
gneo:.5
Gin:30 " g/min
Iin:5
Kdnis:.25
Kdisc:3
Kmnis:.4
Kmisc:.05
Vnis:1
Visc:50
Vl:5
Ki:0.05
Kk:5.5
Ks:1.75
a:.025
h:.1
bo:.05
bm:1
Kld:10
Kgl1:.25
Kgl2:50
Kgl3:3
Kgl4:.2
Kc:1
Vol:75
Kin:.2
point8:.8
zero:0
point9:.9
"
" INIT
g:75 " grams gives pG(0)=1g/l.
gn:1000
gln:.3
"
END
```

A.4   REVISED GLUCOREGULATORY MODEL

CONTINUOUS SYSTEM GLUCOSE6 " Creation date 8/26/97.
Mod.11/05/97.
" Type I diabetic model of 70kg human. **14 States.**
With AEP.
STATE g nisG iscG lG gn gln bI ifI pI v "w
DER dg dnisG discG dlG dgn dgln dbI difI dpI dv "dw

```
TIME t
"
" GLUCOSE PLANT (Because of IPFM insulin input, Must
use Euler
" integration with delT = tau.)
"
dg = -dgu - dgnis - dgisc + nhgb + dGin " dg is rate of glucose
" mass input, g/min.
" dGin is dietary, gluconeogenesis, and gtt sources.
"
" Utilization Uptake by Non-Insulin-Sensitive Cells.
"
dgnis = Kdnis*(rpG - cnisG) " dgnis is rate of pG uptake
by non-
" insulin sens. cells, g/min.
dnisG = -nisG*Kmnis + dgnis " nisG is mass of G inside
NIS cells.
"
cnisG = nisG/Vnis " Conc. Glucose in NIS cells, g/l
"
" Utilization Uptake by Insulin-Sensitive Cells.
"
discG = -iscG*Kmisc + dGisc " iscG is total mass of
glucose inside
" insulin-sensitive cells.
dGisc = Hi*Kdisc*(rpG - ciscG)" rate of mass diffusion
into iscs, g/min.
" from blood.
ciscG = iscG/Visc " Conc. isc Glucose, g/l
"
" OTHER EQUATIONS:
"
Hi = K6 + (K7*Ki*bI)/(1 + Ki*bI) " Hill function modelling
effect of
" insulin on diffusion constant.
R = rpG*Hi " R function.
"
pG = g/Vol " Plasma glucose conc., g/l.
"
rpG = IF pG < 0 THEN 0 ELSE pG
"
dpG = dg/Vol " Rate of change of plasma glucose conc.
"
dgu = IF pG < 1.8 then 0 ELSE Kk*(-1.8 + pG) " Kidney
spillover
```

" nonlinearity
"

" **NHGB EQUATIONS:**
"

NHGB = Hi*Kld*(clG -pG) " NHGB in g/min.
"

dlG = -NHGB - Ks*Hi*clG + fgn*Hgln " lG is hepatic glucose
mass.
"

dgn = Ks*Hi*clG - fgn*Hgln " Glycogen stored in liver,
gms.
"

clG = lG/Vl " Conc. glucose in liver.
"

Hgln = bo + bm*Kgl4*gln/(1 + gln*Kgl4) " Hill fctn for
glucagon action
"

fgn = IF gn < 0 THEN 0 ELSE K16 " Glycogen switch.
"

" **GLUCAGON EQUATIONS:**
"

dgln = - h*gln + Kgl3*fgln " Glucagon release from
pancreatic alpha cells.
"

fgln = Kgl1/(1 + Kgl2*R) " glucagon release rate inverse
to R parameter.
"

"

" **3 INSULIN COMPARTMENTS:**
"

dbI = - (K10 + K11)*bI + K9*ifI + K8*pI + dUxin + eI " Blood
insulin
difI = - (KLifi + K9)*ifI + K11*bI " Interstitial fluid
compartment.
dpI = - (KLpi + K8)*pI + K10*bI " Portal compartment
"

" **CONTROLLER (PI)**
"

"e = pG - SP " System error
"dv = K2*e " v is error integral
"u = K3*(e + v) " P + I
"ru = IF u < 0 THEN 0 ELSE u " u is controller output.
"eI = ru
"

" **CONTROLLER (PID)**

```
"
e = pG - SP
dv = K3 * e
u = K1 * e + v + K2 * dpG
ru = IF u < 0 THEN 0 ELSE u
eI = ru
"
" IPFM INJECTOR
"
"dw = ru - z " Integrator
"p = if w > phi then 1 else 0 " Threshold
"s = DELAY(p, tau) " Pulse generator
"x = p - s
"y = if x > 0 then x else 0 " Pulse rectification
"z = y * phi/tau " Integrator reset feedback.
"eI = Do * y/tau " Insulin bolus, area Do.
"
" OTHER EXOGENOUS INSULIN INPUT:
"
dUxin = IF t > tins THEN Iin * EXP(-(t - tins)) ELSE 0
"
" GLUCOSE IVGTT:
"
dGin = Gin1 + Gin2 + Gneo
Gin1 = IF t > tgluc THEN Gin ELSE 0
Gin2 = IF t > (tgluc + 5) THEN -Gin ELSE 0
"
" CONSTANTS:
"
tins:800
tgluc:400
SP:0.9 " g/liter set point.
Do:1
tau:0.05
phi:1
beta:0.1
Kf:.333
Ksi:1.7
K1:0.2
K2:.01
K3:.1
K4:.001
K5:0.333
K6:.05
K7:15
```

```
K8:.7
K9:.7
K10:.4
K11:.75
KLpi:.2
KLifi:.2
K14:.3
K15:2.
K16:12
gneo:.5
Gin:30 " g/min
Iin:5
Kdnis:.25
Kdisc:3
Kmnis:.4
Kmisc:.05
Vnis:1
Visc:50
Vl:5
Ki:0.05
Kk:5.5
Ks:3
a:.025
h:.1
bo:.05
bm:1
Kld:10
Kgl1:.25
Kgl2:50
Kgl3:3
Kgl4:.2
Kc:1
Vol:75
Kin:.2
point8:.8
zero:0
point9:.9
"
" INIT
g:150 " grams gives pG(0) = 2g/l.
gn:1000
gln:.3
"
```

END

A.5   REIBNEGGER'S HIV MODEL

CONTINUOUS SYSTEM reibsys2
" Rev. 11/30/94, 5/30/98.
" Original 19-state Reibnegger model with R's numbers.
" See: 'Theoretical implications of cellular immune reactions against
" helper lymphocytes infected by an immune system retrovirus.'
" Reibnegger, G. *et al*; *P.N.A.S.* 84: 7270-7274, Oct. 1987.
"

TIME t " Weeks.
"

STATE x1 x2 x3 x4 x5 x6 x7 x8 x9 x10 x11 x12 x13 x14 x15 x16 x17 x18 x19
"

DER dx1 dx2 dx3 dx4 dx5 dx6 dx7 dx8 dx9 dx10 dx11 dx12 dx13 dx14 dx15
DER dx16 dx17 dx18 dx19
"

" STATE EQUATIONS:…………………………………………..
"

" x1 = p.c. pathogen P
dx1 = r*rx1/(1+rx1/kr) - k1*rx1*rx7/(kk+rx1)
-k2*rx1*rx3/(kk+rx1) +pin
"

" x2 = p.c. precursor CTL w/ specificity to P.
dx2 = ic*(1+h*F)/(kf+F)) - A*rx2*rx1 - e1*rx2
"

" x3 = p.c. CTL w/ specificity to P.
dx3 = A*rx2*rx1 - e1*rx3 + p*F*rx3/(kf+F)
"

" x4 = p.c. of precursor Th w/ TCR specificity to P.
dx4 = ih*(1+h*F)/(kf+F)) - A*rx4*Mx -e1*rx4
-a1*rx4*rx10/(ky+rx10)
"

" x5 = p.c. of Th w/ TCR specificity to P.
dx5 = A*rx4*Mx - e1*rx5 + p5*F*rx5/(kf+F) - k5*rx5*rx10/(ky+rx10)
"

" x6 = p.c. of resting macrophages.
dx6 = im*(1+h*F)/(kf+F)) - A7*rx6*F - em*rx6
"

" x7 = p.c. of cytotoxic (angry) macrophages.

$dx7 = A7*rx6*f - zm*rx7$

"

" x8 = p.c. of pathogen debris (Dp).

$dx8 = k1*rx1*rx7/(kk + rx1) + k2*rx1*rx3/(kk + rx1) - ed*rx8$

"

" x9 = p.c. of precursor Th w/ TCR specificity exclusive of pathogen.

$dx9 = iph - e1*rx9 - a9*rx9*rx10/(ky + rx10)$

"

" x10 = p.c. of free HIV virions.

$dx10 = q*(rx17 + rx18) - zy*rx10 - a1*pHt*rx10/(ky + rx10) + addx10$

$addx10 = -(rx5 + rx14)*rx10/(ky + rx10) + av1in + av2in$

"

" x11 = p.c. of preCTL specific for HIV (gp120).

$dx11 = ic*(1 + h*F)/(kf + F)) - A*rx11*(rx17 + rx18) - e1*rx11$

"

" x12 = p.c. of CTL specific for HIV.

$dx12 = A*rx11*(rx17 + rx18) - e1*rx12 + p*F*rx12/(kf + F)$

"

" x13 = p.c. of preTh w/ TCR specific for HIV.

$dx13 = ih*(1 + h*F)/(kf + F)) - Ay*rx13*My - e1*rx13 - a1*rx13*rx10/(ky + rx10)$

"

" x14 = p.c. of Th w/ TCR specific for HIV.

$dx14 = A*rx13*My - e1*rx14 + p14*F*rx14/(kf + F) - rx14*rx10/(ky + rx10)$

"

" x15 = p.c. of HIV-infected, preTh w/ TCR specific for pathogen P.

$dx15 = a1*rx4*rx10/(ky + rx10) - A*rx15*Mx - e1*rx15$

"

" x16 = p.c. of HIV-infected preTh w/ TCR specific for HIV.

$dx16 = a1*rx13*rx10/(ky + rx10) - Ay*rx16*My - e1*rx16$

" x17 = p.c. of HIV-infected Th w/ TCR specific for pathogen P.

$dx17 = k5*rx5*rx10/(ky + rx10) + A*rx15*Mx + p17*F*rx17/(kf + F) + addx17$

$addx17 = - e1*rx17 - k1*rx17*rx7/(kk + rx17) - k2*rx17*rx12/(kk + rx17)$

"

" x18 = p.c. of HIV-infected Th w/ TCR specific for HIV.

```
dx18 = rx14*rx10/(ky + rx10) + Ay*rx16*My + p18*F*rx18/
(kf + F) + addx18
addx18 = - e1*rx18 - k1*rx18*rx7/(kk + rx18) - k2*rx18*rx12/
(kk + rx18)
"
```

" x19 = p.c. of HIV debris (sgp120).
```
dx19 = ( (rx17 + rx18)/(kk + rx17 + rx18) )*(k1*rx7
+ k2*rx12) - ed*rx19
"
```

" INPUTS............
```
av1in = if t <t1avinj then 0 else if t<(t1avinj + dt)
then vinj1 else 0
av2in = if t <t2avinj then 0 else if t<(t2avinj + dt)
then vinj2 else 0
"
pin1 = if t<t1pinj then 0 else if t <(t1pinj + dt) then
pinj1 else 0
pin2 = if t<t2pinj then 0 else if t <(t2pinj + dt) then
pinj2 else 0
pin = pin1 + pin2
q = if t < t1avinj then 0 else qmax
r = if t < t1pinj then 0 else rv
"
```

" QUASI-SS VARIABLES & FUNCTIONS.................................
```
"
Mx = rx8*(rx6 + rx7)/(kd + rx8) "APC's for pathogen.
My = rx19*(rx6 + rx7)/(kd + rx19) "APC's for viral gp-120
protein.
"
```

" Lymphoid "factors" resulting from antigen presentation to Th.
```
F = (rx5 + rx17)*Mx/kf + Mx*(rx5 + rx17)) + (rx14 + rx18)*My/
(kf + My*(rx14 + rx18))
"
pHt = rx4 + rx9 + rx13 " Total in infected pre-Th.
Ht = rx4 + rx5 + rx9 + rx13 + rx14 " Total uninfected CD4+Th
Hit = rx18 + rx17 " Total HIV infected Th.
TOTHt = rx4 + rx5 + rx9 + rx13 + rx14 + rx15 + rx16 + rx17 + rx18
Ratio = Hit/TOTHt " Ratio of HIV-infected Th to Total
Th.
"
```

"...RECTIFICATION OF STATES...
```
rx1 = if x1 < 0 then 0 else x1
rx2 = if x2 < 0 then 0 else x2
rx3 = if x3 < 0 then 0 else x3
```

```
rx4 = if x4 < 0 then 0 else x4
rx5 = if x5 < 0 then 0 else x5
rx6 = if x6 < 0 then 0 else x6
rx7 = if x7 < 0 then 0 else x7
rx8 = if x8 < 0 then 0 else x8
rx9 = if x9 < 0 then 0 else x9
rx10 = if x10 < 0 then 0 else x10
rx11 = if x11 < 0 then 0 else x11
rx12 = if x12 < 0 then 0 else x12
rx13 = if x13 < 0 then 0 else x13
rx14 = if x14 < 0 then 0 else x14
rx15 = if x15 < 0 then 0 else x15
rx16 = if x16 < 0 then 0 else x16
rx17 = if x17 < 0 then 0 else x17
rx18 = if x18 < 0 then 0 else x18
rx19 = if x19 < 0 then 0 else x19
"

" LOGGING STATES & Variables……
lx1 = if x1 < .01 then -2 else log(x1)
lx2 = if x2 < .01 then -2 else log(x2)
lx3 = if x3 < .01 then -2 else log(x3)
lx4 = if x4 < .01 then -2 else log(x4)
lx5 = if x5 < .01 then -2 else log(x5)
lx6 = if x6 < .01 then -2 else log(x6)
lx7 = if x7 < .01 then -2 else log(x7)
lx8 = if x8 < .01 then -2 else log(x8)
lx9 = if x9 < .01 then -2 else log(x9)
lx10 = if x10 < .01 then -2 else log(x10)
lx11 = if x11 < .01 then -2 else log(x11)
lx12 = if x12 < .01 then -2 else log(x12)
lx13 = if x13 < .01 then -2 else log(x13)
lx14 = if x14 < .01 then -2 else log(x14)
lx15 = if x15 < .01 then -2 else log(x15)
lx16 = if x16 < .01 then -2 else log(x16)
lx17 = if x17 < .01 then -2 else log(x17)
lx18 = if x18 < .01 then -2 else log(x18)
lx19 = if x19 < .01 then -2 else log(x19)
lMy = if My < .01 then -2 else log(My)
lMx = if Mx < .01 then -2 else log(Mx)
lf = if f < .01 then -2 else log(f)
lHt = if Ht < .01 then -2 else log(Ht)
lHit = if Hit < .01 then -2 else log(Hit)
lTOTHt = if TOTHt < .01 then -2 else log(TOTHt)
lRatio = if Ratio < .01 then -2 else log(Ratio)
"
```

" CONSTANTS:..

a1:0.01 " conversion rate const., pThp to Thp.
A:0.001 " activition rate const.
Ay:0.1 " conversion rate const. pre to active Thvi
A7:9.E-4 " activation rate for macrophages.
a9:0.1 " viral loss constant for pH cells.
e1:0.02 " efflux of precursor T-cells.
ed:2.0 " efflux of debris, per day.
em:0.05 " efflux of macrophage, M.
h:9.0 " inflammation constant.
ic:0.25 " influx of the precursor T-cells into the
compartment.
ih:5e-2 " influx of pHx.
im:1.25e5 " macrophage influx rate.
inbc:2.0 " influx rate of unactivated (naive) B-Cells.
iph:1e4 " influx of pH.
k1:10 " killing capacity constant.
k2:10 " killing capacity constant.
k5:1.2 " HIV infection r.c., Thp to Thpi
kd:1e7 " presentation saturation.
kf:65. " lymphoid factor saturation.
kk:5e4 " killing saturation constant.
kt:1e3 " restimulation saturation.
ky:1e3 " saturation constant.
kr:1e7 " pathogen repro. saturation const.
p:1 " activated CTL prolif. factor.
p5:1 " activ. Th prolif. factor.
p14:2 " activ. Thv prolif factor.
p17:2 " activ. Thpi prolif. factor.
p18:4 " activ. Thv1 prolif. factor.
qmax:1e3 " IDV production constant, max.
rv:0.95 " pathogen replication rate.
dt:3 " time over which infusion given.
t1avinj:0 " time of first injection of HIV.
t2avinj:1000 " time of 2nd HIV injection.
t1pinj:75 " time of first introduction of pathogen.
t2pinj:300 " time of second pathogen introduction.
zm:1.414 " effector macrophage decay rate, per day.
zy:0.693 " decay constant.
vinj1:100 " First injection of HIV virus.
vinj2:100 " Second injection of HIV virus.
pinj1:1e5 " Injected pathogen cells; first injection.
pinj2:1e5 " Second injection of pathogen cells.
" For actual values used, see parameter file
REIBPAR8.T below.

```
"
" INITIAL CONDITIONS..........................................
" See parameter file REIBPAR8.T below.
"
END
[reibsys2] " Reibpar8.t, 7/28/98 Parameters used.
x1:0, x2:30, x3:0, x4:15, x5:0, x6:2.5E5, x7:0, x8:0,
x9:5.E5, x10:0, x11:30,
x12:0, x13:15, x14:0, x15:0, x16:0, x17:0, x18:0,
x19:0.
kr:1.E6, k1:90, kk:6.E4, k2:90, ic:0.25, h:1, kf:50,
A:5.E-3, e1:0.02, p:1,
ih:1.5, a1:0.2, ky:1000, p5:1, k5:1, im:6.E4,
A7:1.E-3, em:0.05, zm:1, ed:2,
iph:1.E4, a9:0.01, zy:2, Ay:1.E-3, p14:1, p17:1,
p18:1, t1avinj:0, dt:2, vinj1:100, t2avinj:250,
vinj2:0, t1pinj:75, pinj1:3000, t2pinj:300, pinj2:5.
E4, qmax:2500, rv:0.8, kd:1.E7, inbc:2, kt:1000.
```

A.6   HIV/AIDS MODEL

CONTINUOUS SYSTEM AIDS12
```
"
" Program to simulate HIV infection of the immune
system. Includes HIV
" infection of Th and Mp cells, B cells and antibodies
to HIV and viral
" pathogen. Mps and NK cells have Fc receptors. 32
States. rev.6/01/98.
"
STATE P pCx Cx M AM Mi AMi pHx Hx pHxi Hxi pCy Cy pHy
Hy pHyi
STATE Hyi Dv Bx By PBx PBy AbX AbY pNK NK gIFN Y SCp
mBx mBy IL2
"
DER dP dpCx dCx dM dAM dMi dAMi dpHx dHx dpHxi dHxi
dpCy dCy dpHy dHy dpHyi
DER dHyi dDv dBx dBy dPBx dPBy dABx dABy dpNK dNK
dgIFN dY dSCp dmBx dmBy
DER dIL2
"
TIME t
"STATE EQUATIONS......................
"
```

" **P is viral pathogen.**
dP =DELAY(rSCp*fab, 1)/(1+rP/kps) - (k1*rAM
+k2*rAMi)*rP-k3*rP*rAbX +c0
c0 = Pin-k33*rP
"

" **SCp are somatic cells infected by viral pathogen, P.**
dSCp = k88*rP-k33*rSCp-k5*rSCp*rAbX*rNK -
k4*rCx*rSCp-k1*rSCp*rAM "
" **pCx are precursor CTL for pathogen.**
dpCx = ic*ff-k8*rpCx*rSCp*fIL2-k9*rpCx
"

" **Cx are activated CTL for pathogen.**
dCx = k8*rpCx*rSCp*fIL2-k9*rCx+pf*rCx*ff*fIL2
"

" **M are resting macrophages.**
dM = im*ff-k11*rM-k12*rM*fgIFN-k13*rM*satY
"

" **AM are activated macrophages secreting IL1, etc.**
dAM = k12*rM*fgIFN-k15*rAM-k13*rAM*satY
"

" **Mi are resting macrophages infected with HIV.**
dMi = k13*rM*satY-k17*rCy*rMi-k18*rMi-k31*rAbY*rMi*rNK
-k12*rMi*fgIFN
"

" **AMi are activated macrophages infected with HIV.**
dAMi = k12*rMi*fgIFN+k13*rAM*satY -k17*rCy*rAMi
-k31*rAbY*rAMi*rNK+c2
C2 = -k19*rAMi
"

" **Y is HIV virions.**
dY = f20*satHi*fIL2+f21*(satAMi
+satAPi)+avin1+avin2+avin3+avin4 +C3
C3 = avin5-k34*rY -k25*rY*rAbY -k23*(rHx+rHy+rpHx
+rpHy+rpH)*satY +C4
C4 = -k24*rY*sCD4
"

" **pNK are precursor NK cells.**
dpNK = ink*ff+k26*rpNK*fIL2-k28*rpNK-k29*rpNK*fIL2
"

" **NK are active NK cells.**
dNK = k29*rpNK*fIL2-k30*rNK
"

" **pCy are precursor CTL specific for HIV Ag.**
dpCy = ic*ff-k8*rpCy*totVI*fIL2-k9*rpCy
"

" Cy are active CTL specific for HIV Ag.
dCy = k8*rpCy*totVI*fIL2 - k9*rCy + pf*rCy*ff*fIL2
"

" pHx are precursor Th with TCR for pathogen Ag.
dpHx = ih*ff -k37*rpHx*satY -k35*rpHx-
k36*rpHx*(APCxi+APCx)/(1+rDv/k32)
"

" Hx are mature Th with TCR for pathogen Ag.
dHx = k36*rpHx*APCx/(1+rDv/
k32) - k38*rHx - k37*rHx*satY + k39*rHx*fIL2
"

" pHxi are precursor CD4 + Th w/ TCR specific for
patho. infected w/ HIV.
dpHxi = k37*rpHx*satY - k40*rpHxi - k41*rpHxi*rAbY*rNK
"

" Hxi are mature Th w/ TCR for pathogen Ag, infected
with HIV.
dHxi = k39*rHxi*fIL2 + k37*rHx*satY - (k46*rCy + k45*(rAM +
rAMi))*rHxi + c10
c10 = -k41*rHxi*rAbY*rNK + k36*rpHx*APCxi/(1+rDv/
k32) - k43*rHxi
"

" pHy are precursor CD4 + Th w/ TCR specific for HIV
Ag.
dpHy = ih*ff - k36*rpHy*APCy/(1+rDv/k32) -k37*rpHy*satY
-k35*rpHy + c16
c16 = -k36*rpHy*APCyi/(1+rDv/k32)
"

" Hy are mature Th w/ TCR for HIV Ag.
dHy = k36*rpHy*APCy/(1+rDv/
k32) + k39*rHy*fIL2 - k38*rHy - k37*rHy*satY
"

" pHyi are pre-Th w/ TCR for HIV Ag, infected w/ HIV.
dpHyi = k37*rHy*satY - k40*rpHyi - k41*rpHyi*rAbY*NK
"

" Hyi are mature Th w/ TCR for HIV Ag, infected w/
HIV.
dHyi = - k43*rHyi + k37*rHy*satY - k41*rHyi*rAbY*rNK +
k39*rHyi*fIL2 + c13
c13 = -(k46*rCy + k45*(rAM + rAMi))*rHyi + k36*rpHy*APCyi/
(1+rDv/k32)
"

" Dv is HIV debris (e.g., gp120).
dDv = (rMi +APCxi +APCyi +rpHxi +rHxi +rpHyi
+rHyi)*(k50*rCy+k51*rNK*AbY) +c14

```
c14 = - k52*rDV+k34*rY+Dvin
"
```

" gIFN is gamma interferon, units.
```
dgIFN = k82*satHx*(APCx+APCxi)/
(k87+APCx+APCxi)+gIFNin+c21
c21 = - k9*rgIFN+k82*satHy*(APCy+APCyi)/
(k68+APCy+APCyi)
"
```

" Interleukin 2
```
dIL2 = IL2in +k70*fIL1*rHx*(APCx+APCxi)
+k72*fIL1*rHy*(APCy+APCyi)+C22
C22 = - k96*rIL2
"
```

".B-CELL SECTION..............................
```
"
```

" By are inactive B-cells specific for HIV Ag.
```
dBy = k54*rnBy*BCF*satV+k58*rBy*BCFy*fIL2-k56*rBy
-k57*rBy*BCF -k92*rBy
"
```

" Bx are inactive B-cells specific for pathogen Ag.
```
dBx = k55*rnBx*satP*BCF+k58*rBx*BCFx*fIL2-k56*rBx
-k61*rBx*BCF -k92*rBx
"
```

" PBy are plasma B-cells making Ab for HIV Ag.
```
dPBy = k57*rBy*BCF - k59*rPBy+k91*rmBy*(fIL2 - 1)
"
```

" PBx are plasma B-cells making Ab for pathogen Ag.
```
dPBx = k61*rBx*BCF - k59*rPBx+k91*rmBx*(fIL2 - 1)
"
```

" AbX are Abs for pathogen Ag.
```
dAbX = k63*rPBx*BCF - k64*rAbX - k60*rAbX*SatP
"
```

" AbY are Abs for HIV Ag.
```
dAbY = k63*rPBy*BCF - k64*rAbY - k66*rAbY*satV+c15
c15 = - k67*(rHxi+rHyi+rpHxi+rpHyi+rMi+rAMi+APCyi+
APCxi)*rAbY
"
```

" mBx are memory B-cells for pathogen Ag.
```
dmBx = k92*rBx - k62*rmBx - k91*rmBx*(fIL2 - 1)
"
```

" mBy are memory B-cells for HIV Ag.
```
dmBy = k92*rBy - k62*rmBy - k91*rmBy*(fIL2 - 1)
"
```

" QUASI-STEADY-STATE VARIABLES...................
```
"
```

rDx = rP*k49 + rSCp* (rAbX*rNK + rCx) *k89 + rSCp*k90 "
Pathogen debris
"

APCx = rM*satP*fIL2*k76 +k77*rBx*fIL2*satP " Antigen
presenting Mp for patho.
"

APCxi = rMi*satP*fIL2* (k78 +k79*rAbX) " HIV-infected
Antigen presenting Mp
" for patho.
"

APCyi = rMi*satV*fIL2* (k78 +k79*rAbY) " HIV-infected
Antigen presenting Mp
" for HIV.
"

APCy = rM*satV*fIL2*k76 +k77*rBy*fIL2*satV " Antigen
presenting Mp for patho.
"

IL1 = k80* (rM + rMi) *rP/ (1 + rP/
kps) +k81* (rM + rMi) *satHx + c22
c22 = k81* (rM + rMi) *satHy + k80* (rM + rMi) *rY + IL1c
"

fIL1 = 1 + 9*IL1/ (kIL1 + IL1)
"

fIL2 = (1 + 9*rIL2/ (k10 + rIL2))
"

fgIFN = 1 + 9*rgIFN/ (k27 + rgIFN)
"

BCFx = 1 + kbcf*fIL1*rHx/ (k16 + rHx*fIL1)
"

BCFy = 1 + kbcf*fIL1*rHy/ (k83 + rHy*fIL1)
"

BCF = BCFx + BCFy
"

qv1 = k20max - dk20*AZT/ (k85 + AZT) " Functions to model
reduced HIV repro-
" duction rate due to AZT.
qv2 = k21max - dk21*AZT/ (k85 + AZT)
"

f20 = if t < tav1inj then 0 else qv1 " Variable rate
constants to model
" effect of AZT
f21 = if t < tav1inj then 0 else qv2
"

Thtot = (rHx + rHxi + rHy + rHyi)
"

```
f = foo + k84* (rHx + rHy) *IL1/ (kIL1 + IL1)
"
ff = 1 + 9.0*f/ (kf + f)
"
fab = rv/ (1 + ABI/k7)
"
satV = (rY + rDv) / (1 + (rY + rDv) /k73)
"
satP = (rP + rDx) / (1 + (rP + rDx) /k71)
"
satY = rY/ (k14 + rY)
"
satHx = rHx/ (1 + rHx/k69)
"
satHy = rHy/ (1 + rHy/k69)
"
satHi = (rHxi + rHyi) / (1 + (rHxi + rHyi) /k44)
"
satAMi = rAMi/ (1 + rAMi/k32)
"
satAPi = (APCxi + APCyi) / (1 + (APCxi + APCyi) /k44)
"
Htot = rpHx + rpHy + rHx + rHy + rpH
"
Hitot = rpHxi + rpHyi + rHxi + rHyi
"
"Hratio = Hitot/ (Htot + Hitot)
"
"totH = Htot + Hitot
"
"Miratio = (rMi + rAMi) / (rM + rAM + rMi + rAMi)
"
totVI = rMi + rAMi + APCXi + APCyi + rpHxi + rpHyi + rHxi + rHyi
" INPUTS.............................
"
avin1 = if t < tav1inj then 0 else if t < (tav1inj + dt)
then vinj1 else 0
avin2 = if t < tav2inj then 0 else if t < (tav2inj + dt)
then vinj2 else 0
avin3 = if t < tav3inj then 0 else if t < (tav3inj + dt)
then vinj3 else 0
avin4 = if t < tav4inj then 0 else if t < (tav4inj + dt)
then vinj4 else 0
avin5 = if t < tav5inj then 0 else if t < (tav5inj + dt)
then vinj5 else 0
```

```
"
Pin1 = if t < tp1inj then 0 else if t < (tp1inj + dt) then
pinj1 else 0
Pin2 = if t < tp2inj then 0 else if t < (tp2inj + dt) then
pinj2 else 0
Pin3 = if t < tp3inj then 0 else if t < (tp3inj + dt) then
pinj3 else 0
Pin4 = if t < tp4inj then 0 else if t < (tp4inj + dt) then
pinj4 else 0
Pin5 = if t < tp5inj then 0 else if t < (tp5inj + dt) then
pinj5 else 0
Pin = Pin1 +Pin2 +Pin3 +Pin4 +Pin5
"
IL2in = if t<til2inj then 0 else if t< (til2inj + dt1)
then IL2inj else 0
"
gIFNin = if t<tgifinj then 0 else if t< (tgifinj + dt2)
then gIFinj else 0
"
sCD4 = if t<tcd4inj then 0 else if t< (tcd4inj + dt3) then
sCD4inj else 0
"
AZTin1 = if t<taz1inj then 0 else if t< (taz1inj+dtaz)
then AZin1 else 0
AZTin2 = if t<taz2inj then 0 else if t< (taz2inj+dtaz)
then AZin2 else 0
AZTin3 = if t<taz3inj then 0 else if t< (taz3inj+dtaz)
then AZin3 else 0
AZTin4 = if t<taz4inj then 0 else if t< (taz4inj+dtaz)
then AZin4 else 0
AZT = AZTin1 + AZTin2 + AZTin3 + AZTin4
"
ABIin1 = if t<tab1inj then 0 else if t< (tab1inj +dtab)
then ABin1 else 0
ABIin2 = if t<tab2inj then 0 else if t< (tab2inj +dtab)
then ABin2 else 0
ABIin3 = if t<tab3inj then 0 else if t< (tab3inj +dtab)
then ABin3 else 0
ABIin4 = if t<tab4inj then 0 else if t< (tab4inj +dtab)
then ABin4 else 0
ABI = ABIin1 + ABIin2 + ABIin3 + ABIin4
"
Dvin = if t < tdvinj then 0 else if t< (tdvinj + dt6) then
Dvinj else 0
"
```

``` 
"  RECTIFICATION of STATES...............
"
"  LOGGING STATES & VARS...................
"
"  CONSTANTS.............................
"  See PAR727.T file below.
"
"  INPUT CONSTANTS........................
"  See PAR727.T file
"
"  INITIAL CONDITIONS > 0................
"  See PAR727C.T file
"
END
```

[The constant simulation parameters for the program above are:]

```
[AIDS12]  "  PAR727C 7/27/98
P:0, pCx:0.9, Cx:0.7, M:2.45E4, AM:13.5, Mi:0, AMi:0,
pHx:128, Hx:6.2E5, pHxi:0, Hxi:0, pCy:1.65, Cy:0,
pHy:2.47E5, Hy:0, pHyi:0, Hyi:0, Dv:0, Bx:10,
By:0, PBx:0, PBy:0, AbX:100, AbY:0, pNK:0.2, NK:1.3,
gIFN:1.66, Y:0, SCp:0, mBx:100, mBy:0, IL2:0, kps:1.
E9, k1:9.E-6, k2:5.E-5, k3:5.E-5, k33:0.01, k88:0.3,
k5:9.E-6, k4:9.E-6, ic:0.25, k8:3.E-5, k9:0.2,
pf:10.E-5, im:1000,
k11:0.05, k12:2.E-4, k13:5.E-4, k15:1, k17:0.01,
k18:0.05, k31:2.E-5, k19:1,
k34:4.E-3, k25:10.E-7, k23:10.E-7, rpH:1.E6, k24:0.2,
ink:0.5, k26:0.125, k28:0.3, k29:1, k30:0.5, ih:1.E4,
k37:0.05, k35:0.05, k36:4.E-3, k32:1.E7,
k38:0.02, k39:5.E-4, k40:0.1, k41:2.E-5, k46:1, k45:2,
k43:0.2, k50:0.02,
k51:0.02, k52:0.5, k82:2.E-4, k87:1.E5, k68:5000,
k70:10.E-11, k72:10.E-11,
k96:1, k54:0.01, rnBy:10, k58:2.E-3, k56:0.1,
k57:0.05, k92:5.E-5, k55:0.03,
rnBx:10, k61:0.01, k59:0.2, k91:1.E-3, k63:5, k64:0.3,
k60:5.E-4, k66:5.E-3,
k67:5.E-3, k62:3.E-4, k49:0.3, k89:0.1, k90:0.1,
k76:10.E-11, k77:10.E-7, k78:10.E-5, k79:10.E-6,
k80:5.E-7, k81:5.E-7, IL1c:0.2, kIL1:1.E5, k10:10,
k27:10, kbcf:9, k16:3.E5, k83:2.E5, k20max:0.4,
dk20:0.498, k85:10, k21max:0.4, dk21:0.498, foo:0.5,
k84:10.E-5, kf:20, rv:250, k7:10, k73:1.E5,
```

k71:5000, k14:300, k69:5000, k44:200, tav1inj:0, dt:4,
vinj1:0, tav2inj:60,
vinj2:0, tav3inj:120, vinj3:0, tav4inj:150, vinj4:0,
tav5inj:200, vinj5:1000,
tp1inj:0, pinj1:0, tp2inj:30, pinj2:0, tp3inj:60,
pinj3:0, tp4inj:100, pinj4:1.E4, tp5inj:300,
pinj5:100, til2inj:250, dt1:200, IL2inj:100,
tnd1inj:150, dt2:240, ND1inj:0, tcd4inj:150, dt3:180,
CD4inj:0, taz1inj:250, dtaz:200, AZin1:250,
taz2inj:240, AZin2:0, taz3inj:300, AZin3:0,
taz4inj:440,
AZin4:0, tab1inj:300, dtab:300, ABin1:0, tab2inj:240,
ABin2:0. tab3inj:300, ABin3:0, tab4inj:440, ABin4:0,
tdvinj:140, dt6:3, Dvinj:0, kcd4:100, inbc:150, k6:2,
k22:0.1, k42:0.1, k47:0.05, k48:10.E-5, k53:0.1,
k65:2, k74:10.E-5, k75:10.E-10, k86:2.E-3, k93:10.E-5,
k94:100.
k95:10.E-5

A.7   CLAPP ET AL. HIS VS. CANCER MODEL

CONTINUOUS SYSTEM TUMOR5
"Define State Variables: 18 states.
"

pNK=precursor natural killer cells.
NK=natural killer cells.
pTh=precursor T-helper cells.
Th=T-helper cells.
pTc=precursor cytotoxic T-cells.
aTc=1st stage activated CTL.
Tc=fully activated cytotoxic T-cells.
pMp=precursor Macrophages.
aMp=1st stage activated Macrophages.
Mp=fully active Macrophages.
Lys=cellular debris.
CA=Cancer/tumor.
IL2=Interleukin-2.
nB=non-activated B-cells.
aB=Antigen-activated B-cells.
pB=Plasma B-cells specific for CA.
AbCA=Antibodies (free & cellular) for CA.
mBCA=Memory cells for CA.
"

TIME t
"

STATE pNK NK pTh Th pTc Tc pMp aMp Mp Lys CA IL2 nB aB
pB AbCA mBCA
"

DER dpNK dNK dpTh dTh dpTc dTc dPMp dMp dLys dCA dIL2
dnB daB dpB
DER dAbCA dmBCA
"
"************* **STATE EQUATIONS** ***************
"
dpNK = (1 + inflam)*mf1 - mat1*rpNK + ep1*rpNK - rpNK/tpNK
dNK = mat1*rpNK - rNK/tNK
dpTh = (1 + inflam)*mf3 - apf*rpTh - rpTh/tpTh
dTh = rpTh*apf + ep2p*rTh - rTh/tTh
dpTc = (1 + inflam)*mf5 - rpTc*apf - rpTc/tpTc
daTc = rpTc*apf - raTc*(gIFN + rIL2)/(1 + PGE2) - raTc/taTc
dTc = raTc*(gIFN + rIL2)/(1 + PGE2) + rTc*IL2*c6/
(sf10 + rIL2) + addTc
addTc = dmTc - rTc/tTc
dpMp = (1 + inflam/(1 + PGE2))*mf6 - gIFN*rpMp - rpMp/tpMp
daMp = gIFN*rpMp - raMp*rTh*rCA*c7/
(sf11 + rCA*raMp*rTh) + addaMp
addaMp = -raMp/taMp
dMp = raMp*rTh*rCA*c7/(sf11 + rCA*raMp*rTh) - rMp/tMp
dLys = rCA*(rNK*co1 + (rMp + c5*raMp)*co2 + rTc*co3)*c14
- rLys/tLys
dCA = gf*rCA/(1 + rCA/grs) - nvmp*rCA + addCA
addCA = -rCA*(rNK*rAbCa*co1 + raMp*co2 + rTc*co3)*c14
dIL2 = IL2in - lf4*rIL2 + rCA*rTh*c13/(sf26 + rCA*rTh)
dnB = (1 + inflam)*mf7 - c18*rnB - c19*rCA*rnB*fil2
daB = c19*rCA*rnB*fil2 - c20*raB + (c21-
c22)*raB*BCFca*factor - raB*c27
dpB = c22*factor*BCFca*raB - c23*rpB + c29*rmBCA*BCFca
dmBCA = c27*raB - c28*rmBCA - c29*rmBCA*BCFca
dAbCA = c24*rpB*rCA/(sf32 + rpB) - c25*rAbCA
- c26*rAbCA*rCA
"(Note that c26 governs the antigenicity of cancer
surface proteins.)
"

"The eight **immunocytokines** and their functions are:
gIFN = rTh*IL1/(sf9 + rTh) + rNK/(sf23 + rNK) + gIFNin "*Gamma
interferon*
IL1 = c16*(raMp + rMp)*rTh*rCA/((sf19 + rTh)*(sf20 + rCA))
+ IL1in "*Interleukin 1 from Mp*
HPF = c17*factor/(sf27 + factor) "*Hematopoeitic factors*
fil2 = c30*rIL2/(sf29 + rIL2) + 1 "Hill fctn. of IL2

BCFca = IL1*rTh*rCA*c31/(sf30 + rTh) "*B-cell maturation factor*
PGE2 = c8*rMp/(sf12 + rMp) "*Prostaglandin E2*
aIFN = c10* (rMp + raMp)/(sf17 + rMp + raMp) + rNK/(sf24 + rNK) "*Alpha interferon*
factor = rTh*IL1/(sf28 + IL1) "*Interleukin 1 acting on Th.*
"

"Immunocytokines causing B-cell maturation & inflammation
apf = agf* (raMp +rMp)*rLys*(gIFN/(1 +PGE2))/(sf4 +(raMp +rMp)*rLys) "Antigen presentation factor "to activate Th
cxf1 = aIFN*c11/(sf16 + aIFN) + IL2*c12/(sf18 + IL2)
"

"Other functions used in the state equations are:
mat1 = c15*(gIFN + IL2)/(sf1 + gIFN + IL2)
ep1 = c1*(gIFN/(sf2 + gIFN) + IL2)
rp1 = c2*PGE2/(sf3 + PGE2)
ep2p = (IL1/(1 + PGE2))/(sf9 + rTh)
co1 = cxf1/(sf13 + rNK*rCA)
co2 = cxf2/(sf14 + (raMp*c5 + rMp)*rCA)
co3 = cxf3/(sf15 + rTc*rCA)
"

"Constants used:
tpNK:2, tNK:.1, tpTh:15, tTh:5, tpTc:30, taTc:50,
tTc:1, tpMp:40, taMp:50, tMp:5, t2Mp:100, tLys:0.5,
mf1:0.1, mf2:0, mf3:25, mf4:1, mf5:20, mf6:200, mf7:8,
agf:0.2, sf1:1e5, sf2:100, sf3:5000, sf4:3500, sf5:50,
sf6:1000, sf7:10000, sf8:1.0e5, sf9:200, sf10:1000,
sf11:2500, sf12:5000, sf13:500, sf14:500, sf15:2500,
sf16:2000, sf17:1000, sf18:1000, sf19:1000, sf20:1000,
sf21:1000, sf22:1000, sf23:200, sf24:600, sf25:2e3,
sf26:1e3, sf27:15, sf28:5, sf29:20, sf30:100,
sf31:5e3, sf32:500, lf1:0.0001, lf2:0.0001,
lf3:0.0001, lf4:0.2, nvmp:0. 001, gf:1, grs:1e7,
c1:5.E-2, c2:0, c3:1, c4:3, c5:0.5, c6:10, c7:100,
c8:5, c9:2, c10:10, c11:5, c12:2, c13:2, c14:0.2,
c15:1e-7, c16:0.1, c17:20, c18:0.1, c19:5.E-2,
c20:0.1, c21:0.01, c22:0.2, c23:1e-2, c24:4, c25:0.01,
c26:1.E-5, c27:0.1, c28:1e-4, c29:0.1, c30:9,
c31:3.3E-5, cxf2:4, cxf3:1000, dt1:7, dt2:14, dt3:7,
dt5:1, til1inj:14, til2inj:14, tgifinj:14, gifinj:0,
il1inj:0, il2inj:0, mTcinj:0
"

```
"The default initial conditions were:
pNK:2e3, NK:200, pTh:15, Th:10, pTc:25, Tc:0, aTc:10,
pMp:2500, aMp:0, Mp:2, Lys:0
CA:1, IL2:0.5, nB:10.
"Other state ICs are 0.
```
END

[The default constants for the revised Clapp et al. model were derived from the DeBoer papers, the immunology literature, and artful estimates. Note that: tXYZ=time constants in days, sfx=saturation factor, txyzinj=time at which exogenous therapy given, dtx=duration of exogenous therapy, etc. (The colons are the equivalent of "=" in the *Simnon* notation, and an asterisk denotes multiplication. To save space, I have not included the rectification of states or the logging for display purposes. Also, *constants should be put in column form.*)]

Appendix B: How to Use Root Locus to Determine the Stability of SISO Linear Control Systems

Given the knowledge of the positions of the poles and zeros of the loop gain of a linear, SISO, single-loop feedback system, the root locus (RL) technique allows us to predict precisely where the closed-loop poles of the system will be as a function of system scalar gain. Thus RL can be used to predict the conditions for instability, as well as to design for a desired, closed-loop transient response. Many texts on linear control systems treat the generation and interpretation of RL plots in detail (Truxall 1955, Kuo 1982, Ogata 1990, Northrop 2000, Section 3.2).

RL has also been used in the design of electronic feedback amplifiers, including sinusoidal oscillators (Northrop 1990). It should be stressed that physiological systems are in general, nonlinear, and therefore RL techniques strictly cannot be applied. However, if a physiological system is operating linearly (all states positive) and the nonlinearities are linearized, RL can be applied and some insight can be gained about the system's closed-loop behavior.

It is often tedious to construct RL diagrams by hand on paper, except in certain simple cases described below. Detailed, quantitative RL plots can be generated using a MATLAB® subroutine, "RLOCUS." Several of the examples in this section are from MATLAB plots.

The concept behind the RL diagram is simple: Figure B.1 shows a simple one-loop, linear SISO feedback system. The closed-loop gain is

$$\frac{Y}{X}(s) = \frac{C(s)G_p(s)}{1 - A_L(s)} = \frac{C(s)G_p(s)}{1 + C(s)G_p(s)H(s)} = F(s) \tag{B.1}$$

where
  $C(s)$ is the controller transfer function acting on $E(s)$
  $G_p(s)$ is the plant transfer function (input $U(s)$, output $Y(s)$)
  $H(s)$ is the feedback path transfer function

The system loop gain is

$$A_L(s) = -C(s)G_p(s)H(s) \tag{B.2}$$

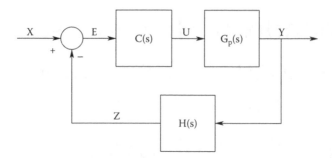

FIGURE B.1 A simple, single-loop, linear negative feedback (NFB) system block diagram. (From Northrop, R.B., *Endogenous and Exogenous Regulation and Control of Physiological Systems*, Chapman & Hall/CRC Press, Boca Raton, FL, 2000.)

The RL technique allows us to plot in the s-plane, those complex **s** values that make the return difference → 0, thus making the closed-loop transfer function, $|\mathbf{F(s)}| \to \infty$. The **s** values that make $|\mathbf{F(s)}| \to \infty$ are of course, by definition, the poles of $\mathbf{F(s)}$. These poles move in the **s**-plane as a function of a gain parameter in a predictable, continuous manner as the gain is changed. Since finding **s** values that set $\mathbf{A_L(s)} = 1 \angle 0°$ is the same as setting $|\mathbf{F(s)}| = \infty$, the RL plotting rules are based on finding **s** values that cause the angle of $-\mathbf{A_L(s)} = -180°$, and $|-\mathbf{A_L(s)}| = 1$. The RL plotting rules are based on satisfying either the angle or magnitude condition on $-\mathbf{A_L(s)}$. We use the vector format of $-\mathbf{A_L(s)}$ to derive the *plotting rules*.

An example of how this is done is given below. First, we assume a feedback system with the loop gain $A_L(s)$ given below. Note that there are three equivalent ways of writing $A_L(s)$:

$$A_L(s) = \frac{-K\beta(s\tau_1 + 1)}{(s\tau_2 + 1)(s\tau_3 + 1)} \quad \text{(time-constant form)} \tag{B.3A}$$

$$-A_L(s) = \frac{K\beta\tau_1}{\tau_2\tau_3} \frac{(s + 1/\tau_2)}{(s + 1/\tau_2)(s + 1/\tau_3)} \quad \text{(Laplace form)} \tag{B.3B}$$

$$-\mathbf{A_L(s)} = \frac{K\beta\tau_1}{\tau_2\tau_3} \frac{(\mathbf{s} - \mathbf{s_1})}{(\mathbf{s} - \mathbf{s_2})(\mathbf{s} - \mathbf{s_3})} \quad \text{(vector form)} \tag{B.3C}$$

where the loop gain poles are $\mathbf{s_1} = -1/\tau_1$, $\mathbf{s_2} = -1/\tau_2$, and $\mathbf{s_3} = -1/\tau_3$.

For the magnitude *criterion*, **s** must satisfy

$$\frac{|\mathbf{s} - \mathbf{s_1}|}{|\mathbf{s} - \mathbf{s_2}||\mathbf{s} - \mathbf{s_3}|} = \frac{\tau_2\tau_3}{K\beta\tau_1} \tag{B.4}$$

For the *angle criterion*, **s** must satisfy $\theta_1 - (\phi_2 + \phi_3) = -180°$.

Nine basic RL plotting rules are derived from the angle and magnitude conditions above, and are used for pencil and paper construction of RL diagrams. They are

1. *Number of branches*: There is one branch for each pole of $A_L(s)$.
2. *Starting points*: Locus branches start at the poles of $A_L(s)$ for the scalar gain $K\beta = 0$.
3. *End points*: The branches end at the finite zeros of $A_L(s)$ for $K\beta \to \infty$. Some zeros of $A_L(s)$ can be at $|s| = \infty$.
4. *Behavior of the loci on the real axis*: (From the angle criterion.) For a negative feedback system, on-real axis locus branches exist to the left of an odd number of on-axis poles and zeros of $A_L(s)$. If the feedback for some reason is positive, then on-axis locus branches are found to the right of a total odd number of poles and zeros of $A_L(s)$. See the examples below.
5. *Symmetry*: RL plots are symmetrical around the real axis in the **s** plane.
6. *Magnitude of gain at a point on a valid locus branch*: From the magnitude criterion, at a vector point **s** on a valid locus branch we have

$$K\beta = \frac{|s - s_2||s - s_3|\tau_2\tau_3}{|s - s_1|\tau_1} \tag{B.5}$$

7. *Points where locus branches leave or join the real axis*: The breakaway or reentry point is algebraically complicated to find; also, there are methods based on both angle and magnitude criteria. For example, in a negative feedback system loop gain function with three real poles, two locus branches leave the two poles closest to the origin and travel toward each other along the real axis as $K\beta$ is raised. At some critical $K\beta$, the branches break away from the real axis, one at $+90°$ and the other at $-90°$. If we examine the angle criterion at the breakaway point, as shown in Figure B.2, it is clear that

$$-(\theta_1 + \theta_2 + \theta_3) = -180° \tag{B.6}$$

From the geometry on the figure, the angles can be written as arctangents:

$$\tan^{-1}\left[\frac{\varepsilon}{(\omega_1 - P)}\right] + \tan^{-1}\left(\frac{\varepsilon}{(\omega_2 - P)}\right) + \left\{\pi - \tan^{-1}\left(\frac{\varepsilon}{(P - \omega_3)}\right)\right\} = \pi \tag{B.7}$$

For small arguments, $\tan^{-1}(x) \cong x$, so

$$\left[\frac{\varepsilon}{(\omega_1 - P)}\right] + \left[\frac{\varepsilon}{(\omega_2 - P)}\right] - \left[\frac{\varepsilon}{(P - \omega_3)}\right] = 0 \tag{B.8}$$

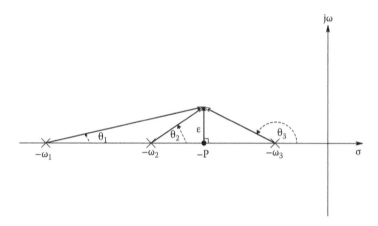

FIGURE B.2 Vectors in the s-plane showing parameters required to calculate the RL branch breakaway point, −P, on the negative real axis. An NFB system loop gain with three, negative real poles is shown. See text for description. (From Northrop, R.B., *Endogenous and Exogenous Regulation and Control of Physiological Systems*, Chapman & Hall/CRC Press, Boca Raton, FL, 2000.)

This equation can be written as a quadratic in P:

$$3P^2 - 2(\omega_1 + \omega_2 + \omega_3)P + (\omega_2\omega_3 + \omega_1\omega_2 + \omega_1\omega_3) = 0 \qquad (B.9)$$

The desired P-root of course lies between $-\omega_2$ and $-\omega_3$.

8. *Breakaway or reentry angles of branches with the real axis*: The loci are separated by angles of 180°/n, where n is the number of branches intersecting the real axis. In most cases, n=2, so the branches approach the axis perpendicular to them.

9. *Asymptotic behavior of the branches for* Kβ → ∞.
   a. The number of asymptotes along which branches approach zeros at |s| = ∞ is $N_A = N - M$. Where N is the number of finite poles and M is the number of finite zeros of $A_L(s)$.
   b. The angles of the asymptotes with the real axis are φ_k, where k = 1, ..., N − M. φ_k is given in Table B.1.
   c. The intersection of the asymptotes with the real axis is a value I_A along the real axis. It is given by

$$I_A = \frac{\sum(\text{real parts of finite poles}) - \sum(\text{real parts of finite zeros})}{N - M} \qquad (B.10)$$

An example of finding the asymptotes and I_A is shown in Figure B.3. This negative feedback system has a loop gain with four poles at s=0, s=−3, and a complex-conjugate (CC) pair at s=−1±j2. There are no finite zeros. Thus $I_A = [(-3-1-1) - (0)]/4 = -5/4$. The angles are, from the table, ±45° and ±135°. The RL plot was done with MATLAB.

TABLE B.1
Asymptote Angles with the Real Axis in the s-Plane

Negative Feedback		Positive Feedback	
N – M	φ_k	N – M	φ_k
1	180°	1	0°
2	90°, 270°	2	0°, 180°
3	60°, 180°, 300°	3	0°, ±120°
4	45°, 135°, 225°, 315°	4	0°, 90°, 180°, 270°

Source: Northrop, R.B., *Endogenous and Exogenous Regulation and Control of Physiological Systems*, Chapman & Hall/ CRC Press, Boca raton, FL, 2000.

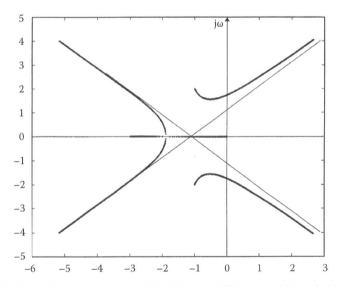

FIGURE B.3   A RL plot done with MATLAB for an NFB system with real poles at $s=0$ and $s=-3$, and CC poles at $s=-1\pm j2$. There are no finite zeros. The asymptote angles are ±45° and ±135°. (From Northrop, R.B., *Endogenous and Exogenous Regulation and Control of Physiological Systems*, Chapman & Hall/CRC Press, Boca Raton, FL, 2000.)

To get a feeling for plotting RL diagrams with the rules above, it is necessary to examine several representative examples. In Example 1, we examine the commonly encountered circle RL. The negative feedback system's loop gain is

$$A_L(s) = \frac{-K_p\beta(s+a)}{(s+b)(s+c)} \tag{B.11}$$

where a>b>c>0. Angelo (1969) gave an elegant proof that this system's RL does indeed contain a circle centered on the zero of $A_L(s)$. The circle's radius R is shown to be the *geometrical mean distance from the zero to the poles*. That is,

$$R = \sqrt{(a-b)(a-c)} \tag{B.12}$$

The break away and reentry points are easily found from a knowledge that the circle has a radius R and is centered at $s=-a$. The circle RL is shown in Figure B.4A. If the loop gain's poles are CC, the RL is also an interrupted circle, shown in Figure B.4B. The poles are at $s=-\alpha\pm j\gamma$, the zero is at $s=-\sigma$. The $A_L(s)$ is

$$A_L(s) = \frac{-K_p\beta(s+\sigma)}{s^2 + s(2\alpha) + (\alpha^2 + \gamma^2)} \tag{B.13}$$

The circle's radius is now found from the Pythagorean theorem:

$$R = \sqrt{(\sigma-\alpha)^2 + \gamma^2} \tag{B.14}$$

Note that as the gain, $K_p\beta$, is increased, the closed-loop system poles become more and more damped, until they both become real, one approaching the zero, the other going to $-\infty$.

The damping of a CC pole-pair in the s-plane can be determined quantitatively by drawing a line from the origin to the upper-half plane pole. The damping factor, ζ, associated with the CC poles can be shown to be the cosine of the angle the line from the origin to the CC pole makes with the negative real axis. That is, $\zeta=\cos(\phi)$. If the poles lie close to the $j\omega$ axis, $\phi \rightarrow 90°$, and $\zeta \rightarrow 0$. The length of the line from the origin to one of the CC poles is the undamped natural frequency, ω_n, of that CC pole-pair. Recall that the CC pole pair is the result of factoring the

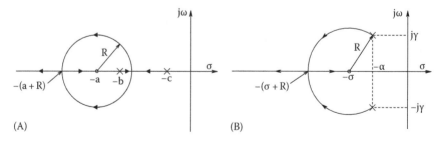

(A)   (B)

FIGURE B.4 (A) Illustration of the circle RL. The NFB system's loop gain has real poles at $-c$ and $-b$, and a zero at $-a$. (B) The interrupted circle RL. The poles are at $s=-\alpha\pm j\gamma$, and the real zero is at $-\sigma$. See text for details. (From Northrop, R.B., *Endogenous and Exogenous Regulation and Control of Physiological Systems*, Chapman & Hall/CRC Press, Boca Raton, FL, 2000.)

quadratic term, $[s^2 + s(2\zeta\omega_n) + \omega_n^2]$. By way of an example, the damping of the open-loop, CC pole pair in Figure B.4B is $\zeta = \cos[\tan^{-1}(\gamma/\alpha)] = \alpha/\omega_n$.

In Example 2, we examine the RL of a simple, type 1 negative feedback system with a transport lag.

The system's loop gain is

$$A_L(s) = \frac{-Ke^{-s\delta}}{(s+a)} \tag{B.15}$$

Let $a = 1$, $\delta = 1$. We write $-A_L(s)$ in vector form, noting that in general, $s = \sigma + j\omega$.

$$-A_L(s) = \frac{Ke^{-\delta\sigma}e^{-j\omega\delta}}{s - s_1} \tag{B.16}$$

By the *magnitude criterion*

$$|-A_L(s)| = \frac{Ke^{-\delta\sigma}}{|s - s_1|} = 1 \quad \text{or} \quad K = |s - s_1|e^{+\delta\sigma} \tag{B.17}$$

The *angle criterion* gives

$$\angle [-A_L(s)] = -\omega\delta R - \theta_1 = -180°(2k+1), \quad k = 0,1,2,\ldots \tag{B.18}$$

The loop gain has one real pole at $s = -1$, and an infinite number of poles at $s = -\infty \pm j2\pi k$, where $k = 0, 1, 2, \ldots$. The complicated RL plot for this simple delay system is shown in Figure B.5.

First, we find the ω_0 value where the first closed-loop, CC pole pair crosses the $j\omega$ axis. The ω_0 value is found by solving the angle criterion by trial and error. That is,

$$\omega_0\delta R + \tan^{-1}(\omega_0) = 180° \tag{B.19}$$

ω_0 is found to be $2.029 \, \text{r/s}$. To find the K value required to put the closed-loop poles at $\pm j\omega_0$, we substitute ω_0 into the magnitude criterion. Thus,

$$K = \sqrt{1 + \omega_0^2 e^{+\sigma\delta}} = 2.262 \ (NB : \sigma = 0 \quad \text{for } s = j\omega_0.) \tag{B.20}$$

The gain at which the primary, closed-loop pole pair becomes CC is found for $s = -2$ is

$$K = 1 \, e^{-2(1)} = 0.1353 \tag{B.21}$$

Because the higher-order, closed-loop pole pairs cross the $j\omega$ axis at progressively higher K values, it is the first-order ($k = 0$) pole pair that is dominant in

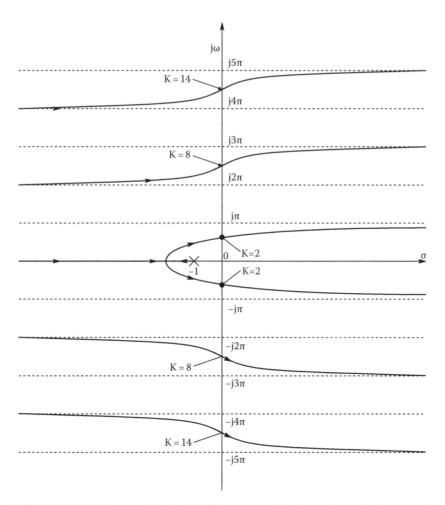

FIGURE B.5 RL diagram of a simple NFB system with a loop gain real-pole at s=−1, a scalar gain K, and a transport lag, e⁻ˢ. See text for discussion. (From Northrop, R.B., *Endogenous and Exogenous Regulation and Control of Physiological Systems*, Chapman & Hall/CRC Press, Boca Raton, FL, 2000.)

determining system instability. In other words, once the first-order, closed-loop, CC pole pair is in the right-half **s** plane, the other locus branches are of academic interest only; the system is already unstable.

For a third example, let us consider a CPK plant with two real poles and direct, negative feedback. The system's loop gain is

$$A_L(s) = \frac{-K_c K_p}{(s+1)(s+5)} \tag{B.22}$$

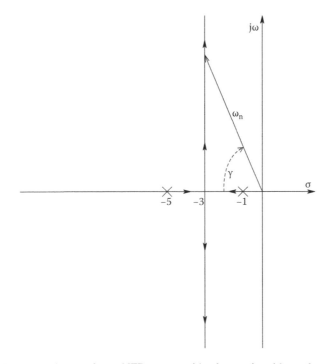

FIGURE B.6   RL diagram for an NFB system with a loop gain with a pole at $s = -1$ and $s = -5$. The breakaway point on the real axis is at $s = -3$. (From Northrop, R.B., *Endogenous and Exogenous Regulation and Control of Physiological Systems*, Chapman & Hall/CRC Press, Boca Raton, FL, 2000.)

This is a type 0 system, and its RL is shown in Figure B.6. We note that the system is never unstable, but that there is a practical limit to the size of $K_c K_p$ in terms of acceptable closed-loop settling time and minimum damping, ζ. To make a useful controller, we wish to be able to raise the gain to reduce settling time while retaining a practical range of closed-loop system pole damping, say $0.5 \le \zeta_{CL} \le 0.707$.

In Example 4, we introduce a proportional plus derivative feedback (PD) for the system of Example 3. We also condition the input (set-point) by a factor of $(5 + aK_p K_c)/K_p$ in order to give the closed-loop system unity dc gain. The PD feedback introduces a zero as $s = -a$ into the loop gain function. If $a > 5$, then the closed-loop system has the circle RL of Example 1. The transfer function of the PD compensator is $C(s) = K_c(s + a)$. The system loop gain is now

$$A_L(s) = \frac{-K_p K_c (s + a)}{(s + 1)(s + 5)}$$ (B.23)

We chose **a** so the zero lies on the real axis at $s = -15$. Figure B.7 illustrates the system's RL diagram. Note that there are two closed-loop system poles;

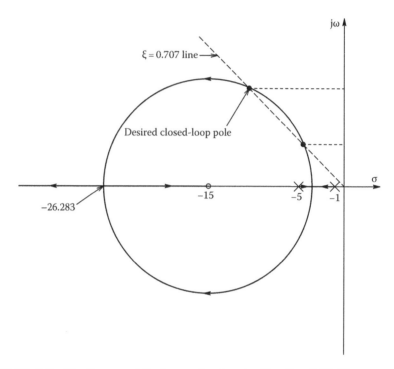

FIGURE B.7 RL diagram of the loop gain given by Equation B.23. The system is given PD compensation. Because the closed-loop system poles' damping $\xi=0.707$ line intersects the circle RL at two points, we choose the higher gain (K_pK_c) that causes the system to have the higher natural frequency, hence step response speed. See text for calculations. (From Northrop, R.B., *Endogenous and Exogenous Regulation and Control of Physiological Systems*, Chapman & Hall/CRC Press, Boca Raton, FL, 2000.)

they start out real then become CC, then real again as the gain K_pK_d increases. One then approaches the zero, the other goes to $-\infty$. The circle radius is $R = \sqrt{(15-5)(15-1)} = 11.832$. Thus, the locus branches leave the real axis at $s = -(15 - 11.832) = -3.168$ and rejoin it at $s = -(15 + 11.832) = -26.832$.

To obtain a good, quick step response for the system with little overshoot, we want the CC poles to have a damping factor $\zeta=0.707$. Recall that ζ is equal to the cosine of the angle a line drawn from the origin to the upper CC pole makes with the negative real axis. In this case, it is a line at 45°, as shown in Figure B.7. Where the 45° line intersects the CC RL branch at its most distant intersection is the desired closed-loop, CC pole position. Once the pole position is known, the RL magnitude criterion can be used to find the required gain, K_pK_d. The length of the line from the origin to the desired, closed-loop, CC pole position is the closed-loop system's undamped natural frequency, ω_{nCL}. Graphical measurement on the plot yields $\omega_{nCL} = 15.5$ r/h. From the RL magnitude criterion, we can find the corresponding gain

$$\left|A_L(\omega_{nCL})\right| = 1 = \frac{K_p K_c R}{AB} \rightarrow$$

(B.24)

$$K_p K_c = AB/R = \sqrt{[(11-5)^2 + 11^2]}\sqrt{[(11-1)^2 + 11^2]} / 11.832 = 15.743$$

Note that the $SP*(5+a*K_p*K_c)/K_p$ term in the program normalizes the steady-state output gain to 1 for any K_c. Figure B.8 illustrates the step response of the PD-compensated system of Example 4. Output plots for K_c values of 1, 3.15, 10, and 50 are shown. The system has the desired closed-loop damping $\zeta = 0.707$ for $K_c = 3.15$, $K_p = 5$.

As a *final example* with the PD compensated system, we use MATLAB's RLOCUS subroutine to plot the system's RL diagram. The dots represent the

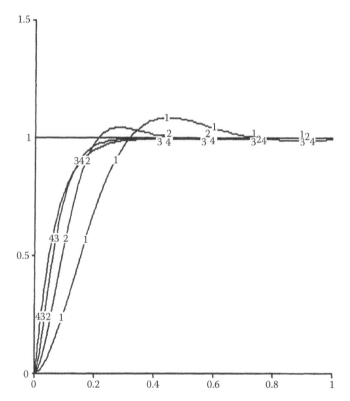

FIGURE B.8 Unit step response of the PD-compensated system of Example 4 (Figure B.7). Vertical axis: closed-loop system output. Horizontal axis: time in seconds. All traces: $K_p = 5$. Trace $1 = K_c = 1$, trace $2 = K_c = 3.15$, trace $3 = K_c = 10$, trace $4 = K_c = 50$. At $K_c = 3.15$, the closed loop $\zeta = 0.707$. Note there is no overshoot for $K_c \geq 10$. (From Northrop, R.B., *Endogenous and Exogenous Regulation and Control of Physiological Systems*, Chapman & Hall/CRC Press, Boca Raton, FL, 2000.)

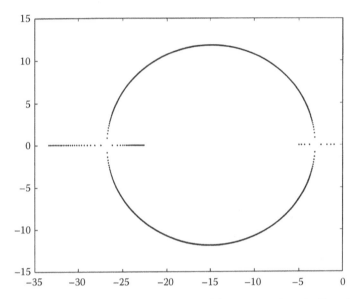

FIGURE B.9   MATLAB RL plot of the system with loop gain given by Equation B.23. $K_p K_c$ ranges from 0 to 50 in increments of 0.10. (From Northrop, R.B., *Endogenous and Exogenous Regulation and Control of Physiological Systems*, Chapman & Hall/CRC Press, Boca Raton, FL, 2000.)

closed-loop system's pole positions for values of $K_p K_c$ ranging from 0 to 50 in increments of 0.10. Note the dot spacing on the real axis and along the circle. Large spacing indicates that the closed-loop pole positions are changing rapidly with gain. Although the closed-loop, CC poles lie on a circular path, they appear elliptical in Figure B.9 because of scale distortion (the real axis spans 0 to −35 while the jω axis spans −15 to +15).

Many other interesting examples of RL plots are to be found in the venerable control systems texts by Kuo (1982) and Ogata (1970). We have seen that circle RL plots are easy to construct graphically by hand. More complex RL plots should be done on a computer.

Appendix C: Bode Plots

A *Bode plot* is a system's frequency response (FR) plot done on *log-linear graph paper*. The (horizontal) frequency axis is logarithmic; on the linear vertical axis, $20\log_{10}|\mathbf{H}(f)|$ is plotted, which has the units of decibels (dB), and the phase angle of $\mathbf{H}(f)$, which is the phase angle between the output and input sinusoids. One advantage of plotting $20\log_{10}|\mathbf{H}(f)|$ vs. f on semilog paper is that plots are made easier by the use of *asymptotes*, giving the FR behavior relative to the system's break frequencies or natural frequencies. Also by using a logarithmic function, products of terms appear graphically as sums. Perhaps the best way to introduce the art of Bode plotting is by examples.

Example C.1

A simple low-pass system: Let us assume a system is described by the first-order ODE:

$$a\dot{\mathbf{y}} + b\mathbf{y} = c\mathbf{u} \tag{C.1}$$

In operator notation this is

$$ap\mathbf{y} + b\mathbf{y} = c\mathbf{u} \tag{C.2}$$

Assume the input $\mathbf{u}(t) = U(t)A\sin(\omega t)$, and the system is in the SSS. Now

$$\mathbf{Y}[aj\omega + b] = c\mathbf{U} \tag{C.3}$$

Writing the phasors as a vector ratio

$$\frac{\mathbf{Y}}{\mathbf{U}}(j\omega) = \frac{c}{aj\omega + b} \tag{C.4}$$

When doing a Bode plot, *the FR function should be put into a time-constant form*. That is, the 0th power of $(j\omega)$ is given a coefficient of 1, in both numerator and denominator: Sic:

$$\frac{\mathbf{Y}}{\mathbf{U}}(j\omega) = \frac{c/b}{j\omega(a/b) + 1} = \mathbf{H}(j\omega) \quad \text{(time-constant form)} \tag{C.5}$$

The quantity (a/b) is the system's time constant; it has the units of time. Significantly, $\tau^{-1} = b/a$ is the break frequency of the system in r/s. One advantage of the time-constant format is that when $\mathbf{u}$ is dc, $\omega = 0$, and the dc gain of the system is simply (c/b). The magnitude of the FR function at dc is just

$$|\mathbf{H}(j\omega)| = \frac{c/b}{\sqrt{\omega^2(a/b)^2 + 1}} \tag{C.6}$$

And its dB logarithmic value is, in general

$$dB = 20\log\left(\frac{c}{b}\right) - 10\log\left[\omega^2\left(\frac{a}{b}\right)^2 + 1\right] \qquad (C.7)$$

For $\omega = 0$, $dB = 20\log(c/b)$. For $\omega = (a/b)^{-1}$ r/s, $dB = 20\log(c/b) - 10\log[2]$, or the dc level minus 3 dB. $\omega_o = (a/b)^{-1}$ r/s is the system's break frequency. For $\omega \geq 10\omega_o$, the amplitude response (AR) is given by

$$dB = 20\log\left(\frac{c}{b}\right) - 20\log[\omega] - 20\log\left[\frac{a}{b}\right] \qquad (C.8)$$

From Equation C.8, we see that the asymptote has a slope of −20 dB/decade of radian frequency, or equivalently, −6 dB/octave (doubling) of frequency. The phase of this system is given by

$$\phi(\omega) = -\tan^{-1}\left(\frac{\omega a}{b}\right) \qquad (C.9)$$

Thus, the phase goes from 0° at $\omega = 0$ to −90° as $\omega \to \infty$. Figure C.1 illustrates the complete Bode plot for this simple first-order, low-pass system.

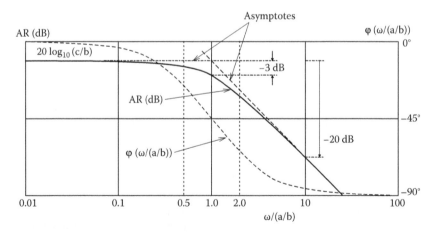

FIGURE C.1 Bode plot (log 20 magnitude and phase vs. f Hz) for the simple low-pass FR function given by Equation C.5. Note the AR is down by −3 dB, and the phase = −45° at the break or corner frequency, $f_c = (b/a)/(2\pi)$ Hz. (From Northrop, R.B., *Endogenous and Exogenous Regulation and Control of Physiological Systems*, Chapman & Hall/CRC Press, Boca Raton, FL, 2000.)

Example C.2

This is an *underdamped, second-order, low-pass system* described by the ODE

$$a\ddot{\mathbf{y}} + b\dot{\mathbf{y}} + c\mathbf{y} = d\mathbf{u}(t) \tag{C.10}$$

As before, $\mathbf{u}(t) = \mathbf{U}(t)\, A\sin(\omega t)$ has been applied to the system for a long time so it is in the steady state. Again, we use the operator notation

$$\mathbf{p}^2 \mathbf{y} a + \mathbf{p}\mathbf{y}b + c\mathbf{y} = d\mathbf{u}(t) \tag{C.11}$$

The output $\mathbf{y}(t)$ will be a sinusoid of some amplitude B and phase ϕ with respect to $\mathbf{u}(t)$ so it can be represented as a vector $\mathbf{Y}$. The input can also be represented as a vector with amplitude A and zero (reference) phase. Again, we let $p = j\omega$.

$$\mathbf{Y}[a(j\omega)^2 + b(j\omega) + c] = d\mathbf{U} \tag{C.12}$$

Thus the FR function is

$$\frac{\mathbf{Y}}{\mathbf{U}}(j\omega) = \frac{d}{a(j\omega)^2 + b(j\omega) + c} = \frac{B}{A}\angle\phi = \mathbf{H}(j\omega) \tag{C.13}$$

To put this FR function in time-constant form to facilitate Bode plotting, we must divide both numerator and denominator by c:

$$\frac{\mathbf{Y}}{\mathbf{U}}(j\omega) = \frac{d/c}{(a/c)(j\omega)^2 + (b/c)(j\omega) + 1} = \frac{d/c}{(j\omega)^2/\omega_n^2 + (2\xi/\omega_n)(j\omega) + 1} \tag{C.14}$$

The constant, $c/a \equiv \omega_n^2$, the system's *undamped natural radian frequency* squared, and $b/c \equiv 2\xi/\omega_n$, where ξ is the system's *damping factor*. The system has complex-conjugate roots to the characteristic equation of its ODE if $0 < \xi < 1$, and real roots if $\xi > 0$. In this *second example*, we have assumed that the system is underdamped, i.e., $0 < \xi < 1$. Now the Bode magnitude plot is found from

$$dB = 20\log\left(\frac{d}{c}\right) - 10\log\left\{\left[1 - \frac{\omega^2}{\omega_n^2}\right]^2 + \left[\left(\frac{2\xi}{\omega_n}\right)\omega\right]^2\right\} \tag{C.15}$$

At dc and $\omega \ll \omega_n$ and $\omega_n/2\xi$, $dB \cong 20\log(d/c)$. The undamped natural frequency is $\omega_n = \sqrt{c/a}$ r/s. When $\omega = \omega_n$, $dB = 20\log(d/c) - 20\log[(2\xi/\omega_n)\,\omega_n] = 20\log(d/c) - 20\log[2\xi]$, and when $\omega \gg \omega_n$, $dB = 20\log(d/c) - 40\log[\omega/\omega_n]$. Thus for $\omega = \omega_n$ and $\xi < 0.5$, the dB curve rises to a peak above the intersection of the asymptotes. The high-frequency asymptote has a slope of -40 dB/decade of radian frequency, or -12 dB/octave (doubling) of radian frequency. These features are

AR (dB)

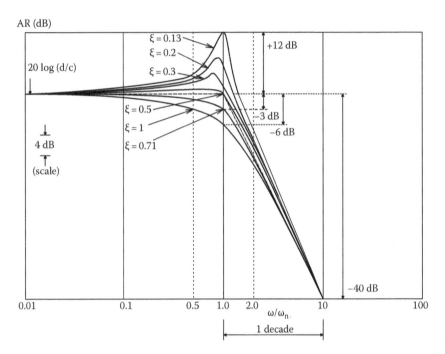

FIGURE C.2   Bode AR plot for a second-order low-pass FR function (see Equation C.14). Different damping factors for the complex-conjugate poles are shown. Note the poles are real for $\xi > 1$. See text for description. (From Northrop, R.B., *Endogenous and Exogenous Regulation and Control of Physiological Systems*, Chapman & Hall/CRC Press, Boca Raton, FL, 2000.)

shown schematically in Figure C.2. The phase of the second-order low-pass system can be found by the inspection of the FR function of Equation C.14. It is simply

$$\phi = -\tan^{-1}\left(\frac{\omega(2\xi/\omega_n)}{1 - \omega^2/\omega_n^2} \right) \tag{C.16}$$

It can be shown (Ogata 1970) that the magnitude of the resonant peak normalized with respect to the system's dc gain is

$$M_r = \frac{|H(j\omega)|_{max}}{|H(j0)|} = \frac{1}{2\xi\sqrt{1-\xi^2}} \quad \text{for } 0.707 \geq \xi \geq 0 \tag{C.17}$$

And the frequency at which the peak AR occurs is given by

$$\omega_p = \omega_n\sqrt{1 - 2\xi^2} \text{ r/s} \tag{C.18}$$

Example C.3

A lead/*lag filter* is described by the ODE

$$\frac{\dot{y}}{\omega_2} + y = \frac{\dot{x}}{\omega_1} + x \tag{C.19}$$

Again, using the derivative operator $\mathbf{p}$, substituting $j\omega$ for $\mathbf{p}$, and treating the output and input like vectors, we can write

$$\frac{Y}{X} = \frac{j\omega/\omega_1 + 1}{j\omega/\omega_2 + 1} = H(j\omega), \quad \omega_2 > \omega_1 \tag{C.20}$$

For the Bode plot, at $\omega = 0$ (dc) and $\ll \omega_1$, ω_2, $dB = 20\log_{10}|H(j\omega)| = 0\,dB$. At $\omega = \omega_1$, $dB \cong 20\log(\sqrt{2}) - 20\log(1) = +3\,dB$. For $\omega \gg \omega_2$, $dB \cong 20\log(\omega_2/\omega_1)$. The phase of the lead/lag filter is given by

$$\phi = \tan^{-1}\left(\frac{\omega}{\omega_1}\right) - \tan^{-1}\left(\frac{\omega}{\omega_2}\right) \tag{C.21}$$

The Bode magnitude response and phase of the lead/lag filter are shown in Figure C.3.

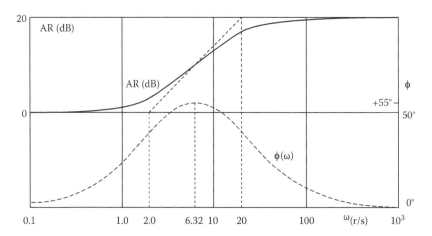

FIGURE C.3   Bode plot for a lead-lag filter given by Equation C.20. Note that the phase reaches a ca. $55°$ lead at $\omega = 6.32\,r/s$, and the AR increases by $20\,dB$ as $\omega \to \infty$. (From Northrop, R.B., *Endogenous and Exogenous Regulation and Control of Physiological Systems*, Chapman & Hall/CRC Press, Boca Raton, FL, 2000.)

Because the SSS FR has traditionally been used as a descriptor for electronic amplifiers and feedback control systems, many texts on electronic circuits and control systems have introductory sections on this topic with examples. (See for example, Schilling and Belove 1979, Section 9.1-2, Northrop 1990, Ogata 1990, Section 6-2, Nise 1995, Chapter 10.) Modern circuit simulation software applications such as MicroCap™, SPICE, and Multisim™ compute Bode plots for active and passive circuits, and MATLAB® and Simulink® will provide them for general linear systems described by ODEs or state equations.

Appendix D: Nyquist Plots

A Nyquist plot is a *polar plot* of the system's steady-state frequency response vector,

$$\mathbf{H}(j\omega) = |\mathbf{H}(j\omega)| \angle \phi(\omega)$$

as $|\mathbf{H}(j\omega)|$ vs. $\phi(\omega)$ for various ω values.

When $|\mathbf{A}_L(j\omega)|$ vs. $\phi(\omega)$ is plotted, its primary use lies in predicting the stability of *nonlinear feedback systems* by the *Popov criterion* (Northrop 2000), or for predicting stability, limit cycle frequency, and amplitude when using the *describing function method* (Ogata 1970, Northrop 2000) on SISO, closed-loop, nonlinear control systems. In the describing function method, the nonlinear feedback system's *loop gain* is partitioned into a nonlinear function and a single-input/single-output (SISO) LTI system. Both the inverse describing function and the Nyquist plot are plotted on polar graph paper.

For a first example of a Nyquist plot, consider the simple, single time constant, low-pass filter:

$$\mathbf{H}(j\omega) = \frac{-1}{j\omega/\omega_o + 1} \tag{D.1}$$

What is plotted in Figure D.1 is

$$|\mathbf{H}(j\omega)| = \frac{1}{\left[\sqrt{\omega^2/\omega_o^2 + 1}\right]} \text{ vs. } \phi = -180° - \tan^{-1}\left(\frac{\omega}{\omega_o}\right) \tag{D.2}$$

Note that this Nyquist plot is a semicircle starting at -1 in the polar plane, and ending at the origin at an angle of $-270°$. At $\omega = \omega_o$, $|\mathbf{H}(j\omega_o)| = 0.707$, and $\phi = -180° - 45° = -225°$.

For a second example of a Nyquist plot, consider the quadratic low-pass frequency response function:

$$\mathbf{H}(j\omega) = \frac{1}{(j\omega)^2/9 + (j\omega)(0.2) + 1} \tag{D.3}$$

Here $\omega_n = 3\,\text{r/s}$, $2\xi/\omega_n = 0.2$, so $\xi = 0.3$. In Figure D.2, we have plotted

$$|\mathbf{H}(j\omega)| = \frac{1}{\sqrt{\{[1 - \omega^2/9]^2 + \omega^2(0.04)\}}} \text{ vs. } -\tan^{-1}\left(\frac{0.2\omega}{1 - \omega^2/9}\right) \tag{D.4}$$

At $\omega = 0$, $|\mathbf{H}(j0)| = 1$ and $\phi = 0°$. When $\omega = \omega_n$, $|\mathbf{H}(j\omega_n)| = 1.667$ and $\phi = -90°$. And as $\omega \to \infty$, $|\mathbf{H}(j\infty)| \to 0$, at an angle of $-180°$.

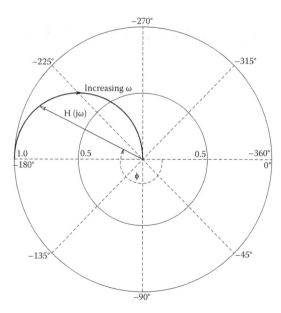

FIGURE D.1 Nyquist (polar) plot of the FR of a simple, inverting low-pass filter (see Equation D.1). (From Northrop, R.B., *Endogenous and Exogenous Regulation and Control of Physiological Systems*, Chapman & Hall/CRC Press, Boca Raton, FL, 2000.)

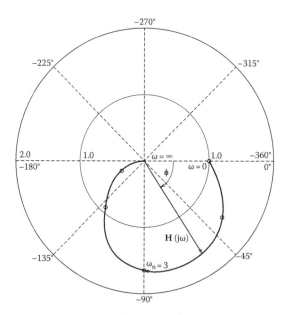

FIGURE D.2 Nyquist plot of the FR of a quadratic low-pass filter with x=0.3 (see Equation D.3). The FR vector **H**(jw) is shown (see Equations D.4). (From Northrop, R.B., *Endogenous and Exogenous Regulation and Control of Physiological Systems*, Chapman & Hall/CRC Press, Boca Raton, FL, 2000.)

Glossary

ABBREVIATIONS AND ACRONYMS

AA	Amino acid
Ab	Antibody
ABM	Agent-based model
ABS	Agent-based simulation
ACS	Adaptive complex system
AcP	Activator protein
ADCC	Antibody-dependent cellular cytotoxicity
Ag	Antigen
AH	Aqueous humor of the eye
AI	Autoinducer, a molecule used to synchronize groups of intracellular biological oscillators by quorum sensing
AMP	Adenosine monophosphate
AMPs	Antimicrobial peptides (or proteins) (in immune systems)
AP	Antigen presentation
APCs	Antigen-presenting cells
BCE	Before Christian era
BCO	Biochemical oscillator
[BG]	Blood glucose concentration [*]
BOLD	Blood oxygen level dependent
bps	(Complementary) base pairs (of DNA)
BQS	Bacterial quasispecies
BSE	Bovine spongiform encephaly
BT or Bt	*Bacillus thuringiensis*
C^3 (in physiology)	Command, communication, and control (of biochemical and physiological processes)
CAbs	Catalytic antibodies
CAIS	Complex adaptive immune system
CAM	Complimentary and alternate medicine
CAMP	Cyclic AMP
CAS	Complex adaptive system
CC	Cyclomatic complexity, or the cell cycle
CD	Cluster of differentiation (CD) antigens (usually cell-surface proteins, e.g., CD8)
CDK	Cyclin-dependent kinases
cDNA	Coding DNA (i.e., genes)
CES	Complex economic system
CIIS	Complex innate immune system

CIP	Cognate immunity protein (in bacteria)
CJD	Creutzfeld–Jakob disease
CNLS	Complex nonlinear system
CS	Complex system
CS	Complement system
CTL	Cytotoxic T-cells (leucocytes)
CWD	Chronic wasting disease (of cervids)
[M]	Concentration of a molecule, M
DE	Difference equation
Dscam	Down syndrome cell adhesion molecule (in humans). A member of the immunoglobulin super-family (IgSF) of proteins. In certain invertebrates, Dscam is associated with immune function
dsDNA	Double-stranded DNA
F1	The first generation following mating
FFL	Feed-forward loop (reaction architecture term coined by Alon 2007b)
FFP	Feed-forward path
fMRI	Functional magnetic resonance imaging
G^-	Gram-negative staining (bacteria)
G^+	Gram-positive staining (bacteria)
GI	Gastrointestinal
GIS	Graphical information systems
GLC	Glucagon
GM	Genetically modified
GM-CSF	Granulocyte-macrophage colony-stimulating factor
GMO	Genetically modified organism
GPS	Global positioning system
gpX	Glycoprotein "X" sgp soluble glycoprotein
HbO	Oxyhemoglobin
HbR	Deoxyhemoglobin
hCAIS	Human complex adaptive immune system
hCNS	Human central nervous system
hCS	Human complement system
HCV	Hepatitis C virus
HGT	Horizontal gene transfer
hIS	Human immune system
ICs	Initial conditions (of an ODE)
IF	Interstitial fluid
IL	Interleukin (e.g., IL2)
IMD	Immune deficiency (pathway)
I/O	Input/output
IOP	Intraocular pressure
IPFM	Integral pulse frequency modulation
IS	Immune system

IVDA	Iterative vector diffusion algorithm (in detecting modularity)
IVGTT	Intravenous glucose tolerance test
IVP	Initial value problem (in an ODE model describing behavior of a CNLS)
LPS	Lipopolysaccharide
LTIS	Linear time-invariant system
LUC	Law of unintended consequences
L–V	Lotka–Volterra
MA	Mass action
MAS	Multi-agent simulation
MB	Memory B-cell
MHC	Major histocompatibility complex molecule
MIMO (in systems theory)	Multiple-input, multiple-output system
Mφ	Macrophage
miRNA	MicroRNA
mRNA	Messenger RNA
MRSA	Methycillin-resistant *Staphococcus aureus*
mya	Million years ago
NA	Nucleic acid
NFB	Negative feedback
NF-κB	Nuclear factor kappa light-chain-enhancer of activated B-cells
NHGB	Net hepatic glucose balance
NK	Natural killer cells in the hIS
NL	Nonlinear
NM	Network motif
Nt	Nucleotide
OECD	The Organization for Economic Cooperation and Development
ODE	Ordinary differential equation
ORF	Open reading frame
ORs	Orthogonal ribosomes
PAF	Platelet activating factor
PAMP	Pathogen-associated molecular pattern
PC	Parametric control
PCD	Programmed cell death (i.e., apoptosis)
pd	Population density (number per volume)
PDF	Probability density function
PETN	Pentaerythritol tetranitrate (an explosive)
PFB	Positive feedback
PG	Plasma glucose, or peptidoglycan
[PG]	Plasma glucose concentration
PHA	Phytohemagglutinin
PR	Parametric regulation (or regulator)

PrP^{Sc}	Misfolded prion protein; the infectious agent in TSEs (PrP or PrP^C=normal prion protein)
QS	Quasispecies; quorum sensing
RAG	Recombination activating gene (in the expression of Abs)
RAMPs	Ribosomally synthesized antimicrobial peptides
RANTES	Regulated on activation, normal T-cell expressed and secreted. Binds with the macrophage CCR5 chemokine receptor
RBC	Red blood cell (erythrocyte)
RNAP	RNA polymerase
RNAi	RNA interference
RP	Regulatory protein
RrP	Repressor protein
RV	Random variable
SFG	Signal flow graph
siRNA	Small interfering RNA
SISO (in systems theory)	Single-input, single-output system
SNP	Single nucleotide polymorphism (a mutation)
SoS	Systems of systems
SRV	Stationary random variable
SS	Steady state
ssRNA	Single-stranded RNA
SV	State variable
TCR	T-cell receptor (cell-surface molecule)
TF	Transcription factor
TL	Transport lag (delay element)
TLR	Toll-like receptor
TP	Tipping point
tRNA	Transfer RNA
TSE	Transmissible spongiform encephaly ("mad cow" disease)
UTR	Untranslated region (of a gene)
vDNA	Viral DNA
VIP	Vasoactive intestinal peptide
VQS	Viral quasispecies
V&V	Verification and validation (of a model)
ya	Years ago

GLOSSARY TERMS

Abzyme: See *catalytic antibody*.

Adjacency matrix, A (in graph theory): This is an $N \times N$ matrix, where N is the number of vertices (nodes) in the graph. If an edge (path) exists from some vertex **p** to some vertex **q**, then the element $m_{p,q} = 1$, else it is 0.

Agent (in CAS modeling): An agent inputs information and outputs behavior and information. They are autonomous decision-making units with diverse characteristics. A network of information flow can exist between agents. They can have "memories." Some agents are adaptive and can learn. Computer models of human agents figure in agent-based models (ABMs) of political, economic, contagion, terrorism, and social CNLSs (complex nonlinear systems), most of which can be treated as CASs. In economic ABMs, human agent models can be programmed to produce, consume, buy, sell, bid, etc. They can be designed to be influenced by the behavior of other, adjacent agents. ABMs can exhibit emergent, collective behavior. Dynamic ABMs have been used to model bacterial chemotaxis (Emonet et al. 2005) and have application in hIS (human immune system) modeling (immune cell trafficking). Electric power companies have used ABMs to better understand the complexities of electric power generation, transmission, and usage. The following species of agents were used in power simulations: customer, physical generator, fuel, transmission line, demand, generation Co., transmission Co., regulatory (power market + ISO + real-time dispatch = administrative), and random event (lightning, failures, etc.) agents. See the ABM tutorial papers by Macal and North (2006) and Bonabeau (2002).

Agent-based model: An ABM is a dynamic system of interacting, autonomous entities. It consists of (1) a set of user-defined agents, (2) a set of agent relationships, and (3) a framework for simulating agent behaviors and interactions (i.e., ABM software; e.g., Swarm, Repast S, Ascape, MASON, MATLAB®, DIAS, NetLogo, StarLogo, SeSAm, Cormas, and VOMAS).

Allele: One of the variant forms of a gene at a particular locus, or location on a chromosome. A single allele for each locus is inherited separately from each parent. Alleles produce variation in inherited characteristics such as eye color or blood type. There are dominant alleles and recessive alleles.

Alternative splicing (in genomics): A process by which primary gene transcripts (pre-mRNA) are reconnected in multiple ways during RNA splicing. The resulting different mRNAs may be translated into different protein isoforms. Thus, a single gene may code for multiple proteins that may have different effects—a form of pleiotropy. Five modes of alternative splicing have been identified: (1) Exon skipping or cassette exon: An exon may be spliced out of the primary transcript or be retained. This is the most common mode in mammalian pre-mRNAs. (2) Mutually exclusive exons: One of two exons is retained in mRNAs after splicing, but not both. (3) Alternative donor site: An alternative $5'$ splice junction (donor site) is used, changing the $3'$ boundary of the upstream exon. (4) Alternative acceptor site: An alternative $3'$ splice junction (acceptor site) is used, changing the $5'$ boundary of the downstream exon. (5) Intron retention: A sequence may be spliced out as an intron, or simply retained.

Anion (in electrochemistry): A negatively charged ion or molecule; one that migrates toward the anode (+) in an electric field.

Anticodon (in genomics): In transfer RNA (tRNA), a sequence of three adjacent nucleotides that binds to a corresponding codon in mRNA. As an example, one codon for the AA (amino acid) lysine is **AAA**, and the anticodon of a lysine tRNA might be **UUU**. Some anticodons can pair with more than one codon due to a phenomenon known as *wobble base pairing*. Frequently, the first Nt of the anticodon is one of two not found on mRNA (inosine and pseudouridine), which can hydrogen-bond to more than one base in the corresponding codon position. In the genetic code, it is possible for a single AA to be specified by all four, third-position possibilities, for example, the AA glycine is coded for by the codons: **GGU**, **GGC**, **GGA**, and **GGG**.

Antiredundancy (in genomic systems): Mechanisms that sensitize cells or individual organisms to genetic damage, and cause them to be preemptively eliminated (as by programmed cell death [PCD], aka apoptosis) from a population. Other genomic mechanisms found in antiredundancy include overlapping reading frames, absence of tRNA suppressor genes, codon bias, loss of DNA repair mechanisms, reduced number of promoters, coordinated expression of genes (operons), and checkpoint genes (Krakauer and Plotkin 2002).

Aquaporins: A unique water transport molecule made up of six, channel-forming, transmembrane protein α-helices, arranged in a right-hand bundle, with the amino and carboxyl termini located on the inside (cytoplasmic side) of the cell membrane. There are over 300 known aquaporins that can be divided into those that are permeable only to water, and those permeable to glycerol as well as water. Aquaporins have been found in all life forms: plants, animals, and bacteria. Their permeabilities to water are regulated by several mechanisms (e.g., hormones, phosphorylation by kinases, and gibberellic acid in plants). AQPs do not actively transport water; they modulate its passive diffusion into and out of cells. There are some 13 known human aquaporins: AQP0–12. The kidneys have no fewer than eight AQPs: AQP1–4, AQP6–8, and AQP11. The genes coding the kidney AQPs are found on chromosomes 7p14, 12q13, 9p13, 18q22, 12q13, 9p13, 16p12, and 11q13, respectively (Chavda and Patel 2008).

Argonaute (in genomics): Argonaute proteins are the catalytic component of the RNA-induced gene-silencing complex (RISC). They are responsible for the gene-silencing process known as RNA interference (RNAi). Argonaute proteins bind to small interfering RNA (siRNA) fragments, and the complex has endonuclease activity against mRNA strands that are complementary to the bound siRNA fragment.

Ashby's law of requisite variety (LRV) (complex systems): "The survival of a system depends on its ability to generate at least as much variety (entropy) within its boundaries as exists in the form of threatening disturbances from its environment." In essence, this means that "...a control[ler or regulatory] system has to be more complicated than the system it is controlling [regulating]" (Smith 2004). Another formulation of Ashby's law: "A model system or controller can only model or control something to the extent that it has sufficient interval variety to represent it. For example, in order to make a choice between two alternatives, the controller must be able to represent at least two possibilities, and thus one distinction" (Heylighen and Joslyn 2001).

Atomic components (in complex systems): A system component that cannot be reduced in terms of other components in the model representation chosen.

Attacins (in invertebrate immunity): Six Attacin proteins have been described (A–F). They have been found in the hemolymph of various moths and flies. Attacins efficiently kill *E. coli* (*Escherichia coli*) and two other Gram-negative bacteria isolated from the guts of silkworms. *B. thuringiensis* (BT) produces an exoprotease that destroys Attacins. In *Drosophila*, Attacin acts synergistically with Cecropin A.

Attractor (in phase plane plots of CNLS behavior): An attractor can be a point, a curve, a closed path, or a set of closed paths in the phase plane to which the phase trajectories of the CNLS converge as the system reaches the steady state.

In the case of a closed path attractor, the system's trajectories can form a stable limit cycle (LC), that is, one in which the system ends oscillating in the steady state, given any set of initial conditions (x_0, y_0). There are also semistable LCs. (An unstable LC is better called a "repeller" than an attractor.) One example of a strange attractor is a pair of connected LC orbits (A and B) that the system follows in the phase plane, switching from A to B and back again in a seemingly chaotic, yet deterministic manner, a behavior noted in the three, coupled Lorenz equations.

Autacoids (in the hIS): (From the Greek *autos* = "self" and *akos* = "medicinal agent" or "remedy.") Locally acting signal molecules secreted by cells of the immune system (IS) and damaged body cells. Some are associated with inflammation. They are used to regulate local IS responses, although they may have more general, hormonal effects once in the circulatory system. Histamine is an autacoid. (See *cytokine*.)

Autocatalytic reaction (in chemistry): Those reactions in which at least one of the products (on the right-hand side of the chemical equation) is a reactant (on the left-hand side of the chemical equation). An example of a simple autocatalytic reaction is $A + B \underset{k_r}{\overset{k_f}{\Longleftrightarrow}} 2B$. An autocatalytic reaction system uses positive feedback, and its mass-action (MA) equation(s) is nonlinear.

The autocatalytic MA ODE is $\dot{b} = k_f ab - k_r b^2$, where a and b are the running concentrations of A and B, respectively. When the simple autocatalytic ODE above is viewed as an IVP, with initial concentrations of $A = a_0$ and $B = b_0$, assuming $k_f \gg k_r$, the running concentrations of A and B can be shown to be given by

$$a(t) = \frac{a_0 + b_0}{\{1 + (b_0/a_0)\exp[+(a_0 + b_0)k_f t]\}} \text{ and } b(t) = \frac{a_0 + b_0}{\{1 + (a_0/b_0)\exp[-(a_0 + b_0)k_f t]\}}$$

Note that $a(\infty) \rightarrow 0$ and $b(\infty) \rightarrow (a_o + b_o)$. Both the $a(t)$ and $b(t)$ curves are sigmoid. An example of an autocatalytic reaction is in the formation of BSE prion protein (PrP^{Sc}) from normal prion protein (PrP_C). The reaction is $PrP^C + PrP^{Sc} \Rightarrow 2\ PrP^{Sc}$.

Autoinducer (AI) molecule (in biochemical oscillators): A synchronizing molecule secreted by groups of like cells having periodic oscillations in gene expression and protein synthesis that is taken up by neighboring cells, ensuring their synchronization with the group.

Autopoiesis (in CASs): Autopoiesis, from the Greek, means self-creation, or self-maintenance. A classical example of an autopoietic system is a eukaryotic cell, which has components including a cell membrane, a nucleus, a DNA genome, a cytoskeleton, and various organelles. Genomic information plus an external source of energy and substrate molecules allow the cell to maintain homeostasis, and reproduce itself. An autopoietic system is autonomous and operationally closed, in the sense that every biochemical process within it directly helps maintaining the whole.

Average geodesic distance (of a graph): The average distance D of each node to any other defines the average geodesic distance of a graph:

$$D = \left(\frac{1}{m} \right) \sum_{i=1}^{n} \sum_{j=1}^{n} d(i, j)$$

where
  n is the total number of nodes
  $d(i, j)$ is the shortest path distance between i and j
  m is the total number of edges in the graph

Azidothymidine (AZT): A drug used to fight HIV (human immunodeficiency virus) infection and AIDS (aka Zidovudine). AZT blocks the action of the HIVs' reverse transcriptase, inhibiting viral replication in infected IS cells. The intracellular half-life of AZT is ca. 3–4 h (Hardman and Limbird 1996).

***Bacillus thuringiensis* (BT)**: *B. thuringiensis* is a G^+, soil-dwelling bacterium whose endospores produce crystals of mixed, proteinaceous, insecticidal, δ-endotoxins. These crystalline endotoxins are produced by the expression of bacterial *cry* genes. BT-toxins have specific activity against Lepidoptera, Diptera, Coleoptera, Hymenoptera, and nematodes. *cry* genes have been inserted into the genomes of GM crops, such as BT-corn and BT-cotton, to make the plants insect resistant. The LUC is seen in the evolution of BT-resistant crop pests due to constant exposure to the endotoxins in host BT-plants (Candas et al. 2002).

Bacteriocin (natural antibiotics from bacteria): Small, heat-stable, protein toxins produced by G^+ bacteria that inhibit the growth of similar or closely related bacterial strains. The producing strains of bacteria are immune to their own bacteriocin. Some bacteriocin AMPs are produced by ribosomes with classical gene transcription, and then posttranscriptionally modified to their active form. Others are produced enzymatically by modular, non-ribosomal peptide sythetases. Colicine is a bacteriocin produced by *E. coli*. Bacteriocins produced by *Staphococcus warneri* are called warnerins or warnericins. Bacteriocins have been grouped into three classes, I, II (a, b, and c), and III, by their molecular structures and method of killing target organisms. For example, class IIa, pediocin-like bacteriocins are produced by strains of the ubiquitous, G+ lactic acid bacteria, *Pediococcus* sp. They contain between 37 and 48 residues. They kill target cells (e.g., *Listeria* sp., *Clostridium* sp., and *Enterococcus* sp.) by making pores in the target cell's membrane. (See the review by Papagianni and Anastasiadou 2009.) Class IIb are two-peptide bacteriocins. The use of bacteriocins to preserve food (e.g., Nisin) and fight tooth decay (e.g., Triclosan) is well established.

Bacteriophage: A virus that reproduces exclusively in a host bacterial species. There generally is a narrow host specificity for bacteriophages.

Baraminology (in creationism): (coined from the Hebrew words *bara* [create] and *min* [kind]). The classification of categories of living creatures according to how they are described in the Bible in *Genesis*, *Leviticus*, and *Deuteronomy*. Baraminology depends on faith in an interpretation of the Bible, and has no basis in the modern theory of evolution, genomics, genetics, etc.

Bifurcation (of a phase portrait): A bifurcation occurs when a small, smooth change made to a parameter (input, initial condition, gain, etc.) in a continuous system causes a sudden topological change in the system's long-term dynamical behavior. In the phase plane, a new attractor appears. Under some conditions, the cause of a bifurcation is called a *tipping point*. (Also see *Hopf bifurcation*.)

Bilateria (in evo-devo): Metazoan animals having bilateral symmetry. They also have a head and tail end, and a dorsal and ventral surface. Jellyfish and starfish are radially symmetrical with dorsal and ventral surfaces; they are not bilateria, however.

Bilateria are triploblastic, that is, they have embryos that develop from three different germ cell layers: the ectoderm, the mesoderm, and the endoderm. Humans are bilateria.

Bottlenecks (in genetics): A genetic bottleneck is a stochastic event that causes a strong reduction in an effective population size. A bottleneck takes place when only a few individuals taken from a large population start a new population. Thus, bottlenecks reduce the genetic diversity of a new population. Bottlenecks occur in many diseases (viral and bacterial) transmitted by respiratory droplets, because the droplets often contain few pathogens. They can accelerate the accumulation of mutations in the consensus sequence of a quasispecies.

Buffy coat (in blood): When a whole blood sample is centrifuged, it separates into three layers: At the bottom, about 45% of the sample is erythrocytes (red). In the middle is a thin whitish layer, <1% of the sample. This is the buffy coat; it consists largely of leukocytes and platelets. The top clear layer (ca. 55%) is plasma. The buffy coat is the source of nucleated blood cells used in the DNA analysis from blood.

Butterfly effect (in CNLSs): A term given to the tipping-point behavior of a CNLS. Often misused in a flawed analogy where the feeble air currents caused by the gentle flapping of a butterfly's wings may lead to a hurricane or tornado at some distant location. The analogy fails to take into consideration the dissipative effect of action at a distance, and that some system parameter must cross a threshold before the CNLS switches behavioral modes.

Campbell's law: "The more any quantitative social indicator is used for social decision making, the more subject it will be to corruption pressures and the more apt it will be to distort and corrupt the social processes it is intended to monitor." Educational achievement tests are cited as subject to Campbell's Law, especially in the context of the No Child Left Behind Act.

Canalization (in genetics; adaptive complex systems): A measure of the ability of a genotype to produce the same phenotype regardless of the variability of its environment (niche), that is, genomic robustness. Canalization can be the result of DNA repair mechanisms and/or chaperone proteins. (See, e.g., Wagner et al. 2007.)

Catalytic antibody (CAb) (in biochemistry): (Also called an Abzyme, or a Catmab [catalytic monoclonal antibody].) A natural or bioengineered antibody that has the property to bind to a specific molecular site and then catalyze a targeted biochemical reaction. Monoclonal CAbs can be used, in conjunction with prodrugs, for targeted destruction of pathogens and tumors.

Cathelicidin (an AMP in the urinary tract): This AMP is continuously synthesized and released without storage from tubular epithelial cells into the tubular lumen. It is also inducible by exposure to *E. coli*. Elevated synthesis continued for hours (Zasloff 2007).

Cation (in electrochemistry): A positively charged ion or molecule; one that migrates toward the cathode (−) in an electric field.

Cecropins (in invertebrate immunity): A family of 3–4 kDa, cationic, linear, amphipatic peptides, first isolated from the hemolymph of the giant silk moth, *Hyalophora cecropia*. Cecropins are positively charged, and are most effective against Gram-negative bacteria.

Chaoplexy: A fusion of chaos theory with complexity mathematics; a term coined by Horgan (2008).

Chaos (in CNLSs): Certain nonlinear dynamic systems described by differential equations exhibit TP behavior, that is, their dynamics are highly sensitive to minute changes in initial conditions, input variables, and/or signal rate parameters. These abrupt, seemingly stochastic changes in behavior are called chaos; however, they are in fact deterministic, and are the result of dynamic interactions in the system. The initiation of turbulence in fluid flow is an example of chaos. Certain simple, nonlinear ODEs, such as the van der Pol equation, exhibit chaos in their behavior in response to small parameter changes.

Chaos theory (in CNLSs): Chaos theory attempts to describe the behavior of nonlinear dynamical systems that under certain conditions (of inputs and initial conditions) exhibit the phenomenon known as *chaos*. Such chaotic behavior is, in fact, deterministic. However, its occurrence may at first appear random. Under appropriate conditions, a chaotic CNLS may exhibit periodic LC oscillations in its states, or unbounded responses. Certain chaotic CNLSs can exhibit tipping point behavior where a combination of initial conditions and/or inputs can trigger an abrupt transition from non-oscillatory, steady-state behavior to LCs, and vice versa. Complex systems exist on a spectrum ranging from equilibrium to complete chaos. A system in equilibrium does not have the internal dynamics to enable it to respond to changes in its environment and will die (or is not alive, e.g., viruses and crystals). A system in chaos ceases to function as an organized system. The most productive state to be in is at the edge of chaos, where there is maximum variety and creativity, leading to new possibilities of behavior. (Fryer.)

Chaperonins and chaperones (in protein chemistry): Chaperones include chaperonins. Chaperones are complex proteins whose function is to ensure that a newly assembled polypeptide is bent, folded, and cross-linked to form the desired, active, 3D protein structure. Some chaperone proteins, the heat-shock proteins, have the role of repairing proteins whose 3D structures are damaged by elevated temperature, chemicals, or ionizing radiation. Other chaperones are involved in the transport of new proteins across membranes (e.g., in mitochondria and the endoplasmic reticulum [ER]). Some chaperones may be involved in protein degradation.

Chaperonins are molecular machines that use chemical energy (ATP) to fold and bond new polypeptides into their mature, functional, 3D protein forms. The structure of a chaperonin is generally in the form of two, stacked, protein tori ("donuts") that form a hollow "barrel" in which the folding occurs. Each chaperonin torus is composed of 7–9 protein subunits, depending on the organism, etc. Group I chaperonins are found in prokaryotes, as well as eukaryotic chloroplasts and mitochondria. Group II chaperonins occur in Archaea and the eukaryotic cytosol.

Chemokines (CKs) (in cell biology): Substances secreted by certain cells that attract certain other IS cells by chemotaxis. There are over 48 human chemokines. Inflammatory chemokines are secreted by distressed tissues to call-in innate and adaptive IS cells to sites of injury or infection to effect healing. One method of classifying chemokines is based on the position of one or two conserved cysteine residues in their AA sequences; four subfamilies are identified: CC-, CXC-, XC, and CX3C-. Another two-group chemokine classification scheme is based on function: inflammatory and homeostatic (lymphoid). In this classification, lymphoid CKs include CCL19, CCL21, CXCL12, CXCL13, CCL17, CCL22, CCL25, CCL27, and CCL28, to name a few. Another important chemokine that causes migrations of dendritic cells (DCs), T-cells, B-cells, and mast cells is sphingosine 1-phosphate (S1P) (Rosen and Goetzl 2005, Maeda et al. 2007, Rivera et al. 2008).

Chemostat (as used in microbiology): A bioreactor chamber in which microorganisms are grown under controlled conditions of chemical and physical parameters. Fresh culture medium is continually added and removed, keeping the culture volume constant. By manipulating the flow-through rate, the nutrients, and physical parameters (T, pH, pO_2, etc.), the growth rate can be easily controlled.

Circadian (in biological clocks): Biological rhythms, when unsynchronized by external cues, have nearly 24 h periods. (From the Latin *circa* = approximately, about; and *dies* = day.)

Circuit rank (in graph theory): (Also known as the cyclomatic number.) The circuit rank of a graph is the minimum number $\mathbf{r}$ of edges that must be removed from a graph to make it cycle-free (i.e., to remove all feedback loops). $\mathbf{r} = \mathbf{e} - \mathbf{n} + \mathbf{c}$, where $\mathbf{e}$ is the number of edges in the graph, $\mathbf{n}$ is the number of nodes in the graph, and $\mathbf{p}$ is the number of disjoint partitions the graph divides into (also given as "separate components").

Cistron (in genomics): A DNA segment coding for a single polypeptide, including its own start and stop codons.

Clade (in evolutionary biology): A taxonomic group that includes a common ancestor and all the ancestor's descendents. Such a clade is considered to be a *monophyletic group*. Because reptiles were supposed to have evolved into birds, the vertebrate phylogenetic tree in which reptiles are grouped with Testudines, Lepidosauria, Crocodylia, Aves, Archosauria, and Diapsida is considered to be a *monophyletic grouping*. A clade that includes warm-blooded vertebrates would include only Mammalia and Aves, and be called a *polyphyletic group*.

Clique (in graph theory): A clique is a subgraph in which every possible pair of nodes is directly connected, and the clique is not contained in any other clique. (Also see *modularity*.)

Clustering coefficient (property of a graph node): The ratio between the number of edges between the graph nodes that the node in question is connected to and the number of all possible edges between them. The CC ranges between 0.0 and 1.0.

Codon (in genomics): A sequence of three consecutive DNA bases in a gene's single DNA strand that codes for one unique AA being incorporated into a linear polypeptide molecule, which goes on to form the protein product of the protein-coding gene. For example, the DNA bases **TGG** code for the AA tryptophan, and the bases **CAC** and **CAT** both code for histidine.

Coevolution (in evo-devo): Seen in van Valen's Red Queen hypothesis (1973). Coevolution is, broadly, the change of a biological object triggered by the change of a related object. Coevolution can occur at all levels of scale, from SNPs in coding genes to covarying traits between two competing species in an ecosystem. The species can be a host animal's IS and a parasite, a host animal's IS and a bacterium or a virus, or a phage and a bacterium, or two different bacterial species. Evolution in response to environmental changes is not coevolution.

Cognate immunity protein (CIP) (in bacteria): A protein synthesized by a bacterial species that makes it immune to its own secreted bacteriocin. G^+ bacteria that produce pediocin-like AMPs also produce an 11 kDa CIP that is a well-structured, four-helix bundle cytosolic protein. The CIP shows a high degree of specificity in protecting the bacteria from its cognate AMP and, in some cases, from a few AMPs that are closely related to the cognate AMP. The C-terminal half of the CIP contains a domain that is involved in the specific recognition of the C-terminal membrane recognition hairpin domain of the cognate AMP (Fimland et al. 2005). Pyogenecin is a CIP found in *Streptococcus pyogenes*, associated with the class IIb bacteriocins secreted by *S. pyogenes* (Chang et al. 2009). The genes for CIPs are often found on the same operon as the cognate AMP.

Compartment (in pharmacokinetic systems): A physical volume throughout which a certain chemical species is found well mixed in constant concentration (e.g., blood volume, cerebrospinal fluid volume, aqueous humor volume, and cell volume). Compartments can be interconnected and interact by diffusion or active transport. An N-compartment system is described by N first-order ODEs.

Competition (in biology): Occurs in a population of individuals given finite resources and/or an environmental challenge. Competition can be for energy resources, mates to breed with, or for surviving an adverse physical or chemical change in the surrounding environment. Survivors are selected to breed their genotypes.

Complex adaptive system (CAS): A CAS has the capacity to change its parameters (nodes and branches) in order to optimize its performance, a form of learning. The term complex adaptive systems was coined at the Santa Fe Institute by Holland, Gell-Mann et al. Examples of CASs include the brain, the IS, ant colonies, and manufacturing businesses. CASs generally develop robustness as they adapt. Holland's definition of a CAS is "A Complex Adaptive System (CAS) is a dynamic network of many agents (which represent cells, species, individuals, firms, nations) acting in parallel, constantly acting and reacting to what the other agents are doing. The control of a CAS tends to be highly dispersed and decentralized. If there is to be any coherent behavior in the system, it has to arise from competition and cooperation among the agents themselves. The overall behavior of the system is the result of a huge number of decisions made every moment by many individual agents."

A key feature of all CASs is that their behavior patterns as a whole are not determined by centralized authorities but by the collective results of interactions among independent entities.

CAS system behavior generally exhibits robustness. The brain is a CAS.

Complexity: (From the Latin, *complexus*, meaning twisted together or entwined.) Broadly stated, complexity is a subjective measure of the difficulty in describing and modeling a system (thing or process). In this respect, complexity lies in the eyes of the beholder. In other words, complexity is a global characteristic of a system; it represents the gap between component knowledge and knowledge of overall behavior. There is a lack of high predictability of the system's output. Complexity is relative, graded, and system dependent. A necessary (but not sufficient) property of a complex system is that it is composed of many components that are interconnected in intricate ways, some nonlinear and/or time variable. The complexity of a system can also be correlated with stochastic parameters, and unspecified connectivity parameters. One measure of a system's complexity is the amount of information necessary to describe the system. A complex system may exhibit hierarchy and self-organization resulting from the dynamic interaction of its parts. It is also sensitive to initial conditions on its states. Complex systems can be subdivided into fixed (time-invariant) and dynamic systems. A major challenge in objectively analyzing and describing a complex system model is the choice of mathematical algorithms to use.

Complex systems include, but are not limited to, biochemical pathways, cells, organs (the liver, the IS, and the CNS), physiological systems, organisms, ecosystems, economic systems, weather, political parties, governments, corporate management, etc.

Connectance (in graph theory): Connectance has been defined as

$$\mathbf{Conn}[\%] = \frac{2\mathbf{E}}{\mathbf{V}(\mathbf{V}-1)}$$

where

$\mathbf{V}$ is the number of graph vertices (nodes)
$\mathbf{E}$ is the number of graph edges (Bonchev 2004)

Connected component (in graph theory): A connected component of an undirected graph is a subgraph in which any two vertices are connected to each other by paths, and to which no more vertices or edges can be added while preserving its connectivity. See also *strongly connected graph*.

Connected graph (in graph theory): A graph is connected if there is a path connecting every pair of nodes. A graph that is not connected can be divided into connected components (disjoint connected subgraphs).

Connectivity (in graph theory): A graph is said to be connected if it is possible to establish a path from any vertex (node) to any other vertex of the graph. Otherwise the graph is disconnected. A graph is totally disconnected if there is no path connecting any pair of vertices.

Controllabilty (of a linear time-invariant system [LTIS]): Given an nth order LTIS characterized by the linear state equations:

$$\dot{x} = Ax + Bu$$

$$y = Cx$$

where
   x are the n states of the LTIS
   u are the inputs
   y are the outputs

The system is controllable if the controllability matrix, C_M, is of rank n. C_M is defined as

$$C_M \equiv \begin{bmatrix} B & AB & A^2B & \cdots & A^{n-1}B \end{bmatrix}$$

Controller (in feedback systems): A controller is a dynamic subsystem of a feedback control system that causes an output to follow a variable input. The departure of the controlled output from the desired value is called an *error*. In a simple feedback controller, the error causes the controller to generate a corrected output. A controller can also act as a regulator, forcing a feedback system output to remain constant in spite of external disturbances and internal system noise.

Creation biology: Biology examined from a faith-based, creationist perspective in which God is assumed to have created all life on the planet, as described in the *Genesis* account of creation.

Cross-reactivity (in the hIS): A property of the hIS. When a single antigen provokes an immune reaction, it activates a limited spectrum of immune effectors (Abs, Ths) rather than a single, specific set of immune effectors. Cross-reactivity is not degeneracy.

Crustins (in invertebrate immunity): A family of AMPs isolated from shrimps and crabs. An 11.5 kDa antibacterial peptide, Crustin Cm1, is found in the crab *Carcinus maenas*.

Cycle (in graph theory): A cycle is a closed walk or path having at least three nodes, in which no edge is repeated. An engineer will recognize a cycle as a feedback loop.

Cyclin (in developmental control): Proteins involved with the control of the cell cycle (CC) by their association with CDKs (see below). cyclins D (D1, D2, and D3) are associated with the G1-phase of the CC, cyclins E and A with the S-phase of the CC, and cyclins B and A with mitosis.

Cyclin-dependent kinase (CDK) (in developmental control): A family of protein kinases involved in regulating the cell cycle, transcription, and mRNA processing. The expression of a CDK is regulated by a cyclin protein, forming a CDK complex. CDKs phosphorylate protein serine and threonine residues on target proteins. CDK4 is involved with the G1-phase, CDK2 with the S-phase, and CDK1 with the M-phase of the CC. CDKs themselves are regulated. For example, the p53 protein can block the activity of CDK2 and halt the CC in G1. *p53* plays the role of a tumor suppressor gene.

Cyclomatic complexity (in computer science): A software metric. CC=**M** is computed using a graph that describes the control flow in a program. The graph's nodes (vertices) correspond to the commands of the program. A directed edge connects two nodes if the second command can be executed immediately after the first. Cyclomatic complexity directly measures the number of linearly independent paths through a program's source code. **M** is given by

$$\mathbf{M} \equiv \mathbf{E} + 2\mathbf{P} - \mathbf{N}$$

where
  E is the number of edges in the graph
  N is the number of nodes in the graph
  P is the number of connected components

An alternative definition is $\mathbf{M} \equiv \mathbf{E} + \mathbf{P} - \mathbf{E}$ (this has also been called "Circuit Rank"). Yet another way of calculating **M** uses the number of closed loops **C** in the graph: $\mathbf{M} \equiv \mathbf{C} + 1$. (See McCabe 1976 and Crawford 1992.)

Cyclomatic number r (in graph theory): See *circuit rank*.

Cytokine (in cell biology): (From the Greek *cyto*=cell, and *kinos*=movement.) Certain proteins and peptides used for intercellular signaling, for example, by the IS. They are secreted by a wide variety of cell types and can have effects on nearby cells with appropriate cell-surface receptors, on the secreting cells, or on cells throughout the body. Autocrine cytokines act on the cells that secrete them. These are also called *autoinducer* molecules. Paracrine cytokines act locally on surrounding cells, and endocrine cytokines act at a distance, carried by blood or lymph. Cytokines are generally smaller, water-soluble glycoproteins having masses between 8 and 30 kDa. Autacoids are cytokines released specifically by cells of the IS. Specialized cytokines have been classified as autacoids, lymphokines, interleukins, and chemokines. Chemokines mediate chemoattraction, aka chemotaxis or trafficking of cells.

Defensins (in invertebrate and innate vertebrate immunology): A family of small, 3–4 kDa, cysteine-rich, cationic peptides found in both vertebrates and invertebrates (insect hemolymph). They contain 18–45 AAs, and are effective against bacteria, fungi, and many viruses. They kill bacteria by making holes in their membranes. In humans, neutrophils secrete HNP1–4 defensins, and Paneth cells in the small intestine secrete defensins HD5 and HD6 into the lumen, protecting intestinal stem cells and influencing the composition of intestinal commensal bacteria. β-Defensins are widely expressed throughout human epithelia.

Degeneracy (in biology): The ability of various elements of a system that are structurally different to perform the same function or yield the same output; a well-known characteristic of the genetic code and the IS. (q.v. Pleiotropy.)

Degeneracy as a term applied to biological systems was first used to describe the relationship between codons and the AAs they specify in protein synthesis.

Degeneracy is required for natural selection because natural selection can only operate on a population of genetically dissimilar organisms. Degeneracy can be viewed as a feature of complexity at molecular biological, genetic, cellular, physiological, and population levels (Edelman and Gally 2001). Degeneracy is not redundancy.

Degree (in graph theory): (A nonnegative integer.) The degree, $d_G(v)$, of a vertex (node) v in a graph G is the number of edges incident to v; loops are counted twice. If $d_G(v) = 0$, node v is isolated in graph G.

Degree distribution (in graph theory): The degree distribution of a graph or network is the probability distribution, $P(k)$, of node degrees over the N nodes in the whole graph or network. $P(k)$ is the fraction of the nodes in the graph/network with degree k. (k is an integer ≥ 0.) Thus, $P(k) = n_k/N$, where n_k of N nodes have degree k.

Dendritic cells (DCs) (in the IS): These are actively motile IS leukocytes that possess many surface projections analogous in form to the dendrites of neurons. They ingest foreign antigens (bacteria, viruses, fungi, and cell debris) by endocytosis, process them, and then display Ag fragments with the MHC (major histocompatibility complex molecule)-peptide complex plus the costimulatory B7 molecules that bind to CD28 on the helper T-cells (Th), enabling Ag presentation and activation of the Th. When DCs ingest self-Ags (e.g., dead body cells and their fragments), they are processed, but are presented to Th without any costimulatory molecules. This causes the Th to divide for several cycles, and then kill themselves by programmed cell death (apoptosis), avoiding autoimmunity. The toll-like receptors (TLRs) on DCs "recognize" (have an affinity for) pathogen-associated molecular patterns (PAMPs). PAMPs are molecular patterns conserved in bacteria and viruses, for example, peptidoglycan of Gram + bacteria, flagellin from bacterial flagellae, and the endotoxin of Gram − bacteria). Stimulated DCs secrete IL-12, IL-23, and also IL-10, which stimulates regulatory (suppressor) T-cells. In summary, DCs present antigen to and stimulate Th and B-cells

(Kimball 2007b). DCs are being used in the development of autologous cancer vaccines (Kochar 2006).

Deuterostome (in evo-devo): A subtaxon of Bilateria animals that are distinguished by their development. The first embryonic opening (the blastopore) becomes the animal's anus. The four living phyla of deuterostomes include Chordata (vertebrates), Echinodermata (sea urchins, starfish, sea cucumbers, etc.), Hemichordata (acorn worms), and Xenoturbellida (two species of wormlike animals).

Dicer (in genomics): A very important cytoplasmic RNase III enzyme that processes the pre-miRNA to produce the mature ca. 22 Nt miRNA duplex. GM mice without the *dicer-1* gene die during gastrulation; mouse embryonic stem cells are impaired in their ability to proliferate (Bushati and Cohen 2007).

Diptericins (invertebrate immunity): A family of antibacterial proteins of ca. 82 AAs found in the adults and larvae of dipteran flies.

Diagraph (in Vester's analysis of CSs): A 2D plot of vectors derived from the "impact matrix" of two-bit, internodal "influences" in the graphical model of a CS. The diagraph indicates approximately the impact of each node on each other, and allows tentative identification of inputs, outputs, and internal (intermediate) nodes (see Section 12.3). A diagraph is not a digraph (see below).

Digraph (in graph theory): A graph with directed branches between its nodes, such as a signal flow graph (SFG). The directed branches indicate the direction of signal (or information) flow between pairs of nodes.

Distance (in graph theory): In a graph, d_{pq} is the smallest number of edges connecting nodes **p** and **q**.

Distance matrix, D (in graph theory): A symmetric $N \times N$ matrix an element of which, d_{pq}, is the shortest path length between nodes **p** and **q**. If there is no path between **p** and **q**, then $d_{pq} = \infty$. If the shortest path between **p** and **q** contains three edges, then $d_{pq} = 3$.

Distributed robustness (DR) (in CNLSs): In DR, many parts of a system contribute to overall system function, but all of these parts have different roles. When one part fails or is changed through mutations, the system can compensate for this failure, but not because a redundant (backup) part takes over for the failed part (Wagner 2005). DR is related to complex adaptive behavior.

Distance (in graph theory): The distance d_G (u, v) between two vertices (nodes), u and v, in a graph **G** is the (integer) number of edges in a shortest path connecting them. (This is also known as the *geodesic distance*.) If u and v are identical, d_G (u, v) = 0. When u and v are unreachable from each other, d_G (u, v) = ∞.

Distance matrix (in graph theory): A symmetrical, $N \times N$ matrix containing the distances, taken pairwise, of a set of points in a graph. Its elements are nonnegative real numbers, given N points in Euclidean space. A distance matrix describes the costs or distances between the vertices (nodes), and can be thought of as a weighted form of an adjacency matrix. The number of pairs of points, $N \times (N-1)/2$, is the number of independent elements in the $N \times N$ distance matrix.

Drosocin (in invertebrate immunity): An antibacterial, proline-rich peptide of 19 AAs (GKPRPYSPRPTSHPRPIRV) found in *Drosophila melanogaster*. It is active mainly against Gram-negative bacteria. Glycosylated Drosocin is active against *E. coli* in the hemolymph of flies. It is one of many invertebrate AMPs.

Drosomycin (in invertebrate immunity): An inducible antibacterial and antifungal peptide of 44 AAs found in *Drosophila*. It shares some molecular structural features with plant antifungal defensins.

Dscam (in invertebrate ISs): Invertebrate Dscam proteins are homologs of vertebrate Down's syndrome cell adhesion molecules (DSCAM). In invertebrates, Dscam isoforms serve the role of vertebrate Abs in the identification of invading pathogens. In the fruit fly *Drosophila*, there are over 38,000 different isoforms of the Dscam protein. In the Dscam of the shrimp *Litopenaeus vannamei*, which has 1587 AAs, there are ca. 8970 possible isoforms of this protein. Dscam isoforms evidently can reside as cell-surface recognition molecules in certain invertebrates' cytophagic hemocytes, or be distributed in the animal's hemolymph.

Eclosion (in infection): Eclosion is the release of a number of virions or of a parasite from within the cell. Eclosion ruptures the cell membrane and kills the cell.

Economic allocation (in complex economic systems): Allocation is the channeling of the various factors (components) of production into the different types of production required to generate the particular mix of products (goods) that households (markets) demand. A market will be allocatively efficient if it is producing the right goods for the right people at the right price, that is, its prices are equal to its marginal costs in a perfectly competitive market.

Economic distribution (in complex economic systems): Distribution describes exactly who gets what goods that have been created. Considered are the particular mix of persons who consume a product by virtue of the incomes they have earned for the factors of production they have contributed to in the prior production of the goods currently being consumed.

Edge (in graph theory): A line connecting two vertices (aka nodes), called *end-vertices* (aka endpoints). An edge is also called a (unidirectional) branch in SFGs.

Eicosanoids (in the IS): A family of autacoids that includes the leukotrienes, thromboxanes, and prostacyclin. Collectively, they are called eicosanoids because they are synthesized from 20-carbon fatty acids that are in turn derived from the 20-carbon arachidonic acid. The metabolites of arachidonic acid are varied, and have diverse pharmacological effects. Receptors for eicosanoids are highly specific for affinity and effect.

Emergence (in CASs): The unpredicted appearance of new characteristics, organization, structures, or behaviors in the course of biological or social evolution. Unexpected emergent characteristics are the result of dynamic interactions between the components of a complex system. Biological complexity may emerge as a response to evolutionary pressure on the encoding of structural features that lead to differential survival.

Endonuclease (aka restriction endonuclease [RE]): A protein enzyme that cleaves a single-stranded DNA molecule at a specific site (recognition sequence of 4, 5, or 6 nucleotides) in its nucleotide sequence. Many different endonucleases exist. REs are used in DNA forensic analysis.

Entity (in systems theory): The independent elements or parts of a system. In an SFG, they are the SFG's input nodes.

Entropy (in thermodynamics, ecology, economic systems, and information theory): In thermodynamic systems, entropy is defined by the equation

$$S_2 - S_1 \equiv \int_2^1 \frac{dQ}{T} \geq 0 \quad \text{(a thermodynamic system going from state 1 to state 2)}$$

where

S_k is the entropy of a system at state k; its unit is J/°K

T is the system's Kelvin temperature

dQ is a differential change in system energy (joules, ergs, calories, BTUs, etc.) going from state 1 to state 2

Note that no real physical process is possible in which the entropy decreases. Entropy also can be thought of as a measure of thermodynamic disorder, which is maximum at thermodynamic equilibrium, where a system has uniform temperature, pressure, composition, etc., at all points in its volume.

The order that living cells produce as they metabolize, grow, and reproduce is more than compensated for by the disorder they create in their surroundings as a result of their lives. Cells preserve their internal order by using stored information (their genomes) to direct their processing of free energy, and returning to their surroundings an equal amount of energy as heat and entropy. Unlike inanimate physical systems, living and economical systems make use of stored information. Some free energy must be used to reproduce or transmit this information to successive generations.

In information theory, entropy is a measure of the amount of information that is missing in a transmitted message. Informational entropy was defined by Claude Shannon at Bell Labs in the late 1940s.

Epistasis (in population genetics): A nonreciprocal interaction between two or more genes in which one gene suppresses the expression of another affecting the same part of an organism. The gene whose phenotype is expressed is called *epistatic*, while the phenotype suppressed is called *hypostatic*.

Epitope (in the IS): A specific structural domain on an antigen (Ag) molecule to which an antibody (Ab) molecule has a high chemical affinity. The paratope region of the Ab or receptor molecule binds strongly chemically to the epitope on the antigen. Certain IS B- and T-cells can also have cell membrane-bound Abs having paratopes to an antigenic epitope.

Ergodic (in noisy signals and systems): A noisy signal is assumed to be ergodic when one sample record taken from an ensemble of records is assumed to be equivalent of any other record taken from the ensemble. Thus time averages can be used in lieu of ensemble or probability averaging to calculate statistics for the signal (e.g., mean, mean square, and autocorrelation).

Eukaryotes (in cell biology): Cells and animals with cells that have true nuclei (dsDNA associated with histones and a bilayer nuclear membrane), plus other cytoplasmic organelles (Golgi apparatus, mitochondria, chloroplasts, etc.), all bounded by a plasma membrane.

Evolutionary landscape (in evolutionary theory): A multidimensional parameter space that organisms occupy. The dependent parameter is usually the fitness of the organism (plotted on the ordinate or "y" axis). When a landscape is "flat," it signifies that changes in the immediate mutational neighborhood are of negligible effect on the fitness function (FF). This seldom-seen situation can be interpreted to mean that the organism is not under any significant selection pressure. A poorly adapted organism exhibits a rough landscape with many peaks and troughs. Although many parameters affect fitness, we humans are most comfortable with 3D plots (x, y, z), or even 4D plots (**r**, x, y, z) (**r** is the radius of a vector from the origin). (See *fitness landscape*.)

Fas ligand (FasL) (in cellular apoptosis): FasL is a type II transmembrane protein that belongs to the tumor necrosis factor (TNF) family. When FasL binds to the Fas receptor (FasR-I), the event initiates apoptosis of the cell on whose membrane FasR is located. FasL–FasR interactions are important in the progression of cancer and in regulating the hIS. FasR is a type I transmembrane protein; its gene is located on human chromosome 10. Note that FasL and FasR can be expressed on the same cell's surface. There is a soluble form of FasL that does not induce FasR trimerization and the induction of the death-inducing signaling complex (DISC).

Fc receptor (FcR) (in the IS): FcR is a cell-surface protein with high affinity to the "constant" or stem region of a "Y"-shaped Ab molecule. Mast cells have FcRs for IgE.

Fitness (in evolutionary biology): (Definitions of fitness may vary.) Fitness is a measure of a species' reproductive success in its niche. Those individuals who leave the largest number of mature, breedable offspring are the fittest. Fitness can be attained in several ways: (1) Selection for longevity (survival). (2) Selection for mating success. (3) Selection for fecundity or family size (e.g., large litters and several breeding seasons). (4) Individuals are robust to adverse conditions in the niche (e.g., sudden changes in temperature, light, salinity, and pO_2, and escape from predators). (5) Ability to find food/energy sources. (6) Ability to pass on advantageous, learned behavior to the next (F1) generation, etc. (7) Any combination of the preceding ways.

If fitness is the capability of an individual with a certain genotype to reproduce, one measure of fitness is the proportion of an individual's genes found in all of the genes of the F1 generation. *Absolute fitness* is given by $f_{abs} \equiv N_{after}/N_{before}$, where N_{before} is the number of individuals with a particular genotype, and N_{after} is the number of individuals with that genotype after selection. Absolute fitness can also be calculated as the product of the proportion of a certain genotype surviving and the average fecundity, and it is equivalent to the reproductive success of a genotype. A fitness $f_{abs} > 1$ indicates that the frequency of that genotype in the population increases, while $f_{abs} < 1$ means that it decreases. If differences in individual genotypes affect fitness, then clearly the frequencies of the genotypes will change over successive generations. In natural selection, genotypes with higher fitness become more prevalent.

Relative fitness is given by $f_{rel} \equiv N_{F1}/N_{compF1}$, where N_{F1} is the average number of surviving progeny of a certain genotype in the F1 generation, and N_{compF1} is the average number of surviving progeny of all competing genotypes in the F1 generation.

The fitness of nonbiological CNLSs can also be defined. In a complex manufacturing process, fitness might be made a function of the rate of production of a good, R, and the number of manufacturing errors per unit, E. Thus, we might use F=R/E as a fitness measure.

Fitness function (FF) (in genetic algorithms [GAs]): A kind of objective function that quantifies the optimality of a solution in a network using a GA. An FF is calculated iteratively until some minimum error criterion is reached.

Fitness landscape (FL) (in evolutionary biology): (aka adaptive landscape.) A poorly defined but much used term in evolutionary biology. One version of the FL is that it is a hypersurface whose height is proportional to a species' reproductive rate (or fitness). The fitness can be plotted as a function of one or more environmental (niche) variables, such as temperature, salinity, availability of a certain energy source (food, light), and density of another species that competes for resources. A population tends to move to the peaks in its FL over a number of generations.

Another view of the FL plots fitness in a population as a function of the frequency of different forms of specific genes (alleles). This produces a contoured landscape in which the peaks represent local optima corresponding to a particular genotype. The niche is assumed to be constant.

Follicular dendritic cells (FDC) (in the IS): A type of leukocyte found in the follicles of the lymphatic system. Morphologically, FDCs look like IS DCs, but, unlike the latter, are probably not of hematopoietic origin. However, FDCs have similar roles in IS function. They ingest antigens (bacteria, viruses, fungi, and cell debris) by endocytosis and present their fragments to B-cells, inducing their proliferation. The role of FDCs in the IS is to amplify and coordinate the adaptive response to infection.

Fractal: A rough or fragmented geometric shape that can be subdivided into parts, each of which is generally a reduced size copy of the whole. The fine structure of a fractal exists at arbitrarily small scales; it is too irregular to be easily described by traditional Euclidean geometric language. Fractals have a simple, recursive, mathematical definition. Approximate fractals occur in nature, for example, snowflakes, blood vessels, and cauliflower. The term was coined by the mathematician Benoit Mandelbrot in 1975, derived from the Latin *fractus*, meaning "broken" or "fractured."

Free energy (in chemistry, physics, and economics): (aka Gibbs free energy, ΔG, Joules.) Also called available energy, or energy set free. In chemical reaction systems, ΔG is the chemical potential that is minimized when a system reaches equilibrium at constant pressure and temperature. If a reaction is spontaneous, $\Delta G < 0$ and energy is given off, usually in the form of heat and/or photons; a non-spontaneous reaction requires energy input (heat, photons, etc.) and $\Delta G > 0$. At equilibrium, $\Delta G = 0$. In general, in chemical thermodynamics, $\Delta G = \Delta H - T\Delta S$, where ΔH = change in enthalpy (J) and ΔS = change in entropy (J/K). For example, when hydrogen is burned in oxygen, water vapor is formed with a release of -54.64 kcal/mol (heat and photons). When CO_2 is formed by burning carbon in oxygen, $\Delta G = -94.26$ kcal/mol.

Functional redundancy (in ecology): In an ecosystem, several species contribute in equivalent ways to the function of the ecosystem such that the removal of one species does not disrupt the ecosystem's function.

Gain (in linear systems [LSs] analysis): Gain is a term derived from electronics; it refers to the amplification factor of an amplifier. More generally, gain is the frequency-domain transfer function (or SS frequency response) between an input signal, V_i, and an output signal, V_o. As a frequency response, the gain can be written as a complex number (vector): $H(j\omega) = V_o/V_i$. Gain can also be expressed in terms of the Laplace complex variable, s: $H(s) = (V_o/V_i)(s)$. $H(s)$ will generally be a rational polynomial in s.

Gamma-aminobutyric acid (in neurophysiology): The chief inhibitory neurotransmitter found in the nervous systems (e.g., the CNS and retinas) of widely divergent species. Also found in pancreatic islet cells and in the kidneys. Synaptic GABA release causes hyperpolarization of membranes by opening chloride channels that allow Cl^- ions to enter the postsynaptic membrane, or K^+ ions to flow outward. These stimulated ion flows cause the postsynaptic membrane potential to be "clamped" to a negative potential, inhibiting postsynaptic neuron firing. The GABA molecule is $H_2N–CH_2–CH_2–CH_2–CO–OH$.

Genetic algorithm (GA) (in computing): A search technique used in computing (and artificial neural network optimization) to find exact or approximate solutions to optimization and search problems. GAs use techniques inspired by evolutionary biology such as inheritance, mutation, selection, and recombination (crossover), as well as synaptic strength in neural networks. GAs have been applied to problems in bioinformatics, chemistry, computer science, economics, engineering, manufacturing, mathematics, physics, and many other fields. A problem-dependent FF is used to measure the quality of the solution calculated iteratively by the GA. The GA initializes a population of solutions randomly, and then improves it by a repetitive application of mutation, crossover, inversion, and selection operators. The generational process continues until a terminating condition has been reached using the FF.

Genetic drift (in population genetics): (Also called allelic drift.) The evolutionary process in which gene (allele) frequencies change from one generation to the next due to purely random events. Thus, certain genes will be carried forward in a reproductive population, while others will disappear. Genetic drift is one of the primary mechanisms of evolution. It should not be confused with natural selection, a nonrandom evolutionary selection process in which the tendency of alleles is to become more or less widespread in a population with time due to the alleles affecting adaptive and reproductive success. (That is, the organism's habitat is involved in natural selection.)

Glycolysis (in cell metabolism): Glycolysis is an anaerobic metabolic pathway in which, through several enzyme-catalyzed steps, 1 glucose molecule is used to form 2 ATP molecules + 2 NADH molecules + 2 pyruvate molecules. The pyruvates can be used to make ethanol, or be inputs to the citric acid (Krebs) cycle. Pyruvic acid can also combine reversibly with NADH to form lactic acid with the aid of lactic dehydrogenase enzyme. Thus, the conversion of pyruvic acid to lactate allows the glycolysis process to run for several minutes, making ATP required for other metabolic processes, even without respiratory O_2.

Good (n.) (in economics): In economics, a good is any object or service that increases utility, directly or indirectly, of the consumer. A good is manufactured, grown, crafted, etc., and generally sold for profit. A good is supplied to meet a general demand.

Graph (in graph theory and systems analysis): A set of objects called nodes, vertices, or points, connected by links, also called branches, edges, or lines. A graph can be used to model a system. Graphs can be *undirected*, in which a line from node **A** to node **C** is the same as a line from node **C** to node **A**. In a *directed graph* (aka *digraph*), all branches leave certain nodes and are directed to other nodes. This directivity implies unidirectional flow of information, signal, mass, etc., from the originating (source) nodes to a sink node. A *mixed graph* contains both directed and undirected edges. A *simple graph* has no multiple edges or loops. An *oriented graph* is one in which all its nodes and branches are numbered, and arbitrary directions are assigned to the branches (i.e., the branches are directed). (For an extensive glossary of terms used in graph theory, see Albert [2005], Caldwell [1995], also http://en.wikipedia.org/wiki/Glossary_of graph_theory.)

Gross domestic product (GDP) (in economic systems): GDP is a measure of national income and output for a country's economy. GDP is the total market value of all final goods and services produced within the country in a year. It can be calculated from $GDP = C + I + G + (X - M)$, where $C = $ consumption, $I = $ gross investment, $G = $ government spending, and $(X - M) = $ (exports – imports).

Half-life (in biochemistry): The time it takes for the static concentration of a drug, hormone, cytokine, etc., to fall to one-half of its concentration at $t = 0$. $T_{1/2}$ is not the same as a time constant, τ.

Hamiltonian path (in graph theory): A Hamiltonian path is a path in an undirected graph that visits each node only once. A Hamiltonian cycle is a closed path in a graph that visits each node only once and also returns to the starting node. A complete graph with more than two nodes is Hamiltonian.

Hapten (in immunology): A hapten is a small, non-peptide molecule that can elicit an immune response only when bound to a large carrier such as a protein molecule; the carrier may be one that also does not elicit an immune response by itself. Once the body has created Abs to the hapten-carrier adduct, usually only the adduct will initiate immune response. Poison ivy contains urushiol, an oil that when absorbed by the skin undergoes oxidation by skin cells that generates the actual hapten, a quinone which then binds to epithelial cell proteins to form an adduct allergen. It is the activated sub- and intradermal T-cells that create the blisters and itching. Other haptens (to some individuals) include fluorescein, cotton-seed oil, and nickel ions.

Hash functions (in computer science and degenerate neural systems): A hash function is basically a hardwired lookup table. It takes a very large input space of possible "words" (keys), and maps N expected words into a hash table of N entries (buckets). A hash function may map two or more keys to the same hash value (called a hash collision), a form of pleiotropy. Hash tables are used to implement associative arrays. Some evidence is emerging that the CNSs of animals use a neural equivalent of hash functions (Leonardo 2005).

Hemimetabolic insects: Insects having incomplete metamorphosis. Their life cycle includes the egg, a nymph, and the adult stage or imago. Representative orders include Blattaria, Hemiptera, Orthoptera, Mantodea, Odonata, etc.

Hematopoiesis: The production of red and white blood cells by specialized stem cells (myeloid and lymphoid) in the bone marrow.

Hepatitis C virus (HCV): A member of the Flaviviridae. A small, ca. 50 nm diameter virus, encoded by a + sense, single-strand, RNA string. Its genome consists of a single ORF (open reading frame) of ca. 9600 nucleoside bases that encode a single 3011 AA protein. At the 5′ end of the RNA is an untranslated region (UTR) that has a ribosome binding site. The 3011 AA protein is cleaved by viral and cellular proteases into three structural and seven nonstructural proteins. Of interest in the study of viral quasispecies (VQS), the HCV has a high rate of replication (10^9/day) and a high mutation rate. The Nt sequence of HCV is highly variable; about 60% Nt sequence homology is found between the most divergent virions around the world. The HCV host cells are hepatocytes. HCV species have been broken down into six genotypes, with several subtypes (a, b, c, ...) within each genotype. The subtypes can be further broken down into quasispecies based on their genetic diversity. Unfortunately, HCV cannot yet be cultured in vitro, and so animal models and human hosts must be used for research.

Hill function (used in modeling dynamic biochemical systems): The **n**th order Hill function is used to model the parametric variation of rate constants in CNLS models. They provide one means of introducing parametric feedback. Rate-constant saturation is modeled by the Hill function:

$$K_S = \frac{K_{So}[M]^n}{\beta + [M]^n}$$

where

K_{So} and β are positive constants
[M] is the concentration of the regulating variable, M
n is an integer ≥ 1
$K_S \rightarrow K_{So}$ for $[M]^n \gg \beta$

Rate-constant suppression by the concentration of M is modeled by the decreasing Hill function:

$$K_N = \frac{K_{So}}{\beta + [M]^n}$$

K_N is a decreasing function of $[M]^n$. $K_N \rightarrow 0$ as $[M]^n \gg \beta$.

Holometabolic insects: Insects of the subclass Endopterygota, which go through distinct larval, pupal, and adult stages. These include, among others, the Diptera (true flies), Lepidoptera (moths and butterflies), Coleoptera (beetles), etc.

***Homeobox* gene** (in eukaryote development): A DNA sequence of about 180 DNA base pairs long, found within genes that are involved with the regulation of development (morphogenesis). The sequence encodes a protein (a homeodomain) that can bind to DNA. The homeodomain proteins are transcription factors (TFs), which in turn switch on cascades of other genes, for example, those required to make an eye, or a leg. Homeodomain proteins generally act in the promoter region of their target genes as complexes with other TFs and/or homeodomain proteins.

Homeobox genes were first found in the fruit fly, and appear ubiquitous in all animals. They have also been found in fungi, yeasts, and plants. In an analogy to computing, a *homeobox* gene is like a call to a subroutine. It switches on the expression of an entire subsystem, the code for which must already be present elsewhere in the genome.

The development-controlling *Hox* genes are a subgroup of *homeobox* genes.

Homeostasis (in physiology): The dynamic maintenance of steady-state conditions in the internal compartment (cytosol) of a cell by the programmed, regulated expenditure of chemical energy. Cellular homeostasis stabilizes the interior of a cell against chemical and physical changes in its external environment as well as from chemical and physical changes arising within the cell from metabolic processes. Homeostasis requires the dynamic equilibria of many biochemical systems, and is generally the result of the action of multiple chemical negative feedback control systems. An example of homeostasis is the maintenance of the concentration of sodium ions in the cytosol of a cell within "normal" bounds against extracellular changes in sodium concentration. Sodium "leaks" into the cell down concentration and membrane potential gradients. Sodium homeostasis in cells is effected by sodium "pumps," specialized proteins fixed in the cell membrane that have the capability of moving sodium ions from the cell's cytosol to the extracellular fluid at the expense of metabolic energy. The sodium ion concentration in a cell determines the osmotic pressure across its cell membrane, hence the amount of water that leaks in or out of the cell. Even water passage through a cell membrane is regulated by transmembrane aquaporin proteins as a part of osmotic homeostasis. Molecular signals that activate transmembrane pumps and gates are generally regulated by the expression of genes by regulated TFs.

Multicellular organisms practice physiological homeostasis. Regulation is generally effected by the action of multicellular organs, such as kidneys, pancreas, and liver, under hormonal and CNS regulatory signals. Homeostasis at all levels of scale leads to oganismal robustness.

Hopf bifurcation (in nonlinear dynamical systems): A local bifurcation in which a fixed point of a dynamical system loses stability as a pair of complex-conjugate eigenvalues of the linearization around the fixed point cross the imaginary axis of the complex (root) plane. Hopf bifurcations occur in the Hodgkin–Huxley model for nerve impulse generation (Northrop 2001), the Lorenz system, and the Oregonator and Brusselator chemical oscillator models. In a CNLS's phase plane, a bifurcation point is one where the phase plane attractor changes as a result of one or more changes in the system's parameters. (See also *tipping point*.)

Horizontal gene transfer (HGT) (in genetics): (aka lateral gene transfer.) HGT occurs when an organism transfers its genetic material to an organism other than one of its (F1) offspring. HGT was first described by Ochiai et al. (1959) and Akiba et al. (1960) for the transfer of antibiotic resistance between different strains of the bacterium *Shigella* and between *Shigella* and *E. coli*. HGT has played a major role in the evolution of bacteria. There are three common means of HGT: (1) Bacterial conjugation, in which bacterial DNA plasmids are exchanged when bacteria temporarily fuse membranes. (2) Transduction, a process in which a bacteriophage virus moves sections of bacterial DNA from one bacteria host to another. (The F1 phages pick up DNA from the initial host, and then infect the second host.) (3) Transformation, which is the genetic alteration of a bacterial cell resulting from the introduction, uptake, and expression of foreign genetic material (DNA or RNA). Transformation can be used in bioengineering/biotechnology applications.

***Hox* genes** (in development): *Hox* genes are a subset of *homeobox* genes. They are essential in the sequential control of embryological development in segments along the body axis; they are the "master regulators" for embryological development. *Hox* genes are highly conserved in all forms of life. Over millions of years of evolution, they have retained their function of assigning particular positions in the embryo. However, the structures (organs, systems) actually built depend on a different set of genes specific for a particular species. An error in the expression of a *Hox* gene generally leads to a nonviable embryo, a major factor in their conservation through natural selection. Exogenous retinoic acid has been shown to cause *Hox* gene mutations and birth defects in mammals.

Hox genes encode TFs, each with a specific, DNA-binding homeodomain protein. They act in sequential zones of the embryo in the same order in which they occur on the chromosome. Humans have 4 *Hox* gene clusters (HOXA–HOXD) with a total of 39 genes. Human *Hox* genes (A–D) are located on chromosomes 7p15, 17q21.2, 12q13, and 2q31, respectively.

Hub (in a graph): A node in a network that has an above-average number of interactions with other nodes in that graph. (In a directed graph, these interactions can be by both input and output edges.)

Immune deficiency (IMD) pathway: An innate immune cascade (of gene expression) triggered by Gram-negative bacterial expression; found in *Drosophila*. G$^-$ bacterial-derived diamino-pimelic acid (DAP)-type peptidoglycan is recognized by upstream recognition proteins LC (PGRP-LC) and LE. This binding leads to the activation of the adaptor protein IMD (Tanji et al. 2007).

Immunoglobulin (in the IS): Antibody proteins made by IS plasma B-cells. Some Igs are circulating, while others are fixed to cell surfaces. Ig antibodies are generally "Y" shaped. Both the stem (Fc region) and arms of the "Y" are paired molecules. The binding site (paratope region) for antigenic Ig molecules lies on or between the arms. There are five classes of Igs: IgA, IgD, IgE, IgG, and IgM. IgE is involved in allergic hypersensitivity and anaphylaxis.

Ig Abs can act in three different ways to fight invading pathogens: (1) They can directly bind to the invading pathogen, inactivating it by changing its gross structure. (2) They can activate the complement system that destroys the invader. Direct binding of Ab to Ag can lead to large clusters of Abs bound to many Ags; such a cluster is said to be an *agglutination*. Agglutinations can precipitate, becoming immobile, and prey for CTLs (cytotoxic T-cells), NKs neutrophils, and macrophages. (3) The Fc regions of Abs bound to Ags on cell surfaces form attachment sites for NK cells, which kill the cells involved.

Incidence matrix (in graph theory): A graph is characterized by a matrix of E (edges) by V (vertices), where [edge, vertex] contains the edges' data (simplest case: 1 if connected, 0 if not connected).

Increasing returns (to scale) (in complex economics): A phenomenon correlated with positive feedback loops in a CES. Goods or profits are seen to increase at a rate larger than the scale of a microeconomic system, that is, goods or profit are proportional to S^k, where S is the size (scale) of the microeconomic system and the exponent $k > 1$. If $k < 1$, the system exhibits decreasing returns to scale; if $k = 1$, there are constant returns to scale.

Inflation (in economic systems): A general rise in the prices of goods and services over a time. It is due to a decline in the purchasing power of money. Inflation is quantified by calculating the Inflation Rate = % change in a price index over time. Price indices include consumer price index (CPI), cost-of-living index, commodity price index, etc.

Infrastructure (in social systems): A society's *infrastructure* is a collective noun denoting its highways, bridges, tunnels, airports, public transportation systems, electric power supply and distribution, water supply, waste removal, communications (newspapers, internet, TV, radio, telephone, cell phone, telecom cable, fiber optic, etc.), dams, inland waterways, levees and flood control, etc. Infrastructure is a mixture of publicly and privately supported works that make our society work smoothly.

Intelligent design (ID): The assertion that certain features of the universe and of living organisms are best explained by an "intelligent cause," and not an unguided process such as the big bang or natural selection in evolution. It is a modern form of the traditional teleological argument for the existence of God, altered to avoid the nature or identity of the "designer." ID deliberately does not try to identify or name the specific agent of design; it merely postulates that one (or more) must exist.

Irreducible complexity (in biology): A faith-based philosophy that argues that life is so complicated that it must be the work of an intelligent designer (aka God) rather than the result of evolutionary processes. Or, more specifically, "... a single system composed of several well-matched, interacting parts that contribute to the basic function, wherein the removal of any one of the parts causes the system to effectively cease functioning" (Behe 1996). Another definition: "... an organism doing something (the function) in such a way that the system (that portion of the organism that directly performs the function) has no more parts that are strictly necessary" (Dunkelberg 2003).

JAK-STAT signaling pathway (in molecular biology): JAK stands for Janus kinase; STAT stands for signal transducers and activators of transcription. The JAK-STAT pathway plays an important role in the regulation of cell proliferation, differentiation, hematopoiesis, and apoptosis. It is found in animals from humans to flies. There are four mammalian JAK members: JAK1, JAK2, JAK3, and tyrosine kinase 2 (TYK2). Each contains a conserved kinase domain and a catalytically inactive, pseudokinase domain at the carboxyl terminus. There are seven mammalian STATs: STAT 1, 2, 3, 4, 5A, 5B, and 6. In a typical JAK-STAT pathway, a cytokine binds to the extracellular receptors on paired, transmembrane JAK molecules. The activated, intracellular JAK molecules lead to the phosphorylation of associated STAT molecules, which then dimerize and translocate to the nucleus where they activate the transcription of certain genes. (See reviews by Shuai and Liu [2003], Rawlings et al. [2004], and Hou et al. [2002].)

Junk DNA (in genomics): (aka intron.) Early in the development of genomic science, it was found that a large percentage of human nuclear DNA did not code proteins. With incredible scientific hubris, this was called "junk DNA." It is now a major scientific challenge to discover what this DNA actually does code for. A large part of intron DNA is composed of transposable elements; parts may code for specialized, active RNAs; parts may be regulatory for gene expression; and other sections are actually noncoding. About 98%–99% of human nuclear DNA is nonprotein coding.

Kinase (protein kinase): A specialized class of protein enzyme that adds phosphate groups ($PO_4^{\equiv}$) to other proteins (phosphorylation). The human genome contains ca. 500 protein kinase genes. Most kinases remove a phosphate group from ATP and attach it covalently to one of three AAs that has a free hydroxyl group (serine, threonine, and tyrosine). (The human genome codes for 90 different tyrosine kinases [TKs].) The phosphorylation of these target AAs alters the balance of non-covalent, intramolecular interactions, which affects tertiary and quaternary protein structures, hence protein function. Phosphorylation can lead to the splitting off or the addition of polypeptide subunits to the phosphorylated protein. Phosphorylation thus can lead to the activation (or inactivation) of a protein. For example, some cytosolic TKs enter the nucleus and transfer their phosphate to TF proteins, activating them and hence the transcription of certain genes.

Knockout (in cell biology): The selective removal of a cytokine, hormone, protein, etc., from an experimental animal in order to study its effect. In animals, knockouts can be accomplished by deleting or silencing genes. In agent-based modeling, a knockout is done in software. (Knockin is also used.)

Kolmogorov complexity (KC) (in computer science and genomics): (aka algorithmic entropy, algorithmic complexity, algorithmic information content, descriptive complexity, Kolmogorov–Chaitin complexity, program-size complexity, or stochastic complexity.) Consider a string of data (e.g., a finite binary string of 0s and 1s) and a finite coding region of DNA (groups of three bases called codons, each of which codes one of 20 AAs in a protein peptide sequence, or words in a paragraph of English text). The NIST (2006) definition of KC is "The minimum number of bits into which a *string* can be compressed without losing information. This is defined with respect to a fixed, but universal decompression scheme, given by a *universal Turing machine*." Funes (2008) states "One problem with AIC is its uncomputability. The fact that no program can be known in advance to halt at some point makes it impossible to know the AIC of a given string." "The KC of a stochastic process can be proved to approach its entropy...." Hutter (2008) gives the history of KC and some interesting inequalities describing the KC of strings.

lac **Operon** (in genomics): An operon is a functional group of protein-coding (structural) genes that are regulated and expressed as a group. In the bacterium *E. coli*, the *lac* operon's linear organization begins with the promoter for the p_i regulatory gene, then the regulatory gene (**i**) followed by the *lac* operon promoter sequence (p_{lac}), then the operator sequence (**o**), and then a group of three structural genes, **z**, **y**, and **a**, coding certain enzymes for lactose metabolism. The regulatory gene (**i**) codes for a repressor protein that obstructs the promoter, and hence inhibits transcription of the structural genes for lactose metabolism (Northrop and Connor 2008).

Lactoferrin: An AMP that restricts the local availability of free iron, an essential microbial nutrient. Lactoferrin is secreted in the distal collecting tubules; it chelates free iron ions and damages bacterial membranes (Zasloff 2007).

Landscape (in evolutionary biology): See *fitness landscape*.

Lanthionine: An AA with empirical formula $C_6H_{12}N_2O_4S$ and MW 208.24. (Two cysteine AAs are joined by a common sulfide bridge.)

```
        NH2            NH2
        |    H    H    |
HO – C – C – C – S – C – C – C – OH
        ||   H H    H H   ||
        O                 O
```

Lanthionine is found in certain bacteriocins called lantibiotics. These include nisin, subtilin, epidermin (an anti-staphylococcus and strepto-coccus AMP), and ancovenin (an enzyme inhibitor). A variant form of lanthionine, β-methyllanthionine, is found in some AMPs.

Lantibiotic: A lanthionine-containing antibiotic peptide (AMP) secreted by cer-tain Gram-positive (G^+) bacteria. These AMPs are ribosomally synthe-sized and then modified posttranscriptionally. Nisin is an example of a lantibiotic; it has been used in food preservation. There are some 11 subgroups of lantibiotics (see Cotter et al. 2005).

Law of requisite variety (LRV) (in control systems): Ashby's LRV states in essence that a feedback system's controller always has to be more com-plicated than the system it is controlling.

Law of unintended consequences (in complex systems): (Actually, more of a maxim.) A property of the behavior of complex systems in which an action or input can lead to an unanticipated, unintended "side-effect," as well as a desired action. A side-effect can be beneficial, as well as bad.

Lectins (in molecular biology): Nonenzymatic proteins found throughout ani-mal and plant kingdoms, in viruses and bacteria, and prokaryotes and eukaryotes, which bind to sugars and carbohydrates. Lectins are found in both soluble and cell-associated forms. Lectins are involved in a variety of cellular processes, including cell adhesion, bacterial infec-tion, erythrocyte agglutination, leukocyte mitosis, lymphocyte homing, cell apoptosis, and innate immunity. Major subclasses of lectin proteins include selectins, collectins, and galectins.

Length of path (in graph theory): Consider nodes j and k in a directed graph. The path length, L_{jk}, is the number of concatenated branches between nodes j and k that do not pass through any intermediate node more than once. There may be more than one distinct path between nodes j and k. The *path gain* is the product of the branch gains in L_{jk}.

Leptin (in fat metabolism): A pleiotropic hormone released mostly by the white adipose tissue (WAT) (fat cells). Leptin plays an important role in the regulation of energy input, output, and storage. High leptin lev-els decrease the rate of production of insulin and also neuropeptide Y (NPY), an appetite stimulant. It also increases energy expenditure through the stimulation of sympathetic nerve activity. Leptin is also produced in small amounts by brown adipose tissue, ovaries, skeletal muscle, stomach, mammary epithelial cells, bone marrow pituitary, and liver.

Limit cycle (LC) (in nonlinear, dynamic, complex systems): Under certain ini-tial conditions and/or inputs, a nonlinear, dynamic, complex system may exhibit bounded, self-sustained oscillations of its states. LCs can fur-ther be characterized as stable, unstable, or conditionally stable. The kth state's stable LCs exhibit a closed path in the ($\dot{x}_k$ x_k) plane (Ogata 1970). Stable LCs in enzymatic biochemical systems may form the basis for biological clocks.

Linear system (LS): An LS obeys all of the following properties; a nonlinear system does not obey one or more of the following properties ($x(t)$ is the input, $y(t)$ is the output, and $h(t)$ is the LS's impulse response, or weighting function):

$$x_1 \to LS \to y_1 = x_1 \otimes h \quad \text{real convolution}$$

$$\frac{\text{if}}{\text{then } a_2 x_2} \to LS \xrightarrow{\begin{array}{l} y_2 = x_2 \otimes h \\ y_2' = a_2(x_2 \otimes h) \end{array}} \text{scaling}$$

$$\frac{a_1 x_1 + a_2 x_2}{} \to LS \xrightarrow{y = a_1 y_1 + a_2 y_2} \text{superposition}$$

$$\frac{x_1(t-t_1)}{+x_2(t-t_2)} \to LS \xrightarrow{\begin{array}{l} y = y_1(t-t_1) \\ +y_2(t-t_2) \end{array}} \text{shift invariance}$$

LINES (in genomics): (A type of retrotransposon.) Long interspersed nuclear elements. Long DNA sequences ($>5\,kb$) that code for two proteins: one that can bind to ssRNA, and the other has reverse transcriptase and endonuclease activity, enabling the LINE to copy both itself and non-coding retrotransposons such as Alu elements. Because LINES move by self-copying instead of moving like transposons do to other sites, they enlarge the genome. The human genome is estimated to contain ca. 9×10^5 LINES, which is ca. 21% of our genome. LINES are used to create genetic "fingerprints."

Lipocalins (an AMP): A broad family of antimicrobial proteins with diverse functions. Lipocalin in the urinary tract is dramatically expressed throughout the kidney following injury or after ischemia-induced, acute tubular necrosis (ATN). Lipocalin captures iron-laden organic siderophores that bacteria utilize to scavenge iron from their environment, depriving them of iron needed for growth and metabolism.

LMC complexity measure (in physical systems): LMC stands for the authors who first described the LMC measure: Lopez-Ruiz, H.L. Mancini, and X. Calbet (1995. *Physics Lett. A.* 209: 321-). The LMC measure is defined in Section 9.4. The LMC measure is applied to complex systems in physics and chemistry.

Loop (in graphs): A collection of branches in a graph (oriented or unoriented) that form a closed path. Loops can be feed-forward or feedback in topology. A loop in an SFG is a directed path that originates at and terminates on the same node, and along which no node is encountered more than once. The product of the branch transmissions in a loop is called the *loop gain.*

Loop gain (in a SISO LTIS): The loop gain, $A_L(s)$, in a linear feedback system is the net gain around a feedback loop, starting at one node and finishing at that node. The zeros of the return difference transfer function,

$F(s) \equiv 1 - A_L(s)$, are the closed-loop system's poles that determine its stability, and, in part, its transient response. A system with negative feedback has a loop gain with a net minus sign, for example, $A_L(s) = -[K_p K_c(s+a)]/[s(s+b)]$.

Lyapunov exponent (in phase plane plots of CNLS behavior): The Lyapunov exponent characterizes the rate of separation of a pair of close trajectories originating at the same time from closely spaced points. The trajectories can be converging on a common attractor, or diverging. Consider two closely spaced points in 2D phase space at $t = t_o$; P_0 on the first (reference trajectory) and P_0' on the second trajectory. Each point will generate a trajectory determined by the system's dynamics and its initial conditions. t seconds later, the vector distance between $P(t)$ on the first trajectory and the corresponding point on the second trajectory is $\rho(t)$.

The Lyapunov exponent λ is defined by the equation $|\rho(t)| = \exp(\lambda t)|\rho_0|$ (absolute values are used because ρ_0 and $\rho(t)$ are 2D vectors), where ρ_0 is the initial, small separation at $t = t_o$ between the points P_0 and P_0' on adjacent trajectories, and $\rho(t)$ is the separation between corresponding points on the trajectories at $t = t_o + t$. Often the maximal Lyapunov exponent (MLE) is used to characterize the phase plane behavior of CNLSs. The MLE is defined by

$$\lambda = \lim_{\substack{t \to \infty \\ \rho_0 \to 0}} \left(\frac{1}{t}\right) \ln \frac{|\rho(t)|}{|\rho_0|}$$

In general, if $\lambda > 0$, then the trajectories diverge from each other rapidly. If $\lambda = 0$, the system is a stable oscillator. If $\lambda < 0$, then the system is focal (trajectories converge on an attractor in the steady state ($t \to 0$). Measuring the LE is only of significance at the attractor.

Lysosomes (in cell biology): An organelle found in both animal and plant (eukaryotic) cells that contains a variety of digestive enzymes (acid hydrolases) used to digest unwanted macromolecules (proteins, lipids, carbohydrates, and nucleic acids).

Lysosomes are assembled in the Golgi apparatus and bud off it. The low pH (4.8) of the lysosome is maintained by proton pumps and Cl⁻ ion channels in its single membrane. Lysosomes have a variety of functions: In embryonic development, they digest unwanted structures (e.g., a tadpole's tail and finger webs in a 3–6 month old fetus). They also digest phagocytosed bacteria and damaged mitochondria, and even break down cellular organelles in autophagic cell death.

Macroeconomics (in economic systems): A branch of economics that deals with the structure, modeling, and behavior of a global, national, or regional economy as a whole. Macroeconomic models seek to describe the relationships between factors such as national income, overall employment/unemployment, investments, debt, savings, and international trade and finance.

Major histocompatibility complex (protein) (in the IS): A cell-surface, transmembrane protein that is required for cell–cell contact in antigen presentation. MHC1 is found on all somatic cells except CNS neurons. MHC1 comes in three different types: HLA-A, HLA-B, and HLA-C. (HLA stands for human leucocyte antigen. It is common for an individual to express two allelic versions of each type, and a person heterozygous for HLA-A, HLA-B, and HLA-C will express six different MHC1 proteins [Kimball 2007a].) MHC1 on antigen-presenting somatic cells binds to the CD8 protein on the surface of a CD8+ CTL. MHC1 is coded on chromosome six.

MHC2 molecules occur on the cell membrane surfaces of macrophages and B-cells in the IS. An MHC2 molecule binds to a CD4 protein on a CD4+ helper T-cell (Th) surface during the antigen presentation process by B-cells and macrophages. Another class of MHC2 on B-cells and macrophages can present an antigenic epitope to CD8+ CTLs. MHC2 is coded by six **D** genes on human chromosome six.

Mass action (MA) (in chemical systems): The basis for writing dynamic models for chemical reactions (also used in population dynamic models). A group of coupled chemical reactions described by MA kinetics can be expressed as a set of first-order ordinary differential equations (ODEs). These ODEs are generally nonlinear, and can have time-varying coefficients. The law of mass action is based on the assumptions that the reacting and product molecular species are well mixed and free to move in a closed compartment (i.e., they are not bound to a surface, like a mitochondrial membrane or a cell surface) and there are no diffusion barriers. Furthermore, it is assumed that the reacting molecules and ions are in weak concentrations, and move randomly due to thermal energy and momentum exchanges due to collisions with themselves and each other. The law of mass action is based on the probabilities that molecular species will collide and react, that is, chemical bonds will be broken and made. Basically, it states that the rate at which a chemical is produced in a reaction (at a constant temperature) is proportional to the product of the concentrations of the reactants.

For example, if two reactants combine and form a product, C: $A + B \rightarrow C \rightarrow D$, then by MA, the rate of appearance and disappearance of C is given by the simple, first-order, nonlinear ODE: $[\dot{C}] = k_1[A][B] - k_2[C]$. Also, it is easy to see that both the reactants behave as $[\dot{A}] = [\dot{B}] = - k_1[A][B]$. In these simple ODEs, brackets [*] denote a concentration, such as mol/L, k_1 is the reaction rate constant, and k_2 is the loss rate constant for C. Now let us examine a reaction in which two molecules of B must combine with one of A to make C:

$$A + 2B \xrightarrow{\ k_1\ } C \xrightarrow{\ k_2\ } D$$

MA tells us that the net rate of increase of the molar concentration of the product, **C**, is given by the ODE:

$$[\dot{C}] = k_1[A][B]^2 - k_2[C]$$

Note that the concentration of **B** is squared in the ODE. More examples of the application of MA can be found in Chapter 4 in Northrop and Connor (2008).

Messenger RNA (mRNA) (in genomics): When a functional gene is expressed, RNAP (RNA polymerase) makes a pre-mRNA complimentary copy of the DNA codons that make up the coding gene, as well as any intron codons in the gene. Editing enzymes in the nucleus remove the noncoding introns and make coding mRNA, which is used by ribosomes in the cell's cytoplasm to assemble the protein's polypeptide chain.

Methycillin-resistant *Staphococcus aureus* (MRSA): *Staphococcus* sp. bacteria that have evolved the ability to produce an enzyme that cleaves and destroys the beta-lactam antibiotics, including cephalosporins and all variations of penicillin.

Microeconomics: A branch of economics that studies how individuals, households, and firms make decisions to allocate limited resources. It considers supply and demand in markets where goods and services are being bought and sold.

Microprocessor (in cell biology): Once transcribed, primary miRNAs are pre-processed by a nuclear protein complex called a microprocessor, which contains the Drosha type III RNase, the double-stranded RNA-binding protein DGCR8, plus some other proteins. The microprocessor cleaves the hairpin-loop pre-miRNAs away from the rest of the primary transcript, permitting the pre-miRNAs to pass through the nuclear membrane into the cytoplasm.

MicroRNA (in genomics): (aka miRNA or µRNA.) Single-strand RNA molecules of 21–23 nucleotides (nt) in length. They are noncoding RNAs formed from µRNA genes in genomic DNA. Their functions are diverse, including regulation of gene expression, and they are implicated in diseases and cancer. Each primary miRNA transcript is processed into a short stem-loop structure called pre-miRNA, and then finally into a functional miRNA. Functional miRNA molecules are partially complementary to one or more mRNA molecules, and their main function is to suppress gene expression.

Modularity (in biology): Modularity is independent of scale. Anatomic modularity generally involves parts of organs, for example, the pancreatic α-, β-, γ-, and δ-cells, the adrenal cortical cells versus the adrenal medullary cells. At the intracellular level, ribosomes, mitochondria, and chloroplasts are seen as functional and structural modules. At the molecular level, we can identify complex reaction networks such as the

Krebs cycle, which has limited interfaces with other reaction networks. Anatomic modularity is expressed in the evolution of organ systems and a segmented body plan. Modules allow evolutionary flexibility; specific features may undergo changes during development without substantially altering the functionality of the entire organism. Each module is thus free to evolve, as long as the metabolic pathways at the edges between modules are relatively fixed. Each organism may have an optimal level of modularity. (See *segmentation*.)

Modular pleiotropy (in genetics): A genetic architecture in which a set of genes tends to have pleiotropic effects on the same set of traits, but few and weaker effects on other traits (Wagner et al. 2007).

Module (in a living organism): A part of an organism that is integrated with respect to a certain kind of process (natural variation, function, development, etc.) and relatively autonomous with respect to other parts (and modules) of the organism (Wagner et al. 2007). A *functional module* is composed of features that act together in performing some discrete physiological function that is semiautonomous in relation to other functional modules.

Module (in a graph or network): Broadly defined as a subnetwork of a graph, the nodes of which have more connections to other nodes within the module than to external nodes.

Modulon (in genomics): An *operon* is a set of structural genes whose expression is controlled by a single DNA-binding protein. Multiple operons controlled by the same repressor are grouped together as a *regulon*. A *modulon* is a regulon concerned with multiple pathways of functions, in which the operons may be under individual controls as well as a common, pleiotropic regulatory protein (RP). For example, the CAP modulon contains all regulons/operons, such as the *lac* operon and the *ara* regulon, that are regulated by CAP/cAMP, but each operon has other regulators as well.

Monocistronic: (*adj.*) (See *cistron*.)

Motif (in molecular biology): Network motifs are functional patterns of interconnections (modules) that occur in molecular biological regulatory networks at higher frequencies than those found in (synthetic) randomized networks. Several motifs have been repeatedly identified in many biosynthetic networks: three-node, feed-forward loop (FFL) motifs; single-input multiple-output motifs; multiple-input multiple-output motifs (also called dense overlapping regulons); "diamond" four-node SISO motifs; and multilayer perceptrons.

Motif (in graph theory): A pattern of interconnections occurring either in an undirected or a directed graph G at a number significantly higher than found in randomized versions of the graph, that is, in graphs with the same number of nodes, links, and degree distribution as the original one, but where the links are distributed at random. The pattern M is usually taken as a subgraph of G.

mRNA (messenger RNA) (in gene transcription): mRNA is synthesized on coding (gene) DNA by RNAP-II. It carries the RNA codon code for a specific protein gene product to the cytoplasm where it is translated by ribosomes that assemble the AA polypeptide string (primary structure) of that protein.

Msx homeobox **genes**: Genes expressed during development that control the induction of the neural crest, patterning in diencephalon midline formation. *hMSX2* is a human gene that encodes a member of the muscle segment *homeobox* gene family. The MSX2 protein is a TF (repressor) that is implicated in the balance between survival and apoptosis of neural crest-derived cells necessary for normal craniofacial development.

MTAN (in bacterial metabolism): 5′-methylthioadenosine/S-adenosylhomocysteine nucleosidase (Kamath et al. 2006). The enzyme MTAN occurs in a variety of bacterial cell types (G^+ and G^-). It acts in the intracellular production of 5′-methylthioribose (MTR) and adenine, and adenine and S-ribosylhomocysteine (SRH). MTANs are directly involved in the biosynthesis of AI molecules; AIs are involved in inter- and intraspecies signaling. Inhibition of MTAN may provide a means of down-regulating AIs, thereby blocking quorum sensing signaling. It may also impact the regulation of bacterial growth and/or their DNA synthesis.

Mutation-selection balance (in population genetics): If a single mutation (variant) appears in a population of N individuals, its initial frequency (rate of occurrence) (f_0) in a diploid population will be 1/2N. With successive generations, f_k, the rate of occurrence of the mutation in the kth generation, will reach a steady-state, equilibrium value (as $k \to \infty$) that reflects the balance between the mutation (increasing f_k) and selection (decreasing f_k). The mutation-selection balance is the result of genetic drift.

NAD⁺ (NAD^+) (in biochemistry): Nicotine adenine dinucleotide (oxidized).

Natural selection: The process whereby favorable traits that are heritable propagate throughout a reproductive population. It is axiomatic that individual organisms with favorable traits are more likely to survive and reproduce than those with unfavorable traits. When these traits are the result of a specific genotype, then the number of individuals with that genotype will increase in the following generations. Over many generations, this passive process results in adaptations to a niche, and eventually to speciation.

Neighbor of a node (in graph theory): The neighbors of node k are those nodes connected directly to it by one edge, or by one or two edges if the graph is a digraph.

Network motifs (in modularity): Recurrent building block patterns seen in complex metabolic networks.

Neuropeptide Y (NPY) (in glycoregulation): NPY is a 36 AA peptide found in the brain and the autonomic nervous system. A pleiotropic hormone, NPY is actually one of a family of three peptides having similar structures and pleiotropic actions in the body. The family consists of NPY, peptide YY

(PYY), and pancreatic polypeptide (PP). NPY itself is highly conserved throughout evolution; it is the most abundant neuropeptide in the mammalian CNS. NPY has a major role in appetite control. Neural tissue is the major source of NPY in the human body; it is widely distributed and released from specific kinds of CNS neurons, and also from peripheral neurons including noradrenergic, non-cholinergic, and sympathetic, as also certain sensory neurons.

NPY is, in general, a pleiotropic, inhibitory neuromodulator/ neurotransmitter. According to Sandeva et al. (2007), NPY exhibits "... anxiolytic, anti-stress, anti-convulsant, and anti-nociceptive actions, in addition to its hypertensive, and potent appetite-stimulating effects and its capacity to shift circadian rhythms." These authors also noted that NPY is a potent down-regulator of alcohol consumption. NPY is found in large vesicles in the endings of sympathetic nerve fibers in the body. NPY from this source evidently acts as a potent, long-acting vasoconstrictor, having synergistic action with norepinephrine (Guyton 1991).

NPY-related peptides bind to a large and heterogeneous family of G-protein-coupled receptors, Y1–Y5 and $y6$. The Y1 receptor is involved with the vascular and antinociceptive effects of NPY, as well as its psychological functions, such as decreased anxiety and depression. Receptor Y1 also figures in the appetite stimulation by NPY. When NPY acts as a neuromodulator, it is through the Y2 receptors that are located presynaptically and inhibit the release of the neurotransmitter. The pleiotropy of NPY (why it is enigmatic) is illustrated by its action on Y2 receptor neurons (increases blood pressure) versus its action on Y1 receptor neurons (decreases BP). Y2 receptors are also associated with angiogenesis (important in tumor proliferation), and the effects of NPY on circadian rhythms.

Neutrophils (in the hIS): These important leucocytes in the hIS were named because of their staining properties with the well-known hematoxylin and eosin (H&E) blood stain. Neutrophils stain pink, while basophilic white blood cells (WBCs) stain dark blue, and eosinophilic WBCs stain bright red. Neutrophils are the most abundant (in number) WBCs (ca. 70%). They have a nucleus with 2–5 lobes, and their cytoplasm is full of secretory granules. Neutrophils are components of both the adaptive and the innate mammalian IS. They traffick to the site of an infection, following signals such as IL8, γIFN, and C5a from the complement system. They are phagocytes, and they can do MHC II antigen presentation. From their granules, they can release the AMPs lactoferrin, cathelicidin, myeloperoxidase, bactericidal/permeability increasing (BPI) protein, various defensins, the serine proteases neutrophil elastase and cathepsin G, and gelatinase. Nonactivated neutrophils have a normal, average half-life of ca. 12 h. Once activated, they can survive ca. 1–2 days.

NF-κB (in the IS): (Nuclear factor kappa-light-chain-enhancer of activated B-cells.) A ubiquitous protein complex that regulates the transcription of genes that lead to proteins that regulate the immune responses to infection. It is a rapid-acting, primary TF. It up-regulates the expression of many genes involved in inflammation. Many bacterial products can activate NF-κB through TLRs. The constitutive expression of NF-κB turns on the expression of genes that keep the cell proliferating and protect it from developing conditions that could lead to apoptosis.

Niche (in evolution and ecology): A species' *fundamental niche* is the full range of environmental conditions (physical and biological) under which it can exist. As a result of competition with other organisms, the species may be forced to occupy a realized niche smaller than the fundamental niche. Hutchinson (1958) defined a niche as an **N**-dimensional hypervolume in N-space. Examples of dimensions might include moisture, pH, solar radiation, salinity, etc. A species can actively alter its niche (e.g., beavers), and generally must compete with other species for resources in their mutual niche-hypervolume. Thus, a fundamental niche can be partitioned by species' competition.

NK system (in complex systems): Stuart Kauffman's NK model is used to describe the fitness landscapes (FLs) of CNLSs. **N** is the number of elements (define or elaborate) that characterize the system. **K** is the average number of other elements that each element is interdependent with. "A certain degree of fitness is associated to every possible configuration of the elements of the system." An FL is usually shown as a 3D surface; however, it can be M dimensional (M > 2). The fitness is shown as the surface height. In an NK system, if $K \rightarrow 0$, each element contributes independently to the system's fitness, and each element must be optimized to get a maximally sharp, single fitness peak. $K \gg 0$ will give a ragged, multi-peaked FL. An NK model allows us to compare model architectures characterized by a different number of elements (**N**) and/or a different level of interdependence (**K**) by changing the two parameters.

Nociceptor (in sensory neurophysiology): A nociceptor is a pain receptor, signaling external or internal tissue damage from physical (mechanical, heat, cold, radiation) or chemical (acids, bases) sources.

Node (in graph theory): A variable or state of a system represented at a point. There are input, output, and system nodes. Also see *vertex*.

Node degree (in graph theory): The *degree of a node or vertex* is the number of edges incident on that node, with loops being counted twice. In directed graphs such as SFGs, the edge count for a node k can be broken down into *indegree(k) and outdegree(k)* numbering the number of directed edges ending on node k and leaving node k, respectively.

Nosocomial infection: One acquired in a hospital or doctor's office.

Nuclease: An enzyme that has the function of cleaving the phosphodiester bonds between the nucleotide subunits of nucleic acids. Nucleases are further categorized as endo- or exonucleases.

Observability (of a MIMO LTIS): Observability is a measure of how well the internal states of a system can be inferred by knowledge of its external outputs. The observability and controllability of an LTIS are mathematical duals. A test for observability is the observability matrix, defined as follows. Given an **n**th order LTIS described in SV form:

$$\dot{x} = Ax + Bu$$

$$y = Cx$$

where
    y is the output column vector
    x is the internal states column vector
    A, **B**, and **C** are matrices

This system is *observable* if the observability matrix, O_M, is of rank **n**. O_M is defined by

$$O_M = \begin{bmatrix} C \\ CA \\ \vdots \\ CA^{n-1} \end{bmatrix}$$

Occam's razor (in complex systems): A principle attributed to the William of Occam (ca. 1285–1349), a friar and logician. This principle has been stated in several forms: (1) The explanation of any phenomenon should make as few assumptions as possible, eliminating or "shaving off" those that make no difference in the observable predictions of the explanatory hypothesis or theory. (2) All things being equal, the simplest solution tends to be the best one. (3) (in Latin) *Lex Parsimoniae: Entia non sunt multiplicanda praeter necessitatem.=Entities should not be multiplied beyond necessity.* The term "razor" refers to the act of shaving away unnecessary assumptions in reaching a parsimonious, simple explanation.

    Sir Francis Crick commented on the potential limitations of using Occam's Razor in biology: "While Occam's Razor is a useful tool in the physical sciences, it can be a very dangerous implement in biology. It is thus very rash to use simplicity and elegance as a guide in biological research." Evolution has led to complex designs.

Ocellus (in invertebrate photoreception): (*pl.* ocelli) A simple, multicellular, rhabdomeric photoreceptor found in a variety of invertebrates (e.g., arthropods and jellyfish). Ocelli are generally found in and around an animal's head (anterior). The horseshoe crab *Limulus polyphemus* has seven ocelli, plus two compound eyes. The dragonfly *Libellula* has three dorsal ocelli between its large compound eyes, so does the lubber grasshopper, *Romalea microptera*.

Ommatidium (*pl.* ommatidia): The basic optical/sensory unit in arthropod compound eyes. It consists of an outer corneal lens derived from chitin, a light-concentrating apparatus that directs photon energy down to the cluster of underlying, photosensitive retinula cells that meet in their rhabdom region along a center axis perpendicular to the lens. In grasshopper compound eyes, there are six retinula cells and two eccentric cells. The high-resolution compound eyes of preying mantisses and dragonflies each have ca. 10^4 ommatidia/eye; the cockroach *Periplaneta americana* has about 2000 ommatidia/eye.

Ontogeny: The course of development of an individual, multicellular (eukaryote) organism.

Operating point (in CNLSs): A set of constant (average), steady-state parameter values around which a stable CNLS exhibits stable, piecewise-linear, input–output behavior.

Operator (in genomics): The site at one end of an operon where an activator or repressor molecule binds to the DNA and thus regulates the transcription process.

Operon (in genomics): Found mostly in prokaryotes, operons represent a genetic "module" in which a group of enzymes and substrates needed for the synthesis of a single product are regulated together. In dsDNA, an operon consists of an upstream common promoter region, a common operator region, and then a series of two or more, linked structural genes. RNAP binds to the promoter. Regulatory TFs bind to the operator that can repress or activate transcription of the operon's genes into mRNA. Operons are found in prokaryote, archaeal, and some eukaryote genomes. In prokaryotes, about half of all protein-coding genes are found in multigene operons. The genes in an operon are transcribed by a single molecule of mRNA. Thus, the coded operon proteins are expressed in 1:1:1:... ratios.

Opportunity cost (in economic systems): The value of a good or service forgone in order to acquire (or produce) another good or service. An opportunity cost is involved when a farmer chooses to plant corn for ethanol instead of beans. Another example: A city decides to build a parking lot on municipal vacant land. The opportunity cost is the value of the next best thing that might have been done with the land and funds used. The city could have built a hospital, a sports center, a golf course, or even have sold the land to reduce debt.

Opsonin (in cell biology): Any molecule that acts as a binding enhancer for the processes of endocytosis and phagocytosis. Most phagocytic binding cannot occur without the opsonization of the Ag. Two important opsonin molecules are IgG and C3b of the complement system. Opsonization includes neutralization of the negative charge (zeta potential) on the phagocytosing cell and its target molecule.

Orthogonal ribosome (in synthetic biology): A human-designed ribosome that only synthesizes polypeptides in response to bioengineered orthogonal mRNAs. o-Ribosomes do not respond to endogenous mRNAs that use 3-base codons. Bioengineered o-ribosomes have been designed to read 4-base codons on bioengineered o-mRNAs. Thus, 265 different codons are available, instead of 64. (See, e.g., Chubiz and Rao [2008], and Rackham and Chin [2006a].)

Pallium (CNS anatomy): In mammals, the outer layer of the cerebral hemisphere that envelops structures in the brain stem. The outer pallium includes the cerebral cortex.

Parametric regulation and control (in CNLSs): A nonlinear, feedback/feed-forward system in which regulation/control is effected by having one or more path gains as a function of one or more state variables. Parametric R and C is found in all living systems, for example, biochemical, molecular biological, and physiological systems. Parametric R and C makes complex adaptive biological systems possible; it permits transcription regulation and homeostasis.

Paratope (in the IS): The paratope region of an antibody (Ab) or receptor molecule binds chemically to the epitope region on an antigen (Ag), a kind of molecular, key-in-lock process. Certain B- and T-cells can have cell membrane-bound Abs having paratopes to a specific antigenic epitope. The fit is by chance. (See *epitope*.)

Pareto distribution (in CNLSs): This power-law probability distribution was named after the economist Vilfredo Pareto. It models the statistical behavior of many types of economic systems, as well as various social, geophysical, and actuarial phenomena, even the sizes of meteorites. If X is a random variable (RV) with a Pareto distribution, then the probability that X is greater than some value x is given by $pr(X > x) = (x/x_m)^{-\gamma}$, where x_m is the minimum possible X and $\gamma > 0$. When this distribution is used to model the distribution of wealth, γ is called the *Pareto index*. By differentiation, the Pareto PDF (probability density function) is given by $p(x; \gamma; x_m) = (\gamma x_m^{\gamma})/x^{\gamma+1}$, for $x \geq x_m$. The expected value of the RV X following the Pareto statistics is $E(X) = \gamma x_m/(\gamma - 1)$. If $\gamma \leq 1$, $E(X) \to \infty$. The variance of X is $var(X) = [\gamma/(\gamma - 2)][x_m/(\gamma - 1)]^2$.

Pareto efficiency (or optimality) (in complex economic systems): Given a set of individuals with alternative economic allocations of goods or income, any movement from one allocation to another that can make at least one individual better off without making any of the others worse off is called a *Pareto improvement*. If the economic allocation in any system is not Pareto efficient, there is a theoretical potential for a Pareto improvement, that is, an increase in Pareto efficiency. An allocation is called (strongly) Pareto optimal when no further Pareto improvements can be made.

Path (in graph theory): A route that does not pass any node (vertex) more than once. If the path does not pass any node more than once, it is a *simple path*. (Also see *walk*.)

Pediocin-like bacteriocins: AMPs produced by G^+ lactic acid bacteria. They contain 37–48 AAs, have similar structures, display anti-listeria activity, and kill target cells by permeabilizing their cell membranes. Over 20 pediocin-like bacteriocins have been characterized (Fimland et al. 2005).

Phase plane (in nonlinear system analysis): A two- (or three)-dimensional parametric plot of two (or three) outputs of a system given certain initial conditions, or a transient input. Often $\dot{x}(t)$ versus $x(t)$ is plotted. As an example, assume that a NL second-order system is modeled by a second-order ODE of the form

$$\ddot{x} + F(\dot{x}, x) = 0$$

Let $x = x(t)$ and $y = x(t)$. Note that the second-order nonlinear system can also be written as two first-order ODEs (state equations):

$$\dot{y} = -F(y, x)$$

$$\dot{x} = y$$

A plot of $x(t)$ versus $y(t)$ in the $y = \dot{x}$ versus x plane for t values ≥ 0, starting at $t = 0$ at some $x(0)$, $y(0)$, is called a phase plane *trajectory*.

The y, x plane is the phase plane. The system's phase plane trajectories provide important information about the equilibrium states, oscillations (LCs), and the stability of the nonlinear system. Any Mth-order NL system modeled by a set of M NL ODEs can be written as $\dot{x}_n = F_n(x_1, x_2, \ldots, x_{n-1}, x_n, x_{n+1}, \ldots, x_M)$, $n = 1, 2, \ldots, M$. Thus, M 2D phase plane plots can be made for the $(\dot{x}_n, x_n)$ trajectories. A phase plane trajectory that converges on a closed path as $t \to 0$ gives a graphic demonstration of stable LC oscillations. Patterned, closed, steady-state phase plane trajectories are called *attractors*.

Phase space: A multidimensional phase plane. In a system that has $N > 1$ variables, the phase space is generally a 2N-dimensional space (the N variables plus their N first time-derivatives).

Pheremone (in ecology): A chemical compound secreted by animals to mark territory or attract mates.

Plectics: A research area, named by Murray Gell-Mann, that is "...a broad transdisciplinary subject covering aspects of simplicity and complexity as well as the properties of complex adaptive systems, including composite complex adaptive systems consisting of many adaptive agents" (Wiki 2007c).

Pleiotropy (in physiology): The condition whereby an intercellular signaling molecule (e.g., a cytokine or hormone) can have different physiological effects on different kinds of target cells having receptors specifically for that signal molecule. (Also defined as a single gene influencing multiple phenotypic traits.) An organ can also behave pleiotropically, for example, the kidneys, which participate in the regulation of several different ion concentrations in the blood. In genetics, pleiotropy is the regulation or determination of more than one characteristic, function, or product by a single gene (or operon).

Pole (in LSs considered in the frequency [Laplace] domain): An LTIS is said to have a pole at some finite, complex frequency value, s_k, if its transfer function magnitude $|H(s_k)| \to \infty$, that is,

$$\frac{Y}{X}(s_k) = H(s_k) = \frac{P(s_k)}{Q(s_k)} = \frac{s_k^m + b_{m-1}s_k^{m-1} + \cdots + b_1 s_k + b_0}{s_k^n + a_{n-1}s_k^{n-1} + \cdots + a_1 s_k + a_0} \to \infty$$

In general, this will occur for some finite root of the denominator polynomial, $Q(s)$, s_k.

Polycistronic (in genomics) (*adj.*): A form of gene organization that transcribes an mRNA that codes for multiple proteins. Polycistronic DNA segments have multiple ORFs in one transcript and also have small regions of untranslated sequences between each ORF.

Power-law distribution (PLD) (in graph theory and CNLSs): A PLD has the general form $p(x) \propto L(x)\, x^{-\gamma}$, where $\gamma > 1$, and $L(x)$ is a slowly varying function that satisfies $\lim_{x\to\infty} L(tx)/L(x) = 1$ with t constant. The form of $L(x)$ only controls the shape and finite extent of the long tail. If $L(x) =$ constant, then the PLD holds for all values of x. If there is a lower bound of x, x_{min}, from which the law holds, then the PLD becomes

$$p(x) = \left[\frac{\gamma - 1}{x_{min}}\right]\left(\frac{x}{x_{min}}\right)^{-\gamma}, \quad (x_{min} \le x \le \infty)$$

The moments of this PDF are given by

$$[x^m] = \int_{x_{min}}^{\infty} x^m p(x)\, dx = \frac{\gamma - 1}{\gamma - 1 - m} x_{min}^m$$

which is only well defined for $m < (\gamma - 1)$. When $2 < \gamma < 3$, the mean exists, but the variance and higher-order moments $\to \infty$.

Prodrug (in pharmacology): A pharmacological substance (drug) that is applied in an inactive form. Once administered, the prodrug is acted on chemically to convert it to its active form. Activation can be by photons, a natural enzyme, or by a CAb. Prodrugs allow a targeted therapy directed to specific pathogens, cells, or cancer cells, eliminating many severe side-effects of "broadside" pharmacology.

Promoter (aka Promoter Sequence): A DNA Nt sequence(s) lying upstream (5′) from a gene, to which the RNAP complex binds. It contains the TATA box surrounded by a large complex of various proteins, including transcription factor IID (TFIID), which is a complex of the TATA-binding protein (TBP) that recognizes and binds to the TATA box. There are

14 other protein factors that bind to the TBP and each other, but not to the DNA. Transcription factor IIB (TFIIB) binds to both the DNA and pol II. (TFII A, D, E, F, and H also participate in gene promotion.) Promoters essentially turn a gene on or off. Transcription is initiated at the promoter.

Proteolytic enzymes: Enzymes that break down the chemical bonds (e.g., peptide, S-H) holding the protein tertiary structure and AA chains together. The pancreas secretes many proteolytic enzymes into the duodenum.

Protostomia (in evo-devo): A taxon of metazoan, bilateral animals, in which during development the anterior blastopore becomes the animal's mouth. (In deuterostome development, the blastopore becomes the animal's anus.) Molecular biological evidence suggests that protostome animals can be divided into two major groups: Ecdysozoa (nematodes and arthropods) and Lophotrochozoa (molluscs and annelids).

Pyrogenic (in immune response): Causes fever.

Q_{10} (in physical biochemistry): Q_{10} (pronounced Q-ten) is a measure of a chemical reaction's temperature sensitivity. It is the ratio of the rate of an isothermal reaction at temperature T_0 to that of the same reaction at $T = T_0 - 10°C$. Mathematically, Q_{10} can be written more generally as

$$Q_{10} \equiv \left(\frac{R_2}{R_1}\right)^{[10/(T_2 - T_1)]}$$

where

R_1 is the reaction rate at $T = T_1$
R_2 is the reaction rate at $T = T_2 > T_1$
Q_{10} for most biochemical reactions ≈ 2

Another way of looking at the temperature sensitivity (S_T) of reaction rate constants is

$$S_T \equiv \frac{\Delta R}{\Delta T}\left(\frac{T_1}{R_1}\right)$$

where $\Delta R = R_2 - R_1$, $\Delta T = T_2 - T_1$, $T_2 > T_1$, $R_2 = R@T_2$, $R_2 > R_1$, etc.

Quantization noise (in numerical simulations): Introduced into digital signals by the process of A/D conversion and computational roundoff. An important source of broadband noise introduced into digitized analog signals resulting from the stepwise relationship between analog-to-digital converter (ADC) output bit words and the continuous analog input. The mean-squared quantization noise error voltage can be shown to be given by $N_q = q^2/12$ msV, where q is the quantization step size in volts. If the dynamic range of the ADC is ±3 standard deviations of the signal, then we can show that the mean-squared noise voltage is approximately

$$N_o = \frac{q^2}{12} \cong \frac{3\sigma_x^2}{(2^N - 1)}$$

where

N is the number of bits in the ADC output word (e.g., 8, 16, and 20)
σ_x^2 is the mean-squared signal value (Northrop 2004)

Digital "noise" can also arise within digital simulations from roundoff errors. In simulations, arithmetic (addition, subtraction, multiplication, and division) is generally carried out with 16, 32, 64, or 128 bit precision, depending on the computer and the application software.

Quasispecies (in viral evolution): Well-mixed "clouds" of genotypes that appear in a population at mutation-selection balance (Bull et al. 2005). More specifically, a term used to describe nonliving organisms such as various RNA viruses that possess high mutation and replication rates. Genomic mutation rates that are nonlethal create a broad, mixed population of genotypes, some of which are more robust to niche changes. Thus, they may be able to (1) escape the immune response, (2) better resist antiviral therapy, and (3) adapt to changes in host cell biochemistry. The quasispecies model is applicable to RNA viruses because they have high mutation rates on the order of 1 mutation per round of replication, and their populations can be very large. Modeling and experimental studies have shown that a broad quasispecies population with a broad distribution of fitness may be a better evolutionary strategy for survival than to have a sharply defined, most fit, stable, single genotype. The complexity of an RNA VQS refers to the amount of nonrepetitive information stored in a genome. The measurement of the quasispecies' complexity includes the number of the percentage of different variants (i.e., mutant frequency) and the polymorphism (phenotypes) among the variants (Gómez et al. 1999).

Quorum sensing (in biological oscillators and certain bacteria): (1) Groups of like cells can synchronize their cyclical efforts of gene expression and protein synthesis, as well as their cell cycles. The ability of these synchronously active cells to have their endogenous oscillations entrained in frequency and phase by a commonly secreted, exogenous AI molecule is called quorum sensing (QS). (2) Bacteria communicate with each other using AIs and QS (Gutierrez et al. 2009). Bacterial QS is used to regulate a variety of synchonized bacterial behaviors: bacteriocin production along with CIPs (Rossi et al. 2008), bioluminescence, expression of virulence factors, biofilm formation, conjugation, and pigment formation (Pai et al. 2009).

Race (in biology): An interbreeding, usually geographically isolated population of organisms differing from other populations of the same species in the frequency of hereditary traits. A race that has been given formal taxonomic recognition is known as a *subspecies*. Race applied to humans is more difficult to describe: One definition is "a local geographic or global human population distinguished as a more or less distinct group by genetically transmitted physical characteristics." A more controversial definition is "a group of people united or classified together on the basis of common history, nationality, or geographic distribution, for example, the Scandinavian race" (cf. http://www.answers.com/topic/race-1).

Rank (in matrix algebra): The rank of a matrix **A** is invariant under the interchange of two rows (or columns), or the addition of a scalar multiple of a row (or column) to another row (or column), or the multiplication of any row (or column) by a nonzero scalar. For an $n \times m$ matrix **A**, rank $\mathbf{A} \le \min(n, m)$. For an $n \times n$ matrix **A**, a necessary and sufficient condition for rank $\mathbf{A} = n$ is that $|\mathbf{A}| \neq 0$ (Ogata 1990).

Recession (in economics): A contraction phase of the business cycle. A recession is said to occur when the real GDP is negative for two or more consecutive quarters. Recessions may include bankruptcies, deflation, foreclosures, unemployment, reduced sales, and stock market crash.

Red Queen dynamics (in genomics): They occur when two, competing, complex biological systems interact, for example, a quasispecies (QS) (viral, bacterial, or parasitic) versus the host's IS. Red Queen Dynamics have been viewed as a type of evolutionary "arms race." *Red Queen Cycling* is when the QS genes that result in infectivity track host IS gene frequencies, leading to high QS fitness and a continuing infection in a sympatric host; infectivity LCs can occur. The Red Queen Scenario, first proposed by van Valen (1973), suggests that genetic diversity within a population is required to counter rapidly evolving pathogens. A Red Queen Scenario exists between bacteria and human use of antibiotics. The *Red Queen Principle* can be stated: "For an evolutionary system, continuing development is needed just in order to maintain its fitness relative to the systems it is co-evolving with." (The Red Queen herself comes from Ch. 2 in Lewis Carroll's *Through the Looking Glass*, 1872.)

Reductionist thinking (in complex systems): Includes the natural human proclivity to try to manipulate a complex system by changing one input state at a time. A complex system model is often partitioned (i.e., reduced) by reductionists in order to simplify analysis or obtain a local solution. This approach destroys the unreduced complex system's counterintuitive behavior. A better approach to manipulating a whole, unreduced complex system model is to use multiple inputs and observe its behavioral responses, and then choose which inputs work best.

Redundancy (in biology): Redundancy is found at the molecular, cellular, physiological, and organismic levels, where multiple elements have the same functions, and operate in parallel (e.g., pancreatic beta cells, hepatocytes, erythrocytes, platelets, hematopoietic stem cells, two lungs, two eyes, two ovaries, and many worker bees). Redundancy is one insurance of system robustness. (See *degeneracy*.)

Regulator (in control theory): A subsystem that senses the departure of certain critical system states from normal set values due to adverse external inputs or internal disturbances, and then takes action to bring the SVs back to the set values. A regulator must effectively store the information about acceptable state values to compare with the perturbed SVs, either implicitly or explicitly. It must then make internal adjustments to certain system states in order to maintain constancy in the regulated parameters (e.g., states and outputs). A parametric regulator adjusts system branch gains to effect critical SV constancy. In general, a regulator responds to disturbances, rather than to system control inputs. However, a system controller can also act as a regulator. Regulation is generally effected by some form of negative feedback.

Regulon (in genomic systems): In prokaryotes, a regulon is a collection of genes that may or may not be grouped together in operons. A regulon's genes are regulated by the same, common RP. An example is quorum sensing in bacterial biological clocks.

Relaxation oscillation (a type of BCO): Relaxation oscillation is a term borrowed from electronics (Wang 1999). A simple relaxation oscillator can be made from a switch, a battery, a resistor, a capacitor, and a neon bulb. The neon bulb is placed across the capacitor, which is connected through the resistor to the battery. The capacitor is initially discharged. The switch is closed and the capacitor charges through the resistor toward the battery voltage, V_B. When the voltage across the capacitor, V_C, reaches the ionization threshold voltage ($V_I < V_B$) for the neon bulb, the sudden increase of conductance of the neon bulb causes the capacitor to discharge rapidly through the high-conductivity neon plasma until it reaches the low extinction voltage, $V_E < V_I$, where the ionization of the neon gas $\rightarrow 0$, and the bulb's conductance $\rightarrow 0$. The capacitor now charges again through the series resistor until $V_C = V_I$, and the relaxation cycle repeats. The voltage across the capacitor, V_C, is seen to have a sawtooth shape, rising slowly from V_E until it reaches V_I, and then decreasing rapidly to V_E again as the capacitor discharges through the neon bulb causing it to flash, and the cycle repeats. Many other types of relaxation oscillators exist; as a rule, they all involve threshold switching of a parameter or device.

Repressilator (in systems biology): A repressilator is a genetically engineered modification of the *Lac* operon in *E. coli* that makes its gene product concentrations oscillate in bounded LCs.

Repressor (in genomics): A molecule that binds to a noncoding sequence of a gene, preventing RNAP from making mRNA on the coding section of DNA.

Resilience (in CASs): Resilience has been defined as the capacity of a system to absorb disturbance and reorganize while undergoing change so as to retain essentially the same function, structure, identity, and feedbacks, that is, remain within one regime—it reflects the degree to which a CAS is capable of self-organization and the degree to which the CAS can build and sustain the capacity for learning and adaptation (Norberg and Cumming 2008). (Compare with *robustness*.)

Retrotransposon (in genomics): A subclass of transposons that can induce mutations by being inserted near or within genes. Retrotransposons copy themselves to an RNA template, which is then copied back to DNA in the host cell by a reverse transcriptase enzyme. These copied oligos are inserted in the host's genome. One type of retrotransposon is the LINE, over 5 kb in length, that codes for two proteins: one that has the ability to bind ssRNA, and the other that has reverse transcriptase and endonuclease activity, enabling them to copy both themselves and noncoding SINES (short interspersed element) such as Alu elements.

Retrovirus: A class of virus (e.g., HIV) in which the information for self-replication is stored in ssRNA. After the virus enters its target cell, it uses its enzyme, reverse transcriptase, to generate a DNA copy of its genome, and then uses the cell's internal biochemical machinery to make copies of its genomic RNA and other viral proteins inside the cell. The cell is destroyed when the daughter virions are released (viral eclosion).

Reverse transcriptase (in genomics): An RNA-dependent, DNA polymerase enzyme that makes a DNA copy from an ssRNA Nt sequence. Found in retroviruses such as HIV, and in LINES.

RNA-induced silencing complex (RISC) (in genomics): The ribonucleoprotein complex required for small RNA-mediated gene suppression.

Robustness (in systems theory and biology): A system is said to be robust (have robustness) when it largely preserves its normal functioning in the presence of external (environmental) disturbance inputs (e.g., temperature changes, changes in salinity, incident light flux, and pO_2) and internal noise, and also shows feature persistence in the presence of internal (parametric) pathway changes or failures, or subsystem (module or motif) failures. Pathway parametric changes can result from photon energy absorption; molecular ageing; chemical damage from peroxides, free radicals, etc.; and metabolic poisons.

Negative feedback can produce robustness, as can redundancy, alternate pathways, and modularity. (See *homeostasis*.) Robustness is graded.

Route (in graph theory): A sequence of edges and nodes from one node to another. Any given edge or node might be used more than once.

Scale (of complex systems): The complexity of biological, economic, weather, ecosystem, political, and government systems, for example, can be studied at several levels of scale. Complex biological system behavior can be observed and modeled at molecular (biochemical), cellular, physiological, and organismic levels. An economic system's complexity can be studied at various scales, such as family, municipal, state, regional, national, and global.

In biophysical sciences, scale is generally defined as a property of the dimensions of space and time. This view of scale has two main components: grain and extent. Grain refers to the resolution of analysis, and extent to the coverage (Cumming and Norberg: *Scale and Complex Systems*. Ch. 9 in Norberg and Cumming 2008).

Scale-free graph (in graph/network theory): A graph/network whose degree distribution follows a power-law model, sometimes asymptotically, that is, $P(K) \cong ak^{-\gamma}$, where the exponent γ typically lies in the range between 2 and 3. Many observed networks appear to be scale-free. These include the World Wide Web, protein–protein interaction networks, citation networks (in scholarly papers), human sexual partners, semantic networks, and some social networks. The high-degree nodes are called *hubs*. Some properties of scale-free networks (SFNs) are as follows: (1) SFNs are more robust against random failures; the network is more likely to stay functional than a random network after removal of randomly chosen nodes. (2) SFNs are more vulnerable against targeted (nonrandom) attacks on their hubs. This means an SFN looses function rapidly when nodes are removed according to their degree. (3) SFNs have short average path lengths. The average path length, L, is proportional to $\log(N)/\{\log[\log(k)]\}$ (Hildago and Barabasi 2008).

Segmentation (in biology): Segmentation refers to the anatomical subdivision of some metazoan bodies into a series of semi-repetitive segments. It can be viewed as a form of anatomical modularity. Segmentation evolved hundreds of million years ago (mya). It is very evident in fossils of trilobites, extinct marine arthropods from the Paleozoic era, 250–450 mya. We currently see it in mature annelid worms, insects, certain plants, and in structures such as the vertebrate spinal column. Segmentation allows for a high degree of specialization in body regions under local developmental control by *homeobox* genes.

Sensitivity (of complex LSs): (aka gain sensitivity.) Assume a large LS with a certain input x_i and an output y_k. The complex gain between this input and output is defined as $M_{ik} = Y_k/X_i$. The sensitivity of M_{ik}, given a change in a certain internal path gain, G_{pq}, is defined as

$$S_{Gpq}^{Mik} = \frac{\partial M_{ik}}{\partial G_{pq}}\left(\frac{G_{pq}}{M_{ik}}\right)$$

Ideally, we want $S_{Gpq}^{Mik} \to 0$ for robustness. For example, consider a linear, SISO feedback system with forward gain, P, and feedback path gain, Q. The system's overall gain is well known: $M = Y/X = P/(1+PQ)$. S_P^M is calculated for this simple system and found to be $1/(1+PQ)$, that is, the effect of ΔP on the gain M is minimized as the return difference

$\mathbf{F}=(1+\mathbf{PQ})\to\infty$. This illustrates an important general property of negative feedback systems, that is, the reduction of gain sensitivities, hence the effect of parameter variations on the overall loop gain.

For a stable, MIMO CNLS, the sensitivity for the small-signal gain between the $\mathbf{i}$th input and $\mathbf{k}$th output, $\mathbf{M_{ik}}$, as a function of a change in the $\mathbf{n}$th parameter, $\mathbf{P_n}$, or another input, $\mathbf{x_j}$, can be written as

$$S_{P_n}^{Mik}=\frac{\partial M_{ik}}{\partial P_n}\left(\frac{P_n}{M_{ik}}\right) \quad \text{or} \quad S_{xj}^{Mik}=\frac{\partial M_{jk}}{\partial x_j}\left(\frac{x_j}{M_{jk}}\right)$$

These sensitivities can be approximated by

$$S_{P_n}^{Mik}\cong\frac{\Delta M_{ik}}{\Delta P_n}\left(\frac{P_n}{M_{ik}}\right) \quad \text{or} \quad S_{xj}^{Mik}\cong\frac{\Delta M_{ik}}{\Delta x_j}\left(\frac{x_j}{M_{ik}}\right)$$

Again, for robustness, we generally desire the nondirect-path parameter sensitivities to be small or zero.

Set point (in a regulated system): The desired, steady-state value of a regulated system's output state.

Shannon Diversity Index (SDI) (applied to complex ecosystems): Sometimes called the Shannon–Wiener diversity index, the SDI ($\mathbf{H'}$) is used as an objective measure of biodiversity in complex ecosystems. (Sometimes workers use $\log_2(*)$ instead of $\ln(*)$.)

$$H'=-\sum_{i=1}^{s}p_i\ln(p_i)$$

where
  $\mathbf{n_i}$ is the number of individuals in each of a total $\mathbf{S}$ species, $\mathbf{i}=1, 2, 3,$ ..., $\mathbf{S}$ (aka $\mathbf{n_i}=$ the abundance of a species)
  $\mathbf{S}$ is the number of different species in the ecosystem (species' richness)
  $\mathbf{N}$ is the total number of all individuals in the ecosystem $=\displaystyle\sum_{i=1}^{s}n_i$
  $\mathbf{p_i}$ is the relative abundance of species $\mathbf{i}$, $\mathbf{p_i}\cong\mathbf{n_i/N}$

Estimates the probability of finding species $\mathbf{i}$ in the ecosystem.
(It can be shown that for any given number of species, there is a maximum possible $\mathbf{H'}=\mathbf{H'_{max}}=\ln\mathbf{S}$, when all $\mathbf{S}$ different species are present in equal numbers ($\mathbf{n_i}=\mathbf{N/S}$) [Beals et al. 2000, Wiki 2007d].)

Shannon's equitability index (applied to complex ecosystems): Defined as $\mathbf{E_H}\equiv\mathbf{H'/H'_{max}}$. ($\mathbf{H'_{max}}=\ln(\mathbf{S})$.) Equitability assumes a value between 0 and 1 with 1 being complete evenness.

Siderophores (in bacterial iron metabolism): Small, high-affinity iron (Fe^{3+}) che-lating compounds employed by bacteria such as *E. coli, Vibrio cholerae, Yersina pestis, B. subtilis,* and *B. anthracis* to chelate and internal-ize Fe^{3+} for use in bacterial metabolism. The normal free [Fe^{3+}] is ca. 10^{-24} mol/L. Once inside the bacterial cytoplasm, the Fe^{3+}-siderophore complex is usually reduced to Fe^{++} to release the iron. Siderophores can also chelate Al, Ga, Cr, Cu, Zn, Pb, Mn, Cd, V, Pu, and U ions.

Signal flow graphs (SFGs): SFGs are used in describing LTI systems in the frequency domain. They are directed graphs in which the nodes (ver-tices) are signal summation points (adding inputs from all incoming branches [edges]). SFG-directed edges are signal-conditioning pathways that multiply (condition) the signal at a source node and input that con-ditioned signal additively to the sink node. Outwardly directed branches that leave a node do not change the signal (state) at that node. They pro-cess the signal at the source node that is the result of summing all input signals at that node. SFGs generally process signals in the frequency domain. SFGs can be used to characterize sets of linear state equations (ODEs). In nonlinear SFGs [NLSFGs], one or more branches operate nonlinearly on node signals. Thus, Mason's rule cannot be used to find the NLSFG's transfer functions.

Simplicity (in systems): The absence of complexity. The property of being simple, having few states. Simple systems are easier to describe, model, explain, and understand than complicated ones. Like complexity, simplicity is a relative, graded property. Simplicity can denote beauty, purity, or clarity.

SINE (in genomics): Short interspersed element. These are Class I transposed DNA sequences of 100–400 bps. The most abundant SINEs are the Alu elements. SINES are nonprotein-coding DNA sequences. SINEs do not encode a functional reverse transcriptase and rely on other mobile ele-ments' enzymes for transposition. The most common SINEs in primates are the Alu sequences of ca. 280 bps. SINEs are noncoding, and make up ca. 13% of the human genome.

Size (in graph theory): The *size* of a graph is the total number of its edges. In complex systems, size matters.

Species (in biology): A basic definition of species is a group of organisms capable of interbreeding and producing fertile offspring. The classification of a species often depends on morphology (phenotype), habitat (niche), behavior, and genotype. The presence of specific local traits or morphol-ogy may allow a species to be divided into subspecies or races (e.g., the races of man, and the breeds of dogs, horses, and cattle). Each species is assigned to a specific genus. Such assignment is based on the hypothesis that the species in that genus are more closely related (e.g., *Homo sapi-ens* and *Homo habilis,* where *Homo* is the genus, *sapiens* is a species).

Specified complexity (SC) (in creationism): A concept promoted by ID propo-nent William Dembski. It singles out patterns that are both specified and complex. Dembski claims that SC (aka complex specified information) is a reliable marker of design by an intelligent agent (God). Dembski developed a test for SC based on the Kolmogorov complexity index (Wiki 2007b). An SC factor was calculated (Dembski 2005).

Stability tests (for an LS): An LS is stable if its transfer function H(s) has no poles (denominator roots) in the right-half **s**-plane. In a SISO, single-loop feedback system, the closed-loop transfer function H(s) is of the form

$$H(s) = \frac{F(s)}{[1 - A_L(s)]}$$

Some tests for stability examine the complex **s** values that make the denominator of H(s) → ∞, that is, values of **s** that make $A_L(s) = 1\angle 0$. This criterion has given rise to the venerable Nyquist stability criterion, the root-locus technique, and the Popov stability criterion.

Stagflation (in economics): An economic condition in which there is a simultaneous inflation and economic stagnation (loss of jobs).

State (in LTISs): The state of an LTIS is the smallest set of variables (called state variables) such that the knowledge of these variables at $t = t_o$ together with the input for $t \geq t_o$ completely determines the behavior of the system for any time $t \geq t_o$ (Ogata 1970).

State variable (in LTISs): The SVs of an LTIS are the smallest set of variables that determine the state of the dynamic system. If at least **n** variables $x_1(t)$, $x_2(t)$, ..., $x_n(t)$ are needed to completely describe the behavior of a dynamic LTIS, then such **n** variables $x_1(t)$, $x_2(t)$, ..., $x_n(t)$ are a set of state variables (Ogata 1970). In a linear SFG, sums of certain SVs are found at certain nodes, and their integrals at other nodes.

Stationary (in systems analysis): A *stationary* signal arises from a stationary system in which no parameters or branch gains are changing in time over the period in which the signal is measured. Stationary also refers to noise produced from within a system.

Step function (in systems analysis): Used as a system input, a unit step function, U(t), is 0 for $t < 0$ and 1 for $t \geq 0$. A delayed step function, $U(t - \tau)$, is 0 for $t < \tau$ and 1 for $t \geq \tau$. Also, $U(t - \tau) = \int_0^\infty \delta(t - \tau)dt.\ \tau > 0.$

Stiff ODEs (in simulating CNLSs): Computer solutions of nonlinear ODEs are done by numerically integrating difference equations. There are many numerical integration routines that are used for these solutions (e.g., rectangular, trapezoidal, Euler, Runge–Kutta/Fehlberg, Dormand–Prince, Gear, and Adams). The use of certain integration algorithms with stiff ODEs can result in numerically unstable or noisy solutions. Such chaotic behavior can often be eliminated by making the step size extremely small, and choosing an integration routine such as Gear or rectangular, which are generally well behaved with stiff ODEs. An example of a stiff, nonlinear ODE was given by Moler (2003): An ODE that models the size of a ball of flame when a match is struck is $dr/dt = r^2 - r^3$, $r(0) = \rho$, where ρ is the initial (normalized) radius r of the flame ball; try 0.0001–0.01. In this initial value problem (IVP), we are interested in the steady-state solution, in this case, $r(\infty) \to 1$, and how the ODE approaches SS ($dr/dt = 0$). Moler shows, using the MATLAB ODE solver ode45, that this ODE approaches its SS in a very long, noisy manner before rapidly converging on $r = 1$.

Stimulon (in genomics): Stimulons are involved in activating global biochemical and physiological responses in response to environmental stress stimuli such as starvation, desiccation, or osmotic stress. A *stimulon* is a collection of genes (which may be in operons and regulons) under regulation by the same stimulus. This term has been used for prokaryotic systems, for example, quorum sensing in bacteria.

Stromal cells (in human anatomy): Stromal cells are connective tissue cells in the loose connective tissues surrounding organs. They are often associated with the uterine endometrium, prostate, bone marrow precursor cells, and the ovaries. Stromal cells make up the support structure of biological tissues and support the parenchymal cells. Endothelial cells, fibroblasts, pericytes, immune cells, and inflammatory cells are examples of the most common types of stromal cells.

Strongly connected graph (in graph theory): A directed graph (e.g., an SFG), directed subgraph, or module is called strongly connected if there is a path from each vertex in the graph or subgraph to every other vertex. In particular, this means a path in each direction—a path from node **p** to node **q**, and also a path from **q** to **p**.

Structural gene (in genomics): A gene that codes for the structure of a protein.

Subgraph (in graph theory): A subgraph of a graph G is a graph whose vertex and edge sets are subsets of the total vertices and edges of G. If a graph G is contained in graph H, H is a supergraph of G.

Sympatric (in ecology): (*adj.*) Occupying the same niche without interbreeding. Said of populations of closely related species, for example, viral QS.

Sympatric speciation (in ecology): Describes the genetic divergence of various populations (derived from a single clone species) inhabiting the same niche. This divergence can presumably lead to the emergence of new species.

Synchronization (of N oscillating systems): Frequency synchronization causes two or more oscillators to oscillate at exactly the same frequency. Phase-lock synchronization causes two or more oscillators to oscillate with the same frequency, in phase.

Synteny (in genetics): The physical co-localization of genetic loci on the same chromosome. *Hox* genes exhibit synteny.

Synthetic biology: An emerging area of biological engineering. The name synthetic biology was coined in 1974 by the Polish geneticist, Waclaw Szybalski. It was used again in 1978 by Szybalski and Skalka. One of the aims of synthetic biology is to create artificial gene circuits that perform designated functions outside the normal capabilities of a host organism (e.g., a bacterium). Synthetic biological gene systems have been designed to be "orthogonal," that is, to be substantially independent of the gene systems of the host organism. This means that orthogonal mRNA (o-mRNA) and o-ribosome pairs are used to make new proteins (and consequently, GMOs) that have utility in various fields of human endeavor (e.g., medicine, agriculture, and materials). o-Ribosomes generally do not recognize endogenous mRNAs.

System: A group of interacting, interrelated, or interdependent elements (also parts, entities, or states) forming or regarded as forming a collective entity. There are many definitions of system that are generally context dependent. (The preceding definition is very broad and generally acceptable.)

System boundary: The notional dividing line between a system and its system environment (and other systems).

System environment: The system's niche. "That set of *entities* outside the *system boundary*, the state of which set is affected by the system or which affects the state of the system itself." Think arctic ice cap and polar bears.

Tamm–Horsfall glycoprotein (THP) (in innate, antimicrobial defense): Secreted by cells lining the thick ascending limb of the loop of Henle in the mammalian kidney. It has a MW of ca. 68 kDa. This protein does not kill invading bacteria directly, rather it interferes with the bacteria's ability to bind to epithelial cells and grow there, for example, *E. coli*. The bacteria covered with THP are washed out by the filtrate and urine stream.

Teratogen (in developmental biology): An exogenous chemical that causes developmental defects in a developing embryo. 13-*cis*-retinoic acid (vitamin A) is an example of a teratogen that can disrupt the expression of *Hox* genes and cause dysmorphology in embyros. Thalidomide is another teratogen; when taken by pregnant human mothers, it caused defects in limb development. Susceptibility to teratogenesis varies with the developmental stage of the embryo at the time of exposure, and also depends on the genotype of the animal. There are many teratogeneic agents, including but not limited to ionizing radiation, certain infections (parvovirus B19, rubella virus, herpes virus, toxoplasmosis, syphilis, etc.), metabolic disturbances (alcoholism, diabetes, folic acid deficiency, etc.), and chemicals and drugs (excess vitamin A, isoretinoin, temazepan, nitrazepam, nimetazepam, aminopterin, androgenic hormones, captopril, chlorobiphenyls [PCBs], dioxin, coumarin, diethylstilbesterol, lithium, tetracyclines, thalidomide, penicillamine, etc.).

Time constant form (of a rational polynomial in s): For example, the transfer function H(s) in Laplace form is written

$$H(s) = \frac{K(s+a)}{s^2 + 2\xi\omega_n s + \omega_n^2}$$

In time constant form, H(s) is written

$$H(s) = \frac{(Ka/\omega_n^2)(s\tau + 1)}{s^2/\omega_n^2 + (s2\xi/\omega_n) + 1}$$

where $\tau = 1/a$. Time constant form is used for Bode plotting; Laplace form is used to find the time functions by inverse Laplace transforms.

Tipping point (TP) (in CNLSs): The TP of a CNLS occurs when a critical set of initial conditions and/or inputs is reached such that the system rapidly changes its input/output dynamic behavior. In the phase plane, this change is indicated by a switch from one attractor to another. One example of a TP is in global warming climate change. The slow buildup of man-made atmospheric greenhouse gasses (e.g., CO_2 and CH_4) has caused a slow increase of global temperature from trapped solar IR radiation. This temperature increase causes the slow melting of arctic permafrost, the arctic ice cover and glaciers. Melting permafrost releases more methane and carbon dioxide into the atmosphere; melting ice sheets and glaciers expose more IR-absorbing ground (causing a decrease in arctic albedo), and so more heat is absorbed, accelerating the melting process. These events can be viewed as local positive feedback (autocatalytic) processes that accelerate the process of global warming, hence global temperature rise. A TP for these processes occurs at some critical atmospheric CO_2 concentration. Of course many other parameters affect global climate, including atmospheric dust and SO_3 gas from volcanism, solar energy output, alteration of heat-carrying ocean currents by fresh melt water, etc. (See also *Hopf bifurcation*.)

Another ecological TP has recently occurred off the coast of Namibia. Namibia once had a large and profitable sardine fishery. Sardines eat plankton. The Namibian sardine population fell abruptly from overfishing, leading to a plankton bloom. The plankton died and their bodies settled on the sea floor. Soon, bacterial decomposition of the dead plankton led to the release of an extensive volume of methane gas and also poisonous hydrogen sulfide gas (H_2S). These gasses dissolved in seawater and created a hostile environment for sardines and plankton, and so the sardine population was prevented from recovering. If the sardine catch had been regulated, there would have been a sustainable catch and no plankton superbloom.

Toll-like receptors (TLRs) (in innate ISs): A family of transmembrane proteins that play a key role in the innate ISs of animals. They are found on the surfaces and inside of macrophages, DCs, and epithelial cells.

They receive their name from their similarity to the products of the *Toll* genes found in *Drosophila*. TLRs recognize structurally conserved molecular components (PAMPs) derived from bacteria, viruses, and mycoplasma. There are at least 12 mammalian TLRs. About half of the mammalian TLRs are located on cell surfaces; the others are in the cytoplasm. The cytoplasmic portion of transmembrane TLRs is highly similar to that of the IL1 receptor family. TLR3, 7, and 9 are associated with endosomes (TLR3 binds to the dsRNA of viruses engulfed in endosomes). The stimulation of TLRs by bacterial components, in a multistep process, triggers expression of several genes involved in IS cytokine activation. Depending on the specific TLR activated, certain cytokine genes (for TNF-α, IL12, various interferons) are expressed. TLR4 is activated by the LPS (lipopolysaccharide) in the outer coats of G$^-$ bacteria such as *Salmonella* and *E. coli* O157:H7. (See the reviews by Kimball [2009b], Takeda and Akira [2005], and Vasselon and Detmers [2002].)

Toll pathway (in fly innate immunity): The Toll pathway regulates the immune response (production of AMPs) in flies in response to a G$^+$ bacterial infection.

Trafficking (in immunology): Trafficking is where immune cells (e.g., naive T-cells, DCs, Mϕ), or stem cells (e.g., hematopoietic stem cells), migrate within the body in response to chemokine gradients and communicate (via cytokines and cell-surface receptors) with many other immune and nonimmune cells. Such trafficking is essential for the function of the adaptive IS and tissue growth and healing. The trafficking of cells is guided by chemokines secreted by other cells. (See Luster et al. [2005] for a comprehensive review of immune cell migration and its regulation, as also papers by Ward and Marelli-Berg [2009] and Rivera et al. [2008].)

Trait (in biology): A genetically inherited property. The alleles for a trait occupy the same locus on homologous chromosomes. Traits can be physical (part of an organism's phenotype), functional, or behavioral. Some examples are brown irises (human), darkly pigmented skin (human), penicillin resistance (bacteria), DDT resistance (insects), dislocating mandibles (snakes), polycystic kidney disease (humans), and the flashing sequence of a species of lightning bug.

Transcription (in genomics): The complicated process whereby a structural gene is expressed: once the gene is activated, RNAP generates pre-mRNA, which is then edited to mRNA that leaves the nuclear volume, and carries the anticodon template to a ribosome where the actual polypeptide assembly is carried out.

Transcription factor (TF) (in molecular biology): A TF is a protein molecule that regulates the binding of RNAP to an "upstream" (5'–3' direction) promoter region of a DNA-coding gene segment. The TF protein may bind directly to the DNA promoter region or bind to other TFs already attached, modulating their action in the initiation of gene transcription. A TF may work to either stimulate or repress gene transcription. The regulation of TFs themselves is a highly complex process involving many molecular biological feedback and feed-forward pathways.

In addition, epigenetic modifications of histone proteins associated with the gene, as well as epigenetic DNA methylation can play an important role in transcriptional regulation, hence gene expression.

Transcriptional repressors (in developmental control): A transcriptional repressor is a TF that vetoes or represses the transcription of a gene, usually in embryonic development. *Msx homeobox* genes are generally transcriptional repressors.

Transition state (in biochemistry): In a chemical reaction, the transition state is the intermediate molecular configuration or state having the highest chemical energy. In an irreversible reaction, colliding reactant molecules in the transition state will always go on to form product(s).

Transport lag (TL) (in systems): (aka delay or dead time.) A TL is when a train enters a tunnel and emerges τ minutes later, unchanged. As a physiological example, a TL can be assigned to the time it takes for a nerve impulse to propagate down an axon, or the time it takes for a hormone molecule released into the blood by the adrenal gland to reach the heart. A TL can be modeled in the time domain as $y(t) = x(t - \tau)$, that is, the TL operator output y is the input x delayed by τ seconds.

Transposon (in genomics): (aka jumping genes, mobile DNA, or mobile genetic elements.) A segment of DNA that can move around (or be moved around) to different positions in the genome of a single cell. In doing so, it may replicate itself and may encode its own promoter sequence. Transposons may cause mutations, destroy or alter the ability of normal genes to be expressed, and increase (or decrease) the amount of DNA in the genome. Transposons are inserted at new locations in a cell's DNA by means of a transposase enzyme that effects enzymatic breakage and strand reunion. This process is called DNA recombination; it produces recombinant DNA. Thus, transposons are mutagens, and in humans they have been linked to certain diseases: Hemophilia A and B, severe combined immunodeficiency, porphyria, and predisposition to certain cancers. When in an organism's germ cells, they may also be linked to evolutionary processes. The most common form of transposon in human intron DNA is the Alu sequence, a form of SINE. The Alu sequence is ca. 300 bases long, and ca. 1000 copies can be found in the human genome (Citizendium 2008b).

There are three classes of transposons: Class I, also called retrotransposons, in which a DNA segment is first transcribed into RNA, and then the enzyme reverse transcriptase makes a DNA copy that is inserted into a new DNA location in the host's genome. (Retroviruses such as HIV are specialized Class I transposable elements. Once inside the host cell, their RNA genome is copied into DNA by reverse transcriptase, and then this viral DNA is inserted into the host's DNA. Once the infected viral DNA is copied, new retrovirions assemble, eclose, and go to infect other host cells.) Class II, in which a DNA segment moves directly from site to site. Class III, which is called miniature inverted-repeat transposable elements (MITEs). MITEs are too small to encode any protein; how they are copied and moved is uncertain.

Trans-splicing (of operons): Trans-splicing is a special form of RNA processing in eukaryotes where exons from two different primary RNA transcripts are joined end to end and ligated. An exon from one mRNA molecule is spliced onto the 5′ end of a completely separate mRNA molecule, post-transcriptionally. In contrast, normal *cis*-splicing produces a single molecule. Trans-splicing makes an RNA transcript that comes from multiple RNAPs on the genome.

Tree (in graph theory): A connected, acyclic, simple graph. A tree has no path loops.

TSE disease: Transmissible spongiform encephaly disease (aka mad cow disease, scrapie [sheep], or the chronic wasting disease of cervids), caused by misfolded prion protein in the CNS.

Ultrastable system (complex interacting systems): A term coined by Ashby (1960). Two complex systems of continuous variables interact (one is an environment, or niche, and the other is an organism or reacting part). The interaction is in the form of two feedback modalities: The first is a rapid, primary feedback (through complex sensory and motor channels). The second feedback works intermittently at a slower speed from the environment to certain continuous variables in the organism. These continuous variables affect the values of step-mechanisms in the organism only when their values fall outside set threshold limits (upper or lower thresholds, or "windows"). The changes in the step-mechanisms determine how the organism reacts to its environment. If environmental conditions become adverse (but not fatal), an ultrastable organism may modify its environment, or move out of it to a more suitable one. (See *complex adaptive system*.)

Unit impulse (a system input): Used to characterize linear and nonlinear systems. Mathematically, a unit impulse, $\delta(t-\tau)$, occurs at $t=\tau$, and is zero elsewhere. As a rectangular pulse, its height is $1/\varepsilon$, and its width is ε, giving it unity area. In the limit as $\varepsilon \to 0$, the impulse becomes infinitely tall and has zero width and its area is unity: $\int_{0}^{\infty} \delta(t-\tau)\,dt \equiv 1.0$, beginning at $t \geq \tau$, else 0. An LS's response to a unit impulse, called its weighting function, completely characterizes its dynamics.

3′ Untranslated region (3′UTR) (in genomics): 3′ untranslated base sequences following the protein-coding ORF in an mRNA molecule.

Utility (in economics): The condition or quality of being useful; usefulness.

Variety (in complex systems theory): "The total number of possible states of a system, or of an element of a system" (Beer 1981). In cybernetics, a term introduced by Ashby (1956) to denote the total number of distinct states (or # of state variables [nodes]) in a system.

Vasoactive intestinal peptide (VIP) (in physiology): VIP is a highly pleiotropic peptide hormone consisting of 28 AAs; it is produced by many cells and organs in the body, and has many effects. Many different cells have receptors for VIP and its analogs. It has a short half-life of ca. 2 min. It is widely distributed in cholinergic presynaptic neurons in the CNS and also in peptidergic neurons innervating many organs (e.g., heart, lungs, digestive system, genitourinary tract, eyes, skin ovaries, and thyroid gland) (Onoue et al. 2008).

In the digestive system, VIP induces smooth muscle relaxation, stimulates secretion of water into pancreatic juice and bile, and inhibits gastric acid secretion and absorption from the intestinal lumen. Also, its role in the intestines is to greatly stimulate secretion of water and electrolytes, as well as dilating peripheral blood vessels, stimulating pancreatic bicarbonate secretion, and generally increasing gut motility. VIP is also a neurotransmitter, found in the CNS and peripheral organs. It is involved in synchronizing the master circadian pacemaker in the suprachiasmatic nucleus. It stimulates prolactin release. In the heart, VIP causes coronary vasodilatation. In the lungs, VIP acts as a vasodilator, bronchodilator, and anti-inflammatory agent. Perhaps the most curious role of VIP lies in the hIS. Certain autonomic nerve fibers innervating organs associated with immune cell growth and maturation (thymus, spleen, lymph nodes, mucosal-associated lymphoid tissue, gut, and Peyer's patches) secrete VIP. Local concentrations of VIP provide a direct causal link between the CNS and the hIS. The immunomodulatory effects of VIP are too numerous to list here. (See Delgado et al. [2004] for a detailed review.)

To summarize some of its effects in the hIS, VIP exerts an anti-inflammatory effect on macrophages and microglia, it up-regulates IL-10 (a potent anti-inflammatory cytokine), it inhibits antigen presentation by macrophages and DCs, it supports the generation and long-term survival of memory Th2 cells, etc. Delgado et al. speculated that VIP could be considered a type 2 cytokine instead of a neuropeptide.

Vertex (in graph theory): (The same as a node.) A fundamental unit out of which graphs are formed. The vertices or nodes in a graph are connected by edges (branches), which can be directed or undirected. In SFGs, the nodes have values representing the states of the system that the graph models. For example, they can represent chemical concentrations in a biochemical system, or voltages and currents in an electronic system. A *source vertex* is a vertex with only directed branches leaving it. A *sink vertex* has no directed branches leaving it.

Vertex distance (in graph theory): See *distance*.

Vibrio: A genus of Gram-negative (G⁻) bacteria having a curved, rodlike shape. Some have a single flagellum, while others have many. *V. cholerae* causes cholera in humans, while other *Vibrio* strains cause disease in fish and shellfish, and are a common cause of mortality in farmed marine life.

Voxel (in 3D imaging): The minimum resolvable (cubic) volume element of a 3D (tomographic) imaging system. A *pixel* is the minimum resolvable area element in a 2D imaging system (e.g., x-ray).

Walk (in graph theory): A walk from node i to node j is an alternating sequence of nodes and edges (a sequence of adjacent nodes) that begins with i and ends with j. The *length l of a walk* is the number of edges in the sequence. A walk is closed if its first and last vertices are the same, and open if they are different. For an open walk, $l = N - 1$, where N is the number of vertices visited (a vertex is counted each time it is visited). $l = N$ for a closed walk. A *cycle* is a closed walk of at least three nodes in which no edge is repeated.

(Compare *Walk* with *SFG path*.)

Wiener index, W (a complexity measure in graph theory): The sum of distances over all pairs of vertices in an undirected graph:

$$W \equiv \frac{1}{2} \sum_{i,j=1}^{N} d_{ij} = d_{11} + d_{12} + d_{13} + \cdots + d_{1N} + d_{21} + d_{22}$$
$$+ d_{23} + \cdots + d_{2N} + \cdots + d_{N1} + d_{N2} + \cdots + d_{NN}$$

where $d_{kk} \equiv 0$, and $d_{jk} = d_{kj}$; $1 \le k, j \le N$.

Zeitgeber: An environmental time cue that can synchronize biological clocks. For example, photoperiod can synchronize and entrain circadian clocks that regulate a number of physiological processes and behaviors.

Zinc finger (Zif) Proteins (in genomics): A component of transcription factor IIIA. Zif proteins can bind to selected sites in dsDNA. Each Zif protein has several "fingers," each containing ca. 30 AAs. Each finger has a pair of cysteine and a pair of histidine molecules that serve as zinc molecule ligands. Fingers typically occur as tandem repeats, with two, three, or more fingers comprising the specific DNA-binding domain of a TF. The "fingers" bind in the major groove of the DNA helix. Synthetic (bioengineered) ZiF nucleases have been used to generate targeted breaks in dsDNA for genomic modifications (see *synthetic biology*). The majority of synthetic Zifs are based on the Zif domain of the murine TF, Zif268. They typically have 3–6 individual ZiF motifs, and bind dsDNA target sites from 9 to 18 bps in length.

Bibliography and Related Readings

ABM Software. 2010. Comparison of agent-based modeling software. http://en.wikipedia.org/wiki/Comparison_of_agent-based_modeling software (accessed on February 11, 2010).

Abta, A., A. Kaddar, and H.T. Alaoui. 2008. Stability of limit cycle in a delayed IS-LM business cycle model. *Applied Mathematical Sciences*. 2(50): 2459–2471.

Ackermann, M. and M. Doebeli. 2004. Evolution of niche width and adaptive diversification. *Evolution*. 58(12): 2599–2612.

Acres, B. and J.Y. Bonnefoy. 2008. Clinical development of MVA-based therapeutic cancer vaccines. *Expert Reviews Vaccines*. 7(7): 889–893.

Agre, P. et al. 2002. Aquaporin water channels—From atomic structure to clinical medicine. *The Journal of Physiology*. 542: 3–16.

Akiba, T. et al. 1960. On the mechanism of the development of multiple-drug-resistant clones of *Shigella*. *Japanese Journal of Microbiology*. 4: 219–227.

Albert, R. 2005. Scale-free networks in cell biology. *Journal of Cell Sciences*. 118: 4947–4957.

Alberts, B. et al. 2007. *Molecular Biology of the Cell*, 5th edn. Garland Publishing, Hamden, CT.

Ali, M., A.G. Hariharan, N. Mishra, and S. Jain. 2009. Catalytic antibodies as potential therapeutics. *Indian Journal of Biotechnology*. 8: 253–258.

Alon, U. 2003. Biological networks: The tinkerer as an engineer. *Science*. 301: 1866–1867.

Alon, U. 2007a. *An Introduction to Systems Biology: Design Principles of Biological Circuits*. Chapman & Hall/CRC, Boca Raton, FL.

Alon, U. 2007b. Network motifs: Theory and experimental approaches. *Nature Reviews/Genetics*. 8: 450–461.

An, W. and J.W. Chin. 2009. Synthesis of orthogonal transcription-translation networks. *PNAS*. 106(21): 8477–8482.

Andrews, P.S. and J. Timmis. 2006. A computational model of degeneracy in a lymph node (a chapter). In *Artificial Immune Systems*. Springer, Berlin, Germany, pp. 164–177. ISBN 978-3-540-37749-8.

Angelo, E.J. 1969. *Electronic Circuits: BJTs, FETs and Microcircuits*. McGraw-Hill, New York.

Arnone, M.I. and E.H. Davidson. 1997. The hardwiring of development: Organization and function of genetic regulatory systems. *Development*. 124: 1851–1864.

Arora, S. and B. Barak. 2009. *Computational Complexity Theory: A Modern Approach*. Cambridge University Press, New York (in press).

Arthur, W.B. 1999. Complexity and the economy. *Science*. 284: 107–109.

Arthur, W.B., S.N. Durlauf, and D.A. Lane. 1997. *The Economy as an Evolving Complex System II*. Addison-Wesley & The Santa Fe institute, Reading, MA.

Ashby, W.R. 1956. *An Introduction to Cybernetics*. Chapman & Hall, London, U.K. ISBN: 0-416-68300-2.

Ashby, W.R. 1958. Requisite variety and its implications for the control of complex systems. *Cybernetica*. 1(2): 83–89. http://pespmc1.vub.ac.be/Books/AshbyReqVar.pdf (accessed on November 20, 2009).

Ashby, W.R. 1960. *Design for a Brain*. Chapman & Hall Ltd., London, U.K.

Axelrod, R. and L. Tesfatsion. 2009. *On-Line Guide for Newcomers to Agent-Based Modeling in the Social Sciences.* 16 pp. http://www.econ.iastate.edu/tesfatsi/abmread.htm (accessed on March 15, 2010).

Ay, N., E. Olbrich, N. Bertschinger, and J. Jost. 2006. A unifying framework for complexity measures of finite systems. http://sbs-net.sbs.ox.ac.uk/complexity_PDFs/ECCS06/Conference_Proceedings/PDF/p202.pdf (accessed on December 3, 2009).

Ayers, R.U. 1999. The second law, the fourth law, recycling and limits to growth. *Ecological Economics.* 29: 473–483.

Bailey, J.J. et al. 1992. A kinetic model of CD4+ lymphocytes with the human immunodeficiency virus (HIV). *BioSystems.* 26: 177–183.

Bailey, A.M. et al. 2009. Agent-based model of therapeutic adipose-derived stromal cell trafficking during ischemia predicts ability to roll on P-selectin. *PLoS Computational Biology.* 5(2): 1–17. http://www.ncbi.nih.gov/pmc/articles/ PMC2636895/pdf/pcbi.1000294.pdf (accessed on December 10, 2009).

Balagaddé, F.K. et al. 2008. A synthetic *Escherichia coli* predator–prey ecosystem. *Molecular Systems Biology.* 4:187 (8 pp.).

Baldwin, R. 2003. Information theory and creationism: William Dembski. Web essay. http://home.mira.net/~reynella/debate/dembski.htm (accessed on December 1, 2009).

Baldwin, R. 2005. Information theory and creationism. Web essay. http://www.talkorigins.org/faqs/information/dembski.html (accessed on November 12, 2009).

Banki, Z., H. Stoiber, and M.P. Dierich. 2005. HIV and human complement: Inefficient virolysis and effective adherence. *Immunology Letters.* 97: 209–214.

Bangham, J., F. Jiggins, and B. Lemaitre. 2006. Insect immunity: The post-genomic era. *Immunity.* 35(1): 1–5.

Barkai, N. and S. Liebler. 2000. Circadian clocks limited by noise. *Nature.* 403: 267–268.

Barnett, W.A., C. Deissenberg, and G. Feichtinger. 2004. *Economic Complexity: Non-Linear Dynamics, Multi-Agents Economics, and Learning.* Elsevier Science, Amsterdam, the Netherlands, 492 pp. ISBN: 9780444514332.

Bartlett, T.C. et al. 2002. Crustins, homologues of an 11.5 kDa antibacterial peptide, from two species of *Penaeid* shrimp, *Litopanaeus vannamei* and *Litopenaeus setiferus*. *Marine Biotechnology.* 4: 278–293.

Bar-Yam, Y. and I.R. Epstein. 2004. Response of complex networks to stimuli. *PNAS.* 101(13): 4341–4345.

Beer, S. 1981. *Brain of the Firm, 2/e.* John Wiley, New York (reprinted in 1986 and 1988).

Behe, M.J. 1996. *Darwin's Black Box: The Biochemical Challenge to Evolution.* Touchstone, New York.

Bell, A. and P.-H. Gouyon. 2003. Arming the enemy: The evolution of resistance to self-proteins. *Microbiology.* 149: 1367–1375.

Bello, G. et al. 2004. Co-existence of recent and ancestral nucleotide sequences in viral quasispecies of human immunodeficiency virus type 1 patients. *Journal of General Virology.* 85: 399–407.

Bellouquid, A. and M. Delitalia. 2006. *Mathematical Modeling of Complex Biological Systems: A Kinetic Theory Approach.* Birkhäuser, Boston, MA.

Bellur, A.S. 2004. Software modeling of the complement system and its role in immune response. MS thesis at University of Colorado, Boulder, CO.

Ben-Shahar, Y., K. Nannapaneni, T.L. Casavant, T.E. Scheetz, and M.J. Welsh. 2007. Eukaryotic operon-like transcription of functionally related genes in *Drosophila*. *PNAS.* 104(1): 222–227.

Berezikov, E. et al. 2005. Phylogenetic shadowing and computational identification of human microRNA genes. Letter to the editor in: *Cell.* 120: 21–24.

Berezikov, E. et al. 2006. Many novel mammalian microRNA candidates identified by extensive cloning and RAKE analysis. *Genome Research*. 16: 1289–1298.

Berger, P. et al. 1998. Immunoglobulin E-induced passive sensitization of human airways. *American Journal of Respiratory and Critical Care Medicine*. 157: 610–616.

Bergsten, P. 2000. Pathophysiology of impaired pulsatile insulin release. *Diabetes Metabolism Research and Reviews*. 16: 179–191.

Beverdam, A. et al. 2002. Jaw transformation with gain of symmetry after Dlx5/Dlx6 inactivation: Mirror of the past? *Genesis*. 34: 221–227.

Binkley, S.A. 1997. *Biological Clocks: Your Owner's Manual*. CRC Press, Boca Raton, FL.

Blaug, M. 2008. Economics. (Article in the online *Encyclopedia Britannica*.) http://www.britannica.com/EBchecked/topic/178548/economics/236771/Fields-of-contemporary-economics (accessed on November 20, 2009).

BLS. 2010. Consumer Price Index—All urban Consumers. U.S. Bureau of Labor Statistics. http://data.bls.gov/PDQ/servlet/SurveyOutputServlet (accessed on February 19, 2010).

Blumenthal, T. 2004. Operons in eukaryotes. *Briefings in Functional Genomics and Proteomics*. 3(3): 199–211.

Blumenthal, T. 2005. Trans-splicing and operons. (June 25, 2005) *WormBook* ed. The *C. elegans* Research Community, Wormbook, http://www.wormbook.org/chapters/www_transplicingoperons/transplicingoperons.pdf (accessed on November 12, 2009).

Blumenthal, T. and K.S. Gleason. 2003. *Caenorhabditis elegans* operons: Form and function. *Nature Reviews/Genetics*. 4: 110–118.

Boccaletti, S. et al. 2006. Complex networks: Structure and dynamics. *Physics Reports*. 424: 175–308.

Boccara, N. 2003. *Modeling Complex Systems*. Springer, Berlin, Germany. ISBN-10: 0387404627.

Boerlijst, M., S. Bonhoeffer, and M.A. Nowak. 1996. Viral quasi-species and recombination. *Proceedings of the Royal Society London B*. 263: 1577–1584.

Bonabeau, E. 2002. Agent-based modeling: Methods and techniques for simulating human systems. *PNAS*. 99(suppl. 3): 7280–7287.

Bonchev, D. 2004. Complexity analysis of yeast proteome network. *Chemistry and Biodiversity*. 1: 312–326.

Bonchev, D. and D.H. Rouvray, eds. 2005. *Complexity in Chemistry, Biology, and Ecology*. Springer, New York.

Bonchev, D. and N. Trinajstić. 1977. Information theory, distance theory and molecular branching. *Journal of Chemical Physics*. 67: 4517–4533.

Bonhoeffer, S. and P. Sniegowski. 2002. The importance of being erroneous. *Nature*. 420: 376–369.

Bonner, J.T. 1988. *The Evolution of Complexity by Means of Natural Selection*. Princeton University Press, Princeton, NJ.

Bourgine, P. and J. Johnson. 2006. *Living Roadmap for Complex Systems Science*, 71 pp. Web essay. http://css.csregistry.org/tiki-download_wiki_attachment.php?attId=123 (accessed on November 6, 2009).

Bowden, L., N.M. Dheilly, D.A. Raftos, and S.V. Nair. 2007. New immune systems: Pathogen-specific host defence, life history strategies and hypervariable immune-response genes of invertebrates. *ISJ*. 4: 127–136. ISSN: 1824-307X.

Bowles, S. and H. Gintis. 2002. *Homo reciprocans. Nature*. 415: 125–128.

Bowling, F.L., E.V. Salgami, and A.J.M. Boulton. 2007. Larval therapy: A novel treatment in eliminating methicillin-resistant *Staphococcus aureus* from diabetic foot ulcers. *Diabetes Care*. 30(2): 370–371.

Boyd, S.D. 2008. Everything you wanted to know about small RNA but ere afraid to ask. *Laboratory Investigation*. 88: 569–578. http://gene-quantification.com/boyd-microrna-review-2008.pdf (accessed on September 2, 2009).

Bratsun, D. et al. 2005. Delay-induced stochastic oscillations in gene regulation. *PNAS*. 102(41): 14593–14598.

Briones, C. and U. Bastolla. 2005. Protein evolution in viral quasispecies under selective pressure: A thermodynamic and phylogenetic analysis. *Gene*. 347: 237–246.

Brissaud, J.-B. 2005. The meaning of entropy. *Entropy*. 7(1): 68–96.

Brock, W.A. and A.G. Malliaris. 1992. *Differential Equations, Stability and Chaos in Dynamic Economics*, 1st repr., North Holland, Amsterdam, the Netherlands.

Bromberg, J. 2002. Stat proteins and oncogenesis. *The Journal of Clinical Investigation*. 109(9): 1139–1142.

Bronnikova, T.V. and W.M. Schaffer. 2005. *Overview: Introduction to Biological Oscillations*. Lecture notes. http://bill.srnr.arizona.edu/NLBchemd/NLBchemdOverview.html (accessed on November 12, 2009).

Brusch, L., G. Cuniberti, and M. Bertau. 2004. Model evaluation for glycolytic oscillations in yeast biotransformations of xenobiotics. *Biophysical Chemistry*. 109: 413–426.

Buchanan, M. 2009. Meltdown modelling: Could agent-based computer models prevent another financial crisis? *Nature*. 460(7256): 680–682.

Büchel, C. et al. 1998. The functional anatomy of attention to visual motion: A functional MRI study. *Brain*. 121: 1281–1294.

Bulet, P. and R. Stöcklin. 2005. Insect antimicrobial peptides: Structures, properties and gene regulation. *Protein and Peptide Letters*. 12: 3–11.

Bull, J.J., L.A. Meyers, and M. Lachmann. 2005. Quasispecies made simple. *PLoS Computational Biology*. 1(6): 0450–0460.

Bullock, T.H. and G.A. Horridge. 1965. *The Structure and Function in the Nervous Systems of Invertebrates*. W.H. freeman & Co., San Francisco, CA.

Burggren, W.W. 2005. Developing animals flout prominent assumptions of ecological physiology. *Comparative Biochemistry and Physiology. Part A*. 141: 430–439.

Burggren, W.W. and M.G. Monticino. 2005. Assessing physiological complexity. *Journal of Experimental Biology*. 208: 3221–3232.

Bürglin, T.R. 2005. *The Homeobox Page*, 8 pp. Web essay. http://homeobox.biosci.ki.se/ (accessed on November 12, 2009).

Bushati, N. and S.M. Cohen. 2007. microRNA functions. *Annual Review of Cell and Developmental Biology*. 23: 175–205.

Business Bestseller. 2000. Review of Frederic Vester's Book: *The Art of Networked Thinking—Ideas and Tools for a New Dealing with Complexity*. Book Review in English from *Business Bestseller*. No. 10/00, 12 pp. www.frederic-vester.de/eng/books/complete-review/ (accessed on November 12, 2009).

Calabretta, R., A. Di Ferdinando, G. Wagner, and D. Parisi. 2003. What does it take to evolve behaviorally complex organisms? *BioSystems*. 69: 245–262.

Calamita, G. 2005. Aquaporins: Highways for cells to recycle water with the outside world. *Biology of the Cell*. 97: 351–353.

Caldwell, C. 1995. *Graph Theory Glossary*. On-line tutorial. www.utm.edu/departments/math/graph/glossary.html (accessed on October 7, 2009).

Cammisotto, P.G. et al. 2005. Regulation of leptin secretion from white adipocytes by insulin, glycolytic substrates, and amino acids. *American Journal of Physiology and Endocrinology Metabolism*. 289: E166–E171.

Campbell, D.T. 1976. Assessing the impact of planned social change. Paper #8, Occasional Paper Series, Dartmouth College Public Affairs Center, Hanover, NH, 70 pp. www.wmich.edu/evalctr/pubs/ops/ops08.pdf (accessed on February 17, 2010).

Candas, M. et al. 2002. Insect resistance to *Bacillus thuringiensis. Molecular and Cellular Proteomics.* 2(1): 19–28. www.mcponline.org (accessed on September 16, 2009).

Carlson, J.M. and J. Doyle. 2000. Highly optimized tolerance: Robustness and design in complex systems. *Physical Review Letters.* 84(11): 2529–2532.

Carlson, J.M. and J. Doyle. 2002. Complexity and robustness. *PNAS.* 99(suppl. 1): 2538–2545.

Carson, E.R., C. Cobelli, and L. Finkelstein. 1983. *The Mathematical Modeling of Metabolic and Endocrine Systems.* John Wiley & Sons, New York.

CCC. 2010. *Civilian Conservation Corps (CCC), 1933–1941.* Web essay. www.u-s-history. com/pages/h1586.html (accessed on January 4, 2010).

Chance, B., B. Schoener, and S. Elaesser. 1964. Control of the waveform of oscillations of the reduced pyridine nucleotide in a cell-free extract. *Proceedings of the National Academy of Sciences (USA).* 52: 337–341.

Chang, T.-C. and J.T. Mendell. 2007. microRNAs in vertebrate physiology and human disease. *Annual Review of Genomics and Human Genetics.* 8: 215–239.

Chang, C. et al. 2009. The structure of pyogenecin immunity protein, a novel bacteriocin-like immunity protein from *Streptococcus pyogenes. BMC Structural Biology.* 9: 75 (9 pp).

Chauhan, A.K. and T.L. Moore. 2006. Presence of plasma complement regulatory proteins clusterin (Apo J) and vitronectin (S40) on circulating immune complexes (CIC). *Clinical and Experimental Immunology.* 145: 398–406.

Chavda, H. and C.N. Patel. 2008. *Aquaporins: The Secret Highway for Water Transport.* Web essay. http://www.pharmainfo.net/reviews/aquaporins-the-secret-highways-water-transport/ (accessed on March 8, 2010).

Chen, K. and N. Rajewsky. 2007. The evolution of gene regulation by transcription factors and microRNAs. *Nature Reviews/Genetics.* 8: 93–103.

Chen, G., Z. Li, D. Yuan, Nimazhaxi, and Y. Zhai. 2006. An immune algorithm based on the complement activation pathway. *International Journal of Computer Science and Network Security.* 6(1A): 147–152.

Cheng-Hua, L., Z. Jian-Min, and S. Lin-Sheng. 2009. A review of advances in research on marine molluscan antimicrobial peptides and their potential application in aquaculture. *Molluscan Research.* 29(1): 17–26.

Chess, L. and H. Jiang. 2004. Resurecting CD8+ suppressor T cells. *Nature Immunology.* 5(5): 469–471.

Chiang, A.S., A.P. Gupta, and S.S. Han. 1988. Arthropod immune system: Comparative light and electron microscopic accounts of immunocytes and other hemocytes of *Blattella germanica* (Dictyoptera: Balttellidae). *Journal of Morphology.* 198: 257–267.

Chin, J.W. 2006. Programming and engineering biological networks. *Current Opinion in Structural Biology.* 16: 551–556.

Chou, P.-H. et al. 2009. The putative invertebrate adaptive immune protein *Litopenaeus vannamei* Dscam (LvDscam) is the first reported Dscam to lack a transmembrane domain and cytoplasmic tail. *Developmental and Comparative Immunology.* 33(12): 1258–1267.

Chubiz, L.M. and C.V. Rao. 2008. Computational design of orthogonal ribosomes. *Nucleic Acids Research.* 36(12): 4038–4046.

Citizendium. 2008a. *Horizontal Gene Transfer.* 7 pp. article. http://en.citizendium.org/wiki/ Horizontal_gene_transfer (accessed on November 12, 2009).

Citizendium. 2008b. *Transposon.* 6 pp. article. http://en.citizendium.org/wiki/Transposons (accessed on November 12, 2009).

Citizendium. 2008c. *Mobile DNA.* 8 pp. article. http://en.citizendium.org/wiki/Mobile_ DNA (accessed on November 12, 2009).

Clairambault, J. 2008. A step toward optimization of cancer therapeutics. *IEEE Engineering in Medicine and Biology Magazine.* 27(1): 20–24.

Clapp, K.P., R.B. Northrop, and Q. Li. 1988. The immune system versus cancer: A modelling study. In *Proceedings of the 10th Annual IEEE/EMBS Conference*, New Orleans, LA, Nov. 4–7, pp. 1023–1024.

Clement, K. and C. Vaisse. 1998. A mutation of the human leptin receptor gene causes obesity and pituitary dysfunction. *Nature*. 392: 398.

Club of Rome. 2008. Homepages. http://www.clubofrome.org/eng/home/ (accessed on November 12, 2009).

CNCS. 2008. *Duke University Center for Nonlinear and Complex Systems Course Listing.* http://www.math.duke.edu/CNCS/couses.html (accessed on October 5, 2009).

Cobelli, C. and A. Mari. 1983. Validation of mathematical models of complex endocrine-metabolic systems. A case study on a model of glucose regulation. *Medical and Biological Engineering and Computing.* 21: 390–399.

Cobelli, C. and A. Mari. 1985. Control of diabetes with artificial systems for insulin delivery—Algorithm independent limitations revealed by a modeling study. *IEEE Transactions on Biomedical Engineering.* 32(10): 840–845.

Cobos, I. et al. 2005. Mice lacking Dlx1 show subtype-specific loss of interneurons, reduced inhibition and epilepsy. *Nature Neuroscience.* 8(8): 1059–1068.

Cohen, M.S. and S.Y. Bookheimer. 1994. Localization of brain function using magnetic resonance imaging. *Trends in Neurosciences.* 17(7): 268–277.

Cohen, I.R., U. Hershberg, and S. Solomon. 2004. Antigen-receptor degeneracy and immunological paradigms. *Molecular Immunology.* 40: 993–996.

Collins, R.D. 1968. *Illustrated Manual of Laboratory Diagnosis.* J.B. Lippincott Co., Philadelphia, PA.

Complexity. 2008. *Courses in Complexity Science.* www.complexity.ecs.soton.ac.uk/courses.php (accessed on November 12, 2009).

Condon, T. 2008. *The Dust Bowl: A Cautionary Tale.* Op ed article in the *Hartford Courant.* August 24, 2008, p. C4.

Considine, R.V. et al. 1996. Serum immunoreactive-leptin concentrations in normal weight and obese humans. *New England Journal of Medicine.* 334: 292.

Cook, D.N. and K. Bottomly. 2007. Innate immune control of pulmonary dendritic cell trafficking. *Proceedings of the American Thoracic Society.* 4: 234–239.

Coon, M. 2002. *Is the Complement System Irreducibly Complex?* Essay in *The Talk Origins.* www.talkorigins.org/faqs/behe/icsic.html (accessed on November 16, 2009).

Corning, P.A. 1998. *Complexity Is Just a Word.* Web paper. http://www.complexsystems.org/commentaries/jan98.html (accessed on November 16, 2009).

Correa, A. et al. 2003. Multiple oscillators regulate circadian gene expression in *Neurospora. PNAS.* 100(23): 13597–13602.

Cotter, P.D., C. Hill, and R.P. Ross. 2005. Bacteriocins: Developing innate immunity for food. *Nature Reviews/Microbiology.* 3: 777–787.

Coyne, J.A. and H.A. Orr. 1998. The evolutionary genetics of speciation. *Philosophical Transactions of the Royal Society London B.* 353: 287–305.

Crawford, D. 1992. Modularization and McCabe's cyclomatic complexity. *Communication of the ACM* 35(12): 17–19.

Cropp, T.A. and J.W. Chin. 2006. Expanding nucleic acid function *in vitro* and *in vivo. Current Opinion in Chemical Biology.* 10: 601–606.

Cruse, G. et al. 2005. Activation of human lung mast cells by monomeric immunoglobulin E. *European Respiratory Journal.* 25: 858–863.

CSE Group. 2008. *Complexity Measures.* 9 pp. Web paper. http://cse.ucdavis.edu/~cmg/Group/group_documents/MeasuresofComplexity.pdf (accessed on December 1, 2009).

CSS. (Complex Systems Society). 2008. *Education.* http://ecss.csregistry.org/tiki-index.php?page=education&bl=y (accessed on November 4, 2009).

Csete, M.E. and J.C. Doyle. 2002. Reverse engineering of biological complexity. *Science.* 295: 1664–1669.

Cui, Q., Z. Yu, E.O. Purisima, and E. Wang. 2006. Principles of microRNA regulation of a human cellular signaling network. *Molecular Systems Biology.* 2: 7 pp. www.nature.com/msb/journal/v2/n1/pdf/msb4100089.pdf (accessed on November 16, 2009).

Culshaw, S. et al. 2008. Murine neutrophils present Class II restricted antigen. *Immunology Letters.* 118: 49–54.

Cunningham, W.J. 1958. *Introduction to Nonlinear Analysis.* McGraw-Hill Book Co. Inc., New York.

Danforth, C.A. 2001. Why the weather is unpredictable, an experimental and theoretical study of the Lorenz equations. BS Honors thesis, Bates College, Lewiston, ME. Mathematics & Physics Departments. March 16, 2001. http://uvm.edu/~cdanforth/research/danforth-bates-thesis.pdf (accessed on January 19, 2010).

Dano, S. et al. 2005. Chemical interpretation of oscillatory modes at a Hopf point. *Physical Chemistry Chemical Physics.* 7: 1674–1679.

Darwin, C. 1859. *On the Origin of Species by Means of Natural Selection, or the Preservation of Favoured Races in the Struggle for Life,* 1st edn. John Murray, London, U.K.

Day, R.H. 1994. *Complex Economic Dynamics: An Introduction to Dynamical Systems and Market Mechanisms,* Vol. I. MIT Press, Cambridge, MA.

de Back, W., E.D. de Jong, and M. Wierling. 2006. Red Queen dynamics in a predator–prey ecosystem. *GECCO '06,* 8–12 July, Seattle, WA [ACM 1-59593-186-4/06/0007].

DeBoer, R.J. et al. 1985. Macrophage T lymphocyte interactions in the anti-tumor immune response: A mathematical model. *Journal of Immunology.* 134(4): 2748–2758.

DeBoer, R.J., S. Michelson, and P. Hogeweg. 1986. Concomitant immunization by the fully antigenic counterparts prevents modulated tumor cells from escaping cellular immune elimination. *Journal of Immunology.* 136(11): 4319–4327.

Decker, T. and P. Kovarik. 1999. Transcription factor activity of STAT proteins: Structural requirements and regulation by phosphorylation and interacting proteins. *CMLS, Cellular and Molecular Life Sciences.* 55: 1533–1546.

de la Rosa, M., S. Rutz, H. Dorninger, and A. Scheffold. 2004. Interleukin-2 is essential for CD4+CD25+ regulatory T cell function. *European Journal of Immunology.* 34: 2480–2488.

Delgado, M., D. Pozo, and D. Ganea. 2004. The significance of vasoactive intestinal peptide in immunomodulation. *Pharmacological Reviews.* 56: 249–290.

de Lorenzo, V. and A. Danchin. 2008. Synthetic biology: Discovering new worlds and new words. *EMBO Reports.* 9(9): 822–827.

Demattei, M.-V. et al. 2000. Features of the mammal mar1 transposons in the human, sheep, cow, and mouse genomes and implications for their evolution. *Mammalian Genome.* 11: 1111–1116.

Dembski, W.A. 1998. *Intelligent Design as a Theory of Information.* Web essay. http://www.arn.org./docs/dembski/wd_idtheory.htm (accessed on November 16, 2009).

Dembski, W.A. 1999. *Intelligent Design.* InterVarsity Press, Downers Grove, IL. Published Nov. 1999, 312 pp.

Dembski, W.A. 2001. *No Free Lunch: Why Specified Complexity Cannot Be Purchased without Intelligence.* Rowman & Littlefield, Plymouth, U.K., Published Dec. 2001.

Dembski, W.A. 2002. *If Only Darwinists Scruitinized Their Own Work as Closely: A Response to "Erik."* Web essay. http://www.designinference.com/documents/2002.08.Erik_Response.htm (accessed on November 16, 2009).

Dembski, W.A. 2004. *Irreducible Complexity Revisited.* Web paper. www.designinference.com/documents/2004.01.Irred_Compl_Revisited.pdf (accessed on November 16, 2009).

Dembski, W.A. 2005. *Specification: The Pattern That Signifies Intelligence.* Web essay. www.designinference.com/documents/2005.06.Specification.pdf (accessed on November 16, 2009).

Deisboeck, T.S. and J.Y. Kresh, eds. 2006. *Complex Systems Science in Biomedicine.* Springer, Berlin, Germany, 846 pp.

Depew, M.J., T. Lufkin, and J.L. Rubenstein. 2002. Specification of jaw subdivisions by Dlx genes. *Science.* 298(5592): 381–385.

Diamond, J. 2005. *Collapse.* Penguin Books Ltd. London, U.K.

Diep, D.B., G. Mathiesen, V.G. Eijsink, and I.F. Nes. 2009. Use of Lactobaccili and their pheremone-based regulatory mechanism in gene expression and drug delivery. *Current Pharmaceutical Biotechnology.* 10: 62–73.

Dodd, D.M.B. 1989. Reproductive isolation as a consequence of adaptive divergence in *Drosophila psuedoobscura. Evolution.* 43: 1308–1311.

Domingo, E. et al. 1998. Quasispecies structure and persistence of RNA viruses. *Emerging Infectious Diseases.* 4(4): 521–527. www.cdc.gov/ncidod/eid/vol4no4/Domingo. html (accessed on November 16, 2009).

Dong, Y., H.E. Taylor, and G. Dimopoulos. 2006. AgDscam, a hypervariable immunoglobulin domain-containing receptor of the *Anopheles gambiae* innate immune system. *PloS Biology.* 4: e229.

Dörner, D. 1997. *The Logic of Failure: Recognizing and Avoiding Error in Complex Situations.* Basic Books, New York, 240 pp. ISBN-10: 0201479486.

Drennan, B. and R.D. Beer. 2006. Evolution of repressilators using a biologically-motivated model of gene expression. *Artificial Life X: 10th International Conference on the Simulation and Synthesis of Living Systems*, Montreal, Quebec, Canada.

Dunkelberg, P. 2003. *Irreducible Complexity Demystified.* Web essay. www.talkdesign.org/faqs/icdmyst/ICDmyst.html (accessed on November 16, 2009).

Dunlap, J.C. 1999. Molecular bases for circadian clocks. *Cell.* 96: 271–290.

Durlauf, S.N. 1997. What should policymakers know about economic complexity. Online paper. http://www.santafe.edu/research/publications/workingpapers/97-10-080.pdf (accessed November 16, 2009).

Dybdahl, M.F. and A. Storfer. 2003. Parasite local adaptation: Red queen versus suicide king. *Trends in Ecology and Evolution.* 18(10): 523–530.

Edelman, G.M. and J.A. Gally. 2001. Degeneracy and complexity in biological systems. *PNAS.* 98(24): 13763–13768.

Edlund, J.A. and C. Adam. 2004. Evolution of robustness in digital organisms. *Artificial Life.* 10: 167–179.

Edmonds, B. 1999a. Syntactic measures of complexity. PhD dissertation, University of Manchester, U.K.

Edmonds, B. 1999b. What is complexity?—The philosophy of complexity *per se* with application to some examples in evolution. In *The Evolution of Complexity*, F. Heylighen and D. Aerts, eds. Kluwer, Dordrecht, the Netherlands. http://cfpm. org/~bruce/evolcomp/ (accessed on December 3, 2009).

Eigen, M. 1971. Self organization of matter and the evolution of biological macromolecules. *Naturwissenschaften.* 58: 465–523.

Eigen, M. 1993. Viral quasispecies. *Scientific American.* 269: 42–49.

Elena, S.F. and R.E. Lenski. 1997. Tests of synergistic interactions among deleterious mutations in bacteria. *Nature.* 390: 395–398.

Elert, G. 2007. *The Chaos Hypertextbook: Measuring Chaos.* 4.3 Lyapunov Exponent. http://hypertextbook.com/chaos/43.shtml (accessed on November 16, 2009).

Elowitz, M.B. and S. Leibler. 2000. A synthetic oscillatory network of transcriptional regulators. *Nature.* 403: 335–338.

Elsberry, W. and J. Shallit. 2003. *Information Theory, Evolutionary Computation, and Dembski's Complex Specified Information*. Web essay. http://talkreason.org/articles/eandsdembski.pdf (accessed on November 16, 2009).

Emonet, T. et al. 2005. Agent cell: A digital single-cell assay for bacterial chemotaxis. *Bioinformatics*. 21(11): 2714–2721.

Epel, E.S. et al. 2000. Stress and body shape: Stress-induced cortisol secretion is consistently greater among women with central fat. *Psychosomatic Medicine*. 62: 623–632.

Ermolaeva, M.D., O. White, and S.L. Salzberg. 2001. Prediction of operons in microbial genomes. *Nucleic Acids Research*. 29(5): 1216–1221.

Esquela-Kersher, A. and F.J. Slack. 2006. Oncomirs-microRNAs with a role in cancer. *Nature Reviews/Cancer*. 6: 259–269.

Eum, S., S. Arakawa, and M. Murata. 2007. *Toward Bio-Inspired Network Robustness-Step 1. Modularity*. Paper presented at *Bionetics'07*, December 10–13, Budapest, Hungary, 4 pp.

Faggini, M. and T. Lux, eds. 2008. *Coping with the Complexity of Economics*. Springer, Berlin, Germany. ISBN: 978-88-470-1082-6.

Faggioni, R. et al. 2001. Leptin regulation of the immune response and the immunodeficiency of malnutrition. *FASEB Journal*. 15: 2565–2571.

Fall, C., E. Marland, J. Wagner, and J. Tyson, eds. 2005. *Computational Cell Biology*. Springer, New York.

Farabee, M.J. 2007. *Control of Gene Expression*. (rev. 3/12/07) On-line book chapter in: Farabee, M.J. *The On-Line Biology Book*. www.emc.maricopa.edu/faculty/farabee/BIOBK/BioBookGENCTRL.html (accessed on November 16, 2009).

Farmer, J.D. and D. Foley. 2009. The economy needs agent-based modelling. *Nature*. 460(7256): 685–686.

Faupel, K. and A. Kurki. 2002. *Biodiesel: A Brief Overview*. (ATTRA document). http://www.attra.ncat.org/attra-pub/PDF/biodiesel.pdf (accessed on November 4, 2009).

Federer, H.M. et al. 2007. A critical appraisal of chronic lyme disease. *New England Journal of Medicine*. 345: 1422–1430.

Félix, M.-A. and A. Wagner. 2008. Robustness and evolution: Concepts, insights and challenges from a developmental model system. *Heredity*. 100: 132–140.

Ferreira, P. 2001. *Tracing Complexity Theory* (on-line notes for ESD.83-Research Seminar in Engineering Systems at MIT). http://web.mit.edu/esd.83/www/notebook/ESD83-Complexity.doc (accessed on November 16, 2009).

Field Museum. 2007. *Gregor Mendel: Planting the Seeds of Genetics*. www.fieldmuseum.org/mendel/story_pea.asp (accessed on November 16, 2009).

Field, R.J. and R.M. Noyes. 1974. Oscillations in chemical systems: IV. Limit cycle behavior in a model of a real chemical reaction. *Journal of Chemical Physics*. 60: 1877–1884.

Field, R.J., E. Körös, and R.M. Noyes. 1972. Oscillations in chemical systems: II. Thorough analysis of temporal oscillation in the bromate-cerium-malonic acid system. *Journal of the American Chemical Society*. 94(25): S. 8649–8664.

Fimland, G., V.G.H. Eijsink, and J. Nissen-Meyer. 2002. Comparative studies of immunity proteins of pediocin-like bacteriocins. *Microbiology*. 148: 3361–3670.

Fimland, G., L. Johnsen, B. Dalhus, and J. Nissen-Meyer. 2005. Pediocin-like antimicrobial peptides (class IIa bacteriocins) and their immunity proteins: Biosynthesis, structure, and mode of action. *Journal of Peptide Science*. 11: 688–696.

First. 2010a. *First Written Language*. Web article. http://mesopotamia.mrdonn.org/cuneiform.html (accessed on January 19, 2010).

First. 2010b. *First Written Language*. Web article. http:www.chevroncars.com/learn/history/first-written language.html (accessed on January 19, 2010).

Fleischmann, W., M. Grassberger, and R. Sherman. 2004. *Maggot Therapy*. Georg Thieme Verlag, New York.

Forrester, J.W. 2003. Dynamic models of economic systems and industrial organizations. *Systems Dynamics Review*. 19(4): 331–345. (Reprinted in *SDR* from the 1956 manuscript of JWF.)

Foster, J. 2004. *From Simplistic to Complex Systems in Economics*. Discussion Paper No. 335, October 2004, School of Economics, The University of Queensland, St. Lucia, Australia.

Foster, R.G. and L. Kreitzman. 2005. *Rhythms of Life: The Biological Clocks That Control the Daily Lives of Every Living Thing*. Yale University Press, New Haven, CT.

Fragkoudis, R. et al. 2009. Advances in dissecting mosquito innate immune responses to arbovirus infection. *Journal of General Virology*. 90: 2061–2072.

Freeman, M. 2000. Feedback control of intercellular signalling and development. *Nature*. 408: 313–319.

Frost, S.D.W. et al. 2005. Neutralizing antibody responses drive the evolution of human immunodeficiency virus type 1 envelope during recent HIV infection. *PNAS*. 102(51): 18514–18519.

Fryer, P. A brief description of complex adaptive systems and complexity theory. Web paper. http://www.trojanmice.com/articles/complexadaptivesystems.htm (accessed on November 16, 2009).

Fugmann, S.D. et al. 2006. An ancient evolutionary origin of the Rag1/2 gene locus. *PNAS*. 103(10): 3728–3733.

Fujimoto, Y., K. Yagita, and H. Okamura. 2006. Does mPER2 protein oscillate without its coding mRNA cycling?: Post-transcriptional regulation by cell clock. *Genes to Cells*. 11: 525–530.

Funes, P. 2008. *Complexity Measures for Complex Systems and Complex Objects* (on-line class notes) 12 pp. http://www.cs.brandeis.edu/~pablo/complex.maker.html (accessed on November 16, 2009).

Gaillard, W.D. et al. 2000. Functional anatomy of cognitive development: fMRI of verbal fluency in children and adults. *Neurology*. 54: 180–185.

Garcia-Ojalvo, J., M.B. Elowitz, and S.H. Strogatz. 2004. Modelling a synthetic multicellular clock: Repressilators coupled by quorum sensing. *PNAS*. 101: 10955–10960.

Gardner, A. and A.T. Kalinka. 2006. Recombination and the evolution of mutational robustness. *Journal of Theoretical Biology*. 241: 705–715.

Gehrman, E. 2007. *The Unintended Consequences of Holding People Accountable*. Web essay. www.iq.harvard.edu/news/unintended_consequences_holding_people_accountable (accessed on February 17, 2010).

Gell-Mann, M. 1995. What is complexity? *Complexity*. 1(1): 16–19.

Genenames. 2006. *Interleukins and Interleukin Receptors*. Web table dated Aug. 1, 2006. http://www.genenames.org/genefamily/il.php (accessed on November 16, 2009).

Georgescu-Roegen, N. 1971. *The Entropy Law and the Economic Process*. Harvard University Press, Cambridge, MA.

Gilbert, N. 2007. *Agent-Based Models*. Sage Publications, London, U.K. ISBN: 978-1-4129-4964-4.

Gilbert, N. and P. Terna. 1999. How to build and use agent-based models in social science. Web paper. http://web.econ.unito.it/terna/deposito/gil_ter.pdf (accessed on December 10, 2009).

Gilon, P., M.A. Ravier, J.-C. Jonas, and J.-C. Henquin. 2002. Control mechanisms of the oscillations of insulin secretion *in vitro* and *in vivo*. *Diabetes*. 51(suppl. 1): S144–S151.

Gladwell, M. 2002. *The Tipping Point*. Little, Brown & Co., New York.

Gleick, J. 1988. *Chaos: Making a New Science*. Penguin Books, New York.

Godfrey, K. 1983. *Compartmental Models and Their Application*. Academic Press, London, U.K.

Goldbeter, A. 1996. *Biochemical Oscillations and Cellular Rhythms: The Molecular Bases of Periodic and Chaotic Behavior*. Cambridge University Press, Cambridge, U.K.

Goldbeter, A. et al. 2001. From simple to complex oscillatory behavior in metabolic and genetic control networks. *Chaos*. 11(1): 247–260.

Goldstone, R. 2002. *Complex Adaptive Systems*. (Course syllabus for P747 (psychology) at U. Indiana.) http://cognitrn.psych.indiana.edu/rgoldsto/complex/p747description. htm (accessed on November 16, 2009).

Golightly, M. and C. Golightly. 2002. Laboratory diagnosis of autoimmune diseases. www. mlo-online.com

Gómez, J., M. Martell, J. Quer, B. Cabot, and J.I. Estaban. 1999. Hepatitis C viral quasispecies. *Journal of Viral Hepatitis*. 6: 3–16.

Grandpierre, A. 2005. Complexity, information and biological organisation. *Interdisciplinary Description of Complex Systems*. 3(2): 59–71.

Granucci, F., I. Zanoni, and P. Ricciardi-Castagnoli. 2008. Review: Central role of dendritic cells in the regulation and deregulation of immune responses. *Cellular and Molecular Life Sciences*. 65: 1683–1697.

Gregersen, P.K. and T.W. Behrens. 2006. Genetics of autoimmune diseases—Disorders of immune homeostasis. *Nature Reviews/Genetics*. 7: 917–928.

Grizzi, F. and M. Chiriva-Internati. 2006. Cancer: Looking for simplicity and finding complexity. *Cancer Cell International*. 6: 1–7.

Gromada, J., I. Franklin, and C.B. Wollheim. 2007. α-Cells of the endocrine pancreas: 35 years of research but the enigma remains. *Endocrine Reviews*. 28(1): 84–116.

Gruber, J. 2005. *Mathematical Immune System Models*. (v. Jan. 3, 2005) (An extensive *Medline* literature survey with 202 citations, 105 pp.) http://www.lymenet.de/ literatur/immunsys.htm (accessed on November 16, 2009).

Guantes, R. and J.F. Poyatos. 2006. Dynamical principles of two-component genetic oscillators. *PLoS Computational Biology*. 2(3): 0188–0197.

Guillemin, E.A. 1949. *The Mathematics of Circuit Analysis*. John Wiley & Sons, New York.

Gupta, D. 2008. Peptidoglycan recognition proteins-maintaining immune homeostasis and normal development. *Cell Host and Microbe*. 3: 273–274.

Gutierrez, J.A. et al. 2009. Transition state analogs of the $5'$-methylthioadenosine nucleosidase disrupt quorum sensing. *Nature Chemical Biology*. 5(4): 251–257.

Guyton, A.C. 1991. *Textbook of Medical Physiology*. 8th edn. W.B. Saunders Co., Philadelphia, PA.

Gylfe, A., S. Bergström, J. Lundström, and B. Olsen. 2000. Reactivation of *Borrelia* infection in birds. *Nature*. 403: 724–725.

Haffler, D.A. 2002. Degeneracy, as opposed to specificity in immunotherapy. *Journal of Clinical Investigation*. 109(5): 581–584.

Hagen, E.H. and P. Hammerstein. 2005. Evolutionary biology and the strategic view of ontogeny: Genetic strategies provide robustness and flexibility in the life course. *Research in Human Development*. 2(1&2): 87–101.

Halder, G., P. Callerets, and W.J. Gehring. 1995. Induction of ectopic eyes by targeted expression of the eyeless gene in *Drosophila*. *Science*. 267: 1788–1792.

Hallinan, J. 2003. Self-organization leads to hierarchical modularity in an internet community. In *Lecture Notes in Computer Science*. Vol. 2773/2003. Springer, Berlin, Germany, pp. 914–920. ISBN: 978-3-540-40803-1. http://www.springerlink.com/ content/82r8ha86g3jluhdl/fulltext.pdf (accessed on November 24, 2009).

Hallinan, J. 2004a. Gene duplication and hierarchical modularity in intracellular interaction networks. *BioSystems*. 74: 51–62.

Hallinan, J. 2004b. Cluster analysis of the p53 genetic regulatory network: Topology and biology. *IEEE 2004 Symposium on Computational Intelligence in Bioinformatics and Computational Biology*. La Jolla, CA, Oct. 7–8, 2004. 8 pp. www.staff.ncl. ac.uk/j.s.hallinan/pubs/CIBCB2004.pdf (accessed on November 16, 2009).

Hallinan, J. and P.T. Jackway. 2005. Network motifs, feedback loops and the dynamics of genetic regulatory networks. In *Proceedings of the 2005 IEEE Symposium on Computational Intelligence in Bioinformatics and Computational Biology*, Nov. 11–15, 2005, San Diego, CA. http://ieeexplore.ieee.org/Stamp/Stamp.jsp?tp=&arnu mber=1594903&isnumber=33563 (accessed on November 16, 2009).

Hallinan, J. and G. Smith. 2002. Iterative vector diffusion for the detection of modularity in large networks. *International Journal of Complex Systems B*. 9 pp. http://research. imb.uq.edu.au/~j.hallinan/ICCS2002.htm (accessed on November 12, 2009).

Hallinan, J. and J. Wiles. 2004. Evolving genetic regulatory networks using an artificial genome. Paper presented at *the Second Asia-Pacific Bioinformatics Conference* (APBC2004), Dunedin, New Zealand. *Conferences in Research and Practice in Information Technology*, Vol. 29. Yi-Ping Phoebe Chen, Ed. pp. 291–286.

Hammerstein, P., E.H. Hagen, A.V.M. Herz, and H. Herzel. 2006. Robustness: A key to evolutionary design. *Biological Theory*. 1(1): 90–93.

Hansen, T.F. 2006. The evolution of genetic architecture. *Annual Review of Ecology, Evolution, and Systematics*. 37: 123–157.

Hansen, T.F. et al. 2006. Evolution of genetic architecture under directional selection. *Evolution*. 60(8): 1523–1536.

Harden, V. and P. D'Souza. 1996. Chemokines and HIV second receptors: A short history of a recent breakthrough. *Nature Medicine*. 2(12): 1293–1300.

Hardman, J.G. and L.E. Limbird, eds. 1996. *Goodman and Gilman's The Pharmacological Basis of Medical Practice*, 9th edn. McGraw-Hill, New York.

Hartwell, L.H. et al. 1999. From molecular to modular cell biology. *Nature*. 402(suppl.): C47–C52.

Hastings, M., J.S. O'Neill, and E.S. Maywood. 2007. Circadian clocks: Regulators of endocrine and metabolic rhythms. *Journal of Endocrinology*. 195: 187–198.

Haygood, R. 2006. Mutation rate and the cost of complexity. *MBE Advance Access*. (Published by Oxford University Press on behalf of the Society for Molecular Biology, Feb. 9, 2006.)

He, L. and G.L. Hannon. 2004. MicroRNAs: Small RNAs with a big role in gene regulation. *Nature Reviews/Genetics*. 5: 522–531.

Heart, S.F. 2008. *Albert Einstein Quotes*. www.sfheart.com/einstein.html (accessed on October 22, 2008).

Heinemann, M. and S. Panke. 2006. Synthetic biology—Putting engineering into biology. *Bioinformatics*. 22(22): 2790–2799.

Held, T.K. et al. 1999. Gamma interferon augments macrophage activation by lipopolysaccharide by two distinct mechanisms, at the signal transduction level and via an autocrine mechanism involving Tumor Necrosis Factor Alpha and Interleukin-1. *Infection and Immunology*. 67(1): 206–221.

Heppenstall, A., A.J. Evans, and M.H. Birkin. 2005. A hybrid multi-agent/spatial interaction model system for petrol price setting. *Transactions in GIS*. 9(1): 35–51.

Heylighen, F. 1996. *What is Complexity?* (Principia Cybernetica web essay, created 1996.) http://pespmc1.vub.ac.be/COMPLEXI.html (accessed on November 16, 2009).

Heylighen, F. and C. Joslyn. 2001. *The Law of Requisite Variety*. Web paper. http://pespmc1. vub.ac.be/REQVAR.htlm (accessed on November 16, 2009).

Hintze, A. and C. Adami. 2008. Evolution of complex modular biological networks. *PLoS Computational Biology*. 4(2): 1–12.

Holland, J.H. 1992. *Adaptation in Natural and Artificial Systems: An Introductory Analysis with Applications to Biology, Control and Artificial Intelligence*. MIT Press, Cambridge, MA. ISBN: 0-262-58111-6.

Holland, J.H. 1995. *Hidden Order: How Adaptation Builds Complexity*. Helix Books (Addison-Wesley), New York.

Holland, J.H. 1999. *Emergence from Chaos to Order*. Perseus Books, Reading, MA. ISBN: 0-7382-0142-1.

Holland, J.H. and J.H. Miller. 1991. Artificial adaptive agents in economic theory. In *AEA Papers and Proceedings*, May 1991, Pittsburgh, PA, pp. 365–370. http://zia.hss.cmu. edu/miller/papers/aaa.pdf (accessed on December 1, 2009).

Holmes, E.C. 2001. On the origin and evolution of the human immunodeficiency virus. *Biological Reviews*. 76: 239–254.

Holt, T.A. and M. Marinker, eds. 2004. *Complexity for Clinicians*. Radcliffe Publishing, Warwick, U.K. ISBN: 1857758552.

Holt, R.D., R. Gomulkiewicz, and M. Barfield. 2003. The phenomenology of niche evolution via quantitative traits in a 'black-hole' sink. *Proceedings of the Royal Society London B*. 270: 215–224.

Horgan, J. May 20, 2008. *Can Chaoplexy Save Economics? The Scientific Curmudgeon* web blog. http://www.stevens.edu/csw/cgi-bin/blogs/csw/?p=150 (accessed on November 16, 2009).

Hou, S.X., Z. Zheng, X. Chen, and N. Perrimon. 2002. The JAK/STAT pathway in model organisms: Emerging roles in cell movement. *Developmental Cell*. 3: 765–778.

Hu, G. et al. 2001. Msx homeobox genes inhibit differentiation through upregulation of *cyclin D1*. *Development*. 128: 2373–2384.

Hultquist, P.A. 1988. *Numerical Methods for Engineers and Computer Scientists*. Benjamin Cummings, Menlo Park, CA.

Hüser, C. 2005. Robustness—A challenge also for the 21st century. Discussion paper. UFZ-Umweltforschungszentrum Leipzig-Halle. Department of Ecological Modeling, 53 pp, August 12, 2005. http://www.ufz.de/data/DP_2006_023868.pdf (accessed on November 17, 2009).

Hutchinson, G.E. 1958. Concluding remarks. *Cold Spring Harbor Symposium on Quantitative Biology*. 22: 415–427.

Hutter, M. 2008. *Algorithmic Complexity*. Web article (v. 19 April 2008). http://www. scholarpedia.org/article/Algorithmic_complexity (accessed on November 20, 2009).

Hynne, F., S. Danø, and P.G. Sørenson. 2001. Full-scale model of glycolysis in *Saccharomyces cerevisiae*. *Biophysical Chemistry*. 94: 121–163.

Ibelgaufts, H. 2008. *Interleukins*. Table: www.copewithcytokines.de/cope.cgi?key= Interleukins (accessed on May 21, 2009).

Ingolia, N.T. 2004. Topology and robustness in the *Drosophila* segment polarity network. *PLoS Biology*. 2(6): 0805–0815.

INCOSE. 2009. Homepage of the International Council on Systems Engineering. http://www.incose.org/practice/guidetosebodyofknow.aspx (accessed on November 30, 2009).

Ioannou, P.A. and A. Pittsillides, eds. 2008. *Modeling and Control of Complex Systems*. CRC Press, Boca Raton, FL.

Ishikawa, A. 2007. Mucosal dendritic cells. *Annual Review of Immunology*. 25: 381–418.

Izzo, J.L. 1975. Pharmacokinetics of insulin: B. Degradation of insulin. In *Handbook of Experimental Pharmacology*, Hasselblatt, A. and F. von Bruchhausen, eds. Springer-Verlag, New York.

Jackson, R.J. and N. Standart. 2007. How do microRNAs regulate gene expression? *Science STKE*. re1: 13 pp. http://gene-quantification.com/jackson-review-microrna-2007.pdf (accessed on September 2, 2009).

Jacob, F. and J. Monod. 1961. Genetic regulatory mechanisms in the synthesis of proteins. *Journal of Molecular Biology*. 3: 318–356.

Jen, E. 2003. Stable or robust? What's the difference? *Complexity*. 8(3): 12–18.

Jin, P. et al. 2004. Biochemical and genetic interaction between the fragile X mental retardation protein and the microRNA pathway. *Nature Neuroscience*. 7: 113–117.

Johnson, C. 2009a. *Feds: Bomb Could Have Downed Plane*. Article in the December 29, 2009 *Hartford Courant*.

Johnson, D. 2009b. Darwinian selection in asymmetric warfare: The natural advantage of insurgents and terrorists. Paper presented to *the Washington Academy of Sciences, Fall, 2009*, Washington, DC, pp. 89–112. http://dominicdpjohnson.com/publications/pdf/2009%20-%20Johnson%20-%20Selection%20in%20War.pdf (accessed on March 22, 2010).

Johnson, L., G. Fimland, D. Mantzilas, and J. Nissen-Meyer. 2004. Structure–function analysis of immunity proteins of pediocin-like bacteriocins: C-Terminal parts of immunity proteins are involved in specific recognition of cognate bacteriocins. *Applied Environmental Microbiology*. 70(5): 2647–2652.

Jones, R.W. 1973. *Principles of Biological Regulation*. Academic Press, New York.

Kacani, L. et al. 2001. C5a and C5a^{desArg} enhance the susceptibility of monocyte-derived macrophages to HIV infection. *Journal of Immunology*. 166: 2410–2415.

Kafri, O. 2008. *Sociological and Economic Inequality and the Second Law*. Web paper. http://mpra.ub.uni-muenchen.de/9175 (accessed on September 3, 2008).

Kamath, V.P. et al. 2006. Synthesis of a potent 5′-methylthioadenosine/S-anenosylhomocysteine (MTAN) inhibitor. *Bio-organic and Medicinal Chemistry Letters*. 16(10): 2662–2665.

Kamiji, M.M. and A. Inui. 2007. Neuropeptide Y receptor selective ligands in the treatment of obesity. *Endocrine Reviews*. 28(6): 664–684.

Kamp, C. and S. Bornholdt. 2002a. Coevolution of quasispecies: B-cell mutation rates maximize viral error catastrophes. *Physical Review Letters*. 88(6): 068104-1-4.

Kamp, C. and S. Bornholdt. 2002b. From HIV infection to AIDS: A dynamically induced percolation transition? *Proceedings of the Royal Society London B*. 269: 2035–2040.

Kamp, C., C.O. Wilke, C. Adami, and S. Bornholdt. 2003. Viral evolution under the pressure of an adaptive immune system—Optimal mutation rates for viral escapes. *Complexity*. 8(2): 28–33.

Kandel, E.R., J.H. Schwartz, and T.M. Jessell. 1991. *Principles of Neural Science*, 3rd edn. Appleton & Lange, Norwalk, CT.

Kaneko, K. 2006. Life: *An Introduction to Complex Systems Biology*. Springer, New York.

Kaufman, S. 1955. *At Home in the Universe: The Search for the Laws of Self-Organization and Complexity*. Oxford University Press, New York.

Kauffman, S. 1993. *The Origins of Order*. Oxford University Press, Oxford, U.K. ISBN 0-19-505811-9.

KEGG: Kyoto Encyclopedia of Genes and Genomes. KEGG Data-Oriented Entry Points include: KEGG Atlas, KEGG PATHWAY, KEGG BRITE, KEGG ORTHOLOGY, KEGG GENES, KEGG LIGAND. Organism-specific entry points: KEGG organisms. Subject-specific entry points: KEGG DISEASE, KEGG DRUG, KEGG GLYCAN, KEGG COMPOUND, KEGG REACTION. http://www.genome.jp/kegg/ (accessed on November 17, 2009).

Kelly, K. 2009a. *The Arc of Complexity*. Web essay. www.kk.org/thetechnium/archives/2009/05/the_arc_of_comp.php (accessed on January 4, 2010).

Kelly, K. 2009b. *Upcreation*. Web essay. www.kk.org/thetechnium/archives/2009/05/upcreation.php (accessed on January 10, 2010).

Kim, J. and T. Wilhelm. 2008. What is a complex graph? *Physica A: Statistical Mechanics and Its Applications*. 387(11): 2637–2652.

Kimball, J. 2007. *The Complement System*. (v. Sept. 22, 2007) http://users.rcn.com/jkimball.ma.ultranet/BiologyPages/C/Complement.html (accessed on November 17, 2009).

Kimball, J. 2009a. *Dendritic Cells*. (v. Nov. 3, 2009) http://users.rcn.com/jkimball.ma.ultranet/BiologyPages/D/DCs.html (accessed on November 17, 2009).

Kimball, J. 2009b. *Innate Immunity*. (v. Nov. 13, 2009) http://users.rcn.com/jkimball.ma.ultranet/BiologyPages/I/Innate.html (accessed on November 17, 2009).

Kimball, J. 2009c. *Allergies*. (v. Aug. 30, 2009) http://users.rcn.com/jkimball.ma.ultranet/BiologyPages/A/Allergies.html (accessed on November 17, 2009).

Kimball, J. 2009d. *Histocompatibility Molecules*. (v. Aug. 30, 2009) http://users.rcn.com/jkimball.ma.ultranet/BiologyPages/H/HLA.html (accessed on November 17, 2009).

Kimball, J. 2009e. *Dendritic Cells*. (v. Nov. 13, 2009) http://users.rcn.com/jkimball.ma.ultranet/BiologyPages/D/DCs.html (accessed on November 17, 2009).

Kimball, J. 2009f. *Embryonic Development: Putting on the Finishing Touches*. (v. Nov. 7, 2009) http://users.rcn.com/jkimball.ma.ultranet/BiologyPages/H/HomeoboxGenes.html (accessed on November 17, 2009).

Kimball, J. 2009g. *Speciation*. (v. March 30, 2009) http://users.rcn.com/jkimball.ma.ultranet/BiologyPages/S/Speciation.html (accessed on November 17, 2009).

Kimball, J. 2009h. *Antigen Receptor Diversity*. (v. Oct. 4, 2009) http://users.rcn.com/jkimball.ma.ultranet/BiologyPages/A/AgREceptorDiversity.html (accessed on November 17, 2009).

Kimball, J. 2009i. *Homeobox Genes*. http://users.rcn.com/jkimball.ma.ultranet/BiologyPages/H/HomeoboxGenes.html#hox_cluster (accessed on November 30, 2009).

Kimball, J. 2009j. *The Operon*. (v. June 3, 2006) http://users.rcn.com/jkimball.ma.ultranet/BiologyPages/L/LacOperon.html (accessed on November 30, 2009).

Kirilyuk, A. 2006. *Universal Science of Complexity: Consistent Understanding of Ecological, Living and Intelligent System Dynamics*. Invited talk at *DECOS 2006*. (*International Conference Describing Complex Systems 2006*, Brijuni Islands, Croatia, June 12–14, 2006.) http://arxiv.org/abs/0706.3219v1 (accessed on November 18, 2009).

Kirk, C.J. and J.J. Mulé. 2000. Gene-modified dendritic cells for use in tumor vaccines. *Human Gene Therapy*. 11: 797–806.

Kirkpatrick, M. and V. Ravigné. 2002. Speciation by natural and sexual selection: Models and experiments. *The American Naturalist*. 159: S22–S35.

Kitano, H. 2004. Biological robustness. *Nature Reviews/Genetics*. 5: 826–837.

Klevescz, R.R., C.M. Li, I. Marcus, and P.H. Frankel. 2008. Collective behavior in gene regulation: The cell is an oscillator, the cell cycle a developmental process. *FEBS Journal*. 275: 2372–2384.

Kochar, P.G. 2006. *Cancer Vaccines*. 18 pp. review article. www.csa.com/discoveryguides/cancer/review.php (accessed on May 12, 2009).

Kaufman, S. 1995. *At Home in the Universe. The Search for Laws of Self-Organization and Complexity*. Viking Press, Oxford, NY.

Komarova, N.L. 2006. Oscillations in population sizes—From ecology to history. *Structure and Dynamics: eJournal of Anthropological and Related Sciences*. 1(1): Article 8. 5.

Korthof, G. 1998, 2009. Kauffman at home in the universe. The secret of life is autocatalysis. (A review of Kauffman 1995). http://home.planet.nl/~gkorthof32.htm (accessed on December 8, 2009).

Kostianovsky, M. 2000. Evolutionary origin of eukaryotic cells. *Ultrastructural Pathology*. 24: 59–66 (Review Article).

Kovacs, K., L.D. Hurst, and B. Papp. 2009. Stochasticity in protein levels drives colinearity of gene order in metabolic operons of *Escherichia coli*. *PLoS Biology*. 7(5): 9.

Krakauer, D.C. and V.A.A. Jansen. 2002. Red queen dynamics of protein translation. www.santafe.edu/research/publications/wpabstract/200205025 (accessed on November 17, 2009).

Krakauer, D.C. and J.B. Plotkin. 2002. Redundancy, antiredundancy, and the robustness of genomes. *PNAS*. 99(3): 1405–1409.

Krakauer, D.C., V.A.A. Jansen, and M. Nowak. 2002. Red queen dynamics and the evolution of translational redundancy and degeneracy. Book chapter in: *Lecture Notes in Physics*. Vol. 585/2002. Springer, Berlin/Heidelberg, Germany. ISBN: 978-3-540-43188-6. www.springerlink.com/content/v263766g415k8q78/ (accessed on November 17, 2009).

Kreimer, A., E. Borenstein, U. Gophna, and E. Ruppin. 2008. The evolution of modularity in bacterial metabolic networks. *PNAS*. 105(19): 6976–6981.

Krugman, P. 1994. Complex landscapes in economic geography. *The American Economic Review*. 84(2): 412–416.

Kuo, B.C. 1967. *Linear Networks and Systems*. McGraw-Hill Book Co., New York.

Kuo, B.C. 1970. *Modern Control Engineering*. Prentice-Hall, Inc., Englewood Cliffs, NJ, p. 06632.

Kuo, B.C. 1982. *Automatic Control Systems*, 4th edn. Prentice-Hall, Inc., Englewood Cliffs, NJ, p. 06632.

Kuo, L. et al. 2007. Neuropeptide Y acts directly in the periphery on fat tissue and mediates stress-induced obesity and metabolic syndrome. *Nature Medicine*. 13(7): 803–811.

Kurtz, J. and S.A.O. Armitage. 2006. Alternative adaptive immunity in invertebrates. *Trends in Immunology*. 27(11): 4.

Labossiere, R. 2009. *Yale Scientist's Quest: Search for Fuel's Gold*. Feature article in *The Hartford Courant*, November 30, 2009.

LaCourse, J.R. and R.B. Northrop. 1983. Eye of the mussel, *Mytilus edulis* Linnaeus: electrophysiological investigations. *The Veliger*. 25(3): 225–228.

Lacroix-Desmazes, S. et al. 2005. High levels of catalytic antibodies correlate with favorable outcome in sepsis. *PNAS*. 102(11): 4109–4113.

Lansing, J.S. 2003. Complex adaptive systems. *Annual Review of Anthropology*. 32:183–204.

Lappin, T.R.J. et al. 2006. HOX GENES: Seductive science, mysterious mechanisms. *Ulster Medical Journal*. 75(1): 23–31.

Latchman, D.S. 2010. *Gene Control*. Garland Science, London, U.K. ISBN: 978-0-8153-6513-6.

Lathi, B.P. 1974. *Linear Networks and Systems*. McGraw-Hill, New York.

Lawrence, J.G. 2003. Gene organization: Selection, selfishness, and serendipity. *Annual Review of Microbiology*. 57: 419–440.

Lawrie, J. and J. Hearne. 2007. Reducing model complexity via output sensitivity. *Ecological Modelling*. 207: 137–144.

Lee, R.C., R.L. Feinbaum, and V. Ambros. 1993. The *C. elegans* heterochronic gene *lin-4* encodes small RNAs with antisense complimentarity to *lin-14*. *Cell*. 75: 843–854.

Leloup, J.-C. and A. Goldbeter. 1999. Chaos and bi-rythmicity in a model for circadian oscillations of the PER and TIN proteins in *Drosophila*. *Journal of Theoretical Biology*. 198: 445–459.

Le Moigne, J.-L. 1995. On theorizing the complexity of economic systems. *Journal of Socio-Economics*. 24(3): 477–499.

Lemaitre, B., J.-M. Reichart, and J.A. Hoffmann. 1997. *Drosophila* host defense: Differential induction of antimicrobial peptide genes after infection by various classes of microorganisms. *PNAS*. 94: 14614–14619.

Lempradl, A. and L. Ringrose. 2008. How does noncoding transcription regulate Hox genes? *Bioessays*. 30(2): 110–121.

Lenski, R.E. and M. Travisano. 1994. Dynamics of adaptation and diversification: A 10,000-generation experiment with bacterial populations. *PNAS*. 91: 6808–6814.

Lenski, R.E., J.E. Barrick, and C. Offria. 2006. Balancing robustness and evolvability. Essay in: *PLoS Biology*. 4(12): 6. www.plosbiology.org/article/info:doi/10.1371/journal.pbio.0040428 (accessed on November 18, 2009).

Leonardo, A. 2005. Degenerate coding in neural systems. *Journal of Comparative Physiology A*. 191: 995–1010.

Lerner, R.A., S.J. Benkovic, and P.G. Schultz. 1991. At the cross roads of chemistry and immunology: Catalytic antibodies. *Science*. 252: 659–667.

Lévi, F.A. 2008. The circadian timing system: A coordinator of life processes. *IEEE Engineering in Medicine and Biology Magazine*. 27(1): 17–19.

Lewin, R. 2000. *Complexity: Life at the Edge of Chaos*. University of Chicago Press, Chicago, IL.

Lewison, E.F., ed. 1976. *Conference on Spontaneous Regression of Cancer*. DHEW Publ. No. (NIH) 76–1038, NCI Monograph 44.

Li, C., L. Chen, and K. Aihara. 2006. Synchronization of coupled nonidentical genetic oscillators. *Physical Biology*. 3: 37–44.

Li, C., L. Chen, and K. Aihara. 2007. Stochastic synchronization of genetic oscillator networks. *BMC Systems Biology*. 1: 6. 11 pp.

Liang, H. and W.-H. Li. 2009. Lowly expressed human microRNA genes evolve rapidly. Letter in: *Molecular Biology and Evolution*. 26(6): 1195–1198.

Lipson, H., J.B. Pollack, and N.P. Suh. 2002. On the origin of modular variation. *Evolution*. 56(8): 1549–1556.

Litman, G.W., J.P. Cannon, and L.J. Dishaw. 2005. Reconstructing immune phylogeny: New perspectives. *Nature Reviews/Immunology*. 5: 866–879.

Little, T.J., D. Hultmark, and A.F. Read. 2005. Invertebrate immunity and the limits of mechanistic immunology. *Nature Immunology*. 6(7): 651–654.

Liu, J.K., I. Ghattas, S. Liu, S. Chen, and J.L. Rubenstein. 1997. Dlx genes encode DNA-binding proteins that are expressed in an overlapping and sequential pattern during basal ganglia differentiation. *Developmental Dynamics*. 210(4): 489–512.

Liu, K. et al. 2009. In vivo analysis of dendritic cell development and homeostasis. *Science*. 324: 392–397.

Lloyd, S. Aug. 2001. Measures of complexity: A nonexhaustive list. *IEEE Control Systems Magazine*. 21: 7–8. Web paper. http://web.mit.edu/esd83/www/notebook/Complexity.PDF (accessed on November 12, 2009).

Lloyd, S. 2006. *Programming the Universe*. Knopf, New York.

Loker, E.S., C.M. Adema, S.-M. Zhang, and T.B. Kepler. 2004. Invertebrate immune systems—Not homogeneous, not simple, not well understood. *Immunological Reviews*. 198: 10–24.

Loo, D.D.F., E.M. Wright, and T. Zeuthen. 2002. Water pumps (Topical Review). *Journal of Physiology*. 542(1): 53–60.

Lopes, J.C., J. Dias, and J. Ferreira do Amaral. 2008. Assessing economic complexity in some OECD countries with input-output based measures. Paper presented at the *ECOMOD2008-International Conference on Policy Making*, July 2–4, 2008. Berlin, Germany, 25 pp. http://www.ecomod.org/files/papers/746.pdf (accessed on November 18, 2009).

Lorenz, E.N. 1963. Deterministic non-periodic flow. *Journal of the Atmospheric Sciences*. 20: 130–141.

Lorenz, E.N. 1972. Predictability: *Does the Flap of a Butterfly's Wings in Brazil Set off a Tornado in Texas*? Paper delivered to the *American Association for the Advancement of Science*, Washington, DC, December 29, 1972.

Lorenz, E.N. 1993. *The Essence of Chaos*. University of Washington Press, Seattle, WA.

Lotka, A.J. 1920. Undamped oscillations derived from the law of mass action. *Journal of American Chemical Society*. 42: 1595–1599.

Lu, T.K., A.S. Khalil, and J.J. Collins. 2009. Next-generation synthetic gene networks. *Nature Biotechnology*. 27(12): 1139–1150.

Lucas, C. 2006. *Quantifying Complexity Theory*. Web paper, v. 4.83. http://www.calresco.org/lucas/quantify.htm (accessed on November 4, 2009).

Luster, A.D., R. Alon, and U.H. von Andrian. 2005. Immune cell migration in inflammation: Present and future therapeutic targets. *Nature Immunology*. 6(12): 1182–1190.

Ma, H.W. and A.P. Zheng. 2003. The connectivity structure, giant strong component and centrality of metabolic networks. *Bioinformatics*. 19: 1423–1430.

Ma, W., L. Lai, Q. Ouyang, and C. Tang. 2006. Robustness and modular design of the *Drosophila* segment polarity network. *Molecular Systems Biology*. http://arxiv.org/ftp/q-bio/papers/0610/0610028.pdf (accessed on November 18, 2009).

Mabbott, N.A. 2004. The complement system in prion diseases. *Current Opinion in Immunology*. 16: 587–593.

Mabbott, N.A. and M. Turner. 2005. Prions and the blood and immune systems. *Haematoligica*. 90: 545–548.

Macal, C. and M. North. 2006a. Tutorial on agent-based modeling and simulation. Part 2: How to model with agents. In *Proceedings of the 2006 Winter Simulation Conference*, Perrone, L.F. et al. eds. Monterey, CA, 3–6 Dec. pp. 73–83. http://www.informs-sim.org/wsc06papers/008.pdf (accessed on December 10, 2009).

Macal, C. and M. North. 2006b. *Introduction to Agent-Based Modeling and Simulation* (37 presentation slides). *MCS LANS Informal Seminar*. Argonne National Laboratory. http://www.cas.anl.gov/ (accessed on March 15, 2010).

Mackay, C.R. 2008. Moving targets: Cell migration inhibitors as new anti-inflammatory therapies. *Nature Immunology*. 9(9): 988–998.

Mackay, I.R. and F.S. Rosen. 2001. Autoimmune diseases. *New England Journal of Medicine*. 345(5): 340–350 (Review Article).

Mackenzie, D. 2002. *The Science of Surprise*. Can complexity theory help us understand the real consequences of a convoluted event like September 11? Web paper. http://discovermagazine.com/2002/feb/featsurprise (accessed on November 18, 2009).

Madsen, M.F., S. Danø, and P.G. Sørensen. 2005. On the mechanisms of glycolytic oscillations in yeast. *FEBS Journal*. 272: 2648–2660.

Maeda, Y. et al. 2007. Migration of CD4 T cells and dendritic cells toward sphingosine 1-phosphate (S1P) is mediated by different receptor subtypes: S1P regulates the functions of murine mature dendritic cells via S1P receptor type 3. *Journal of Immunology*. 178: 3437–3446.

Maillet, F. et al. 2008. The *Drosophila* peptidoglycan recognition protein PGRP-LF blocks PGRP-LC and IMD/JNK pathway activation. *Cell Host and Microbe*. 3: 293–303.

Malik mzsg. 2006. *Sensitivity Model Prof. Vester®*. The Computerized System Tools for a New Management of Complex Problems. 12 pp. www.frederic-vester.de/uploads/InformationEnglishSM.pdf (accessed on November 18, 2009).

Mann, D.L. et al. 1990. HIV-1 transmission and function of virus-infected monocytes/macrophages. *Journal of Immunology*. 144(6): 3152–2158.

Manrubia, S.C. and C. Briones. 2007. Modular evolution and increase of functional complexity in replicating RNA molecules. *RNA*. 13: 97–107.

Manrubia, S.C., S. Escarmís, E. Domingo, and E. Lázaro. 2005. High mutation rates, bottlenecks, and robustness of RNA viral quasispecies. *Gene*. 347: 273–282.

Maron, M. 2003. *Modelling Populations*. University of Sussex, Brighton, U.K. Web paper. http://brainoff.com/easy/report.pdf (accessed on November 18, 2009).

Marshall, S.H. and G. Arenas. 2003. Antimicrobial peptides: A natural alternative to chemical antibiotics and a potential for applied biotechnology. *Electronic Journal of Biotechnology*. 6(2): 271–284. ISSN: 0717-3458.

Mason, S.J. 1953. Feedback theory—Some properties of signal flow graphs. *Proceedings of the IRE*. 41(9): 1144–1156.

Mason, S.J. 1956. Feedback theory—Further properties of signal flow graphs. *Proceedings of the IRE*. 44(7): 920–926.

Matthias, R., B. Hannon, and J.W. Forrester. 1997. *Modeling Dynamic Economic Systems*. Springer, New York.

Matsuo, T. et al. 2003. Control mechanism for the circadian clock for timing of cell division in vivo. *Science*. 302: 255–259.

McCabe, T. 1976. A complexity measure. http://www.literateprogramming.com/mccabe.pdf (accessed on April 10, 2010).

McCauley, J.L. and C.M. Küffner. 2004. Economic system dynamics. *Discrete Dynamics in Nature and Society*. 1: 213–220. http://mrpa.ub.uni-muenchen.de/2158/1/MRPA_paper_2158.pdf (accessed on November 18, 2009) (MPRA-Munich Personal RePEc Archive.)

McGinnis, W. et al. 1984. A conserved DNA sequence in homeotic genes of the *Drosophila* Antennapedia and Bithorax complexes. *Nature*. 308: 428–433.

McMillen, D., N. Kopell, J. Hasty, and J.J. Collins. 2002. Synchronizing genetic relaxation oscillators by intercell signaling. *PNAS*. 99(2): 679–684.

Medoff, B.J., S.Y. Thomas, and A.D. Luster. 2008. T cell trafficking in allergic asthma: The ins and outs. *Annual Review of Immunology*. 26: 205–232.

Mendao, M., J. Timmis, P.S. Andrews, and M. Davies. 2007. The immune system in pieces: Computational lessons from degeneracy in the immune system. In *Proceedings of 2007 IEEE Symposium on Foundations of Computational Intelligence* (FOCI 2007), Honolulu, HI, pp. 394–400.

Mendelson, T.C. and K.L. Shaw. 2005. Rapid speciation in an arthropod. *Nature*. 433: 375–376.

Merck. 2007. *Osteoporosis. Merck Manual Home Edition*, 7 pp. http://www.merck.com/mmhe/print/sec05/ch060/ch060a.html (accessed on November 18, 2009).

Metcalfe, J.S. and J. Foster, eds. 2004. *Evolution and Economic Complexity*. Cheltenham, U.K. ISBN: 1843765268.

Meyer, A., R. Pellaux, and S. Panke. 2007. Bioengineering novel *in vitro* metabolic pathways using synthetic biology. *Current Opinion in Microbiology*. 10: 246–253.

Mihacescu, I., W. Hsing, and S. Liebler. 2004. Resilient circadian oscillator revealed in individual cyanobacteria. *Nature*. 430(8995): 81–85.

Miller, J.H. and S.E. Page. 2007. *Complex Adaptive Systems: An Introduction to Computational Models of Social Life*. Princeton University Press, Princeton, NJ. ISBN-10: 0691127026

Milo, R. et al. 2002. Network motifs: Simple building blocks of complex networks. *Science*. 298: 824–827.

Mitleton-Kelly, E. ed. 2001. *Complex Systems and Evolutionary Perspectives: The Application of Complexity Theory to Organisations*. Elsevier, London, U.K.

Mittmann, B. 2002. Early neurogenesis in the horseshoe crab, *Limulus polyphemus* and its implication for arthropod relationships. *Biological Bulletin*. 203: 221–222.

Moler, C. 2003. *Stiff Differential Equations*. MATLAB News & Notes, May 2003. www.mathworks.com/company/newsleters/news_notes/clevescorner/may03_cleve.html (accessed on November 13, 2009).

Montague, C.T. et al. 1997. Congenital leptin deficiency is associated with severe early-onset obesity in humans. *Nature*. 387: 903.

Morgan, S.M. et al. 2005. Sequential actions of the two component peptides of the lantibiotic Lacticin 3147 explain its antimicrobial activity at nanomolar quantities. *Antimicrobial Agents and Chemotherapy*. 49(7): 2606–2611.

Moss, R. 2009. *Provenge Fights Prostate Cancer*. News bulletin. http://chetday.com/provengeprostatecancer.htm (accessed on November 18, 2009).

Mukherji, S. and A. van Oudenaarden. 2009. Synthetic biology: Understanding biological design from synthetic circuits. *Nature Reviews/Genetics*. 10: 859–871.

Müller, G.B. and G.P. Wagner. 1996. Homology, *Hox* genes, and developmental integration. *American Zoology*. 36: 4–13.

Nagarajan, R. 2002. Quantifying physiological data with Lempel-Ziv complexity—Certain issues. *IEEE Transactions on Biomedical Engineering*. 49(11): 1371–1373.

Nair, S.V., A. Ramsden, and D.A. Raftos. 2005. Ancient origins: Complement in invertebrates. *ISJ*. 2: 114–123.

Nan, F. and D. Adjeroh. 2004. On complexity measures for biological sequences. In *Proceedings of the 2004 IEEE Computational Systems Bioinformatics Conference* (CSB 2004), Stanford, CA, 5 pp.

Nanavati, R.P. 1975. *Semiconductor Devices*. Intext Educational Publishers, New York, Chap. 10, *Some Additional Semiconductor Devices*.

Neck, R. ed. 2003. *Modeling and Control of Economic Systems*. Elsevier Science, Oxford, U.K., 442 pp.

Nevinsky, G.A., T.G. Kanyshkova, and V.N. Buneva. 2000. Natural catalytic antibodies (abzymes) in normalcy and pathology. *Biochemistry Moscow*. 65(11): 1245–1255.

Nibbering, P. 2004. Effects of maggot excrete on human endothelial cells. Line Symposium Biotherape Conference, Neu-Ulm, Germany.

Nigam, Y. et al. 2006a. Maggot therapy: The science and implication for CAM. Part I—History and bacterial resistance. *eCAM*. 3(2): 223–227.

Nigam, Y. et al. 2006b. Maggot therapy: The science and implication for CAM. Part II—Maggots combat infection. *eCAM*. 3(3): 303–308.

Nise, N.S. 1995. *Control Systems Engineering*, 2nd edn. Benjamin Cummings, Redwood City, CA.

Nissen-Meyer, J., P. Rogne, C. Oppegård, H.S. Haugen, and P.E. Kirstiansen. 2009. Structure–function relationships of the non-lanthionine-containing peptide (class II) bacteriocins produced by Gram-positive bacteria. *Current Pharmaceutical Biotechnology*. 10: 19–37.

NIST. 2006. *Kolmogorov Complexity* by (CRC-A). http://www.nist.gov/dads/HTML/kolmogorov.html (accessed on November 18, 2009).

Norberg, J. and G.S. Cumming, eds. 2008. *Complexity Theory for a Sustainable Future.* Columbia University Press, New York.

Northrop, R.B. 1990. *Analog Electronic Circuits: Analysis and Applications.* Addison-Wesley, Reading, MA.

Northrop, R.B. 2000. *Exogenous and Endogenous Regulation and Control of Physiological Systems.* Chapman & Hall/CRC Press, Boca Raton, FL.

Northrop, R.B. 2001. *Dynamic Modeling of Neuro-Sensory Systems.* CRC Press, Boca Raton, FL.

Northrop, R.B. 2003. *Signals and Systems Analysis in Biomedical Engineering.* CRC Press, Boca Raton, FL.

Northrop, R.B. 2004. *Analysis and Application of Analog Electronic Circuits to Biomedical Instrumentation.* CRC Press, Boca Raton, FL.

Northrop, R.B. and A.N. Connor. 2008. *Introduction to Molecular Biology, Genomics and Proteomics for Biomedical Engineers.* Taylor & Francis/CRC Press, Boca Raton, FL.

Northrop, R.B., X.-Z. Liu, and Q. Li. 1989. A modelling study of human immune deficiency disease. In *Proceedings of the 15th Annual Northeast Bioengineering Conference,* S. Buus, ed. IEEE Press, New York, pp. 171–172.

Noyes, R.M. and R.J. Field. 1974. Oscillatory chemical reactions. *Annual Review of Physical Chemistry.* 25: 95–119.

Obuchowicz, E., R. Krysial, and Z.S. Herman. 2004. Does neuropeptide Y (NPY) mediate the effects of psychotropic drugs? *Neuroscience and Biobehavioral Reviews.* 28: 595–610.

Ochiai, K., T. Yamanaka, K. Kimura, and O. Sawada. 1959. Inheritance of drug resistance (and its transfer) between *Shigella* strains and between *Shigella* and *E. coli* strains. *Hihon Iji Shimpor.* 1861: 34 (in Japanese).

Odell, J. 2003. Between order and chaos. *Journal of Object Technology.* 2(6): 45–50.

Ogata, K. 1970. *Modern Control Engineering.* Prentice-Hall Inc., Englewood Cliffs, NJ.

Ogata, K. 1990. *Modern Control Engineering,* 2nd edn. Prentice-Hall Inc., Englewood Cliffs, NJ.

Olenev, N. 2007. A normative balance dynamic model of regional economy for study economic integrations. Paper presented at *International Conference on Economic Integration Competition and Cooperation,* Session 6, Opatija, Croatia, April 19–20, 2007. http://mpra.ub.uni-muenchen.de/7823/ (accessed on November 18, 2009).

Onoue, S., S. Misaka, and S. Yamada. 2008. Structure–activity relationship of vasoactive intestinal peptide (VIP): Potent agonists and potential clinical applications. *Archives of Pharmacology.* 377: 579–590.

Ormerod, P., H. Johns, and L. Smith. 2001. *An Agent-Based Model of the Extinction Patterns of Capitalism's Largest Firms.* Web paper. 15 pp. http://www.complexity-society.com/papers/model_of_ capitalism.pdf (accessed on December 10, 2009).

Orr, H.A. 1996. Darwin v. intelligent design (again). *Boston Review.* Dec. 1996/Jan. 1997 issue.

Ottawa. 2009. *Complexity Theory.* (On-line course notes for POP8910, Scientific Paradigms in Population Health.) http://courseweb.edteched.uottawa.ca/POP8910/Notes/Complexity.htm (accessed on November 18, 2009).

Pace, J.K. II and C. Fescotte. 2007. The evolutionary history of human DNA transposons: Evidence for intense activity in the primate lineage. *Genome Research.* (Article published online 3/05/07 before print.) 11 pp. http://genome.cshlp.org/content/17/4/422 (accessed on November 18, 2009).

Pai, A., Y. Tanouchi, C.H. Collins, and L. You. 2009. Engineering multicellular systems by cell–cell communication. *Current Opinions in Biotechnology.* 20: 461–470.

Pan, W., X. Liu, F. Ge, J. Han, and T. Zheng. 2004. Perinerin, a novel anntimicrobial peptide purified from the clamworm *Perinereis aibuhitensis* Grube and its partial characterization. *Journal of Biochemistry*. 135: 297–304.

Panganiban, G. and J.L.R. Rubenstein. 2002. Developmental functions of the *Distal-less/Dlx* homeobox genes. *Development*. 129: 4371–4386 (Review Article).

Papagianni, M. and S. Anastasiadou. 2009. Pediocins: The bacteriocins of Pediococci. Sources, production, properties and applications. *Microbial Cell Factories*. 8: 3 (16 pp.). www.microbialcellfactories.com/content/8/1/3 (accessed on February 22, 2010).

Pape, H.-C. 1995. Nitric oxide: An adequate modulatory link between biological oscillators and control systems in the mammalian brain. *Seminars in the Neurosciences*. 7: 329–340.

Parnes, O. 2004. From interception to incorporation: Degeneracy and promiscuous recognition as precursors of a paradigm shift in immunology. *Molecular Immunology*. 40: 985–991.

Paul, S. et al. 1995. Catalytic antibodies to vasoactive intestinal peptide. *Chest*. 107: 125S–126S.

Paul, S. et al. 2003. Specific HIV gp120-cleaving antibodies induced by covalently reactive analog of gp120. *The Journal of Biological Chemistry*. 278: 20429–20435.

Pauza, C.D. 1988. Commentary: HIV persistence in monocytes leads to pathogenesis and AIDS. *Cellular Immunology*. 112: 414–424.

Pedrazzini, T., F. Pralong, and E. Grouzmann. 2003. Neuropeptide Y: The universal soldier. *Cellular and Molecular Life Sciences*. 60: 350–377.

Petrosky, H. Sept.–Oct. 2009. Infrastructure. *American Scientist*. 97(5): 370–375.

Pham, L.N. et al. 2007. A specific primed immune response in *Drosophila* is dependent of phagocytes. *PLoS Pathogens*. 3(3): 8.

Phillipson, M., B. Heit, P. Colarusso, L. Liu, C.M. Ballantyne and P. Kubes. 2006. Intraluminal crawling of neutrophils to emigration sites: A molecularly distinct process from adhesion in the recruitment cascade. *Journal of Experimental Medicine*. 203(12): 2569–2575.

Pikovsky, A., M. Rosenblum, and J. Kurths. 2001. *Synchronization: A Universal Concept in Nonlinear Science*. Cambridge University Press, Cambridge, U.K.

Piqueira, J.R.C. 2008. A mathematical view of biological complexity. *Communications in Nonlinear Science and Numerical Simulation*. 14(6): 2581–2586.

Price, M.N., A.P. Arkin, and E.J. Alm. 2006. The life-cycle of operons. *PLoS Genetics*. 2(6): 859–873.

Prigogine, I. 1980. *From Being to Becoming: Time and Complexity in the Physical Sciences*. W.H. Freeman, San Francisco, CA. ISBN: 0-7167-1107-0.

Proteinogenic. 2010. Proteinogenic amino acid. http://en.wikipedia.org/wiki/ Proteinogenic (accessed on February 18, 2010).

Pryor, F.L. 1996. *Economic Evolution and Structure: The Impact of Complexity on the US Economic System*. Cambridge University Press, New York. ISBN: 0-521-55097-1.

Purnick, P.E.M. and R. Weiss. 2009. The second wave of synthetic biology: From modules to systems. *Nature Reviews/Molecular Cell Biology*. 10: 410–422.

Quastler, H. 1958. A primer on information theory. In *Symposium on Information Theory in Biology*. Yockey, H.P., R.L. Platzman, and H. Quastler, eds. Pergamon Press, New York, pp. 3–49.

Rackham, O. and J.W. Chin. 2005a. A network of orthogonal ribosome·mRNA pairs. *Nature Chemical Biology*. 1(3): 159–166.

Rackham, O. and J.W. Chin. 2005b. Cellular logic with orthogonal ribosomes. *Journal of American Chemical Society*. 127(50): 17584–17585.

Rackham, O. and J.W. Chin. 2006. Synthesizing cellular networks from evolved ribosome-mRNA pairs. *Biochemical Society Transactions*. 34(part 2): 328–329.

Rackham, O., K. Wang, and J.W. Chin. 2006. Functional epitopes at the ribosome subunit interface. *Nature Chemical Biology*. 2(5): 254–258.

Radzicki, M.J. 2003. Mr. Hamilton, Mr. Forrester, and a foundation for evolutionary economics. *Journal of Economic Issues*. 37(1): 133–173.

Railsback, S.F., S.L. Lytinen, and S.K. Jackson. 2006. Agent-based simulation platforms: Review and development recommendations. *Simulation*. 82(9): 609–623. http://sim.sagepub.com/cgi/reprint/82/9/609 (accessed on February 11, 2010).

Raine, A., J. Foster, and J. Potts. 2006. The new entropy law and the economic process. *Ecological Complexity*. 3: 354–360.

Raff, E.C. and R.A. Raff. 2000. Dissociability, modularity, evolvability. *Evolution and Development*. 2(5): 235–237.

Rameshwar, P. and A. Bardaguez. 2009. Molecular basis of cytokine function. Chapter in: *The Neuroimmunological Basis of Behavior*. Siegel, A. and S. Zalcman, eds. Springer Science, New York, pp. 59–70. ISBN: 978-0-387-84850-8 (on-line).

Ramos, C. and B. Robert. 2005. *msh/Msx* gene family in neural development. *Trends in Genetics*. 21(11): 624–632.

Rast, J.P., L.C. Smith, M. Loza-Coll, T. Hibino, and G.W. Litman. 2006. Genomic insights into the immune system of the sea urchin. *Science*. 314: 952–956.

Rawlings, J.S., K.M. Rosler, and D.A. Harrison. 2004. The JAK/STAT signaling pathway. *Journal of Cell Science*. 117: 1281–1283.

Reiber, C.L. and S.P. Roberts. 2005. Ontogeny of physiological regulatory mechanisms: Fitting into the environment. Introduction to the symposium. *Comparative Biochemistry and Physiology Part A*. 141: 359–361.

Reibnegger, G. et al. 1987. Theoretical implications of cellular immune reactions against helper lymphocytes infected by an immune system retrovirus. *PNAS*. 84: 7270–7274.

Reis e Sousa, C. 2006. Dendritic cells in a mature age. *Nature Reviews/Immunology*. 6(6): 476–483. www.nature.com/nri/journal/v6/n6/full/nri1845.html (accessed on May 19, 2009).

Remesar, X. et al. 1996. Is leptin an insulin counter-regulatory hormone? *FEBS Letters*. 402: 9–11.

Remondino, M. and A. Cappellini. 2006. Agent based simulation in biology: The case of periodical insects as natural prime number generators. *Vet On-Line*. http://priory.com/vet/cicada.pdf (accessed on February 11, 2010).

Rice, W.R. and G.W. Salt. 1988. Speciation via disruptive selection on habitat preference: Experimental evidence. *The American Naturalist*. 131: 911–917.

Richardson, R.C. 2001. Complexity, self-organization and selection. *Biology and Philosophy*. 16: 655–683. http://www.springerlink.com/content/l72422h001m73p43/fulltext.pdf (accessed on December 8, 2009).

Ridley, M. 2000. *Mendel's Demon: Gene Justice and the Complexity of Life*. Weidenfield & Nicolson, London, U.K.

Rinn, J.L. et al. 2007. Functional demarcation of active and silent chromatin domains in human HOX loci by noncoding RNAs. *Cell*. 129(7): 1311–1323.

Rivera, J., R.L. Proia, and A. Olivera. 2008. The alliance of sphingosine 1-phosphate and its receptors in immunity. *Nature Reviews Immunology*. 8(10): 753–763.

Rizza, A., L.J. Mandarino, and E. Gerich. 1982. Cortisol-induced insulin resistance in man: Impaired suppression of glucose production and stimulation of glucose utilization due to a postreceptor defect of insulin action. *Journal of Clinical Endocrinology and Metabolism*. 54: 131–138.

Robalino, J. et al. 2005. Double-stranded RNA induces sequence-specific antiviral silencing in addition to nonspecific immunity in a marine shrimp: Convergence of RNA interference and innate immunity in the invertebrate antiviral response? *Journal of Virology*. 79: 13561–13571.

Robertson, H.H. et al. 1996. Reconstruction of the ancient 'mariners' of humans. *Nature Genetics*. 12: 360–361.

Robison, K. 1996. *Darwin's Black Box: Irreducible Complexity or Irreproducible Irreducibility?* Essay in *The Talk Origins Archive*. www.talkorigins.org/faqs/behe/review.html (accessed on November 18, 2009).

Roche Diagnostics Corporation. 2004. *Biochemical Pathways*, 3rd edn. (Parts 1 and 2). (Two complex metabolic charts from Roche Diagnostics Corp., Indianapolis, IN.)

Rollins, M. 2010. *'Theory' Has a Different Meaning in Science*. Letter to the editor. *The Hartford Courant*, March 22, 2010.

Rosen, H. and E.J. Goetzl. 2005. Sphingosine 1-phosphate and its receptors: An autocrine and paracrine network. *Nature Reviews/Immunology*. 5: 560–570. http://www.nature.com/nri/journal/v5/n7/nri1650.pdf (accessed on December 15, 2009).

Rosser, J.B. Jr. 2008. *Econophysics and Economic Complexity*. Web essay. http://cob.jmu.edu/rosserjb/ECONOPHYSICS%20AND%20ECONOMIC%20COMPLEXITY.doc (accessed on November 18, 2009).

Rossi, L.M. et al. 2004. *Epistemological Implications of Economic Complexity*. James Madison University. Web paper. http://cob.jmu.edu/rosserjb/EPISTEMOLOGY.eco.complex.doc (accessed on November 18, 2009).

Rossi, L.M. et al. 2008a. Research advances in the development of peptide antibiotics. *Journal of Pharmaceutical Sciences*. 97(3): 1060–1070.

Rossi, L.M. et al. 2008b. *Complex Dynamics in Ecologic-Economic Systems*. James Madison University. http://cob.jmu.edu/rosserjb/Handbook.%20Complex%20Dynamics%20%20IN%20ECOLOGIC.doc (accessed on November 18, 2009).

Rössler, O.E. 1976. An equation for continuous chaos. *Physics Letters*. 35A: 397–398.

Rowley, A.F. and A. Powell. 2007. Invertebrate immune systems-specific, quasi-specific, or non-specific. *Journal of Immunology*. 179: 7209–7214.

Rugh, W.J. 1996. *Linear System Theory*, 2nd edn. Prentice-Hall, Inc., Upper Saddle River, NJ. ISBN: 0-13-441205-2.

Ruiz-Argüelles, A. and L. Llorente. 2006. The role of complement regulatory proteins (CD55 and CD59) in the pathogenesis of autoimmune hemocytopenias. *Autoimmunity Reviews*. 6(3): 155–161.

Ruiz-Jarabo, C. et al. 2000. Memory in viral quasispecies. *Journal of Virology*. 74(8): 3543–3547.

Ruth, M. and B. Hannon. 1997. *Modeling Dynamic Economic Systems*. Springer, New York, 339 pp.

Ryan, D. 2004. *Biodiesel—A Primer*. ATTRA document. http://attra.ncat.org/attra-pub/PDF/biodiesel.pdf (accessed on November 4, 2009).

Sadd, B.M. and P. Schmid-Hempel. 2006. Insect immunity shows specificity in protection upon secondary pathogen exposure. *Current Biology*. 16: 1206–1210.

Sagués, F. and I.R. Epstein. 2003. Nonlinear chemical dynamics. *Dalton Transactions*. (*Royal Society of Chemistry*). 7: 1201–1217.

Salzet, M. 2005. Neuropeptide-derived antimicrobial peptides from invertebrates for biomedical applications. *Current Medicinal Chemistry*. 12: 2663–2681.

Saltzman, B. 1962. Finite amplitude free convection as an initial value problem. *Journal of Atmospheric Sciences*. 19: 329–341.

Samuelson, P.A. and W.D. Nordhaus. 2001. *Economics*, 17th edn. McGraw-Hill, New York, p. 157.

Sandeva, R., Sv. Dimova, and Zh. Tsokeva. 2007. Neuropeptide Y family of peptides: Structure, anatomical expression, regulation, receptors and physiological functions. *Trakia Journal of Sciences.* 5(2): 8–15. ISSN: 1312-1723.

Sandilands, G.P. et al. 2005. Cross-linking of neutrophil CD11b results in rapid cell surface expression of molecules required for antigen presentation and T-cell activation. *Immunology.* 114: 354–368.

Santa Fe Institute. 2008. http://www.santafe.edu/about/ (accessed on November 18, 2009).

Sardanyés, J. and R.V. Solé. 2006. Red Queen strange attractors in host–parasite replicator gene-for-gene coevolution. *Chaos, Solitons and Fractals.* 32(5): 1666–1678.

Sasaki, K., K. Doh-ura, J.W. Ironside, N. Mabbott, and T. Iwaki. 2006. Clusterin expression in follicular dendritic cells associated with prion protein accumulation. *Journal of Pathology.* 209: 484–491.

Saslow, W.M. 1999. An economic analogy to thermodynamics. *American Journal of Physics.* 67(12): 1239–1247.

Scarlat, E. 2001. Complexity and chaos in economic cybernetic systems. Chaotic models of economic systems functionality. *Economy Informatics.* 1: 78–83.

Schaffer, W.M. 2001. *Quantized Dynamics in the Lorenz Equations.* Web essay. http://bill.srnr.arizona.edu/Quantized/quantized.htm (accessed on November 18, 2009).

Schaffer, A. et al. 2003. L-Arginine: An ultradian-regulated substrate coupled with insulin oscillations in healthy volunteers. *Diabetes Care.* 26(1): 168–171.

Schilling, D.L. and C. Belove. 1979. *Electronic Circuits, Discrete and Integrated.* McGraw-Hill, New York.

Schneider, E. and J. Kay. 1994. Life as a manifestation of the second law of thermodynamics. *Mathematical and Computer Modeling.* 19(6–8): 25–48.

Schlosser, G. and G.P. Wagner, eds. 2004. *Modularity in Development and Evolution.* University of Chicago Press, Chicago, IL, 600 pp.

Schlosser, R. et al. 1998. Functional magnetic resonance imaging of human brain activity in a verbal fluency test. *Journal of Neurosurgery and Psychiatry.* 64: 492–498.

Schuster, H.G. 2005. *Complex Adaptive Systems: An Introduction.* Scator Verlag, Saarbrücken, Germany. ISBN: 3-9807936-0-5. Book description. http://www.theophysik.uni-kiel.de/theo-physik/schuster/cas.html (accessed on November 19, 2009).

Schwartz, M. 1959. *Information, Transmission, Modulation and Noise.* McGraw-Hill, New York.

Scott, M.O. and A.J. Weiner. 1984. Structural relationships among genes that control development: Sequence homology between the Antennapedia, Ultrabithorax, and fushi tarazu loci of Drosophila. *PNAS.* 81(13): 4115–4119.

Searchinger, T. et al. 2008. Use of U.S. croplands for biofuels increases greenhouse gasses through emissions from land use change. *Science.* 319: 1238–1240. http://www.sciencemag.org/cgi/reprint/319/5876/1238.pdf (accessed on November 19, 2009).

Sercarz, E.E. and E. Maverakis. 2004. Recognition and function in a degenerate immune system. *Molecular Immunology.* 40(14–15): 1003–1008.

Shannon, C. 1948. A mathematical theory of communication. *Bell System Technical Journal.* 7: 535–563.

Shannon, C. and W. Weaver. 1949. *Mathematical Theory of Communications.* University of Illinois Press, Urbana, IL.

Sharma, V.K. and M.K. Chandrashekaran. 2005. Zeitgebers (time cues) for biological clocks. *Current Science.* 89(7): 1136–1146.

Sheard, S.A. 2007. Principles of complex systems for systems engineering. *Symposium on Complex Systems Engineering*, Jan. 11–12, 2007. Rand Corp., Santa Monica, CA, 15 pp. http://cs.calstatela.edu/wiki/images/8/84/sheard.doc (accessed on November 20, 2009).

Sheard, S.A. and A. Mostashari. 2008. Principles of complex systems for systems engineering. Published online in Wiley InterScience, New York, 17 pp. http://www3.interscience.wiley.com/cgi-bin/fultext/121519813/PDFSTART (accessed on November 20, 2009).

Shen-Orr, S.S., R. Milo, S. Mangan, and U. Alon. 2002. Network motifs in the transcriptional regulation network of *Escherichia coli*. *Nature Genetics*. 31: 64–68.

Sherman, R.A. 2003. Maggot therapy for treating diabetic foot ulcers unresponsive to conventional therapy. *Diabetes Care*. 26(2): 446–451.

Sherwin, R.S. et al. 1974. A model of the kinetics of insulin in man. *Journal of Clinical Investigations*. 53: 1481–1492.

Shi, L. and S.M. Paskewitz. 2006. Proteomics and insect immunity. *ISJ*. 3: 4–17.

Shimamoto, T. et al. 1997. Inhibition of Dlx7 homeobox genes causes decreased expression of GATA-1 and c-myc genes and apoptosis. *PNAS*. 94(7): 3245–3249.

Shuai, K. and B. Liu. 2003. Regulation of JAK-STAT signalling in the immune system. *Nature Reviews/Immunology*. 3: 900–911.

Silva-Rocha, R. and V. de Lorenzo. 2008. Mining logic gates in prokaryotic regulation networks. *FEBS Letters*. 582: 1237–1244.

Simnon™. 1998. *Simnon*™ for Windows. v. 3.0. © 1998 SSPA Maritime Consulting, Göteborg, Sweden.

Simon, H.A. 1962. The architecture of complexity. *Proceedings of the American Philosophical Society*. 106(6): 567–482.

Simon, S. and G. Brandenberger. 2002. Ultradian oscillations of insulin secretion in humans. *Diabetes*. 51(suppl. 1): S258–S261.

Śladowski, D. et al. 2006. Expression of the membrane complement regulatory proteins (CD55 and CD59) in human thymus. *Folia Histochemica et Cytobiologica*. 44(4): 263–267.

Slobin, L.I. 1966. Preparation and some properties of antibodies with specificity toward ortho nitrophenol esters. *Biochemistry*. 5: 2836–2841.

Smith, D.J. 2004. *Systems Thinking: The Knowledge Structures and the Cognitive Process*. (Online essay). 33 pp. www.smithsrisca.demon.co.uk/systems-thinking.html (accessed on November 30, 2009).

Solé, R.V. and B. Luque. 1999. Statistical measures of complexity for strongly interacting systems. (Submitted to *Physical Review E*, Aug. 27, 1999. 12 pp.) http://arxiv.org/PS_cache/adap-org/pdf/9909/9909002v1.pdf (accessed on November 30, 2009).

Sornette, D. 2003. *Why Stock Markets Crash*. Princeton University Press, Princeton, NJ. ISBN: 0-691-09630-9.

Spangler, R.A. and F.M. Snell. 1961. Sustained oscillations in a catalytic chemical system. *Nature*. 191(4787): 457–458.

Stagner, J.I., E. Samols, and G.C. Weir. 1980. Sustained oscillations of insulin, glucagon, and somatostatin from the isolated canine pancreas during exposure to a constant glucose concentration. *Journal of Clinical Investigations*. 65: 939–942.

Stebbing, A.R.D. 2006. Genetic parsimony: A factor in the evolution of complexity, order and emergence. *Biological Journal of the Linnean Society*. 88: 295–308.

Steenkamp, E.T. et al. 2006. The protistan origins of animals and fungi. *Molecular Biology and Evolution*. 23: 93–106.

Suerbaum, S. and C. Josenhans. 2007. *Helicobacter pylori* evolution and phenotypic diversification in a changing host. *Nature Reviews/Microbiology*. 5: 441–452.

Sullivan, B.A. et al. 2002. Positive selection of a Qa-1-restricted T cell receptor with a specificity for insulin. *Immunity*. 17(1): 95–105.

Sussman, G.J. 2007. *Building Robust Systems*. Web essay. http://groups.csail.mit.edu/mac/users/gjs/6.945/readings/robist-systems.pdf (accessed on November 19, 2009).

Sustainable. 2008. *Paper Computer*. Web description. http://sustainable.a.wiki-site.com/index.php/Paper_Computer (accessed on November 19, 2009).

Szybalski, W. and A. Skalka. 1978. Nobel prizes and restriction enzymes. *Gene*. 4: 181–182.

Tabarrok, A. 2008. *The Law of Unintended Consequences*. Blog. www.marginalrevolution.com/marginalrevolution/2008/01/the-law-of-unin.html (accessed on November 19, 2009).

Tagkopoulos, I., Y.-C. Liu, and S. Tavazoie. 2008. Predictive behavior within microbial genetic networks. *Science*. 320: 1313–1317. www.sciencmag.org/cgi/reprint/320/5881/1313.pdf (accessed on November 19, 2009).

Takeda, K. and S. Akira. 2005. Toll-like receptors in innate immunity. *International Immunology*. 17(1): 1–14.

Talk Origins Archive. 2009. www.talkorigins.org/faqs/behe.html (accessed on November 19, 2009).

Tanaka, F. 2002. Catalytic antibodies as designer proteases and esterases. *Chemical Reviews*. 102: 4885–4906.

Tanji, T. et al. 2007. Toll and IMD pathways synergistically activate an innate immune response in *Drosophila melanogaster*. *Molecular and Cellular Biology*. 27(12): 4578–4588.

Tesfatsion, L. 2002. Agent-based computational economics: Growing economies from the bottom up. *Artificial Life*. 8: 55–82.

Tesfatsion, L. 2005a. *Syllabus of Readings for Complex Adaptive Systems and Agent-Based Computational Economics. 4. Biological Evolution*. Web paper. http://www.econ.iastate.edu/tesfatsi/bioevol.htm (accessed on November 19, 2009).

Tesfatsion, L. 2005b. *Agent-Based Computational Economics: A Constructive Approach to Economic Theory*. pp. 1–55. Preprint of a chapter that appeared in Tesfatsion, L. and K.L. Judd, eds. 2006. *Handbook of Computational Economics, Vol. 2: Agent-Based Computational Economics*. Elsevier/North Holland, Amsterdam, the Netherlands. http://www.econ.iastate.edu/tesfatsi/hbintlt.pdf (accessed on December 10, 2009).

Tesfatsion, L. 2009. Agent-based computational economics (ACE) home page. http://www.econ.iastate.edu/tesfatsi/ace.htm (accessed on December 10, 2009).

Tesfatsion, L. and K.L. Judd, eds. 2006. *Handbook of Computational Economics, Vol. 2: Agent-Based Computational Economics*. Elsevier/North Holland, Amsterdam, the Netherlands. (Preface, topics, contributors & chapter abstracts. http://www.econ.iastate.edu/tesfatsi/hbace.htm, accessed on December 11, 2009.)

Theobald, D. 2007. *The Mullerian Two-Step: Add a part, make it necessary* or, *Why Behe's "Irreducible Complexity" is silly*. v. 1.1. Essay in *The Talk Origins Archive*. www.talkorigins.org/faqs/comdesc/ ICsilly.html (accessed on November 19, 2009).

Tincu, J.A. and S.W. Taylor. 2004. Antimicrobial peptides from marine invertebrates. (Minireview.) *Antimicrobial Agents and Chemotherapy*. 48(10): 3645–3554.

Tomović, R. 1966. *Introduction to Nonlinear Automatic Control Systems*. John Wiley & Sons, Inc., New York.

Tononi, G., O. Sporns, and G.M. Edelman. 1994. A measure for brain complexity: Relating functional segregation and integration in the nervous system. *PNAS*. 91: 5033–5037.

Tononi, G., O. Sporns, and G.M. Edelman. 1999. Measures of degeneracy and redundancy in biological networks. *PNAS*. 96: 3257–3262.

Tools. 2010. *Tools for Agent-Based Modelling. SwarmWiki*. http://www.swarm.org/index.php/Tools_for_Agent_Based_Modeling/ (accessed on March 15, 2010).

Trachtulec, Z. 2007. Eukaryotic operon genes can define highly conserved syntenies. *Folia Biologica (Praha)*. 50: 1–6.

Truxal, J.G. 1955. *Automatic Feedback Control System Synthesis*. McGraw-Hill Book Co., New York.

Turner, J.S. 2007. *The Tinkerer's Accomplice*. Harvard University Press, Cambridge, MA.

TVA. 2010. From the new deal to a new century: A short history of TVA. Web article. www. tva.gov/about/history.htm (accessed on January 4, 2010).

Tyson, J.J. 2002. Biochemical oscillators. Ch. 9 in *Computational Cell Biology: An Introductory Text on Computer Modeling in Molecular and Cell Biology*, C. Fall, E. Marland, J. Wagner, and J.J. Tyson, eds. Springer-Verlag, New York.

Ulrich, W. 2005. Can nature teach us good research practice? A critical look at Frederic Vester's bio-cybernetic systems approach. *Journal of Research Practice*. 1(1): Article R2. 10. http://jrp.icaap.org/index.php/jrp/article/view/1/1 (accessed on November 19, 2009).

UNHCSRC. 2005. The Complex Systems Research Center. www.csrc.sr.unh.edu/about/index.html (accessed on November 4, 2009).

Universe. 2008. *Chaos Theory: Non-Linear Equations*. Web paper. http://universe-review. ca/R01–09-chaos.htm (accessed on November 19, 2009).

UCMP-University of California Museum of Paleontology. 2007. *Morphology of the Tetrapods*. UCMP-University of California Museum of Paleontology, Berkeley, CA. www.ucmp.berkeley.edu/vertebrates/tetrapods/tetramm.html (accessed on November 19, 2009).

Valentine, J.W. 2000. Two genomic paths to the evolution of complexity in body plans. *Paleobiology*. 26(3): 513–519.

van Hoek, M.J.A. and P. Hogeweg. 2006. *In silico* evolved *lac* operons exhibit bistability for artificial inducers, but not for lactose. *Biophysical Journal*. 91: 2833–2843.

van Regenmortel, M.H.V. 2004. Reductionism and complexity in molecular biology. *EMBO Reports*. 5(11): 1016–1020.

van Valen, L. 1973. A new evolutionary law. *Evolutionary Theory*. 1: 1–30.

Variano, E.A., J.H. McCoy, and H. Lipson. 2004. Networks, dynamics and modularity. *Physical Review Letters*. 92(18): 188701-1–188701-4.

Vasselon, T. and P.A. Detmers. 2002. Toll receptors: A central element in innate immune responses. *Infection and Immunity*. 70(3): 1033–1041.

Venter, J.C., S. Levy, T. Stockwell, K. Remington, and A. Halpern. 2003. Massive parallelism, randomness and genomic advance. *Nature Genetics Supplement*. 33: 219–227.

Verkman, A.S. and A.K. Mitra. 2000. Structure and function of aquaporin water channels. *American Journal of Physiology Renal Physiology*. 278: F13–F28.

Verstraete, W. 1999. Energy and the entropy business. *Nature Biotechnology*. 17(suppl.): BV27–BV28.

Vester, F. 1999. *Die Kunst vernetzt zu Denken: Ideen und Werkzeuge für neuen Umgang mit Komplexität*. [*The Art of Networked Thinking: Ideas and Tools for a New Way of Dealing with Complexity*] (6th edn., 2000. In German.) Deutsche Verlags-Anstalt. Stuttgart, Germany. ISBN: 3-421-05308-1. (Amazon.de offers an English translation of this book published by MCB Verlag, München, Germany. Nov. 2007. ISBN-10: 3939314056.) http://www.frederic-vester.de/eng/books/preface-of-the-book/ (Abstract accessed on November 19, 2009).

Vester, F. 2004. Sensitivity Model/Sensitivitäts modell Prof. Vester®. v. SMW 5.0e. Commercial software package in English, German or Spanish language. (Orig. version 1991.) For Windows 95, 98, NT, 2000 & XP. Munich, Germany. Frederic Vester GmbH. http://www.frederic-vester.de/eng/sensitivity-model/ (accessed on December 8, 2009).

Via, S. 2002. The ecological genetics of speciation. *The American Naturalist*. 159(suppl.): S1–S7.

Viana, M.P., E. Tanck, M.E. Beletti, and Lda F. Costa. 2009. Modularity and robustness of bone networks. *Molecular BioSystems*. 5: 255–261.

Vignuzzi, M. et al. 2006. Quasispecies diversity determines pathogenesis through cooperative interactions in a viral population. *Nature*. 439(19): 344–348.

Vilar, J.M.G., H.Y. Kueh, N. Barkai, and S. Liebler. 2002. Mechanisms of noise-resistance in genetic oscillators. *PNAS*. 99(9): 5988–5992.

Villadangos, J.A. and L. Young. 2008. Antigen-presenting properties of plasmacytoid dendritic cells. *Immunity*. 29: 352–361.

Vinogradov, A.E. 2008. Modularity of cellular networks shows general center-periphery polarization. *Bioinformatics*. 24(24): 2814–2817.

Vizioli, J. and M. Salzet. 2002. Antimicrobial peptides from animals: Focus on invertebrates. *Trends in Pharmacological Science*. 23(11): 494–496.

Vogel, C. and C. Chothia. 2006. Protein family expansions and biological complexity. *PLoS Computational Biology*. 2(5): 0370–0382.

Volterra, V. 1926. Variations and fluctuations of the number of individuals in animal species living together. In *Animal Ecology*, R.N. Chapman, ed. McGraw-Hill, New York, pp. 409–448.

von Andrian, U.H. 2003. Introduction: chemokines—Regulation of immune cell trafficking and lymphoid organ architecture. Editorial in: *Seminars in Immunology*. 15(5): 239–241. http://labs.idi.harvard.edu/vonandrian/Pages/sem_immun_uva.pdf (accessed on December 15, 2009).

von Dassow, G. and E. Munro. 1999. Modularity in animal development and evolution: Elements of a conceptual framework for Evo-Devo. *Journal of Experimental Zoology* (*Molecular and Developmental Evolution*) 285: 307–325.

Wagner, A. 2005. Distributed robustness versus redundancy as causes of mutational robustness. *BioEssays*. 27.2: 176–188.

Wagner, G.P., M. Pavlicev, and J.M. Cheverud. 2007. The road to modularity. *Nature Reviews/Genetics*. 8: 921–931.

Waldrop, M.M. 1992. Complexity: *The Emerging Science at the Edge of Order and Chaos*. Simon & Schuster, New York.

Wall, J.T., J. Xu, and X. Wang. 2002. Human brain plasticity: An emerging view of the multiple substrates and mechanisms that cause cortical changes and related sensory dysfunctions after injuries of sensory inputs from the body. *Brain Research Brain Research Reviews*. 39(2–3): 181–215.

Walls, G.L. 1967. *The Vertebrate Eye and Its Adaptive Radiation*. Haffner, New York.

Wallinga, D. and M. Mellon. 2008. *Factory Farms Feeding Antibiotic Crisis*. OP-ed essay in the July 14, 2008 *Hartford (CT) Courant*.

Walsh, C.T. and M.A. Fischbach. 2009. Squashing superbugs—The race for new antibiotics. *Scientific American*. 13 July.

Wang, D. 1999. Relaxation oscillators and networks. In *Wiley Encyclopedia of Electrical and Electronics Engineering*, vol. 18. J.G. Webster, ed. J. Wiley & Sons, New York.

Wang, R. and L. Chen. 2005. Synchronizing genetic oscillators by signaling molecules. *Journal of Biological Rhythms*. 20(3): 257–269.

Wang, Z. and J. Zhang. 2007. In search of the biological significance of modular structures in protein networks. *PLoS Computational Biology*. 3(6): 1011–1021.

Wang, L., J. Xie, and P.G. Schultz. 2006. Expanding the genetic code. *Annual Review of Biophysics and Biomolecular Structure*. 35: 225–249.

Wang, Y., H.M. Stricker, D. Gou, and L. Liu. 2007a. MicroRNA: Past and present. *Frontiers in Bioscience*. 12: 2316–2329.

Wang, K., H. Neumann, S.Y. Peak-Chew, and J.W. Chin. 2007b. Evolved orthogonal ribosomes enhance the efficiency of synthetic genetic code expansion. *Nature Biotechnology*. 25(7): 770–777.

Wang, Q., A.R. Parrish, and L. Wang. 2009. Expanding the genetic code for biological studies. *Chemistry and Biology*. 16: 323–336.

Ward, S.G. and F.M. Marelli-Berg. 2009. Mechanisms of chemokine and antigen-dependent T-lymphocyte navigation. (Review Article.) *Biochemistry Journal*. 418: 13–27.

Weitz, J.S., H. Hartman, and S.A. Levin. 2005. Coevolutionary arms races between bacteria and bacteriophage. *PNAS*. 102(27): 9535–9540.

Welch, J.J. and D. Waxman. 2003. Modularity and the cost of complexity. *Evolution*. 57(8): 1723–1734.

Wentworth, P., Jr. 2002. Antibody design by man and nature. *Science*. 296: 2247–2249.

Wentworth, P. Jr. and K.D. Janda. 2001. Catalytic antibodies: Structure and function. *Cell Biochemistry and Biophysics*. 35: 63–87.

Whitacre, J. and A. Bender. 2009. Degeneracy: A design principle for achieving robustness and evolvability. Complexity Digest, Issue 2009.5. arXiv:0907.0510, 2009/07/03. http://arxiv.org/arxiv/papers/0907/0907.0510.pdf (accessed on August 26, 2009).

Whitaker, I.S. et al. 2007. Larval therapy from antiquity to the present day: Mechanism of action, clinical applications and future potential. *Postgraduate Medical Journal*. 83: 409–413.

Whitworth, J.A. and J.J. Kelly. 1995. Evidence that high dose cortisol-induced Na⁺ retention in man is not meditated by the mineralocorticoid receptor. *Journal of Endocrinology Investigation*. 18: 586–591.

Whitworth, J.A., G.J. Mangos, and J.J. Kelly. 2000. Cushing, cortisol, and cardiovascular disease. *Hypertension*. 36: 912–916.

Widrow, B. and S.D. Stearns. 1985. *Adaptive Signal Processing*. Prentice-Hall, Englewood Cliffs, NJ.

Wieder, E. 2003. *Dendritic Cells: A Basic Review*. www.celltherapysociety.org/files/PDF/Resources/OnLine_Dendritic_Education_Brochure.pdf (accessed on May 14, 2009).

Wiki. 2009a. *Schools of Economic Thought*. (v. Nov. 16, 2009) http://en.wikipedia.org/wiki/Schools_of_economics (accessed on November 19, 2009).

Wiki. 2009b. *Speciation*. (v. Oct. 30, 2009) http://en.wikipedia.org/wiki/Speciation (accessed on November 19, 2009).

Wiki. 2009c. *Specified Complexity*. (v. Nov. 11, 2009) http://en.wikipedia.org/wiki/Specified_complexity (accessed on November 19, 2009).

Wiki. 2009d. *Consumer Price Index*. (v. Nov. 16, 2009) http://en.wikipedia.org/wiki/Consumer_price_index (accessed on November 19, 2009).

Wiki. 2009e. *Glossary of Graph Theory*. (v. Oct. 31, 2009) http://en.wikipedia.org/wiki/Glossary_of_graph_theory (accessed on November 19, 2009).

Wiki. 2009f. *Plectics*. (v. Oct. 27, 2009) http://en.wikipedia.org/wiki/Plectics (accessed on November 19, 2009).

Wiki. 2009g. *Shannon Index*. (v. Oct. 24, 2009) http://en.wikipedia.org.wiki/Shannon_index (accessed on November 19, 2009).

Wiki. 2010. *Interleukin*. (v. Feb. 21, 2010) http://en.wikipedia.org/wiki/Interleukin (accessed on March 22, 2010).

Wilke, C.O. 2003. Probability of fixation of an advantageous mutant in a viral quasispecies. *Genetics*. 163: 467–474.

Wilke, C.O. 2005. Quasispecies theory in the context of population genetics. *BMC Evolutionary Biology*. 5: 44.

Win, M.N., J.C. Liang, and C.D. Smolke. 2009. Frameworks for programming biological function through RNA parts and devices. *Chemistry and Biology*. 16: 298–310.

Witteveldt, J. et al. 2004. Protection of *Panaeus monodon* against white spot syndrome virus by oral vaccination. *Journal of Virology*. 78: 2057–2061.

Wojas, K., J. Tabarkiewicz, and J. Roliński. 2003. Dendritic cells in cancer immunotherapy—A short review. *Folia Morphology*. 62(4): 317–318.

Wolf, J. et al. 2000. Transduction of intracellular and intercellular dynamics in yeast glycolytic oscillations. *Biophysical Journal.* 78: 1145–1153.

Wooldridge, M. 2002. *An Introduction to Multi-Agent Systems.* Wiley & Sons, New York. ISBN: 0-471-49691-X.

Wray, G.A. et al. 2003. The evolution of transcriptional regulation in eukaryotes. *Molecular Biology And Evolution.* 20: 1377–1419.

Xu, P., M. Shi, and X.-X. Chen. 2009. Antimicrobial peptide evolution in the Asiatic honey bee *Apis cerana. PLoS ONE.* 4(1): 9.

Yamaguchi, S. et al. 2003. Synchronization of cellular clocks in the suprachiasmatic nucleus. *Science.* 302: 1408–1412.

Young, B.M. and G.M. Young. 2002. Ypla is exported by the Ysc, Ysa and flagellar type III secretion systems of *Yersina enterocolitica. Journal of Bacteriology.* 184(5): 1324–1334.

Zadeh, LA., G.J. Klir, and B. Yuan, eds. 1996. *Fuzzy Sets, Fuzzy Logic, and Fuzzy Systems: Selected Papers by Lofti A. Zadeh.* World Scientific, Singapore.

Zarei, M. et al. 2006. Functional anatomy of interhemispheric cortical connections in the human brain. *Journal of Anatomy.* 209: 311–320.

Zasloff, M. 2007. Antimicrobial peptides, innate immunity, and the normally sterile urinary tract. *Journal of American Society Nephrology.* 18: 2810–2816.

Zeuthen, T. 2007. The mechanism of water transport in Na^+-coupled glucose transporters expressed in *Xenopus* oocytes. (Comment to the Editor.) *Biophysical Journal.* 93: 1423–1416.

Zhang, W.-B. 2005. Differential equations in economics. Ch.1 in *Differential Equations, Bifurcations and Chaos in Economics.* World Scientific Pub. Co., Inc. ISBN-13: 978-9812563330.

Zhao, C.-X., H.-B Shao, and L.-Y. Chu. 2008. Aquaporin structure–function relationships: Water flow through plan living cells. *Colloids and Surfaces B: Biointerfaces.* 62: 163–172.

Zhou, W. 2005. Immunosuppressive networks in the tumor environment and their therapeutic relevance. *Nature Reviews/Cancer.* 5: 263–273.

Zimmer, C. 1999. Complex systems: Life after chaos. *Science.* 284(5411): 83–86.

SOME COMPLEX SYSTEMS JOURNALS

Advances in Complex Systems. Print ISSN: 0219-5259, Online ISSN: 1793-6802. Frank Schweiter, Editor-in-Chief. http://www.worldscinet.com/acs/ (accessed on November 20, 2009).

Chaos. ISSN: 1054-1500. David K. Campbell, Editor-in-Chief, BU, Boston, MA. Publisher: American Institute of Physics. http://ojps.aip.org/chaos/about_the_journal (accessed on November 20, 2009).

Complexity Digest. Carlos Gershenson, Editor. http://turing.iimas (accessed on November 20, 2009). unam.mx/~comdig

Complexity International. ISSN: 1320-0682. David Green, Executive Editor. An electronic journal published by Monash University, Australia. http://www.complexity.org.au/ci/info-journal.html (accessed on November 20, 2009).

Complex Systems. ISSN: 0891-2513. Todd Rowland, Managing Editor. Published Quarterly. http://www.complex-systems.com/ (accessed on November 20, 2009).

Journal of Complexity. ISSN: 0885-064X. J. Traub, Editor-in-Chief. http://www.elsevier.com/wps/find/journaldescription.cws_home/622865/description#description (accessed on November 20, 1998).

Journal of Economic Issues. ISSN: 0021-3624. Glen Atkinson, Editor, Department of
    Economics, U. Nevada, Reno, NV 89557. http://diglib.lib.utk.edu/utj/jei-home.php
    (accessed on November 20, 2009).
Journal of Mathematical Economics. ISSN: 0304-4068. Bernard Cornet, Editor. http://www.
    elsevier.com/wps/find/journaldescription.cws_home/505577/description#description
    (accessed on November 20, 2009).

Index

A

Abzymes, *see* Catalytic antibodies
Acinar cells, 138
Acquired immune deficiency syndrome
    (AIDS)
  antibodies, 285
  Bailey model, 288
  CD4+ T-cells, 282
  chemokines, 284–285
  CXCR4 receptor, 284
  eclosion, HIV, 284
  five Bailey state equations, 288–290
  hCS, 286–287
  HIV binding, 284
  mathematical model, 287
  MΦs, 282–283
  opportunistic infection, 282
  PHA, 287
  putative process, 285–286
  retroviruses, 281–282
  soluble CD4 molecule, 284–285
  "typical" HIV virion, 282–283
  Visna virus, 282
Active variables, 372–374
Adaptive immune system (AIS), 140
Adrenocorticotropic hormone (ACTH),
    233, 313
Advanced rate elimination method
    (AREM), 331
Agent-based computational economics
    (ACE), 366
Agent-based models (ABMs), 5, 256
  ACE, 366
  characteristics, 365
  dynamic model, ES, 357
  hASC behavior, 366–367
  properties, 366
  PS agent, 365–366
  software applications, 367
Agent-based simulation (ABS), 5
Aldosterone (ALDO) hormone, 313
AMPs, *see* Antimicrobial peptides;
    Antimicrobial proteins
Antibody-dependent cellular cytotoxicity
    (ADCC), 213
Antigen presentation, 140
Antimicrobial peptides, 181, 184
Antimicrobial proteins (AMPs), 219–220
Apoptosis process, *see* Programmed cell death

Aquaporins (AQPs) protein, 316
Atrial fibrillation (AFib), 6–7
Autocatalytic reaction (ACR), 78
Autoimmune (AI) disease, 317
Average vertex (node) degree, 327
Average vertex distance, 328

B

Bacterial quasispecies (BQS) model,
    242–243
Bacteriocins, 188–189
Belousov–Zhabotinski reaction (BZR), 91, 93
Biochemical oscillators (BCOs)
  biochemical systems, endogenous feedback
    loops, 108
  biological clocks, 121
  $[Ca^{++}]_i$ oscillations, 111
  dynamic models, 105
  eight state equations, 112
  eukaryote, 85, 105
  glycolysis metabolic pathway, 108, 110
  insulin oscillations, 111
  intracellular process, Per oscillator,
    105–106
  limit-cycle behavior, 108–109
  linearization function, 118
  mass action–diffusion model, 107
  NADH concentration, 108–109
  ODEs, 107
  relaxation oscillations, 109
  repressilator oscillator, 115–118
  root-locus technique, 107, 118
  SCN circadian oscillator, 105
  Simnon phase-plane plots, 113–114
  snell2.t program, 113–114
  Spangler and Snell oscillating system,
    113–115
  synchronization, 85
  Vilar's genetic oscillator, 118–120
Biological oscillators synchronization, *see*
    Quorum sensing
Blood oxygen level-dependent (BOLD) fMRI,
    140–141
Bode plots
  circuit simulation software
    applications, 424
  lead/lag filter, 423
  logarithmic function, 419
  simple low-pass system, 419–420